Knowledge and Space

Volume 15

Series editor
Johannes Glückler, Department of Geography, Heidelberg University, Heidelberg, Germany

Knowledge and Space

This book series entitled "Knowledge and Space" is dedicated to topics dealing with the production, dissemination, spatial distribution, and application of knowledge. Recent work on the spatial dimension of knowledge, education, and science; learning organizations; and creative milieus has underlined the importance of spatial disparities and local contexts in the creation, legitimation, diffusion, and application of new knowledge. These studies have shown that spatial disparities in knowledge and creativity are not short-term transitional events but rather a fundamental structural element of society and the economy.

The volumes in the series on Knowledge and Space cover a broad range of topics relevant to all disciplines in the humanities and social sciences focusing on knowledge, intellectual capital, and human capital: clashes of knowledge; milieus of creativity; geographies of science; cultural memories; knowledge and the economy; learning organizations; knowledge and power; ethnic and cultural dimensions of knowledge; knowledge and action; and mobilities of knowledge. These topics are analyzed and discussed by scholars from a range of disciplines, schools of thought, and academic cultures.

Knowledge and Space is the outcome of an agreement concluded by the Klaus Tschira Foundation and Springer in 2006.

More information about this series at http://www.springer.com/series/7568

Johannes Glückler • Gary Herrigel
Michael Handke
Editors

Knowledge for Governance

Springer

Klaus Tschira Stiftung
Gemeinnützige GmbH

KTS

Editors
Johannes Glückler
Department of Geography
Heidelberg University
Heidelberg, Germany

Gary Herrigel
Department of Political Science
University of Chicago
Chicago, IL, USA

Michael Handke
Department of Geography
Heidelberg University
Heidelberg, Germany

ISSN 1877-9220
Knowledge and Space
ISBN 978-3-030-47149-1 ISBN 978-3-030-47150-7 (eBook)
https://doi.org/10.1007/978-3-030-47150-7

This Springer imprint is published by the registered company Springer Nature Switzerland AG
The registered company address is: Gewerbestrasse 11, 6330 Cham, Switzerland

Acknowledgements

The editors thank the Klaus Tschira Stiftung for funding the symposia and the Springer book series on Knowledge and Space. The team of the Klaus Tschira Stiftung and the Studio Villa Bosch have been contributing greatly to the success of the symposia for more than a decade. Together with all the authors in this volume, we are grateful to Anna Mateja Punstein, Leslie Ludwig and Katrin Janzen for their superb assistance to the editors as well as to the technical editing team for their tireless dedication. Volker Schniepp of the Department of Geography at Heidelberg University has generously helped us to get figures and maps into shape for publication. We also thank all student assistants and colleagues from the Department of Geography who have helped accomplish the symposium as well as this 15th volume in this book series. We are particularly grateful to Tobias Friedlaender, Alaina Marangos, Johannes Nützel, Sandy Placzek and Marius Zipf.

Contents

Contributors

Claudia N. Avellaneda O'Neill School of Public and Environmental Affairs, Indiana University Bloomington, Bloomington, IN, USA

Stephen Bell School of Political Science and International Studies, University of Queensland, Brisbane, QLD, Australia

Ricardo Andrés Bello-Gómez O'Neill School of Public and Environmental Affairs, Indiana University Bloomington, Bloomington, IN, USA

Christian Binz Department of Environmental Social Sciences, Eawag-Swiss Federal Institute of Aquatic Science and Technology, Dübendorf, Switzerland

Centre for Innovation, Research and Competence in the Learning Economy CIRCLE, Lund University, Lund, Sweden

Alex Clapp Department of Geography, Simon Fraser University, Burnaby, BC, Canada

Richard C. Feiock Askew School of Public Administration and Policy, Florida State University, Tallahassee, FL, USA

Cláudia Florêncio Stockholm Resilience Centre, Stockholm University, Stockholm, Sweden

Johannes Glückler Department of Geography, Heidelberg University, Heidelberg, Germany

Michael Handke Department of Geography, Heidelberg University, Heidelberg, Germany

Jarrod Hayes Department of Political Science, University of Massachusetts, Lowell, MA, USA

Roger Hayter Department of Geography, Simon Fraser University, Burnaby, BC, Canada

Gary Herrigel Department of Political Science, University of Chicago, Chicago, IL, USA

Andrew Hindmoor Department of Politics and International Relations, University of Sheffield, Sheffield, UK

Erik-Logan Hughes Department of Urban Studies and Planning, Massachusetts Institute of Technology, Cambridge, MA, USA

Terry L. Hunt Honors College, University of Arizona, Tucson, AZ, USA

Patrick Kenis Tilburg School of Economics and Management, Tilburg University, Tilburg, The Netherlands

Serena Y. Kim School of Public Affairs, University of Colorado Denver, Denver, CO, USA

Janelle Knox-Hayes Department of Urban Studies and Planning, Massachusetts Institute of Technology, Cambridge, MA, USA

Marleen Kraaij-Dirkzwager Department for Environmental Health, Aftercare and Security (VLH), Dutch National Institute for Public Health and the Environment (RIVM), Bilthoven, The Netherlands

Carl P. Lipo Department of Anthropology, Binghamton University, Binghamton, NY, USA

Pamela Mischen Department of Public Administration, Binghamton University, Binghamton, NY, USA

Simon Niemeyer Centre for Deliberative Democracy and Global Governance, University of Canberra, Canberra, ACT, Australia

Johabed G. Olvera O'Neill School of Public and Environmental Affairs, Indiana University Bloomington, Bloomington, IN, USA

Markus Pohlmann Max-Weber-Institute of Sociology, Heidelberg University, Heidelberg, Germany

Jörg Raab Department of Organization Studies, Tilburg University, Tilburg, The Netherlands

Ortwin Renn Institute for Advanced Sustainability Studies, Potsdam, Germany

Lisen Schultz Stockholm Resilience Centre, Stockholm University, Stockholm, Sweden

Michael Scott College of Humanities, Arts and Social Sciences, Flinders University, Adelaide, SA, Australia

Nico Stehr Karl Mannheim Chair for Cultural Studies, Zeppelin University, Friedrichshafen, Germany

William L. Swann School of Public Affairs, University of Colorado Denver, Denver, CO, USA

Aura Timen National Coordination Centre for Outbreak Management (LCI), Dutch National Institute for Public Health and the Environment (RIVM), Bilthoven, The Netherlands

Nebahat Tokatli Milano School of Policy, Management, and Environment, The New School, New York, NY, USA

Bernhard Truffer Department of Environmental Social Sciences, Eawag-Swiss Federal Institute of Aquatic Science and Technology, Dübendorf, Switzerland

Copernicus Institute of Sustainable Development, Utrecht University, Utrecht, The Netherlands

Elizangela Valarini Max-Weber-Institute of Sociology, Heidelberg University, Heidelberg, Germany

Simon West Stockholm Resilience Centre, Stockholm University, Stockholm, Sweden

Chapter 1
On the Reflexive Relations Between Knowledge, Governance, and Space

Johannes Glückler, Gary Herrigel, and Michael Handke

Governance is both a contested concept and an increasingly empirical concern. On the one hand, it has become an almost universal and all-encompassing concept, which has attracted scholarly interest from many disciplines and over many decades to tackle the dilemmas of collective action and to facilitate effective coordination of interests and resources toward commonly accepted goals. On the other hand, conceptions of governance vary considerably, and their meanings depend on disciplinary perspectives, theoretical traditions, and empirical focus. Although it is easy to agree what governance is not, it is more challenging to create broad consensus on what it is and how it works effectively. Most capaciously, governance denotes the coordination of collective action. These actions take place in institutionally or physically organized spaces of interaction, where knowledge is needed to shape governance appropriately. Organized spaces and knowledge are both conditions as well as consequences of the governance process.

Contributions to theories and perspectives of governance originate from diverse disciplinary fields, including political science, economics, organization studies, sociology, geography, and planning studies. The authors of recent handbooks provide comprehensive overviews of the interdisciplinary breadth of ideas and debates of governance (e.g., Ansell & Torfing, 2016; Bache & Flinders, 2004; Bevir, 2010; Chhotray & Stoker, 2009; Levi-Faur, 2012). They portray the multitude of applications and concepts, ranging from practical fields such as corporate, contract, project, public, private, or nonprofit governance, to internet, land, urban, risk, environmental, and climate governance, to conceptual or normative approaches of network,

J. Glückler (✉) · M. Handke
Department of Geography, Heidelberg University, Heidelberg, Germany
e-mail: glueckler@uni-heidelberg.de; handke@uniheidelberg.de

G. Herrigel
Department of Political Science, University of Chicago, Chicago, IL, USA
e-mail: g-herrigel@uchicago.edu

J. Glückler et al. (eds.), *Knowledge for Governance*, Knowledge and Space 15,
https://doi.org/10.1007/978-3-030-47150-7_1

collaborative, participative, fair or good governance, and to various levels ranging from local to regional and national to global and multilevel governance.

Sometimes, it is helpful for an emergent concept not to be defined too narrowly in order to attract continuous attention and research effort and to stimulate scholarly conversation. Yet such diversity may also make it difficult for researchers to exchange, compare, and assess empirical studies on common ground. Therefore, governance research often remains locked within a discipline's cognitive boundaries. Governance debates always comprise a selection of perspectives. Debaters view the complex reality in a certain way, concentrating on selected characteristics and neglecting other ones (Benz & Dose, 2010). The polysemy of the concept of governance confronts scholars with the difficult task of coherently theorizing the social, political, environmental, institutional, and economic challenges to collective action.

We are positioning our book at the neglected intersection between disciplines, conceptual perspectives, and geographical scales to offer room for conversation on the role of knowledge and space in governance theory and practice. We are bringing together conceptual and empirical work on governance from varying fields, and the contributions' authors not only illustrate the multidisciplinary character of governance research, but they also point to new opportunities for interdisciplinary exchange. We argue for—and the chapters each speak to—particular aspects of the reflexive nature of the relationships between governance, knowledge, and space. In the next section, we briefly define an inclusive framing of some of the characteristics of governance, before we elaborate on the reflexive relations between governance, knowledge and space. By doing so, we highlight the contributions of the individual chapters of this volume to each of these relations, which also provide the rationale for structuring the book into distinct parts.

Framing Governance

Rather than unnecessarily restricting the concept of governance, we propose carving out some of its commonly shared characteristics by means of comparison with related notions in social science, such as management, government, and institutions. All three concepts relate to mechanisms that yield and sustain some degree of social order. Utilizing the governance perspective sheds new conceptual light on these concepts by emphasizing the process character of coordinating and moderating organized social interactions. Governance represents relational practices. There is governance inside firms and across firms, and there is governance within governments and across government boundaries. Governance also coordinates between firms and governments. However, unlike management or government, governance is not an actor capable of making decisions. Rather, it offers a reliable structure for interaction, which is something that it has in common with institutions (Glückler, Suddaby, & Lenz, 2018). In many cases, these structures have yet to be established and institutionalized by the actors involved in governance practices, which once again leads back to the process character of governance (Pierre & Peters, 2000).

To capture the specific meaning of governance, we place it on a conceptual continuum that expands along the dimensions of time (from long-term to short-term convertibility) and of authority (from heteronomy or interdependence to autonomy). Concretely, we conceive governance as lying between the concepts of institutions on one end of longer term societal coordination, and of management as well as of government on the other end of shorter term coordination and decision making within the scope of a single authority (organization, state).

Researchers use the concept of *institutions* to describe relatively stable patterns of social interactions that are based on legitimate mutual expectations and that are enforced by social sanctions (Barley & Tolbert, 1997; Glückler et al., 2018; Hodgson, 2006; Scott, 2007). Institutions refer to legitimate beliefs and practices that are shared to varying degrees by society, guarantee social order, offer behavioral orientation, and facilitate individual action within society. It is, therefore, difficult for individuals or organizations to actively change or overturn existing institutions (Lawrence, Leca, & Zilber, 2013), and if so only over longer periods. Institutional change often unfolds incrementally and only in exceptional cases can it be interpreted ex post as the result of specific moments that disrupted the reproduction of institutional practices, and which then gave rise to collective strategies to recompose rules and institutional features (Glückler & Lenz, 2018; Herrigel, 2010; Streeck & Thelen, 2005).

Whereas institutions are characterized by relative temporal stability and usually slow rates of transformation, the other end of the continuum comprises shorter term, current forms of coordination by means of management. Researchers use "management" to describe the administration of an organization, including the choice of a strategy and the coordination of all the resources in pursuit of the organization's objective. Consequently, management is a coordination mode bounded within the scope of and legitimized by one single authority. Managers are entitled to take decisions within their organizational jurisdiction and thus are empowered to allocate resources, design organizational structures and processes, to assign roles and responsibilities, to solve conflicts, and allocate resources in pursuit of the organization's goals.

Governance should also not be confused with government (Osborne & Gaebler, 1992; Rosenau & Czempiel, 1992), which also provides hierarchical order. A republican form of government, for example, is composed of temporary legitimated representatives who execute the power of the state and, similarly to management, internally assign roles and allocate resources towards collective goals. These goals have grown in number and diversity in modern societies, disproportionately increasing intervention costs for the state. However, being a prominent object of study in conceptual government research (Kooiman, 1993), hierarchical structures often become encrusted and retain decision making in the silos of specific knowledge domains. Government thus risks failing to appropriately address contemporary collective problems, especially if they transcend different functional and territorial jurisdiction (Mayntz, 2003). Meanwhile, modern governments are increasingly confronted with the idea that sovereignty lies with the people and not with their representatives—what role, then, can a government play as an actor in governance processes?

In contrast to longer term and less dirigible institutions on the one hand, and hierarchically controllable shorter term management or temporarily elected governments on the other, we conceive governance as the coordination of distinct stakeholders toward the achievement of consensual goals. These stakeholders might be legally independent of each other, but interdependently linked by a collective problem. In contrast to management or government, governance extends beyond the scope of the rule enforcement of a single authority, and, hence, requires negotiation between different interests (Rhodes, 1996). In governance, those affected by rules are themselves involved in the design of regulation and their institutional enforcement (Marin & Mayntz, 1991). In this way, governance differs from social institutions in that creators of governance processes actively and purposively tackle collective action problems to find acceptable solutions. Such a broad understanding of interdependent controllability or "heterarchy" (Jessop, 1998), including negotiation, collaboration, and legitimacy, is supported in most governance approaches.

Knowledge and Governance

Governance attracts much of its attention as an analytical perspective because it is used to address learning processes in changing environments (Stoker, 1998). The relationship between knowledge and governance is reflexive in nature: Knowledge is a prerequisite for governance as much as governance affects how one creates or shares knowledge. Being at the core of this book series, the concept of knowledge denotes the human understanding of concrete and abstract phenomena. Whereas people can keep "stocks" and exchange "flows" of data and information, the creation of knowledge is initially bound to the individual mind. However, individuals are not atomistic, isolated actors, but embedded in social and institutional contexts. People constantly compare and align their understandings with the understandings of others, and engage in imitation and communication to challenge or confirm common understanding. When framing knowledge as a relational rather than a substantive concept of human understanding, knowledge is difficult to produce in isolation, to store or protect from spillovers, or to copy and reuse in other contexts (Bathelt & Glückler, 2011; Glückler, Meusburger, & El Meskioui, 2013). Whereas knowledge represents the practice of knowing, knowledge creation emerges from the practice of learning, the circulation and interpretation of information, cumulative experience, and cognition (Amin & Cohendet, 2004; Borgatti & Cross, 2003). In this sense, the dynamics of knowing and learning are fundamentally social and most often interactive processes (Lundvall & Johnson, 1994). It includes the collective creation of knowledge as well as its imitation through others. Researchers have acknowledged learning as a reflexive, interactive, and continuous process of recombining knowledge towards new understanding. However, learning is focused on more than just the recombination of existing knowledge. Bridging the barriers to communication between different communities is important (Nooteboom, 2000), but so is collective learning across the boundaries of thought collectives to generate new knowledge for new solutions (Fleck, 1935; Herrigel, Wittke, & Voskamp, 2013; Punstein & Glückler, 2020). Knowledge in governance can be seen as both an input and a collective outcome that is newly created (Hess & Ostrom, 2007). The joint learning process reinforces the positive aspects of interdependence among otherwise loosely coupled actors.

Both knowledge and learning are contextual rather than universal (Bathelt & Glückler, 2011; Meusburger, 2008), which makes effective governance processes a truly complex challenge. Knowledge is hard to store or accumulate and difficult to value. The production of knowledge and recognizing its value (utility) in common depend on the social and spatial context. Governance is always an arranged relational practice. The authors of much of the economic governance literature refer to governance arrangements as "mechanisms." These include markets, hierarchies, and relational contracts (Williamson, 2005). Others also include networks as governance arrangements (Keast, 2016; Powell, 1991). The economic arguments, however, often remain very abstract and are therefore difficult to relate to the common usage of the concept in other disciplines. In the following, we elaborate on the reflexive relationship between knowledge and governance in more detail and divide the book into three parts. The authors of the first two parts refer to the role of knowledge for governance. Whereas those of the first section discuss how knowledge enables governance organization, those of the second analyze how knowledge for governance influences its impact and efficacy. In turn, the third part is a collection of conceptual and empirical contributions that discuss how governance affects learning processes and how it generates new knowledge to solve collective problems. The individual book chapters complement each other thematically and with their analytical emphasis on specific problems of coordinating collective actions.

How Knowledge Enables Governance

The development and implementation of effective governance solutions often requires bringing together distinct knowledge about the collective problem. A precise definition of the problem is often not possible from the very beginning and thus becomes an explicit goal of the governance process. Knowledge can be mobilized and recombined either by drawing on the experience of the actors involved in governance, which often includes those affected by the collective problems, or it can be incorporated as proven expertise provided by outsiders. Expert knowledge is particularly important for governance if the actors involved have not yet made it clear how their different and sometimes conflicting interests relate to each other. This is particularly evident, for example, in the principles Ostrom (2005) has suggested for the appropriate design of the governance of collective goods. The character of collective goods becomes particularly evident with many natural, renewable resources, which are threatened by overexploitation, that is, when actors prioritize the pursuit of individual profit over collective gains (Hardin, 1968). Knowledge about the properties of these resources, the periods of their renewal, their location in space, and their accessibility for exploitation are a prerequisite of resource governance. In addition, knowledge about the constellation of actors involved in the use of the resources as well as knowledge about the possibilities to monitor and sanction those who direct their behavior against the collective gain relevance. It helps actors to assign roles and responsibilities in governance processes. Knowing the institutional context (Glückler, 2020; Glückler, Punstein, Wuttke, & Kirchner, 2020) allows actors to establish governance practices that are coherent or complementary with legitimate mutual expectations (Boyer, 2005). Institutions provide and impart knowledge for governance in their own specific way.

The authors of the book's first part illustrate and discuss the extent to which knowledge is a prerequisite for governance. Questions of the sources and the quality of available knowledge inputs pave the way to assess the appropriateness of different types of knowledge for a specific governance case. Not all knowledge is suitable for usage and valorization in governance. Knowledge for governance is often subject to doubts on the part of certain participating actors, a circumstance that must be taken seriously in the design of a governance process. In the five chapters forming this part of the book, the contributors tackle four important yet empirically contested ways in which knowledge becomes a crucial resource and prerequisite for governance: (i) the availability and control of knowledge for governance (see Chap. 2 by Lipo, Mischen, and Hunt), (ii) the manipulation of the content and interpretation of knowledge (see Chap. 3 by Scott), (iii) the varying relevance of different types of knowledge for governance (see Chap. 4 by Stehr and Chap. 5 by Renn), and (iv) the differences in legitimacy of those who know (see Chap. 6 by Bell and Hindmoor). These five sets of contributors deal with these aspects from different perspectives of disciplinary governance research.

In the second chapter, Carl P. Lipo, Pamela Mischen, and Terry L. Hunt emphasize the importance of having sophisticated knowledge about a resource's availability and the characteristics of its place of origin in order to be able to manage it sustainably. They use the example of prehistoric Rapa Nui to demonstrate how humankind wrestles place dependently over issues of governance by developing capabilities to survive on an isolated island with scarce resources. Ever since their arrival on Easter Island (now belonging to Chile), the Rapa Nui have been confronted with uncertainties about global climate change and fluctuating resource stores. Lipo, Mischen, and Hunt argue that whereas the case of prehistoric Rapa Nui has often been treated as a warning about human-caused ecological catastrophe, new archaeological and multidisciplinary findings indicate that land use on Easter Island had been sustainable during its prehistory until the Europeans arrived. In retrospect, the governance practices collectively applied at Easter Island appear to have been highly innovative and locally appropriate. These findings point to the potential of alternative action models and new governance structures. The authors further show that scientists who analyze governance cases must constantly reassess their own knowledge.

Actors utilize different techniques to control knowledge for governance. Discursive knowledge (van Dijk, 2014), for example, influences the contents and contested negotiations in governance processes. Knowledge can be manipulated with certain modes of communication. Rule-based routines (Becker & Knudsen, 2005; Ostrom, 2000) are a way of collectively preserving knowledge and facilitating decision making in governance. However, they must first often be discovered and adjusted to each other. In the third chapter, Michael Scott analyzes how knowledge as well as particular opinions can be specifically controlled in governance processes in order to reach agreement and influence results. Drawing on empirical cases of coastal property developments in South Australia, he investigates how key actors in land-use governance—such as developers, planners, politicians, and scientists—reflexively deploy "techniques of neutralization" to deflect critiques and manage opposition to contentious new developments. Scott explores how actors use these techniques to draw on particular spatial metaphors and images to suggest that, somewhat ironically, a tacit metatechnique is to neutralize the projected environmental risks to coastal space through narratives of time. Awareness of the usage of these techniques is a valuable knowledge input for governance on its own.

The manner in which knowledge is made relevant, structured, communicated, and exploited is shaped by its source. The contributors of the two subsequent chapters focus on the way in which governance processes become both a competition among ideas and a contest regarding what may count as legitimate. In the book's fourth chapter, Nico Stehr highlights how important scientific knowledge is a prerequisite for effective climate policy. He critically illustrates how local, regional, and national actions related to climate deliberately ignore the expertise of scientists and unacceptably limit the effectiveness of climate policy. On this basis, he diagnoses the failure of large social organizations to respond in a timely fashion to the progress of climate change knowledge, an observation that he denotes as "inconvenient institutions." The sense of political ineffectiveness felt especially among climate scientists provokes a strong disenchantment with democratic governance. As a result, he proposes that political action based on principles of democratic governance be abandoned. Stehr concludes that such a view is mistaken and calls instead for better democratic processes.

In a similar direction and with regard to risk governance, Ortwin Renn argues in the fifth chapter that scientific knowledge does not provide answers to everything. Instead, in many cases it is ambivalent and uncertain and cannot grasp every aspect of a collective action problem at once. In his conceptual contribution, he argues that risk governance is above all about organizing communication processes so that constant knowledge inputs can feed and reproduce an ongoing learning process. Renn introduces the concept of risk governance developed by the International Risk Governance Council in Geneva, which provides guidance for constructing comprehensive assessment and management strategies to cope with risk. It integrates three types of scientific input: classic, curiosity-driven research; strategic, goal-oriented research; and catalytic, process-related investigations. He demonstrates how these three knowledge pools can help risk assessors and managers to better understand complex risk situations. In many governance cases, expert and scientific knowledge seem to have interpretive sovereignty. The authors of recent studies illustrate, however, that governance failure is likely if social stakeholders, who are invited to participate in governance processes alongside experts from scientific fields, find that their practical knowledge is ignored or undervalued (e.g., in cases where their experience is affected by unsolved collective goods problems; cf. also Chap. 18 by Herrigel in the third part of this book).

For some collective action problems, (scientific) knowledge does not exist in the first place, either because the collective problem represents a new phenomenon or because it is one that only evolves progressively. In this case, actors must rely on experiential knowledge. In Chap. 6, Stephen Bell and Andrew Hindmoor discuss how failures incurred by previous decisions are an important source of learning in governance that can make it possible to improve the coordination of collective action. Drawing on the phenomenon of systemic risk in financial markets that occurs when financial actors collectively (if inadvertently) bring on a major financial crisis through the withholding of credit and asset fire sales, they focus on coordinated efforts to prevent such calamities. Using the European debt crisis as an empirical example, they show that where appropriate knowledge and governance arrangements can be put in place, collective action may be arranged to help prevent the uncertain crystallization of systemic risk.

The authors of the five chapters in the first part of this book show not only how important knowledge is as a prerequisite for governance; they also demonstrate that knowledge, and its interpretation, is not free of mistakes. Actors often purposively manipulate or unintentionally modify knowledge by discourse and normative intervention.

How Knowledge Drives the Effectiveness of Governance

Due to the context-specificity of knowledge and the many alternatives for organizing governance over time, similar and comparable knowledge inputs can often yield different results. The second part of this book, therefore, contains a series of five chapters that, despite differences in their research questions, all help reveal how knowledge drives the very efficacy of governance. These chapters' authors focus on the way in which the level of knowledge and expertise effects governance outcomes (see Chap. 7 by Avellaneda, Bello-Gómez, and Olvera). Contributors variously focus on the type of knowledge, such as restriction on only one domain (see Chap. 8 by Handke); the adaptation of knowledge to local contexts (see Chap. 9 by Knox-Hayes, Hayes, and Hughes); the limitations of knowledge in breaking up ineffective governance (see Chap. 10 by Pohlmann and Valarini); and the way in which knowledge of governance structures can improve legitimacy relations among governance actors (see Chap. 11 by Glückler).

In Chap. 7, Claudia N. Avellaneda, Ricardo Andrés Bello-Gómez, and Johabed G. Olvera assess the impact of differential levels of knowledge on governance. To this end, they look at the local level of communal politics in Mexico and

Colombia, where recent efforts to decentralize responsibilities have produced diverse results. Empirically, they are interested in how mayors' wide range of technical training (codified knowledge) and experience (uncodified knowledge) steer official action towards locally improved development. Depending on the context, decision makers' specialist knowledge and experience can have very different effects. Indeed, sometimes their actions make no difference whatsoever. With their findings, they caution governance researchers from generalizing results. What drives performance in one country may not have the same explanatory power in another country.

In Chap. 8, Michael Handke looks at how the use of knowledge from just one rigid domain can blind actors to possible new solutions for solving collective action dilemmas. He argues that unilaterally quantified knowledge about forest fire risks in Chile fails to capture the spatiotemporal diversity of this very context-specific risk. Whereas some big private companies consider hierarchical risk control and making use of the insurance market to be cost efficient, small forest owners in Chile do not have the same access to this form of risk management due to its high demand on technical knowledge and organizational competences. On the regional scale, where actors perceive forest fires as a collective problem, this leads to negative external effects and social conflicts. Handke assesses the strengths and weaknesses of interacting hierarchical and market forms of risk management and pleads for a geographical approach to risk governance. He demonstrates that current risk management practices explicitly decontextualize detailed geographical knowledge of forest fire risks. Localized knowledge of the causes and effects of forest fire risks thus gets lost in the process.

In Chap. 9, Janelle Knox-Hayes, Jarrod Hayes and Erik-Logan Hughes suggest that knowledge inputs in governance must be adapted to local regulatory and institutional conditions, questioning the doctrine of placeless perfect markets. Empirically, they illustrate how markets have to be contextually designed in order to be effective. Using the example of markets for CO2 certificates that are established around the world and that the Kyoto Protocol incentivizes as the dominant mechanism for mitigating climate change, they analyze how textbook knowledge about the functioning of markets needs to be given context-specific value. They conclude that international efforts to promulgate market mechanisms run up against local cultures of markets that shape economic practices and knowledge to different degrees. Markets are enacted via political processes entailing different amounts of public, stakeholder, and expert involvement and varying levels of trust in technocratic government agencies, private firms, and scientific authority. Market cultures highlight issues at the interface of political and economic governance, including issues of citizen, state, and industry participation, and the materiality of economic and financial productivity.

In some circumstances, however, any governance effort to solve problems of collective action may fail completely. Long-established practices of coordinating interests in society can turn out to be institutionally encrusted so that new impulses for governance remain ineffective. In Chap. 10, Markus Pohlmann and Elizangela Valarini analyze anticorruption governance in Brazil. By carrying out content analyses of court files on corporate crime and the involvement of the public sector, they show how systemic institutional encrustation undermines changes in governance practice. Even with new knowledge for governance—in this case the implementation of rigid international standards for good governance—corruption persists, with far-reaching consequences for Brazil's development.

Finally, knowledge about governance itself helps actors assess, improve, or adapt governance models in a particular context. If governance refers to the purposive coordination of collective action between interdependent stakeholders, then methods of social network analysis can be helpful to map and analyze the structure of social interactions that reflect the "lived" practices of coordination and decision making. At this intersection, governance theory and network theory offer space for crossfertilization (Glückler, Lazega, & Hammer, 2017). Not only do social network researchers offer useful methods to identify the patterns of relations among sets of actors; they also provide valuable theories on how specific positions within a network, and how specific formations of whole networks, facilitate certain social outcomes such as innovation, social support, and solidarity or other types of problem solving (e.g., Kilduff & Tsai, 2003; Knoke, 2012). In Chap. 11, Johannes Glückler goes beyond viewing networks as a governance mode (Podolny & Page, 1998; Williamson, 1991) to examine how the lived practice of governance in a network actually conveys legitimate and acceptable collective coordination. He advances the concept of lateral network governance in the empirical context of organized networks, in which firms pool resources and join their interests in the pursuit of common goals. To solve the puzzle of how independent equals commit themselves to coordinating their actions, Glückler proposes overcoming the traditional dualism between formal and informal mechanisms of governance. He conceives lateral network governance as a structure for the legitimate delegation of decision making and develops a social network analytic approach to assessing the relational distribution of legitimacy, utilizing two case studies of interfirm network organizations to illustrate the extent to which the actual legitimacy distribution diverges from formal governance authority.

How Governance Affects Learning and Innovation

In part three of this volume, the contributors invert the perspective given in the previous chapters and reveal how governance also affects the creation and adoption of knowledge. In five chapters, each from a different angle, they show that learning in governance takes time and needs to be actively organized (see Chap. 12 by Niemeyer), that it requires a minimum degree of flexibility to allow learning from mistakes during the organization of the process (see Chap. 13 by Schultz, West, and Florêncio), and that the outcome of learning processes depends in part on whether organizations have the capacity to act under fragmented responsibilities (see Chap. 14 by Raab, Kenis, Kraaij-Dirkzwager, and Timen) or are able to link collective learning to institutional collective action (see Chap. 15 by Kim, Swann, and

Feiock). In addition, fragmentation seems to facilitate learning outcomes. Learning also occurs in the course of the renegotiation of power, which may in turn change the meaning of knowledge (see Chap. 16 by Hayter and Clapp).

Contemporary collective challenges, such as global environmental change, cannot be dealt with by drawing on the knowledge of one generation alone. In Chap. 12, Simon Niemeyer argues that in order to overcome global dynamics, individuals must adopt an open and humble deliberative attitude to be able to accelerate the assessment of uncertain and complex collective issues. He draws evidence from small-scale settings characterized by deliberative minipublics, in other words, organized communication processes at the microlevel of governance. In group meetings and organized face-to-face situations, actors incrementally pass on knowledge and views on climate change to other governance participants. Perceptions change depending on the group affiliation. The observed mechanisms can be "scaled up" to inform possibilities for wider reform of the processes governing the uptake and use of knowledge.

Flexibility in the organization and coordination of governance stakeholders can accelerate the learning process. In Chap. 13, Lisen Schultz, Simon West, and Cláudia Florêncio argue that actors in learning situations must be allowed to make mistakes in order to adapt to heterogeneous contexts. They call this adaptive governance. Focusing on the people, practices, and politics involved with adaptive governance in the Global South, they use the administration of the Kruger to Canyons Biosphere Region in South Africa to show that the practices for generating knowledge, sharing information, collaborating, and responding to change emerge as players navigate tensions between diverse values, norms, and routines. Focusing on the way that people, practices, and politics monitor and prevent poaching highlights how adaptive governance is situated and involves agency, meaning, and creativity. In this respect, successful governance often requires changes to existing institutional structures: "The governance concept points to the creation of a structure or an order which cannot be externally imposed but is the result of the interaction of a multiplicity of governing and each other influencing actors" (Kooiman & van Vliet, 1993, p. 64).

Although actors involved in governance orient their actions towards a common goal, this does not rule out the possibility that their interests diverge. The manner in which individual interests are reflected in governance can influence the effectiveness of learning processes. In Chap. 14, Jörg Raab, Patrick Kenis, Marleen Kraaij-Dirkzwager, and Aura Timen examine organizations' capacity to act under fragmented responsibility. They focus on the risk of epidemic catastrophes and demonstrate that involved actors perceive knowledge and timely information about the spread of viruses differently, ultimately hindering interorganizational learning. They demonstrate how the organizational network governance approach can generate information necessary for specific organizational players to limit the transmission of a virus and its impact.

In contrast to this, Serena Y. Kim, William L. Swann, and Richard C. Feiock demonstrate how the capacity for organizational learning can be collectively supported even in situations of conflicting interests. In Chap. 15, they argue that greater knowledge about collaboration and more information about partners enables actors to better deal with collaboration problems, heterogeneous preferences, and weak institutions for collaboration. In their conceptual contribution, they posit three pathways that link collective learning to institutional collective action and put forward propositions as to how such learning can reduce collective action dilemmas and enhance future collaboration. In the first path, collaborative choices and outcomes affect collective learning. In the second path, collective learning directly mitigates collaboration risks and in turn alters the choice of integrative mechanism. In the third path, collective learning moderates the relation between the collaboration situation and risks—that is, collective learning has greater impact in situations characterized by highly fragmented, specialized, and multifaceted contexts than in low-complexity situations.

The reflexive relationship of knowledge and governance is particularly evident in those cases in which knowledge enters both as an input to governance and evolves as an improved outcome in the course of the governance process. In Chap. 16, Roger Hayter and Alex Clapp look at conflicts over forest and timber resources in British Columbia, analyzing how stakeholders with different geographical presence and influence continuously renegotiate the societally accepted values of these resources. Negotiations are part of the governance process and lead to a remapping of the relationship between economy and society. Hayter and Clapp interpret remapping in British Columbia as an attempt to transform the commodity-driven and shareholder-oriented forest management associated with Fordism into more locally diverse forms of governance as part of a post-Fordist, techno-economic paradigm. This remapping goes hand in hand with institutional thickening, in other words, a process to bring together opposing parties to exchange views and develop respectful relationships and to implement new forms of governance. Stakeholder remapping is not just a practice, but also a result of governance. The authors of this book's last three contributions all place industrial innovation at the center of their analysis as the result of planned governance efforts. In Chap. 17, Christian Binz and Bernhard Truffer suggest that technological innovation increasingly depends on multiscalar actor networks and institutions. They criticize perspectives on innovation that bracket the problem of scale and focus exclusively on discrete spatial units (regions or countries) that both act as agents structuring innovation governance and serve effectively as containers providing institutional conditions for success. Instead, they elaborate on the recently formulated Global Innovation Systems approach, which enables researchers to capture the emergence of system resources across spatial scales. With this framework, Binz and Truffer emphasize that beyond the focus on knowledge generation, a better understanding of "valuation" processes is necessary to guide governance structures for generating new technologies and products.

Complementarily to this, Gary Herrigel in Chap. 18 explores a particular form of MNC governance practices within interlinked global production clusters producing identical end products in different markets. Because diffusion of these clusters is accompanied by significant operational uncertainty, Herrigel claims that many emergent MNC governance practices have an experimentalist character. Stakeholder inclusive teams at the center provisionally set product standards and performance metrics that are then appropriately and transparently modified by local teams, often resulting, through formal justification procedures, in the modification of the initial central standards and metrics. Recursivity in knowledge flow and practice diffuses learning and innovation throughout the MNC. Movement towards experimentalist governance in MNCs exists but is neither seamless nor uncomplicated. Three sorts of obstacles are most common—hierarchical insulation, stakeholder exclusion, and inadequate empowerment resources for participants. These obstacles exist not only ex ante, as firms attempt to construct formal experimentalist systems and implement them throughout their global operations; they also are continually regenerated by the experimentalist dynamics themselves. In order to prevent such obstacles from paralyzing the global process of recursive learning, MNCs are developing an array of destabilization mechanisms that systematically undermine insulation and exclusion strategies within the global firm and reconstitute the deliberative experimentalist learning process.

In the final chapter of this volume, Nebahat Tokatli explores the innovation processes in the flat-glass industry, questioning the extent to which interfirm networks facilitate innovation over a long period of time. Ultimately, Toklati argues that networks lose influence on innovations in the course of the evolution of an industry. When it comes to the secondary processing of flat glass, the assets of innovation are now much more dispersed (locally and globally) than they were before. In addition, contexts in which individual products and processes draw on multiple internal and external sources of technology are now becoming more and more pervasive in the secondary processing of glass.

Governance and Geography

Governance fulfills tasks in a wide variety of fields and across a variety of spatial scales. One of this book's key arguments is that governance cannot be conceived as a placeless category (Glückler, Rehner, & Handke, 2019). Geography and space are important in at least two respects, again reflecting the logic of reflexivity: First, space is a context for governance. Material conditions as well as social relations and institutions are often place-specific and thus pose particular conditions and require specific adaptation for governance to be effective. A seemingly universal, a-spatial governance blueprint will yield differential effect in different places. Therefore, we claim that governance theory is unlikely to yield general solutions that convey optimal outcomes at any place and any time. Second, and conversely, space is an object of governance, and governance creates spaces. Governance has implications for the geographical boundedness (jurisdiction) as well as the quality of opportunities and constraints of actors. Governance may also generate geographical spillover effects on other actors not directly involved in the process, both locally and in other places (e.g., climate change, financial crises, forest fire risk, or environmental pollution). The relation to space is not only about locality but also about relations between regions and across scales. For instance, governance can resolve local dependencies and combine them with action in other places, as in a local environmental movement's cooperation with a global NGO, in order to develop greater impact on the ground.

The first relationship between governance and space becomes clear simply by the fact that actors, collective action, and collective problems are situated and bounded in geographical contexts. Governance actors must respond to the specific physical, social, political, and institutional contexts in particular places in order to address conflicts of interest and collective action problems for a common good. The traditions of governance research differ fundamentally in the way in which they conceive the actors and the roles ascribed to them (Knill & Lehmkuhl, 2002; Peters, 2014). Governance can be driven forward by self-organizing, autonomous stakeholders (Gardner, Ostrom, & Walker, 1992; Rhodes, 1996) who coordinate their collective actions in a targeted manner, especially at the local or regional level, where interactions are particularly visible and manageable. Governance is also about differently organizing the relations between the state and civic actors (Bell & Hindmoor, 2009). The role of states in governance is no longer limited to regulation in the legal territorial area that they supervise. It ranges from direct alliances with multiple stakeholders (Osborne, 2000) to its action as the "shadow of the hierarchy" (Héritier & Lehmkuhl, 2008). Governance is acknowledged to create its own institutional space.

Several contributors to this volume take geographical location as the starting point for their governance analysis. They either refer to problems in dealing with the accessibility and exploitability of spatially distributed collective resources, such as agricultural land in the confined space of Easter Island (see Chap. 2 by Lipo et al.),[1] or they highlight the spatial dimensions of risks, as in the case of forest fires in Chile, which threaten larger groups in society (see Chap. 8 by Handke).[2] Space is a suitable category for locating collective problems, even where many problems transgress the conventional distinction between local and global. Researchers use the geography of governance to not only look at the diversity and relations between places, but also at the interdependencies across spatial scales. Often, governance cannot be restricted to just one scale but requires the analysis of several scales and of the interrelations across them. Several contributors to our book, for example, address the challenges of mastering global climate change in context-specific ways (see Chap. 9 by Knox-Hayes et al., Chap. 12 by Niemeyer, and Chap. 4 by Stehr).

Other contributors elaborate on the second relationship between governance and geography, how collective interactions make and transform geography. Space is the place in and through which knowledge is generated, to which certain knowledge is bound, and from which it is intentionally exploited. Situated action and spatial social relationships influence the functionality of governance (see Chap. 12 by Niemeyer and Chap. 10 by Pohlmann and Valarini). Individual actors and their

[1] Also forest landscapes and timber resources in Canada (see Chap. 16 by Hayter and Clapp).

[2] Also the spread of infectious viruses (see Chap. 14 by Raab et al.) or coastal regions affected by rising sea levels in Australia (see Chap. 3 by Scott).

levels of impact can be taken as the starting point for a deeper analysis. Influences on their decisions and actions, however, come from many other dimensions that governance research must take into account. Several authors analyze how localized collective coordination problems affect innovations for global markets and recognize the disadvantages of location-bound and one-sided hierarchical or market-based forms of governance (see Chap. 17 by Binz and Truffer, Chap. 18 by Herrigel, and Chap. 19 by Tokatli).

A comparative empirical approach on governance research creates new analytical opportunities. Researchers should put particular emphasis on analyzing governance cases to assess the utility of and fit with fundamental governance models, and on designing and adapting appropriate contextual solutions (Glückler et al., 2019). In addition, by constant collection and comparison of governance cases, researchers are sensitized to the experience that general models perform differently in different situations, that similar governance challenges can be met with different models, and that well-designed governance processes can fail or have unintended consequences.

Conclusion and Questions Ahead

Governance emerged as a scholarly preoccupation, historically, as the post-World-War-II global order began to fall into crisis in the 1990s. Bureaucratic mechanisms deployed by the state (planning) and by corporations (managerial hierarchies/vertical integration) were revealed to be overly rigid and unsustainable. Global competition and pressures for nearly permanent innovation as well as natural hazards and growing environmental impact of social and economic activities together destabilized roles, industries, and regions, while simultaneously placing a high value on flexibility. Traditional top-down command and control bureaucracy was simply overwhelmed by the new conditions. At the same time, despite great ideological enthusiasm, it quickly became clear that market mechanisms alone were nearly always incapable of maintaining stable and equitable coordination. Competition frequently broke down as players undercut one another to improve (or maintain) their position, whereas social and environmental collateral damage from market action generated pushback from many affected quarters. The result has been pervasive exploration of modes of coordination that subsume, abandon, and sometimes transcend both bureaucratic and market forms of social ordering. The authors of this volume demonstrate quite clearly how governance arrangements today can come in a remarkable array of guises and in ways that transgress traditional, formerly very reliable, analytical oppositions between public and private, bureaucracy and market, national and transnational, political and economic—and so on.

Although we do not pretend to present a comprehensive overview of governance in this volume, the essays here do show how plastic and wide ranging the problem of governance has become. The chapters' authors discuss a remarkably diverse range of governance concepts and techniques. These concepts include corporate governance (see Chap. 17 by Binz and Truffer), network governance (see Chap. 11 by Glückler), adaptive governance (see Chap. 13 by Schultz et al.), deliberative governance (see Chap. 12 by Niemeyer), risk governance (see Chap. 6 by Bell and Hindmoore, Chap. 8 by Handke, and Chap. 5 by Renn), and environmental governance (see Chap. 4 by Stehr, Chap. 3 by Scott, and Chap. 9 by Knox-Hayes et al.), governance by experimentalism (see Chap. 18 by Herrigel) as well as good governance (see Chap. 7 by Avellaneda et al. and Chap. 10 by Pohlman and Valarini). In the governance cases presented, civil society (see Chap. 16 by Hayter and Clapp and Chap. 2 by Lipo et al.), the state (see Chap. 15 by Kim et al. and Chap. 14 by Raab et al.) and the private sector (see Chap. 18 by Herrigel and Chap. 19 by Tokatli) perform different roles with different levels of divisions of labor.

Very broadly, the essays in this volume permit a number of theoretical and loosely empirical observations that can be explored in future research. To start with, its contributors indicate that governance can be profitably arranged along two dimensions. First, governance refers to efforts on the part of interdependent actors (however constituted) to resolve social, economic, or political problems that have been either jointly defined or commonly agreed upon. Second, space and knowledge are two crucial parameters for governance. Knowledge gives content to and drives practice, whereas all forms of action both unfold in and construct space. In future, researchers need to explore the relationship between power and knowledge within governance practices. In particular, to what degree is "jointness" in the identification of coordination difficulties compatible with power asymmetries? Can collaborative governance, dependent upon input from affected stakeholders, be imposed from above, either by states or corporate management (say, in the governance of their supply chains)? To what degree are the more traditional governance mechanisms of hierarchy and market still salient in contemporary governance arrangements and coordination practices?

Along the same lines, the essays here not only serve as testament to the centrality of knowledge in contemporary economic and social life, but also reveal considerable ambiguity about the scope and limits of knowledge for governance. How much knowledge is enough? Can coordination be sustained while important knowledge carrying players are excluded from the design and practice of governance arrangements? What are the limits of inclusivity for the identification of joint problems and goals for associated actors? More broadly, must researchers explore the extent to which governance arrangements can exclude forms of knowledge from entering into participant governance deliberations? How can actors make governance practices more robust and sustainable, capable of accommodating a broad range of challenges and innovations? Similar questions can be posed about space: How fungible are the spatial dimensions of governance? Researchers must more thoroughly examine the extent to which space acts as a constraint on the construction of governance arrangements, and the extent to which the search for stable coordination recasts social interaction spatially.

Finally, two other aspects of the governance discussion in this volume emerge as crucial for future work. First, *uncertainty*, driven by competition, on-going organizational and technological innovation and collateral political and social recomposition, is a crucial environmental condition for understanding contemporary governance efforts. To what extent do those making governance arrangements seek to accommodate themselves to this uncertainty? That is, how are practices arranged in such a way that continuous innovation and "dynamic capabilities" (Teece et al., 1997) are the outcome of governance? Alternatively, is it possible for interconnected players to construct governance arrangements that insulate them from uncertainty? Secondly, reflexivity, in other words, the organized practice of observing the outcomes of coordinated action and adjusting rules and roles in social coordination in light of those observations, is an emergent and crucially important area for analytical attention. Crises come when old arrangements for governing coordination no longer function and alternative forms of coordination have yet to be created. What kinds of creative search and recombinatory processes do actors deploy to overcome crisis and reestablish stable governance? Can such search and learning processes be organized in a systematic manner, such that disruptive and paralyzing crises can be avoided? These and other questions indicate some of the directions for future research in a promising and growing field of transdisciplinary scholarship whose participants aim to help solve collective action dilemmas in environmental, social, political, and economic contexts.

References

Amin, A., & Cohendet, P. (2004). *Architectures of knowledge: Firms, capabilities, and communities*. Oxford, UK: Oxford University Press. https://doi.org/10.1093/acprof:oso/9780199253326.001.0001

Ansell, C., & Torfing, J. (2016). *Handbook on theories of governance*. Cheltenham: Edward Elgar.

Avellaneda, C. N., Bello-Gómez, R. A., & Olvera, J. G. (2020). Explaining subnational governance: The role of governors' codified and uncodified knowledge. In J. Glückler, G. Herrigel, & M. Handke (Eds.), *Knowledge for governance* (pp. 131–159). Knowledge and Space: Vol. 15. Cham: Springer. https://doi.org/10.1007/978-3-030-47150-7_7

Bache, I., & Flinders, M. (2004). *Multi-level governance*. Oxford, UK: Oxford University Press. https://doi.org/10.1093/0199259259.001.0001

Barley, S. R., & Tolbert, P. S. (1997). Institutionalization and structuration: Studying the links between action and institution. *Organization Studies, 18,* 93–117. https://doi.org/10.1177/017084069701800106

Bathelt, H., & Glückler, J. (2011). *The relational economy: Geographies of knowing and learning*. Oxford, UK: Oxford University Press. https://doi.org/10.1093/acprof:osobl/9780199587384.001.0001

Becker, M. C., & Knudsen, T. (2005). The role of routines in reducing pervasive uncertainty. *Journal of Business Research, 58,* 746–757. https://doi.org/10.1016/j.jbusres.2003.10.003

Bell, S., & Hindmoor, A. (2009). *Rethinking governance: The centrality of the state in modern society*. Cambridge, UK: Cambridge University Press. https://doi.org/10.1017/CBO9780511814617

Bell, S., & Hindmoor, A. (2020). Knowledge and governance: Can systemic risk in financial markets be managed? The case of the Euro crisis. In J. Glückler, G. Herrigel, & M. Handke (Eds.), *Knowledge for governance* (pp. 113–128). Knowledge and Space: Vol. 15. Cham: Springer. https://doi.org/10.1007/978-3-030-47150-7_6

Benz, A., & Dose, N. (2010). *Governance—Regieren in komplexen Regelsystemen: Eine Einführung* [Governance—Governing in complex control systems: An introduction]. Wiesbaden: VS Verlag für Sozialwissenschaften.

Bevir, M. (2010). *The SAGE handbook of governance*. London: SAGE.

Binz, C., & Truffer, B. (2020). The governance of global innovation systems: Putting knowledge in context. In J. Glückler, G. Herrigel, & M. Handke (Eds.), *Knowledge for governance* (pp. 397–414). Knowledge and Space: Vol. 15. Cham: Springer. https://doi.org/10.1007/978-3-030-47150-7_17

Borgatti, S. P., & Cross, R. (2003). A relational view of information seeking and learning in social networks. *Management Science, 49,* 432–445. https://doi.org/10.1287/mnsc.49.4.432.14428

Boyer, R. (2005). How and why capitalisms differ. *Economy and Society, 34,* 509–557. https://doi.org/10.1080/03085140500277070

Chhotray, V., & Stoker, G. (2009). *Governance theory and practice: A cross-disciplinary approach.* Houndsmill: Palgrave Macmillan. https://doi.org/10.1057/9780230583344

Fleck, L. (1935). *Entstehung und Entwicklung einer wissenschaftlichen Tatsache: Einführung in die Lehre vom Denkstil und Denkkollectiv* [Origin and development of a scientific fact: Introduction into the doctrine of thinking style and thought collective]. Basel: B. Schwabe.

Gardner, R., Ostrom, E., & Walker, J. (1992). Covenants with and without a sword: Self-governance is possible. *American Political Science Review, 86,* 404–417. https://doi.org/10.2307/1964229

Glückler, J. (2020). Institutional context and place-based policy: The case of Coventry & Warwickshire. *Growth & Change, 51,* 234–255. https://doi.org/10.1111/GROW.12362

Glückler, J. (2020). Lateral network governance. In J. Glückler, G. Herrigel, & M. Handke (Eds.), *Knowledge for governance* (pp. 243–265). Knowledge and Space: Vol. 15. Cham: Springer. https://doi.org/10.1007/978-3-030-47150-7_11

Glückler, J., Meusburger, P., & El Meskioui, M. (2013). Introduction: Knowledge and the geography of the economy. In P. Meusburger, J. Glückler, & M. El Meskioui (Eds.), *Knowledge and the economy* (pp. 3–14). Knowledge and Space: Vol. 5. Dordrecht: Springer. https://doi.org/10.1007/978-94-007-6131-5_1

Glückler, J., Lazega, E., & Hammer, I. (2017). *Knowledge and networks.* Knowledge and Space: Vol. 11. Cham: Springer. https://doi.org/10.1007/978-3-319-45023-0

Glückler, J., & Lenz, R. (2018). Drift and morphosis in institutional change: Evidence from the 'Walz' and public tendering in Germany. In J. Glückler, R. Suddaby, & R. Lenz (Eds.), *Knowledge and institutions* (pp. 111–134). Knowledge and Space: Vol. 13. Cham: Springer. https://doi.org/10.1007/978-3-319-75328-7_6

Glückler, J., Punstein, A. M., Wuttke, C., & Kirchner, P. (2020). The 'hourglass' model: An institutional morphology of rural industrialism in Baden-Württemberg. *European Planning Studies, 28,* 1554–1574. https://doi.org/10.1080/09654313.2019.1693981

Glückler, J., Rehner, J., & Handke, M. (2019). Gobernanza, redes y territorio [Governance, networks, and territory]. *Revista de Geografía Norte Grande, 74,* 5–20. http://dx.doi.org/10.4067/S0718-34022019000300005

Glückler, J., Suddaby, R., & Lenz, R. (2018). *Knowledge and institutions.* Knowledge and Space: Vol. 13. Cham: Springer. https://doi.org/10.1007/978-3-319-75328-7

Handke, M. (2020). The (de-)contextualization of geographical knowledge in forest-fire risk management in Chile as a challenge for governance. In J. Glückler, G. Herrigel, & M. Handke (Eds.), *Knowledge for governance* (pp. 161–191). Knowledge and Space: Vol. 15. Cham: Springer. https://doi.org/10.1007/978-3-030-47150-7_8

Hardin, G. (1968). The tragedy of the commons. *Science, 162,* 1243–1248. https://doi.org/10.1126/science.162.3859.1243

Hayter, R., & Clapp, A. (2020). The remapping of forest governance: From shareholder to stakeholder. In J. Glückler, G. Herrigel, & M. Handke (Eds.), *Knowledge for governance* (pp. 375–395). Knowledge and Space: Vol. 15. Cham: Springer. https://doi.org/10.1007/978-3-030-47150-7_16

Héritier, A., & Lehmkuhl, D. (2008). Introduction: The shadow of hierarchy and new modes of governance. *Journal of Public Policy, 28,* 1–17. https://doi.org/10.1017/S0143814X08000755

Herrigel, G. (2010). *Manufacturing possibilities: Creative action and industrial recomposition in the USA, Germany and Japan.* Oxford, UK: Oxford University Press. https://doi.org/10.1093/acprof:oso/9780199557738.001.0001

Herrigel, G. (2020). Experimentalist systems in manufacturing multinationals: Recursivity and continuous learning through destabilization. In J. Glückler, G. Herrigel, & M. Handke (Eds.), *Knowledge for governance* (pp. 415–439). Knowledge and Space: Vol. 15. Cham: Springer. https://doi.org/10.1007/978-3-030-47150-7_18

Herrigel, G., Wittke, V., & Voskamp, U. (2013). The process of Chinese manufacturing upgrading: Learning in supply chains and through the dynamics of FDI. *Global Strategy Journal, 3,* 109–125. https://doi.org/10.1111/j.2042-5805.2012.01046.x

Hess, C., & Ostrom, E. (2007). *Understanding knowledge as a commons: From theory to practice.* Cambridge, MA: MIT Press.

Hodgson, G. M. (2006). What are institutions? *Journal of Economic Issues, 40,* 1–25. https://doi.org/10.1080/00213624.2006.11506879

Jessop, B. (1998). The rise of governance and the risks of failure: The case of economic development. *International Social Science Journal, 50,* 29–45. https://doi.org/10.1111/1468-2451.00107

Keast, R. (2016). Network governance. In C. Ansell & J. Torfing (Eds.), *Handbook on theories of governance* (pp. 442–453). Cheltenham: Edward Elgar.

Kilduff, M., & Tsai, W. (2003). *Social networks and organizations.* London: SAGE.

Kim, S. Y., Swann, W. L., & Feiock, R. C. (2020). Collective learning and institutional collective action in fragmented governance. In J. Glückler, G. Herrigel, & M. Handke (Eds.), *Knowledge for governance* (pp. 351–373). Knowledge and Space: Vol. 15. Cham: Springer. https://doi.org/10.1007/978-3-030-47150-7_15

Knill, C., & Lehmkuhl, D. (2002). Private actors and the state: Internationalization and changing patterns of governance. *Governance: An International Journal of Policy, Administration, and Institutions, 15,* 41–63. https://doi.org/10.1111/1468-0491.00179

Knoke, D. (2012). *Economic networks.* Cambridge: Polity.

Knox-Hayes, J., Hayes, J., & Hughes, E.-L. (2020). Carbon markets, values, and modes of governance. In J. Glückler, G. Herrigel, & M. Handke (Eds.), *Knowledge for governance* (pp. 193–224). Knowledge and Space: Vol. 15. Cham: Springer. https://doi.org/10.1007/978-3-030-47150-7_9

Kooiman, J. (1993). Governance and governability: Using complexity, dynamics and diversity. In J. Kooiman (Ed.), *Modern governance: New government-society interactions* (pp. 35–50). London: SAGE.

Kooiman, J., & van Vliet, M. (1993). Governance and public management. In K. Eliassen & J. Kooiman (Eds.), *Managing public organizations: Lessons from contemporary European experience* (pp. 58–72). London: SAGE.

Lawrence, T. B., Leca, B., & Zilber, T. B. (2013). Institutional work: Current research, new directions and overlooked issues. *Organization Studies, 34,* 1023–1033. https://doi.org/10.1177/0170840613495305

Levi-Faur, D. (2012). *The Oxford handbook of governance.* Oxford, UK: Oxford University Press. https://doi.org/10.1093/oxfordhb/9780199560530.001.0001

Lipo, C. P., Mischen, P., & Hunt, T. L. (2020). Lessons from Rapa Nui (Easter Island, Chile) for governance in conditions of environmental uncertainty. In J. Glückler, G. Herrigel, & M. Handke (Eds.), *Knowledge for governance* (pp. 25–49). Knowledge and Space: Vol. 15. Cham: Springer. https://doi.org/10.1007/978-3-030-47150-7_2

Lundvall, B.-A., & Johnson, B. (1994). The learning economy. *Journal of Industry Studies, 1*(2), 23–42. https://doi.org/10.1080/13662719400000002

Marin, B., & Mayntz, R. (1991). *Policy networks: Empirical evidence and theoretical considerations.* Public Policy and Social Welfare: Vol. 9. Frankfurt: Campus.

Mayntz, R. (2003). Governing failures and the problem of governability: Some comments on a theoretical paradigm. In J. Kooiman (Ed.), *Modern governance: New government-society interactions* (pp. 9–20). London: SAGE.
Meusburger, P. (2008). The nexus of knowledge and space. In P. Meusburger, M. Welker, & E. Wunder (Eds.), *Clashes of knowledge* (pp. 35–90). Knowledge and Space: Vol. 1. Dordrecht: Springer. https://doi.org/10.1007/978-1-4020-5555-3_2
Niemeyer, S. (2020). Knowledge and the deliberative stance in democratic systems: Harnessing scepticism of the self in governing global environmental change. In J. Glückler, G. Herrigel, & M. Handke (Eds.), *Knowledge for governance* (pp. 269–292). Knowledge and Space: Vol. 15. Cham: Springer. https://doi.org/10.1007/978-3-030-47150-7_12
Nooteboom, B. (2000). Learning by interaction: Absorptive capacity, cognitive distance and governance. *Journal of Management and Governance, 4,* 69–92. https://doi.org/10.1023/A:1009941416749
Osborne, S. P. (2000). *Public-private partnerships: Theory and practice in international perspective*. London: Routledge.
Osborne, D., & Gaebler, T. (1992). *Reinventing government: How the entrepreneurial spirit is transforming the public sector.* Reading: Addison-Wesley.
Ostrom, E. (2000). Collective action and the evolution of social norms. *The Journal of Economic Perspectives, 14*(3), 137–158. https://doi.org/10.1257/jep.14.3.137
Ostrom, E. (2005). *Understanding institutional diversity.* Princeton, NJ: Princeton University Press.
Peters, B. G. (2014). Is governance for everybody? *Policy and Society, 33,* 301–306. https://doi.org/10.1016/j.polsoc.2014.10.005
Pierre, J., & Peters, B. G. (2000). *Governance, politics and the state*. London: Macmillan.
Podolny, J. M., & Page, K. L. (1998). Network forms of organization. *Annual Review of Sociology, 24,* 57–76. https://doi.org/10.1146/annurev.soc.24.1.57
Pohlmann, M., & Valarini, E. (2020). The fight against corruption in Brazil: A case of good governance? In J. Glückler, G. Herrigel, & M. Handke (Eds.), *Knowledge for governance* (pp. 225–241). Knowledge and Space: Vol. 15. Cham: Springer. https://doi.org/10.1007/978-3-030-47150-7_10
Powell, W. W. (1991). Neither market nor hierarchy: Network forms of organization. In G. Thompson, J. Frances, R. Levacic, & J. Mitchell (Eds.), *Markets, hierarchies & networks: The coordination of social life* (pp. 265–276). London: SAGE.
Punstein, A. M., & Glückler, J. (2020). In the mood for learning? How the thought collectives of designers and engineers co-create innovations. *Journal of Economic Geography, 20,* 543–570. https://doi.org/10.1093/jeg/lbz019
Raab, J., Kenis, P., Kraaij-Dirkzwager, M., & Timen, A. (2020). Ex ante knowledge for infectious disease outbreaks: Introducing the organizational network governance approach. In J. Glückler, G. Herrigel, & M. Handke (Eds.), *Knowledge for governance* (pp. 319–349). Knowledge and Space: Vol. 15. Cham: Springer. https://doi.org/10.1007/978-3-030-47150-7_14
Renn, O. (2020). Risk governance: From knowledge to regulatory action. In J. Glückler, G. Herrigel, & M. Handke (Eds.), *Knowledge for governance* (pp. 93–111). Knowledge and Space: Vol. 15. Cham: Springer. https://doi.org/10.1007/978-3-030-47150-7_5
Rhodes, R. A. W. (1996). The new governance: Governing without government. *Political Studies, 44,* 652–667. https://doi.org/10.1111/j.1467-9248.1996.tb01747.x
Rosenau, J. N., & Czempiel, E.-O. (1992). *Governance without government. Order and change in world politics.* Cambridge, UK: Cambridge University Press. https://doi.org/10.1017/CBO9780511521775
Schultz, L., West, S., & Florêncio, C. (2020). Nurturing adaptive governance through environmental monitoring: People, practices, politics in the Kruger to Canyons Biosphere Region, South Africa. In J. Glückler, G. Herrigel, & M. Handke (Eds.), *Knowledge for governance* (pp. 293–318). Knowledge and Space: Vol. 15. Cham: Springer. https://doi.org/10.1007/978-3-030-47150-7_13

Scott, M. (2020). Knowledge of governance as knowledge for governance: Spatialized techniques of neutralization. In J. Glückler, G. Herrigel, & M. Handke (Eds.), *Knowledge for governance* (pp. 51–67). Knowledge and Space: Vol. 15. Cham: Springer. https://doi.org/10.1007/978-3-030-47150-7_3

Scott, W. R. (2007). *Institutions and organizations: Ideas, interests, and identities.* Thousand Oaks: SAGE.

Stehr, N. (2020). The atmosphere of democracy: Knowledge and political action. In J. Glückler, G. Herrigel, & M. Handke (Eds.), *Knowledge for governance* (pp. 69–91). Knowledge and Space: Vol. 15. Cham: Springer. https://doi.org/10.1007/978-3-030-47150-7_4

Stoker, G. (1998). Governance as theory: Five propositions. *International Social Science Journal, 50,* 17–28. https://doi.org/10.1111/1468-2451.00106

Streeck, W., & Thelen, K. (2005). *Beyond continuity: Institutional change in advanced political economies.* Oxford, UK: Oxford University Press.

Teece, D. J., Pisano, G., & Shuen, A. (1997). Dynamic capabilities and the strategic management. *Strategic Management Journal, 18,* 509–533. https://doi.org/10.1002/(SICI)1097-0266(199708)18:7<509::AID-SMJ882>3.0.CO;2-Z

Tokatli, N. (2020). Networks as facilitators of innovation in technology-based industries: The case of flat glass. In J. Glückler, G. Herrigel, & M. Handke (Eds.), *Knowledge for governance* (pp. 441–460). Knowledge and Space: Vol. 15. Cham: Springer. https://doi.org/10.1007/978-3-030-47150-7_19

van Dijk, T. A. (2014). *Discourse and knowledge: A sociocognitive approach.* Cambridge, UK: Cambridge University Press. https://doi.org/10.1017/CBO9781107775404

Williamson, O. E. (1991). Comparative economic organization: The analysis of discrete structural alternatives. *Administrative Science Quarterly,* 36, 269–296. https://doi.org/10.2307/2393356

Williamson, O. E. (2005). The economics of governance. *American Economic Review, 95*(2), 1–18. https://doi.org/10.1257/000282805774669880

Part I
How Knowledge Enables Governance

Chapter 2
Lessons from Rapa Nui (Easter Island, Chile) for Governance in Conditions of Environmental Uncertainty

Carl P. Lipo, Pamela Mischen, and Terry L. Hunt

> It amazes me how people are often more willing to act based on little or no data than to use data that is a challenge to assemble. (Robert J. Shiller, Economist (Heins, 2010))

Over the past several years, a group of islanders living on a remote and tiny island in the corner of the southeastern Pacific Ocean have been wrestling with complex but fundamental issues of governance. The island of Rapa Nui (Easter Island, Chile) is just 161 km^2 with a resident population of about 6600 people as of 2016 (see Figs. 2.1 and 2.2). After more than 130 years of Chilean rule—located more than 3600 km away across the sea—islanders have been working to establish a governance system to equitably manage the island's cultural and natural resources while also addressing the overlapping sets of authority that stem from family groups, a series of 10 clans (*mata*), resident-elected town government officials, a provincial governor appointed by the Chilean President, and numerous Chilean agencies at the provincial and national levels. Although the island has been a sovereign territory of Chile since 1888, challenges to the overarching colonial structure of governance have been growing over the past several decades, as islanders have moved from addressing voting rights in the 1960s (Tector, 2014), to referendums for decolonization in the 1980s (Delaune, 2012, p. 129), to the first native governor appointment in 1984, to its status as a special territory in 2007, to calls for complete autonomy,

C. P. Lipo (✉)
Department of Anthropology, Binghamton University, Binghamton, NY, USA
e-mail: clipo@binghamton.edu

P. Mischen
Department of Public Administration, Binghamton University, Binghamton, NY, USA
e-mail: pmischen@binghamton.edu

T. L. Hunt
Honors College, University of Arizona, Tucson, AZ, USA
e-mail: tlhunt@email.arizona.edu

J. Glückler et al. (eds.), *Knowledge for Governance*, Knowledge and Space 15,
https://doi.org/10.1007/978-3-030-47150-7_2

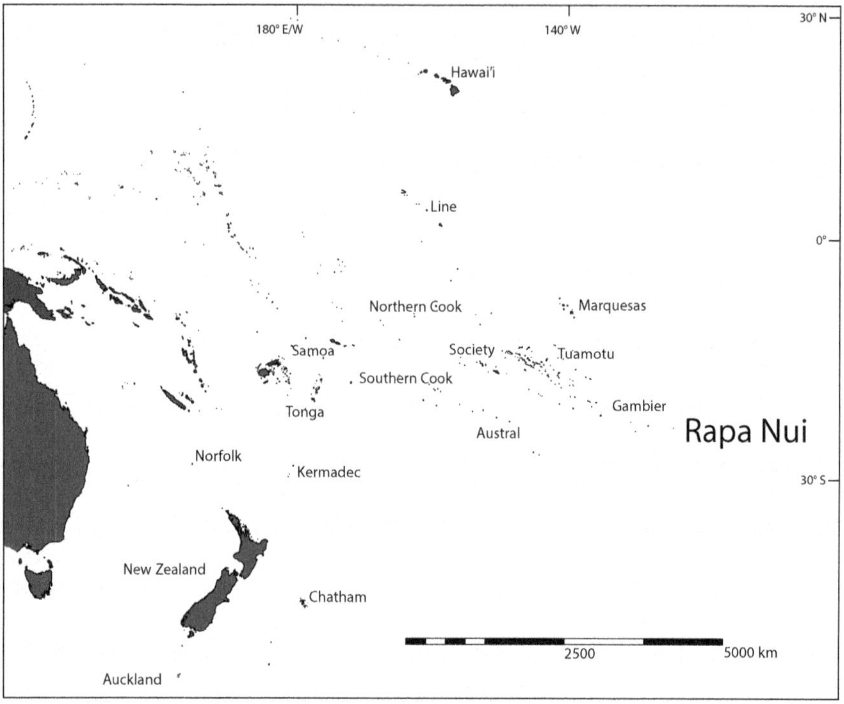

Fig. 2.1 Location of Rapa Nui in East Polynesia. Source: Design by authors

demonstrations, and open conflict in 2010–2012 (COHA, 2011; Delaune, 2012; Warren, 2011). The most recent step towards reshaping island governance has been the establishment of *Ma'u Henua*, a Native Rapa Nui organization that recently assumed administration of the large portion of the island comprising the National Park and formerly under the Chilean National Forest Corporation (i.e., CONAF).

Efforts to change Rapa Nui's governance structure are particularly timely given present and growing threats to the island's resources. With increasing numbers of flights and cruise ship visits, the annual number of visitors to the island has surged to more than 60,000 (CODEIPA, 2015), with expectations of significant increases. Such increased numbers have resulted in greater use of the natural landscape, with widespread impacts to the archaeological record. In 2008, for example, a Finnish tourist broke a portion of an ear from one of the island's iconic statues, a *moai,* creating an international incident (Barfelz, 2011). These kinds of events have led to greater efforts to restrict access to archaeological features and keep tourists from damaging the cultural resources they come to see. In addition to rising visitor numbers, the island's residents grew from about 3000 to nearly 4000 between 1992 and 2002, and the current 6600 has surpassed predictions of just a few years ago (Biblioteca del Congreso Nacionale de Chile, 2015). The population growth has resulted in many new houses in the town of Hanga Roa and its surrounding area, concerns over the growing number of cars and traffic, as well as an expansion of

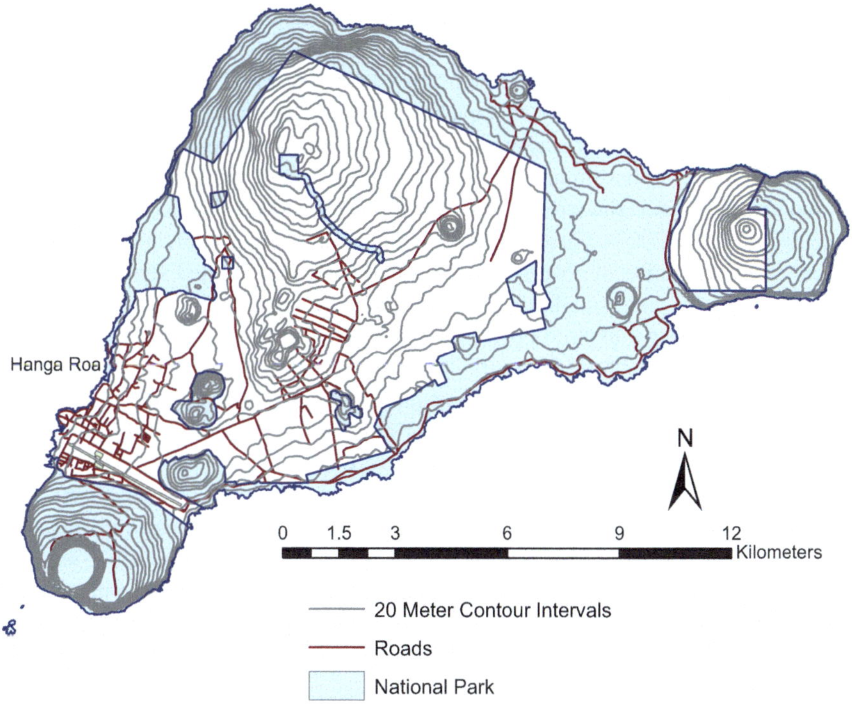

Fig. 2.2 The island of Rapa Nui (Easter Island, Chile). Source: Design by authors

hotels and businesses to support tourism. From a resource perspective, the population growth places an ever-increasing burden on the freshwater resources that come from wells, garbage that must be removed or put in landfills, septic systems, energy requirements, as well as imported fuel, food, and building materials. The impacts to the island's resources are occurring at a time when island residents are increasingly worried about the predicted effects of global climate change (Carabine & Dupar, 2014). To meet such challenges, governance structures must operate efficiently, effectively, and equitably.

Rapa Nui is a particularly notable case when it comes to identifying strategies for living on a remote island in the face of tremendous adversity. Over the course of the twentieth century, Rapa Nui has been promulgated as a case of an island community whose poor decisions ultimately led to environmental and demographic catastrophe. Although this perspective is derived from narratives of the earliest Europeans who visited the island in the eighteeth century (Hunt & Lipo, 2011), the idea that Rapa Nui's past represents a case of failure of governance remains strong in popular culture. Part of this popularity comes from the work of Jared Diamond via his various essays (e.g., Diamond, 1995) and his book *Collapse: How Societies Choose to Fail or Succeed* (2005). The popularity of this notion has also led many scholars to offer Rapa Nui as an exemplar case for potential future human population growth coupled with dwindling natural resources (e.g., Erickson & Gowdy, 2000; Foot, 2004; Nagarajan, 2006). In these cases, Rapa Nui is used to illustrate what happens if communities fail to consider the long-term consequences of their actions, and that if we are to avoid the same fate as Rapa Nui, we (i.e., contemporary Western societies) need to avoid making the same kinds of mistakes. One key assumption embedded in this logic is that prehistoric Rapa Nui was a *failed* governance system—a community that made the wrong choices and thus provides a warning for the future (Flenley & Bahn, 2003). Given that the governance of Rapa Nui is moving toward more traditional forms of governance, the veracity of our knowledge of Rapa Nui's past is critical for understanding the possibility of success in the future.

In this chapter, we explore how knowledge informs governance in general and why understanding the foundations for knowledge is critical to effective governance systems. We present the changing face of modern Rapa Nui governance that points toward a move to more traditional forms of governance. Based on this discussion, we examine the logic that supports the notion of Rapa Nui as a case of environmental and demographic catastrophe. We then review how new research has drastically reshaped our understanding of Rapa Nui prehistory. Based on this new information, we reconsider how the people of Rapa Nui are changing their governance to suit the community's new challenges. We suggest that a better understanding of the role of knowledge in governance potentially re-shapes our assumptions about strategies for the future and how we can engineer governance systems to consider new and changing information.

Governance and Knowledge

As social communities, we are challenged to make effective decisions that have future impacts. Much of the difficulty that surrounds systems of decision-making comes from the uncertainty associated with the knowledge upon which decisions might be based: The greater the uncertainty, the more difficult the decision. People often use policy to guide their decisions and by doing so assume a degree of veracity of a priori knowledge with the hope that their decisions will have a greater likelihood of achieving some desired outcome. For example, policymakers use experience with the incidences of forest fires and the conditions that cause them to guide policy about building requirements and fire break maintenance. The degree to which one can make decisions based on knowledge comes from one's understanding of risk (i.e., events for which one can calculate the odds) and minimizing uncertainty (i.e., events for which one lacks sufficient information to accurately calculate the odds (Knight, 1921)). Uncertainty arises in situations where the number of factors leading to an outcome are unknown, too numerous, and/or too complex based on current systems understanding. Although decisions based on an assessment of risk are typically made by balancing the odds, costs, and potential returns, uncertainty can only be mitigated through the generation of knowledge. The more one knows about a phenomenon and its explanation, the lower the degree of uncertainty and the better one's decision-making can be.

It is in this way that knowledge and its creation are critical features in the operation of effective governance. Traditionally, governance is viewed as "the process by which a society or organization steers itself" (Rosell, 1999, p. 1). For the purposes of this chapter and to accommodate contemporary and archaeological contexts, we must establish a definition for governance that contains the necessary and sufficient conditions for all forms (i.e., observed and inferred from archaeological evidence). Here, we combine a cultural framework from cultural inheritance theory (e.g., Eerkens & Lipo, 2007; Laing, 2008; Richerson & Boyd, 2005) with mechanisms that favor pro-social behaviors on the scale of groups (Bowles, Choi, & Hopfensitz, 2003; Henrich, 2004). For our discussion, governance is thus defined as the *culturally inherited rule set for individual and group behaviors that serve to benefit group level unit of organization*. Based on this definition, the rule set can be explicit (e.g., written) and/or implicit (e.g., culturally inherited) and can take a form that is formal (e.g., laws) and/or informal (e.g., customs). The rule sets are also cultural and contingency-bound, as governance depends on contingent history as well as the combination of information used to assemble the rule set. Because information can change over time, so can governance structures.

Decision-making can happen in one of four contexts: simple, complicated, complex, or chaotic (Snowden & Boone, 2007). Modern day governance occurs largely within a complex context, which is characterized by flux and unpredictability, unknown unknowns, many competing ideas, and the need for creative and innovative approaches (Snowden & Boone, 2007, p. 7). The governance of Rapa Nui is no exception. Governing in complex contexts means recognizing many bases of knowledge. Clarke et al. (2013) argue that in situations such as coastal adaptation to climate change, science and technical knowledge alone are inadequate to deal with the system uncertainties and that participatory, local (particularly indigenous) knowledge and a networked approach to governance is preferred.

Governance Structure

Within contemporary forms of governance, knowledge is managed in ways that depend on the underlying philosophy of governance. The public administration regime, which emerged during industrialization, produced what we now recognize as the large governmental bureaucracy. According to Max Weber, "the more complicated and specialized modern culture becomes, the more its external supporting apparatus demands the personally detached and strictly 'objective' *expert*" (Weber, 1978, p. 216). Other influential perspectives during this time period were scientific management (Taylor, 1911), which prized efficiency over all else and the separation of politics and administration (Goodnow, 1900). Taken together, the focus of these theories was the objectivity of knowledge and its place within large bureaucratic organizations. Toward the end of this period, theorists came to recognize the issue of equity (Frederickson, 1971), but remained focused on governmental institutions as having primary responsibility for the creation of a more equitable state. As time passed and large bureaucracies became known as inflexible bastions of red tape, the public sector began looking to the private sector for answers. Under this New Public Management (NPM) regime (sensu Osborne, 2010), knowledge shifted from the bureaucracy to the "customer." Osborne and Gaebler (1992) argued that as society became one of knowledge workers, the "one-size-fits-all services" of government were no longer satisfactory. They called for government agencies to "reinvent" themselves by being entrepreneurial and listening to their customers. The market became the mechanism by which knowledge exerted its impact on the public sector. The lasting effects of NPM were outsourcing and decentralization (Alonso, Clifton, & Díaz-Fuentes, 2015), both of which served to broaden the base of knowledge from mainly within a large federal bureaucracy to include the private and nonprofit sectors as well as local governments.

Since the late 1990s, western public administration has been moving to yet another governance regime: New Public Governance. To understand the New Public Governance approach to knowledge, one must look at the networks and collaboration literature. As early as 1997, O'Toole (1997, p. 45) urged the field of public administration to "treat networks seriously" as "[p]ublic administration increasingly takes place in settings of networked actors who necessarily rely on each other and cannot compel compliance on the part of the rest." Since that time, there has been an explosion of literature regarding networks and collaboration. Emerson and Nabatchi (2015, p. 25) created an integrative framework for collaborative governance that "attempts to identify and arrange the concepts needed to study and understand collaborative governance regimes." They view knowledge as critical for creating the capacity for joint action. Just as knowledge was the currency of the market in the New Public Management approach, Emerson and Nabatchi (2015, p. 71) argue:

> In many ways, knowledge is the currency of collaboration. Knowledge, once guarded, must be shared with others; and knowledge jointly needed must be generated by participants working together. Contested knowledge requires full consideration, and incomplete knowledge must be balanced and enhanced with new knowledge. In essence, collaboration requires the aggregation, division, and reassembling of data and information, as well as the generation of new, shared information.

One can also look at the roles that information and knowledge play from an organizational perspective. Wei Choo (2006, p. ix) argues that organizations use information for sense making, knowledge creation, and decision-making. A "knowing organization" can anticipate environmental changes, learn and innovate, and "take timely, purposive action."

The key here is that as groups begin to confront new social, economic, and environmental challenges such as those produced by climate change, there will be pressures on governance systems to change to ones better suited to managing a multiplicity of issues and voices. Additionally, those governance structures best suited to a radically uncertain future are those that are most able to attend to change and that can adapt to new knowledge.

Governance of Modern-Day Rapa Nui

When one examines the recent historical governance structures of Rapa Nui, one finds systems that were imposed by an external entity (Chile) to operationalize a colonial structure. This structure emphasized external knowledge in the actions of governance. For example, the Rapa Nui National Park (RNNP) was created in 1935. In 1973, the park administration was officially given to the National Forestry Corporation (CONAF). In addition to managing erosion and deforestation, CONAF was charged with the management of the park as a cultural resource and source of cultural tourism (CODEIPA, 2015, p. 10). CONAF is comprised of three departments. The first, Park Administration, oversees park operations, supervizes the park rangers, and includes one archaeological expert. The second department is Natural Resources, which is in charge of the nursey and the Forest Fire Brigade. The third is the Administration and Finance Department.

Several criticisms have arisen regarding CONAF's management of the Park and resulted in the creation of a proposal for a new park administration (CODEIPA, 2015). The charges can be divided into two categories. The first category addresses inadequacies in CONAF's current management of the park, seen as resulting in the damage of archaeological artifacts, including an inadequate number of park rangers for the protection of the archaeological and cultural artifacts, inadequate collection of entry fees, and an insufficient number of archaeological experts.

The second category is longer-standing and relates to the history of how lands have been acquired, used, and allocated. In 1988, 36 five-hectare parcels of land (*parcelas*) were taken from the National Park and repatriated to islanders (Ramirez, 2000). Later, the Comisión de Desarrollo de Isla de Pascua, created by the Indigenous Law in 1993 and launched in June 1999, began distributing *parcelas* to 267 families, land that was owned by the state development corporation Corfo (IWGIA, 2012). Although this process has rightly resulted in the return of land to the island's native community, the transfer of property to individuals has resulted in substantial conflict over land received (given disparities in economic value), leading to greater community strife and loss of protection for the archaeological record (Ramirez, 2000).

In 2014, the Rapa Nui Commissions of CODEIPA asked CONAF to establish a wholly indigenous management system for the park. CONAF instead presented a co-administration plan (known as GOSPAN), in which the Rapanui People's role was merely consultative (CODEIPA, 2015, p. 2). CODEIPA responded with its own proposal for an organization called *Ma'u Henua*, to be implemented in three phases. The first phase, viewed as a transition phase, is the GOSPAN proposal, which allowed for co-administration between CONAF and the Rapanui People represented by the *Ma'u Henua* Council. In this phase, operations continued to be the responsibility of CONAF, but the voice of the Rapanui People was enhanced in strategic decision-making and the management of a Reinvestment Fund. During this phase, the emphasis was on adequate representation of the Rapanui and the building of organizational capacity to better manage the archaeological and cultural resources.

The second phase, beginning in June 2018, was called the Law Stage. In this phase, the authors of the *Ma'u Henua* proposal argued for the creation of a Public Law Corporation, managed by the members of the Rapanui. In this phase, roles were reversed and CONAF became subsumed by a Public Law Corporation responsible for the management of the park, with state officials acting as advisors. Additionally, a Rapa Nui Park Law established new park boundaries under the administration of the Corporation. The Board of the Corporation is constituted by four members of Rapanui, elected by the 36 traditional families, four Rapa Nui professionals, elected by popular election by members of the Rapanui, one representative of the Council of Elders, one representative of the Provisional Government, one representative of the Municipality of Isla de Pascua, and one representative of CONAF. In addition to the Board of Directors, there is a Technical Advisory Board that includes members from the Provincial Government, Municipality of Easter Island, CONAF, Council of Monuments, CAM, Ministry of National Assets, CONADI, and Sernatur (CODEIPA, 2015, pp. 35–36). The Corporation will be comprised of seven (rather than three) departments: Operations, Archaeology and Heritage, Natural Resources, Planning and Development, Administration, Communications, and Financing.

The final stage, proposed for the year 2025, is called the Consolidation Stage. During this stage, the vision is for the Board of the Corporation to be comprised only of members of the Rapanui People (the four members elected by families, four professionals, and one representative of the Council of Elders).

Analysis of Governance Regimes: From CONAF to Ma'u Henua

How are the theories of governance and knowledge reflected in the past, present, and future governance of the park? Beginning with CONAF administration of the park, it is evident that the initial approach was the public administration approach. Complete control of the park rested within this bureaucracy and knowledge was largely provided from external sources in a colonialist fashion. Over time, with indigenous claims to land and occupations of the park, one could argue that CONAF moved to more of a New Public Management style, dominated by market mechanisms and private property rights rather than collective decision-making. The GOSPAN proposal represents a move toward a New Public Governance approach to the park. However, by continuing to hold the reins and cede no real control of the park to the Rapanui People, CONAF set the stage for the *Ma'u Henua* proposal, which continues organizational (rather that network) governance of the park, but under the control of the Native Rapanui.

The second stage of the *Ma'u Henua* proposal most closely resembles the New Public Governance regime. It is plural and pluralist in its approach, with representatives of the Rapanui, local elected officials (which may or may not be Rapanui), and

CONAF on the Board of Directors, as well as a more expansive Technical Advisory Board that includes representatives of Chilean government organizations. The focus is not solely on the organization, but the organization and its environment, and will undoubtedly deal with the negotiation of values, meanings, and relationships.

The type of governance that will emerge during the Consolidation Stage is unclear. The *Ma'u Henua* proposal includes an emphasis on moving toward a traditional (Polynesian) governance approach. Therefore, to understand what the governance approach of the future will be, one must look to the governance of the past.

How do the authors of the *Ma'u Henua* proposal handle the role of changing knowledge? One type of knowledge is technical. The *Ma'u Henua* proposal's authors recognize this as a lack of concern for archaeological expertise as traditionally embraced by CONAF. Although the new organization will certainly provide direct contribution by local community members over the protection and preservation of the archaeological record, it is not clear from the proposal, however, just how the new organization will foster new forms of expertise that would come from a blend of local knowledge combined with the potential for external technological contributions. Educational programs that train community members, however, will certainly result in significant contributions in this area.

The second is cultural. In the new *Ma'u Henua* structure, local knowledge is now embedded explicitly into governance through the pluralist organization. This structure will potentially go a long way to enabling the community to respond to local needs and in ways that are consistent with local values. But is this governance structure sufficiently adaptive to cope with the magnitude and uncertainty of impacts of events that are associated with future climate change? From the traditional narrative of the island, one might conclude that the answer is no. The "collapse" accounts that are often assumed to be (e.g., Diamond, 2005; Flenley & Bahn, 2003) suggest that the island's populations, acting on individual maximizing strategies, tend to overexploit resources and produce their own demise. Thus, one might argue that the degree to which local knowledge is incorporated into governance, as was entirely the case in prehistoric times, will be correlated with the likelihood of failure. Anti-colonial sentiments aside, wouldn't the island do better with a smartly designed, top-down governance structure?

This claim and the potential that *Ma'u Henua* and local information play in guiding the future of the island requires an evaluation of the assumptions built into traditional ideas about Rapa Nui. Is Rapa Nui's past as solid an example of ecological destruction? Did local strategies for managing limited resources and environmental uncertainty ultimately fail, leading to "collapse"? To develop an understanding of how reliance on traditional governance structures will impact the future sustainability of Rapa Nui, one must reconsider what one knows about Rapa Nui's past.

Knowledge of the Past

Based on popular media such as the 1994 movie *Rapa Nui* (Reynolds & Rose-Price, 1994) and Diamond's 2005 book *Collapse*, the prehistory of Rapa Nui is commonly considered to represent "the" canonical example of a population that failed catastrophically through its own actions. In this account and with steadily increasing intensity, past populations are assumed to have engaged in an island-wide cult of massive statue construction and transport that required tremendous resources to support. Ultimately, due to the resources required to sustain the outsized population and their outlandish behavior, the island's originally abundant natural resources became depleted. Consequently, the people of Rapa Nui suffered the inevitable consequences of their actions: ecological failure, warfare, starvation, cannibalism, depopulation, and societal collapse. The remaining people of the island were left in an environment that was forever degraded relative to previous times of ecological abundance.

Until relatively recently, Rapa Nui's tale has gone largely unchallenged. For much of the twentieth century, this kind of story was taken as simple fact, for its logic seems unassailable. Researchers' assumptions about Rapa Nui were reinforced general assumptions about humans and their behavior. Indeed, the notion that a "Paradise Lost" parable could have transpired on Rapa Nui is certainly not far-fetched when one examines the environment and geography of the island. For instance, the island is remarkably small—just 161 km^2 in overall surface area, a size that allows one to walk across in a single leisurely day. The island is also located in a remote part of the southeastern Pacific, more than 3500 km from the coast of South America, 4000 km from Tahiti, and almost 2000 km from Pitcairn Island, the nearest other inhabited island. And if the small size and remote location were not enough to make inhabitation unlikely, the island is exceedingly poor in natural resources. There are no permanent streams and only limited terrestrial and marine resources. The volcanic soils that comprise the entire island have experienced millennia of weathering and, as a result, have low agricultural productivity, the island lacks a productive fringing reef, and the climate is subtropical with seasonally variable rainfall and droughts. At first glance, Rapa Nui is an island that would present a challenge to simple human habitation, much less monumental architecture.

Paradoxically and despite the limited abundance of natural resources, the island boasts some of the most dramatic examples of prehistoric monuments and statues in the world. European visitors were repeatedly astonished to find that islanders had created more than 1000 massive statues, known as *moai,* with hundreds transported many kilometers across the volcanic terrain and placed atop massive stone-constructed platforms, or *ahu*. This record stands in stark contrast with the island's natural setting.

One way in which the contradiction between the island and its limited natural resources has been reconciled is to presume that the number of statues and monuments can only have been constructed if there was a time in which resources were more plentiful. Speculations about the cause of the island's deforestation and cultural ruin began with one of the island's early European visitors. From a single day's visit in April 1786, French explorer Jean-François de Galaup La Pérouse speculated that Rapa Nui's past inhabitants decimated the island's trees and that the present inhabitants were "indebted to the imprudence of their ancestors for their present unfortunate situation" (La Pérouse, 1797, pp. 318–319). This speculation forms much of the basis for the assumptions made by later authors. For example, many speculate what the likely population must have been (e.g., Bologna & Flores, 2008; Brander & Scott Taylor, 1998; Puleston et al., 2017; Reuveny & Decker, 2000) and then use this to model the environment, assuming that so many statues could not possibly have been made on such a tiny island in any other way. As Diamond (1995, p. 62) reasons: "[T]he statues imply a society very different from the one that Roggeveen saw in 1722. Their sheer number and size suggest a population much larger than 2000 people."

The logic is simple: Large statues dictate that more massive numbers of populations of people once existed and that they were fueled by an environment that must have been more abundant than what was observed at contact. This logic is certainly seductive, in part due to the deep-rooted nature of the idea that humans tend to despoil the world in which they live.

Ecology and Rapa Nui

Drawing on growing ecological awareness, William Mulloy (1974) published an account of pre-contact Rapa Nui society in which the population invested in spectacular constructions, statues, and ceremonial activities leading to over-exploitation of the island's fragile resources and devastating warfare. Mulloy's narrative gained additional support with the documentation that the island lost a once-extensive palm forest through studies of sediment cores taken from the island's volcanic lakes (e.g., Flenley, 1979; Flenley et al., 1991; Flenley & King, 1984). Kirch (1984, p. 264) echoed this story, writing that by the time of European contact the island had "already begun a downward spiral of cultural regression" and "crashed devastatingly." Bahn and Flenley (1992) followed this thread and argued for "collapse" scenario, suggesting Rapa Nui served as a microcosm of the Earth's impending resource and population crisis.

The biogeographer and popular science author Jared Diamond (1995, p. 63) later adopted these accounts and widely popularized them as a moral for our time: "In just a few centuries, the people of Easter Island wiped out their forest, drove their plants and animals to extinction, and saw their complex society spiral into chaos and cannibalism. Are we about to follow their lead?" Diamond (2005, p. 118) asserts that Rapa Nui is "the clearest example of a society that destroyed itself by over exploiting its own resources" and that the consequences of deforestation "start with starvation, a population crash, and a descent into cannibalism." Diamond (2005) goes on to argue that for Rapa Nui the efforts required to carve and transport the giant statues eventually led the population to deplete their own natural resources and plunge into crisis induced by overpopulation and environmental destruction. In other words, people willingly destroyed their island and, in turn, destroyed themselves, thus committing "ecocide."

Diamond and other researchers (e.g., Flenley & Bahn, 2003) tell a story that provides a powerful warning for today's potential destruction of the global environment. With scientific recognition that human industrial practices are resulting in rapid climate change with radical impacts to habitat, rainfall patterns, storms, and sea-level, there are compelling reasons to accept the Rapa Nui "ecocide" narrative as simple fact (even if it is not). Flenley and Bahn (2007, p. 13) argue that "the point about the present ecological prognoses for the world is not that they are absolutely proven, but that they may well happen, and therefore we must take evasive action before it is too late." Apparently, the facts do not matter as much as the need to act quickly.

From this perspective, it is not surprising that Rapa Nui has served as an exemplar of the consequences of ignoring the impacts that humans make on their environment. As of December 5th, 2017, for example, the topic of "Easter Island," "environment," and "warning" currently appears on more than 1,470,000 websites and countless blogs (e.g., Busch, 2016; Hari, 2005). The warning narrative of Rapa Nui has spread through popular culture while also providing rationale for governance decisions—around the globe but also on Rapa Nui itself, as islanders are presently considering the best strategies for local governance of the island's limited resources.

Given the importance of society's understanding of the consequences of its actions relative to the future, if one hopes to use Rapa Nui as a case study that leads to behavior change, one should be particularly concerned about the veracity of current understanding of the island's prehistory. Although the idea that Rapa Nui's history demonstrates the consequences of unbounded growth is consistent with general and widespread ideas about human behavior as well as contemporary ecological fears, does the evidence found in the archaeological record of the island support these assumptions? The answer to this question is significant, as the effectiveness of governance will depend on the degree we have well-documented and thoroughly researched evidence.

Questioning Assumptions of Rapa Nui Governance Failure

Proponents of the "collapse" narrative for Rapa Nui make the critical assumption that the governance structures (i.e., the cultural traditions connected to individual and group-level organization) resulted in actions inconsistent with the island's empirical constraints and conditions. The carving and transport of massive statues and documented loss of a palm forest with the assumed consequences would appear to be inconsistent with resource management of the island, leading one to think that the Rapanui lacked a governance system that accounted for the long-term effects of their actions. But given the fact the islanders lived in a remote and isolated location where their actions (e.g., how much food to grow, where to plant, how much land to clear, how many children to have, how much to share, how to compete) directly determined their survival on an seasonal basis, it is reasonable to question the idea that the islanders engaged in activities without some reason to believe that their efforts would prove beneficial—even they were ultimately disastrous. From an islander's perspective, one should wonder if local conditions led the population to expect long-term benefits from their behavior.

The "collapse" narrative assumes that statue-making was fueled by a population of organizational "complexity" and then, in the words of Kirch (1984), experienced a "downward spiral of cultural regression." The assumption is based in *orthogenesis*, the idea that societies progress in their development and reach "peaks," as indicated by the level of apparent organization exhibited by the cultural achievements and driven by progress (variously defined, if at all) as an inherent mechanism of change. Extrinsic changes, then, necessarily result in failure and regression to earlier simpler states. Orthogenesis—and the related assumptions about the way which societies change—provides much of the theoretical warrant for a concept of "collapse." Overall, however, empirical studies of the nature of change reject orthogenesis. As the many contributions to McAnany and Yoffee's (2009) book *Questioning Collapse* indicate, population "collapse" is often just change in the way populations are organized without any "failure" or "cultural regression." In these views, local populations change with innovation in new solutions for success, even though those new solutions might appear as "collapse." Here, we might ask ourselves whether Rapa Nui statue manufacture and the loss of palm forest might have been solutions to the island's constraints, rather than wanton behavior leading to destruction.

There are multiple reasons to challenge the collapse narrative and to look carefully at the evidence about the conditions leading to monumental architecture, environmental change, and the observations made by early Europeans as they arrived on the island in the eighteenth century.

A New Understanding of Rapa Nui Prehistory: Five Things Now Known About the Island and Its Past

Although the "collapse" story of Rapa Nui fits many cultural expectations about human behavior, researchers conducting fresh studies have greatly reshaped what is now known about the island (e.g., Cauwe, 2011; Cauwe & de Dapper, 2015; Hunt, 2007; Hunt & Lipo, 2006, 2008, 2011; Lipo & Hunt, 2009; Lipo, Hunt, Horneman, & Bonhomme, 2016; Lipo, Hunt, & Rapu Haoa, 2013; Morrison, 2012; Mulrooney, 2012, 2013; Mulrooney, Ladefoged, Stevenson, & Rapu Haoa, 2009). Based on excavations, extensive surface surveys, remote sensing of island structure, revaluation of chronological evidence and detailed examinations of attributes related *moai* transport, the new findings can be summarized in these five categories: (1) the empirical basis for prehistoric "collapse," (2) post-European-contact events, (3) prehistoric population size and structure, (4) the island's natural resources, and (5) *moai* transportation.

Collapse

Most significantly, there is simply no empirical evidence for a prehistoric demographic catastrophe. Instead, notions of a "collapse" prior to European contact can be traced to the misconceptions of these early visitors (Hunt & Lipo, 2011) and the perpetuation of historic myths (Hunt & Lipo, 2010, 2011; Lipo & Hunt, 2009; Mulrooney, 2012; Mulrooney et al., 2009). Archaeological evidence cited for a pre-contact "collapse" reveals just the opposite: steadily expanding landscape use (Stevenson et al., 2015) until the arrival of Europeans followed by well-documented impacts due to European contact (e.g., see Lipo & Hunt, 2009; Mulrooney, 2012). Fundamentally, there are few archaeological indications that the population of Rapa Nui was ever much larger than the estimated 3000 witnessed at European contact (Boersema, 2017; Corney, 1908; Hunt, 2007; Morrison, 2012) or that it had ever been substantially larger in the past and then declined. Although researchers continue to make claims of much larger populations (e.g., Diamond, 2005; Puleston et al., 2017), they base these claims on conjecture or preconceptions of what "could have" happened without linking the claims to any empirical evidence for population size. For example, there is currently no evidence for a hiatus in the archaeological record that might signal a massive population decline (Mulrooney, 2013; Mulrooney et al., 2009; Stevenson et al., 2015), which would be required if a large population once existed. There is also little evidence for the level of conflict associated with the "collapse" narrative, including that for lethal skeletal trauma, mass graves, systematic production of lethal weapons, or fortifications (DiNapoli, Morrison, Lipo, Hunt, & Lane, 2018; Gill & Stefan, 2016; Lipo et al., 2016; Lipo & Hunt, 2009; Owsley, Barca, Simon, & Gill, 2016). A population at contact of about 3000, as Spanish observers reported in 1770 (Boersema, 2017), is consistent with archaeological studies whose authors demonstrate a low-density and dispersed settlement/land-use pattern (Morrison, 2012).

Post-European Contact Events

Historians document dramatic population decline resulting from the impacts of European contact and the introduction of Old-World diseases, slave raiding, and other calamities (Fischer, 2005; Hunt & Lipo, 2011). The case for European-caused population loss is unquestionable: It is documented in historic accounts (see Fischer, 2005) with the population ultimately declining to just 111 people in 1877. Early observers, however, were largely unaware of the effects of disease produced by contact, leading them to interpret the island's state as the result of the "imprudence of the ancestors" (e.g., La Pérouse, 1797, p. 319). Sadly, this confusion has produced the collapse narrative, in which the victims of European contact have been blamed for their own demise (Hunt & Lipo, 2010, 2011; Rainbird, 2002).

Prehistoric Population Structure

Uncovering new evidence, researchers have now documented that the island was always resource poor and that small numbers of people could easily have carved and transported the *moai* (Hunt & Lipo, 2011; Lipo et al., 2013). This observation is supported by studies of archaeological community patterning and structure: There is no evidence of large, dense settlements indicative of large populations. Instead, archaeological data from extensive field surveys and satellite image analysis of rock mulch (Ladefoged, Flaws, A., & Stevenson, 2013; Kovalchik, 2014) and *manavai* gardening (Ayala-Bradford, Lipo, & Hunt, 2005) suggest that the island's communities consisted of distinct groups arrayed along the coast in dispersed settlement patterns (Morrison, 2012; Stevenson, 1984). Rather than living in nucleated villages, communities consisted of family groups living at low density interspersed with areas of cultivation. *Ahu* and *moai* served as central locations for episodic gatherings that served to bind communities in activities and resource sharing (Hunt & Lipo, 2011).

Natural Resources of Rapa Nui

Vast areas of the island were transformed into rock mulch gardens (e.g., Bork, Mieth, & Tschochner, 2004; Hunt & Lipo, 2011; Stevenson, Wozniak, & Rapu Haoa, 1999; Wozniak, 1998, 1999). These gardens' remains can be seen across the island as artificial rock concentrations on the surface. Although European visitors have often viewed these rocky landscapes as the result of "ecocide," such mulch formed a critical dimension to survival. Rapa Nui's soils are derived from heavily weathered volcanic rocks. Given their age, these soils are relatively nutrient poor. Adding broken rock to the soil (i.e., "lithic mulching") served to enrich nutrient-leached soils. Soil samples taken from rock mulch areas show elevated levels of nitrogen, phosphorus, and potassium, key nutrients for cultivation of plants such as taro and sweet potato (Hunt & Lipo, 2011; Ladefoged et al., 2010; Ladefoged, Stevenson, Vitousek, & Chadwick, 2005). In this way, a key dimension to the island's productivity was the area covered by rock mulch gardening. More than 10% of the island's total land surface may have been devoted to lithic mulch cultivation (Ladefoged et al., 2013). Small walled gardens known as *manavai* also contributed to food production but were likely used to grow plants such as taro, banana, and sugar cane that needed additional protection and care to flourish (Ayala-Bradford et al., 2005).

Recognizing the significance of rock-mulch gardening has played a central role in rethinking Rapa Nui's prehistory. Contrary to early observations, rock mulch formed the basis of a productive agricultural system key to the population's success. Second, replacing the now-extinct palm trees with gardens increased agricultural potential and was not a catastrophe, as traditionally assumed. Third, cultivation was widely dispersed, and no single part of the island provided an abundance of crops. Dispersed cultivation coincides with a relatively small population living at low density. Thus, the population size observed by the first European observers, of about

3000 individuals, reflects a likely stable population size and not a remnant population that survived "post-collapse" (Hunt & Lipo, 2011; Morrison, 2012).

Moai Transportation

The question of how the multi-ton statues (*moai*) of Easter were transported has puzzled visitors and researchers for centuries, and for some it even played a role in deforestation. No visitors to the island ever witnessed the process, leaving much to an array of speculations. The islanders' oral traditions have long recounted simply that the statues "walked" (e.g., Thomson, 1889). Modern attempts to explain *moai* transport, however, have focused on experiments that began with Heyerdahl's efforts in the 1950s that involved simply dragging them (Heyerdahl, 1989). To resolve problems of friction and damage to statues, later efforts employed wooden sledges, pods, rollers, and sliders in various configurations (Hunt & Lipo, 2011). The idea that wood contraptions were used fits pre-existing notions of statue transport contributing to deforestation, and researchers have thus rarely questioned it.

New field research and experimentation has resolved the question of statue transport. A central finding of the research of Hunt and Lipo (2011; Lipo et al., 2013) is that the statues found along prehistoric roads have shapes that distinguish them from those statues erected on platforms (*ahu*). The road *moai* have statistically wider bases when measured relative to shoulder width than *ahu moai* (see Lipo et al., 2013, Fig. 3). Once statues arrived on platforms, prehistoric carvers modified the statues to decrease the width of the base relative to the shoulders. In addition, although *ahu moai* stand in an upright fashion with their mass located well over their base, road *moai* show a distinctive angled base that would cause the statue to lean significantly forward, often well over 10 degrees. The pronounced forward lean of the road *moai* points to how they were "walked" in an upright position with little wear to the base. *Moai* "walking" is achieved by ropes tilting the body from side to side, while allowing it to fall forward, controlled by a rope to the rear. This arrangement minimizes friction between the base and the ground, allowing for conservation of energy, increasing overall efficiency, and removing the potential for damage as the statue "walks" (Lipo et al., 2013). This means of transportation is only possible, however, because the statute is carefully shaped to move in this fashion.

Apart from labor and engineering expertise, *moai* transport required only ropes; few if any trees were required in statue transport. A woody shrub (*hau hau, Triumfetta semitrioba)* provided abundant materials for making rope (Metraux, 1940; Skottsberg, 1920). Thus, *moai* carving and transport did not contribute to deforestation, nor can one argue that forests were cleared for extensive cultivation of surplus crops to feed thousands of statue workers, as some have supposed (see Diamond, 2005; van Tilburg & Ralston, 2005, p. 299). Instead, the evidence for *moai* carving and transport points to activities by small-scale social groups rather than the product of laborers unified under a powerful centralized chiefdom.

Explaining the Success of Rapa Nui

Rather than a story of catastrophe and collapse, Rapa Nui prehistory is a case study of success on a remote, resource-poor island. Polynesians populated Rapa Nui around AD 1200 as part of rapid expansion throughout the remote Pacific (Hunt & Lipo, 2006; Wilmshurst, Hunt, Lipo, & Anderson, 2011). Colonists brought a roster of plants (taro, sweet potato, banana, sugar cane, etc.) and animals (rats, chickens) along with a variety of knowledge about subsistence strategies (fishing, cultivation) and cultural practices (statue and monument construction). Starting with these variables, Rapanui populations quickly grew in number as the island was transformed from a palm forest into an agricultural and human landscape.

Polynesian rats, as hitchhikers or an intentional introduction, rapidly spread across the island, potentially reaching numbers in the millions in a short time (Hunt, 2007). Rats would prey upon native plant seeds—especially the nuts of a dominant palm forest—contributing to depressed recruitment and ultimately the forest's demise. Forest removal would make way for cultivation, with workers using slash-and-burn cultivation practices common in Polynesian food production. The nutrients released from burning vegetation would have been key to making the relatively poor soils temporarily more productive. Thus, given rats' predilection for palm nuts, the slow rate of growth of the native *Jubaea* palm, and on-going land clearance with fire, the palms went extinct over several centuries. Importantly, no carrying capacity calamity befell the island when the forests were cleared. Clearing the landscape for cultivation and nutrients released from the burned trees created opportunities for at least short-term soil enrichment and cultivation as the island was transformed from a natural to an agricultural landscape.

From the available archaeological evidence, populations resided in multiple, functionally redundant dispersed communities, but groups benefited from interaction through activities at large *ahu* (Hunt & Lipo, 2011). The benefits of interaction among dispersed communities likely explain why investment in monuments, although present elsewhere across the Pacific, took such an exuberant form in this location. On Rapa Nui, monument construction provided advantages to individuals and communities, serving to provide individuals with ways of competing while also mitigating problems of resource uncertainty (through sharing) and reducing inevitable intergroup competition as populations grew (see Hunt & Lipo, 2011 for a more in-depth discussion).

The benefits of *moai* and *ahu* construction on Rapa Nui allow one to understand these phenomena as the products of effective governance, even if such things are not consistent with common assumptions about what "successful" societies should do. *Moai* making and transport appear incongruous with the island's resource limitations and remote isolation, and thus stand apart from what one might assume as central to survival. On this island, however, *moai* werethe key to long-term sustainability. Although activities and forms of investments varied over time, the Rapanui successfully persisted. Populations remained stable and reasonably healthy until 1722 and the arrival of the Europeans. Rapa Nui's success over its pre-European history is tied directly to the cultural practices involved in *moai* and how these practices structured and supported the island's communities.

From the Past and Looking at the Future: Governance on Rapa Nui

This new understanding of the way prehistoric people on Rapa Nui managed uncertainty offers a means of evaluating governance structures from the recent past as well as what might best be implemented for the future. From the archaeological record, one can see that the island's long-term success came from governance structures comprised of multiple local groups that cooperate as well as compete with one another. Such a structure allowed for variability in local knowledge to feed back into practice and spread across the island, thus leading to a system that would dynamically accommodate changes in environmental conditions. This system worked until the Europeans' arrival, when extrinsic changes resulted in social disruption and massive population loss due to disease and other European-driven catastrophes (Hunt & Lipo, 2011).

An important question for Rapa Nui's future is how well this traditional governance structure will work under vastly different circumstances. Cooperation and competition between local groups must be managed in ways that foster multi-level governance, support a population twice the size of the prehistorical carrying capacity, and enable flows of people and goods between the island and the rest of the world. The knowledge of how to live sustainably on the island will need to evolve and adapt to these changes in circumstances. Furthermore, the pluralist form is not necessarily as adaptive as it was in the past, as it still retains a strong top-down structure, a legacy of CONAF. This structure, therefore, might limit the ability for local innovations to emerge and spread across the island, constraining the community's ability to respond to changing and uncertain conditions. So, although *Ma'u Henua* represents a significant positive step forward toward decolonization and the embracing of local knowledge, it still potentially suffers from over-emphasis on top-down governance. Based on the factors that appear to explain Rapa Nui's sustained prehistorical success, future policymakers might well consider adding dimensions of polycentric governance (e.g., Ostrom, 2010; Waring et al., 2015) that combine cooperation and competition at more local levels than what is currently envisioned. For example, the establishment of events that encourage groups to work together while simultaneously competing in some capacity can have tremendously beneficial effects that result in increased prosociality. Wilson (2011), for example, has demonstrated that group-level competition can increase within group cooperation while also increasing global levels of cooperation across a population. The basis of such mechanisms already exists on the island in the form of *Tapati*, a festival that was created in 1968 in which clan groups compete in a series of cultural and athletic events. Like the cooperative efforts involved in making and transporting *moai* that were the foundation of prehistoric Rapa Nui society and governance, events like *Tapati* have the potential to enhance the island's ability to govern effectively in the face of future uncertainty.

Conclusions

Overall, our new understanding of Rapa Nui challenges the idea that traditional, local-based governance systems are inherently flawed. New knowledge, however, brings with it new productive areas in which to consider governance and knowledge systems. Our studies of the archaeological record demonstrate that Rapa Nui's success in the face of uncertainty and constraints (socially, environmentally, and geographically) derived from governance structures in which individuals and local communities shared information and resources. Innovations spread easily through the interaction of multiple, dispersed communities, and resource shortfalls were met by embedded collaborative efforts marked by activities involved in the construction and transport of *moai*. Following the example provided by Rapa Nui prehistory, we suggest that any regular activities (i.e., competitions, rituals, gatherings) that bring communities together and promote within-group cooperation will provide a direct benefit to everyone who participates. With local structures, those groups that cooperate more strongly will do better, particularly in the context of uncertainty. On the island scale, then, competition between these groups who also participate in between-group cooperation will produce general sustainability. The constraints and local conditions faced by populations living on Rapa Nui give an example of resilient and adaptive governance at its finest.

Our understanding has many implications for thinking about governance structures and how their members consider knowledge under conditions of uncertainty. Researchers must first carefully—and constantly—evaluate the assumptions made about the nature of social change. Although long-cherished beliefs might fit contemporary perspectives, they need to explore where these believes come from and to assess their empirical warrant, and must distinguish between the social acceptability of a conclusion versus its empirical support. This concern is why science matters so greatly, particularly when the stakes of failure grow in magnitude. Those in the field must fight the tendency to view knowledge generation as producing "alternative facts," but instead see knowledge as a process by which one continually and critically re-evaluates information from as many sources as possible.

Given that much knowledge about social and cultural change comes from an understanding of the past, science-based archaeology that demands falsifiability in any claims is particularly required. This suggestion does not prioritize science over locally generated knowledge. Instead, one must see knowledge generation as an iterative process in which we continually evaluate all sources of information. Despite having a "scientific" pedigree, the lack of such critical evaluation has contributed significantly to the erroneous acceptance of the "collapse" narrative. There are likely other areas where the field's perceived knowledge must be closely examined for empirical warrant. Likewise, one must not simply rely on "facts" as the basis for knowledge. One's knowledge comes not only from observations, but also the way in which one generates those observations. In this sense, theory is paramount. The idea that human societies will inevitably result in environmental destruction is not only unsupported by the evidence, but also violates a basic understanding about the relations between organisms and resources: There are conditions in which balances can be reached and sustainability obtained. Researchers must forge their observations from explicit and robust theory (Lewontin, 1974).

In cases of uncertainty such as that posed by imminent climate change, an additional imperative exists to incorporate processes that accommodate dynamic knowledge change into one's governance systems. As society begins to encounter environmental conditions that radically deviate from those upon which its actions have been traditionally based, governance must become increasingly adaptive and dynamic. The systems required must be more like that of prehistoric Rapa Nui: local and pluralist. Following the quote by Robert Shiller that begins this chapter, society requires dynamic and adaptive governance systems that accommodate the iterative process of knowledge generation—rather than those that act on traditional assumptions. These kinds of adaptive management systems (Williams & Brown, 2014) are particularly well-suited to the changing nature of knowledge, as innovation in one area can be evaluated locally and then shared across communities– a process that is difficult to implement in top-down governance models. In this way, Rapa Nui stands to serve once again as an exemplary cultural system, though one of success and promise rather than of collapse and catastrophe.

References

Alonso, J. M., Clifton, J., & Díaz-Fuentes, D. (2015). Did new public management matter? An empirical analysis of the outsourcing and decentralization effects on public sector size. *Public Management Review, 17,* 643–660. https://doi.org/10.1080/14719037.2013.822532

Ayala-Bradford, I., Lipo, C. P., & Hunt, T. L. (2005). The use of satellite imagery to study prehistoric agricultural features (*manavai*) on Rapa Nui. In A. Seelendfreund (Ed.), *Reñaca Papers: VI International conference on Rapa Nui and the Pacific* (pp. 113–123). Los Osos: Easter Island Foundation.

Bahn, P., & Flenley, J. (1992). *Easter Island, Earth Island.* London: Thames and Hudson.

Barfelz, A. L. (2011). The little island that could: How reforming cultural preservation policies can save Easter Island and the world's heritage. *Michigan State International Law Review, 20,* 149–177. Retrieved from https://digitalcommons.law.msu.edu/ilr/vol20/iss1/12

Biblioteca del Congreso Nacionale de Chile. (2015). Reportes Estadísticos Comunales 2015 [Report of the communal statistics 2015]. Retrieved from http://reportescomunales.bcn.cl/2015/index.php/Isla_de_Pascua

Boersema, J. J. (2017). An earthly paradise? Easter Island (Rapa Nui) as seen by the eighteenth-century European explorers. In S. H. Cardinali, K. B. Ingersoll, D. W. Ingersoll Jr., & C. M. Stevenson (Eds.), *Cultural and environmental change on Rapa Nui* (pp. 156–178). Abingdon: Routledge.

Bologna, M., & Flores, J. C. (2008). A simple mathematical model of society collapse applied to Easter Island. *EPL (Europhysics Letters), 81,* 48006. https://doi.org/10.1209/0295-5075/81/48006

Bork, H.-R., Mieth, A., & Tschochner, B. (2004). Nothing but stones? A review of the extent and technical efforts of prehistoric stone mulching on Rapa Nui. *Rapa Nui Journal, 18,* 10–14. Retrieved from https://kahualike.manoa.hawaii.edu/rnj/vol18/iss1/5

Bowles, S., Choi, J.-K., & Hopfensitz, A. (2003). The co-evolution of individual behaviors and social institutions. *Journal of Theoretical Biology, 223,* 135–147. https://doi.org/10.1016/S0022-5193(03)00060-2

Brander, J. A., & Scott Taylor, M. (1998). The simple economics of Easter Island: A Ricardo-Malthus model of renewable resource use. *American Economic Review, 88*(1), 119–138. Retrieved from https://www.jstor.org/stable/116821

Busch, T. (2016, January 6). Collapse issues potent warning for the future. *The Taber Times.* Retrieved from http://www.tabertimes.com/blog/2016/01/06/collapse-issues-potent-warning-for-the-future/

Carabine, E., & Dupar, M. (2014). The IPCC's fifth assessment report: What's in it for small island developing states? Retrieved from https://cdkn.org/wp-content/uploads/2014/08/IPCC-AR5-Whats-in-it-for-SIDS_WEB.pdf

Cauwe, N. (2011). *Easter Island: The great taboo.* Brussels: Versant Sud.

Cauwe, N., & de Dapper, M. (2015). *Easter Island: Collapse or transformation? A state of the art. Proceedings of the International Conference of Brussels (9–10/11/2012).* Brussels: The Royal Academy for Overseas Sciences.

Clarke, B., Stocker, L., Coffey, B., Leith, P., Harvey, N., Baldwin, C., Baxter, T., Bruekers, G., Galano, C. D., Good, M., Haward, M., Hofmeester, C., De Freitas, D. M., Mumford, T., Nursey-Bray, M., Kriwoken, L., Shaw, J., Shaw, J., Smith, T., Thomsen, D., Wood, D., & Cannard, T. (2013). Enhancing the knowledge-governance interface: Coasts, climate and collaboration. *Ocean and Coastal Management, 86,* 88–99. https://doi.org/10.1016/j.ocecoaman.2013.02.009

CODEIPA. (2015). Ma'u Henua: Propuesta de nueva administracion Parque Nacional Rapa Nui [Ma'u Henua: Proposal for a new administration of the Rapa Nui National Park]. Retrieved from http://ciperchile.cl/wp-content/uploads/PROPUESTA-MAU-HENUA.pdf

Council on Hemispheric Affairs. (2011). Difficult times for Chile's Easter Island. Retrieved from http://www.coha.org/difficult-times-for-chiles-easter-island/

Corney, B. C. (1908). *The voyage of Captain Don Felipe Gonzalez: In the ship of the line San Lorenzo with the frigate Santa Rosalia in company, to Easter Island in 1770–1771.* Cambridge: Hakluyt Society.

Delaune, G. (2012). Rapa Nui on the verge: Easter Island's struggles with integration and globalization in the information age. *Berkeley Planning Journal, 25,* 126–139. https://doi.org/10.5070/BP325112098

Diamond, J. (1995). Easter's end. *Discover, 16*(8), 63–68.

Diamond, J. (2005). *Collapse: How societies choose to fail or succeed.* New York: Viking.

DiNapoli, R. J., Morrison, A. E., Lipo, C. P., Hunt, T. L., & Lane, B. G. (2018). East Polynesian Islands as models of cultural divergence: The case of Rapa Nui and Rapa Iti. *The Journal of Island and Coastal Archaeology, 13*(2), 206–223. https://doi.org/10.1080/15564894.2016.1276490

Eerkens, J. W., & Lipo, C. P. (2007). Cultural transmission theory and the archaeological record: Providing context to understanding variation and temporal changes in material culture. *Journal of Archaeological Research, 15,* 239–274. https://doi.org/10.1007/s10814-007-9013-z

Emerson, K., & Nabatchi, T. (2015). *Collaborative governance regimes.* Washington, D.C.: Georgetown University Press.

Erickson, J. D., & Gowdy, J. M. (2000). Resource use, institutions, and sustainability: A tale of two Pacific Island cultures. *Land Economics, 76,* 345–354. https://doi.org/10.2307/3147033

Fischer, S. R. (2005). *Island at the end of the world: The turbulent history of Easter Island.* London: Reaction.

Flenley, J. R. (1979). Stratigraphic evidence of environmental change on Easter Island. *Asian Perspectives, 22*(1), 33–40. Retrieved from https://www.jstor.org/stable/42929142

Flenley, J. R., & Bahn, P. (2003). *The enigmas of Easter Island* (2nd ed.). Oxford, UK: Oxford University Press.

Flenley, J. R., & Bahn, P. (2007). Conflicting views of Easter Island. *Rapa Nui Journal, 21,* 11–13.

Flenley, J. R., & King, A. S. M. (1984). Late quaternary pollen records from Easter Island. *Nature, 307*(5946), 47–50. https://doi.org/10.1038/307047a0
Flenley, J. R., King, A. S. M., Jackson, J., Chew, C., Teller, J. T., & Prentice, M. E. (1991). The late quaternary vegetational and climatic history of Easter Island. *Journal of Quaternary Science, 6*(2), 85–115. https://doi.org/10.1002/jqs.3390060202
Frederickson, H. G. (1971). Toward a new public administration. In F. Marini (Ed.), *Toward a new public administration* (pp. 309–331). Scranton: Chandler.
Foot, D. K. (2004). Easter Island: A case study in non-sustainability. *Greener Management International, 2004*(48), 11–20. https://doi.org/10.9774/GLEAF.3062.2004.wi.00004
Gill, G. W., & Stefan, V. H. (2016). Rapanui origins, relationships, and warfare: A summary in theoretical context. In V. H. Stefan & G. W. Gill (Eds.), *Skeletal biology of the ancient Rapanui (Easter Islanders)* (pp. 286–302). Cambridge, UK: Cambridge University Press.
Goodnow, F. J. (1900). *Politics and administration: A study in government.* New York: Macmillan.
Hari, J. (2005). Is Easter Island a warning for us all? Don't be fooled: Advanced and rational societies can commit environmental suicide. *The Independent.* Retrieved from https://www.emfacts.com/2005/06/is-easter-island-a-warning-for-us-all/
Heins, J. (2010). Unconventional wisdom: An interview with Robert Shiller. Retrieved from https://www.advisorperspectives.com/newsletters10/pdfs/Unconventional_Wisdom-An_Interview_with_Robert_Shiller.pdf
Henrich, J. (2004). Cultural group selection, coevolutionary processes and large-scale cooperation. *Journal of Economic Behavior & Organization, 53,* 3–35. https://doi.org/10.1016/S0167-2681(03)00094-5
Heyerdahl, T. (1989). *Easter Island: The mystery solved.* New York: Random House.
Hunt, T. L. (2007). Rethinking Easter Island's ecological catastrophe. *Journal of Archaeological Science, 34,* 485–502. https://doi.org/10.1016/j.jas.2006.10.003
Hunt, T. L., & Lipo, C. P. (2006). Late colonization of Easter Island. *Science, 311,* 1603–1606. https://doi.org/10.1126/science.1121879
Hunt, T. L., & Lipo, C. P. (2008). Evidence for a shorter chronology on Rapa Nui (Easter Island). *The Journal of Island and Coastal Archaeology, 3,* 140–148. https://doi.org/10.1080/15564890801990797
Hunt, T. L., & Lipo, C. P. (2010). Ecological catastrophe, collapse, and the myth of "ecocide" on Rapa Nui (Easter Island). In P. A. McAnany & N. Yoffee (Eds.), *Questioning collapse: Human resilience, ecological vulnerability, and the aftermath of empire* (pp. 21–44). New York, NY: Cambridge University Press.
Hunt, T. L., & Lipo, C. P. (2011). *The statues that walked: Unraveling the mystery of Easter Island.* Riverside: Free.
IWGIA. (2012). The human rights of the Rapa Nui people on Easter Island: Report of the International Observers' Mission to Rapa Nui 2011. Retrieved from https://www.iwgia.org/images/publications//0597_Informe_RAPA_NUI_IGIA-Observatorio_English_FINAL.pdf
Kirch, P. V. (1984). *The evolution of Polynesian chiefdoms.* New York, NY: Cambridge University Press.
Knight, F. H. (1921). *Risk, uncertainty and profit.* Hart, Schaffner and Marx Prize Economic Essays: Vol. 31. Boston: Mifflin.
Kovalchik, J. (2014). *Remote sensing and the assessment of prehistoric productivity in cultivation practices of Rapa Nui, Chile* (Master's thesis). Department of Anthropology, California State University Long Beach, Long Beach.
La Pérouse, J.-F. de Galaup. (1797). *Voyage de La Pérouse autour du monde* [The voyage of La Pérouse round the world] (M. L. A. Milet-Mureau, Ed.). Paris: De l'Imprimerie de la République.
Ladefoged, T. N., Flaws, A., & Stevenson, C. M. (2013). The distribution of rock gardens on Rapa Nui (Easter Island) as determined from satellite imagery. *Journal of Archaeological Science, 40,* 1203–1212. https://doi.org/10.1016/j.jas.2012.09.006

Ladefoged, T. N., Stevenson, C. M., Rapu Haoa, S., Mulrooney, M., Puleston, C., Vitousek, P. M., & Chadwich, O. A. (2010). Soil nutrient analysis of Rapa Nui gardening. *Archaeology of Oceania, 45,* 80–85. https://doi.org/10.1002/j.1834-4453.2010.tb00082.x

Ladefoged, T. N., Stevenson, C. M., Vitousek, P., & Chadwick, O. (2005). Soil nutrient depletion and the collapse of Rapa Nui Society. *Rapa Nui Journal, 19,* 100–105. Retrieved from https://kahualike.manoa.hawaii.edu/rnj/vol19/iss2/4

Laing, B. D. (2008). Dual inheritance theory, contract law, and institutional change: Towards the co-evolution of behavior and institutions. *German Law Journal, 9,* 491–514. https://doi.org/10.1017/S2071832200006556

Lewontin, R. C. (1974). *The genetic basis of evolutionary change*. New York, NY: Columbia University Press.

Lipo, C. P., & Hunt, T. L. (2009). A.D. 1680 and Rapa Nui prehistory. *Asian Perspectives, 48,* 309–317. https://doi.org/10.1353/asi.2009.0001

Lipo, C. P., Hunt, T. L., Horneman, R., & Bonhomme, V. (2016). Weapons of war? Rapa Nui mata'a morphometric analyses. *Antiquity, 90*(349), 172–187. https://doi.org/10.15184/aqy.2015.189

Lipo, C. P., Hunt, T. L., & Rapu Haoa, S. (2013). The 'walking' megalithic statues (*moai*) of Easter Island. *Journal of Archaeological Science, 40,* 2859–2866. https://doi.org/10.1016/j.jas.2012.09.029

McAnany, P. A., & Yoffee, N. (2009). *Questioning collapse: Human resilience, ecological vulnerability, and the aftermath of empire*. New York, NY: Cambridge University Press.

Metraux, A. (1940). *Ethnology of Easter Island*. Bernice P. Bishop Museum Bulletin: Vol. 160. Honolulu: The Museum.

Morrison, A. E. (2012). *An archaeological analysis of Rapa Nui settlement structure: A multi-scalar approach* (Doctoral dissertation). University of Hawaii, Department of Anthropology, Manoa.

Mulloy, W. T. (1974). Contemplate the navel of the world. *Americas, 26*(4), 25–33.

Mulrooney, M. A. (2012). *Continuity or collapse? Diachronic settlement and land use in Hanga Ho'onu, Rapa Nui (Easter Island)* (Doctoral dissertation). University of Auckland, Department of Anthropology, Auckland.

Mulrooney, M. A. (2013). An island-wide assessment of the chronology of settlement and land use on Rapa Nui (Easter Island) based on radiocarbon data. *Journal of Archaeological Science, 40,* 4377–4399. https://doi.org/10.1016/j.jas.2013.06.020

Mulrooney, M. A., Ladefoged, T. N., Stevenson, C. M., & Rapu Haoa, S. (2009). The myth of A.D. 1680: New evidence from Hanga Ho'onu, Rapa Nui (Easter Island). *Rapa Nui Journal, 23,* 94–105.

Nagarajan, P. (2006). Collapse of Easter Island: Lessons for sustainability of small islands. *Journal of Developing Societies, 22,* 287–301. https://doi.org/10.1177/0169796X06068032

Osborne, D., & Gaebler, T. (1992). *Reinventing government: How the entrepreneurial spirit is transforming the public sector.* Reading: Addison-Wesley.

Osborne, S. P. (Ed.). (2010). *The new public governance? Emerging perspectives on the theory and practice of public governance*. Abingdon: Routledge.

Ostrom, E. (2010). Beyond markets and states: Polycentric governance of complex economic systems. *American Economic Review, 100,* 641–672. https://doi.org/10.1257/aer.100.3.641

O'Toole, L. J. (1997). Treating networks seriously: Practical and research-based agendas in public administration. *Public Administration Review, 57,* 45–52. https://doi.org/10.2307/976691

Owsley, D. W., Barca, K. G., Simon, V. E., & Gill, G. W. (2016). Evidence for injuries and violent death. In V. H. Stefan & G. W. Gill (Eds.), *Skeletal biology of the ancient Rapanui (Easter Islanders)* (pp. 222–252). Cambridge, UK: Cambridge University Press.

Puleston, C. O., Ladefoged, T. N., Rapu Haoa, S., Chadwick, O. A., Vitousek, P. M., & Stevenson, C. M. (2017). Rain, sun, soil, and sweat: A consideration of population limits on Rapa Nui (Easter Island) before European contact. *Frontiers in Ecology and Evolution, 5,* 69. https://doi.org/10.3389/fevo.2017.00069

Rainbird, P. (2002). A message for our future? The Rapa Nui (Easter Island) ecodisaster and Pacific island environments. *World Archaeology, 33,* 436–451. https://doi.org/10.1080/00438240120107468

Ramirez, J. M. (2000). Rapa Nui land management: A personal chronicle. *Rapa Nui Journal, 14,* 47–48. Retrieved from https://kahualike.manoa.hawaii.edu/rnj/vol14/iss2/3

Reuveny, R., & Decker, C. S. (2000). Easter Island: Historical anecdote or warning for the future? *Ecological Economics, 35,* 271–287. https://doi.org/10.1016/S0921-8009(00)00202-0

Reynolds, K., & Rose-Price, T. (1994). *Rapa-Nui: The Easter Island legend on film.* New York: Newmarket.

Richerson, P. J., & Boyd, R. (2005). *Not by genes alone: How culture transformed human evolution.* Chicago, IL: University of Chicago Press.

Rosell, S. A. (1999). *Renewing governance: Governing by learning in the information age.* New York, NY: Oxford University Press.

Skottsberg, C. (1920). *The natural history of Juan Fernández and Easter Island.* Uppsala: Almqvist & Wiksells Boktryckeri.

Snowden, D. J., & Boone, M. E. (2007). A leader's framework for decision making: Wise executive tailor their approach to fit the complexity of the circumstances they face. *Harvard Business Review, 85*(11), 68–76.

Stevenson, C. M. (1984). *Corporate Descent Group Structure in Easter Island Prehistory* (Doctoral dissertation). The Pennsylvania State University.

Stevenson, C. M., Puleston, C. O., Vitousek, P. M., Chadwick, O. A., Rapu Haoa, S., & Ladefoged, T. N. (2015). Variation in Rapa Nui (Easter Island) land use indicates production and population peaks prior to European contact. *Proceedings of the National Academy of Sciences of the United States of America, 112,* 1025–1030. https://doi.org/10.1073/pnas.1420712112

Stevenson, C. M., Wozniak, J., & Rapu Haoa, S. (1999). Prehistoric agricultural production on Easter Island (Rapa Nui), Chile. *Antiquity, 73,* 801–812. https://doi.org/10.1017/S0003598X00065546

Taylor, F. W. (1911). *The principles of scientific management.* New York: Harper & Brothers.

Tector, A. (2014). The delightful revolution: Canada's medical expedition to Easter Island, 1964–65. *British Journal of Canadian Studies, 27,* 181–193. https://doi.org/10.3828/bjcs.2014.12

Thomson, W. J. (1889). *Te Pito te Henua, or Easter Island.* Washington, D.C.: U.S. Government Printing Office.

van Tilburg, J. A., & Ralston, T. (2005). Megaliths and mariners: Experimental archaeology on Easter Island. In K. L. Johnson (Ed.), *Onward and upward! Papers in honor of Clement W. Meighan* (pp. 279–303). Chico: Stansbury.

Waring, T. M., Kline, M. A., Brooks, J. S., Goff, S. H., Gowdy, J., Janssen, M. A., Smaldino P. E., & Jacquet, J. (2015). A multilevel evolutionary framework for sustainability analysis. *Ecology and Society, 20*(2), 34. https://doi.org/10.5751/ES-07634-200234

Warren, M. (2011, February 7). Easter Island cops evict aboriginal squatters at luxury hotel [Article]. *The Denver Post.* Retrieved from http://www.denverpost.com/nationworld/ci_17314100

Weber, M. (1978). *Economy and society: An outline of interpretive sociology* (Vol. 1). Berkeley, CA: University of California Press.

Wei Choo, C. (2006). *The knowing organization: How organizations use information to construct meaning, create knowledge, and make decisions.* Oxford, UK: Oxford University Press.

Williams, B. K., & Brown, E. D. (2014). Adaptive management: From more talk to real action. *Environmental Management, 53,* 465–479. https://doi.org/10.1007/s00267-013-0205-7

Wilmshurst, J. M., Hunt, T. L., Lipo, C. P., & Anderson, A. J. (2011). High-precision radiocarbon dating shows recent and rapid initial human colonization of East Polynesia. *Proceedings of the National Academy of Science of the United States of America, 108,* 1815–1820. https://doi.org/10.1073/pnas.1015876108

Wilson, D. S. (2011). *The neighborhood project: Using evolution to improve my city, one block at a time.* New York: Little, Brown and Company.

Wozniak, J. A. (1998). Settlement patterns and subsistence on the northwest coast of Easter Island: A progress report. In P. Vargas Casanova (Ed.), *Easter Island and East Polynesian prehistory* (pp. 171–178). Santiago: Universidad de Chile, Facultad de Arquitectura y Urbanismo, Instituto de Estudios Isla de Pascua.

Wozniak, J. A. (1999). Prehistoric horticultural practices on Easter Island: Lithic mulched gardens and field systems. *Rapa Nui Journal, 13,* 95–99. Retrieved from https://kahualike.manoa.hawaii.edu/rnj/vol13/iss4/1

Chapter 3
Knowledge of Governance as Knowledge for Governance: Spatialized Techniques of Neutralization

Michael Scott

Governance is a key concept in the social sciences (see Bevir, 2011). It refers to processes of intersectoral, interorganizational modes of networked steering that lead to collaborative and negotiated decision making (Rhodes, 1997). For sociospatial researchers, knowledge for governance evokes images of governmental knowledge of populations and their attitudes, technical knowledge of space and its physical properties, formal and informal bureaucratic processes of problem definition, and the creation of knowledge through stakeholder engagement to address pressing social issues. This rendering of governance has a normative inflection. When done well, governance generates more equitable, socially just, and consensus-derived decision making through the integration of diverse stakeholder voices and technical expertise.

Nevertheless, behind governance lurks government (MacLeod & Goodwin, 1999). Legal systems, public funding, and bureaucratic rationalities often cast a shadow over noble normative aims. Governance as an activity then emerges as a domain of negotiation and contest within state-managed systems. Here, actors aim to advance interests through reference to state power, which they call upon to enact, arbitrate, or legitimate governance-derived decisions. Therefore, governance is performed at the cultural-institutional interstices of, in Weber's (1978) terms, authority and legitimation. Outcomes must be enacted (via state authority), and they must be seen as just (legitimate). In these settings, actors assume that the best technical or scientific knowledge enables authoritative action and is woven into narratives, arguments, and framings to legitimate decisions (Hajer, 2001).

M. Scott (✉)
College of Humanities, Arts and Social Sciences, Flinders University, Adelaide, SA, Australia
e-mail: michael.scott@flinders.ed.au

J. Glückler et al. (eds.), *Knowledge for Governance*, Knowledge and Space 15,
https://doi.org/10.1007/978-3-030-47150-7_3

But is knowledge for governance primarily technical or expert? Or does governance require managing representations, stakeholder voice, and impressions? How might knowledge of how to speak and debate within governance processes emerge as a form of *knowledge of governance, for governance*? How can such knowledge *of* how to argue be deployed to legitimate socially contentious decisions? Although there is an extensive literature on knowledge uses in governance (e.g., Nursey-Bray et al., 2014), which includes critical studies of ways of knowing and the governance of knowledge (van Buuren, 2009) as well as the uses of nonknowledge and the unknown in governance (McGoey, 2012), the following further problematizes the image of *knowledge for governance* as a consensual input into collaborative decision making.

In this chapter, I investigate how governance actors legitimate contentious coastal land developments. I do so using qualitative data and the articulation of two contrasting literatures: techniques of neutralization from the sociology deviance (Sykes & Matza, 1957) and theorizations of spatiality—social produced space—from human geography (e.g., Massey, 2005; Soja, 1989). As Boiral (2016, p. 754) notes, techniques of neutralization involve "the release of information aimed at rationalizing and legitimizing, through different types of socially acceptable arguments." From an argumentative perspective, these techniques are a form of *knowledge of governance* used to debate, deflect criticism, and neutralize opposition (Fischer, 1990). Researchers using techniques of neutralization in sociospatial research have recently considered ethical and sustainable consumption practices (Antonetti & Maklan, 2014; Hansmann, Bernasconi, Smieszek, Loukopoulos, & Scholz, 2006; Harris & Daunt, 2011; Johnstone & Tan, 2015; Yeow, Dean, & Tucker, 2014), wildlife crime (Enticott, 2011), and corporate social responsibility and industrial production (Boiral, 2016; Fooks, Gilmore, Collin, Holden, & Lee, 2013; Meyer & Höllerer, 2016; Stuart & Worosz, 2012; Talbot & Boiral, 2015; Teh, Ahmed, & D'Arcy, 2015). Yet the intersections of knowledge, governance, and techniques of neutralization in land use governance have yet to be considered. This is puzzling because techniques of neutralization provide a frame to explore how legitimating rationalizations are deployed without questioning the environmental impacts of the land use planning.

Crucially, the coastal zone is an inherently contested space and thus an exemplary site to explore the spatialized techniques of neutralization in governance settings. As the meeting point of land and sea, the coast affords multiple human uses including recreation, conservation, and aesthetics. The coast is also habitat and ecosystem linked to species reproduction. Moreover, the coast is a site for capital valorization through land development underscored by the cultural draw of living by the sea. Yet, under anthropogenically accelerated climate change and sea level rise (SLR), storm surges now impact the coast in greater frequency and intensity, reclaiming private and public property, while threatening infrastructure and ecosystems. In such locales, the interaction of state control of coastal land use (through planning institutions) and diverse stakeholders offers insights into the enactment of knowledge *of* governance, *for* governance.

I develop this analysis over four sections. First, I review the qualitative methods and the South Australian cases. Second, I outline land-use planning as a governance institution, the techniques of neutralization, and spatiality. In the following section,

I illustrate how actors adapt the techniques of neutralization using spatial metaphors that refer to the political-juridical, economic, and ideological dimensions of coastal land use. In the final section, I propose a metatheme in these spatialized techniques of neutralization: the ongoing privileging of time over space.

Methodology

This research was a component of an Australia-wide CSIRO project (2010–2013) investigating the social and cultural barriers to the uptake of scientific knowledge into costal decision making (Clarke et al., 2013). As part of the South Australian research cluster, my focus lay on land-use planning institutions, social networks, and the cultural narratives operating within these (Scott, Balaev, & Clarke, 2018; Scott & Harvey, 2016). South Australia has a long history of employing land-use planning to protect and manage its coastline, which varies from high energy Southern Ocean zones to low energy gulf waters. Not only was it the first Australian state to account for SLR in its development setbacks (the permitted distance of the built environment to the coastal zone), it is also recognized as having transparent and well-governed planning systems (Harvey & Caton, 2010).

South Australia's planning framework regulates coastal land use. Under the "Development Act" (Government of South Australia, 2014), a state agency, the Coast Protection Board (CPB), oversees coastal planning, development, protection, and restoration. It provides advice and direction on coastal matters to the Minister for the Environment, the state planning authority (the Development Assessment Commission or DAC), and local government authorities (LGAs or local councils). In general, LGAs assess coastal development proposals with reference to the relevant section of the Development Plan. Coastal LGAs must also give regard (but not legally adhere) to CPB policy on setbacks and coastal flooding hazards (Coast Protection Board, 1991). These scalar planning arrangements guided the research design. In June 2011, a focus group with the Department of Environment and Natural Resources (DENR) identified cases that were:

- Examples of key coastal development decisions
- An initial list of the key actors and organisations involved in these planning decisions
- Examples where scientific knowledge of coastal processes influenced the decision
- Examples where scientific knowledge on coastal processes where not accounted for in the decision

Following the focus group, researchers selected four case studies of coastal developments exposed to SLR risks: a coastal retirement apartment complex and a surf club redevelopment on Adelaide's metropolitan coast (whose extensive white sand beach is managed by costly sand carting), a regional coastal resort, and a large peri-urban coastal township development on a flood plain adjacent the Gulf of

St Vincent. They selected all cases on the basis that they were relatively recent (within the past six years), so that interviewees could recall the events and processes. Moreover, these examples of coastal development are paradigmatic cases (Flyvbjerg, 2001, p. 80). They highlight the more general characteristics of land-use governance and the uses of knowledge. Here, contests over coastal development and the projected risks from SLR provide an *exemplar* of the narratives circulating within land use governance. If the techniques of neutralization were deployed in the South Australian case—where the planning system is robust and transparent—they would be likely to be deployed elsewhere.

Data collection took place between July 2011 and May 2012. From the list of cases, I identified potential respondents in three ways. First, the DENR focus group provided a list of potential interview subjects. Second, a Factiva search of media articles on the selected cases highlighted political and community actors. Third, I identified additional individuals through a search of secondary data: publicly available policy documents, development plans, CPB Reports, council submissions to developers, the South Australian Hansard, surf lifesaving club annual reports, draft plans, architectural plans, and submissions to the state's DAC.

With this snowballing process (Noy, 2008), I was able to identify 47 actors, of which only two declined to participate. I conducted semistructured interviews lasting between 30 and 90 min, allowing respondents to extensively recount their subjective experiences in coastal governance. I then transcribed and analysed the interviews using NVivo, and deductively coded the data along the themes of the techniques of neutralization (discussed below) and further analyzed them through reference to spatial concepts.

Land-Use Planning, Techniques of Neutralization, and Spatiality

Land-Use Planning

Land-use planning requires elaboration, as it is a pivotal institution in the governance of coastal developments. Theoretically, land-use planning and its attendant mapping create property rights. Following Polanyi (1957), land is one of the fictitious commodities (the others being labour and money). These are fictitious for they do not easily lend themselves to commodification, and thus exchange in the market economy. Furthermore, land cannot be expropriated without the risk of short- or long-term degradation, negative externalities, or unintended consequences on adjacent areas. Formalization in property law and planning bureaucracies is necessary to secure land's continued social uses and future valorisation via market exchange (Harvey, 1978). In urban contexts, land development often occurs through *growth coalitions* between property developers and state or municipal power (Logan & Molotch, 2007). Crucially, land use planning undergirds urbanisation and is one of the few economic development levers available to Australian states operating under the pressures of competitive federalism, ongoing deindustrialization, and the need to attract investment.

Carving up the earth's surface under state authority planning creates different zones. Land-use planning is about "what goes where" (Taussik, 2007). Zones regulate what development constructions—or protections—occur in that space. Still, zoning and development is not a technocratic procedure of high modernist planning. Instead, this institutionalization of land means planning is called upon to meet "concrete, multiply-determined objectives that could contain various social processes simultaneously" (Krippner, 2002, p. 804), including the triple bottom line of social, economic, and environmental benefits (Elkington, 1999). As such, planning does not necessarily predetermine development ends. New developments involve governance and networked decision-making that brings together legal processes assigning rights with social assessments of the proposed development's risks and benefits. Here state actors and LGA planners coordinate a range of knowledge holders: engineers, scientists, politicians, community groups, NGOs, and other stakeholders with an interest in the site and its adjacent uses (see Scott et al., 2018).

State planning organizations also establish due processes, which are a game and an incentive structure to be engaged (North, 1990). Land's fictitious nature and its multiple uses means no planning policy can cover every contingency a development proposal might present, nor account for changing political economic or environmental conditions under which developments are proposed. There is an art to aligning, negotiating, and, importantly, legitimating land-use development at controversial sites. This requires knowledgeable and reflexive actors engaging in negotiations and the deft marshalling of various forms of evidence to enact authority over the use of space. Planning's legitimacy then rests upon "a belief in the legality of enacted rules and the right of those elevated to authority under such rules to issues commands" (Weber, 1978, p. 215). Because the future is (relatively/somewhat) unknown, development proposals in contentious locations are as much about the production of belief in the authority of the planning system as they are of technical knowledge. Development decisions can therefore be justified in multiple ways—economic, sociocultural, or environmental—with no form of knowledge as ultimate arbiter; what planning assessors call "a balanced decision" (Scott & Harvey, 2016). In opening this discursive space for heterogeneous, yet socially acceptable narratives, the techniques of neutralization emerge as a form of knowledge *of* governance.

Techniques of Neutralization and Spatiality

Sykes and Matza (1957) developed "techniques of neutralization" in the sociology of deviance to differentiate the narratives "delinquents" used to justify their norm- and law-breaking behaviour. Sykes and Matza argue that techniques of neutralization are a learnt response allowing delinquents to reconcile the conflicting demands of their primary subculture with the external demands of the norm- and law-abiding community. As part of deviant subgroups, delinquents recognise the social pressures to conform but also possess a willingness to break laws in which they too believe. Therefore, if society's formal rules are qualified and flexible, how can they be bent in some ways but not broken? Here the techniques of neutralization are a means to deflect, remould, and recast criticism from the justice system and society. They include:

1. Denial of responsibility: the appeal to external forces beyond the actor's control. Injuries are produced by circumstance and the actor denies personal accountability by claiming to have been "hopelessly propelled" into a situation; one is "acted upon rather than acting" (Sykes & Matza, 1957, p. 667).
2. Denial of injury: pivots on the legal distinction between "acts which are wrong in themselves and acts which are illegal but not immoral" (Sykes & Matza, 1957, p. 667). The question is has anyone been physically hurt by the deviant acts, such as graffiti or brawls between willing parties? Although counter to the law, the action does not cause significant harm.
3. Denial of the victim: when the delinquent accepts responsibility for their actions the victim can be denied. Any injury sustained is downplayed, for the victim might have deserved it or becomes transformed into a wrong doer. Awareness of the victim is further weakened if the deviant behaviour is against property. Where there is no immediate physical harm to owners, there is diminished awareness of the victim.
4. The condemnation of the condemners: involves "a rejection of the rejecters" (Sykes & Matza, 1957, p. 668). This rationalization deflects claims against the deviant by shifting doubt on to the motives and behaviours of those who disapprove. Here cynicism is expressed towards those upholding society's norms, casting them as "hypocrites, deviants in disguise, or impelled by personal spite" (p. 668). The delinquent, in effect, has changed the subject of the conversation in the dialogue between his own deviant impulses and the reactions of others; and by attacking others, the wrongfulness of his own behaviour is more easily repressed or lost from view (p. 668).
5. Appeal to higher loyalties: a process of articulating a devotion to smaller subgroups who are claimed to be more important than wider society: family, friends, and cliques. The actor does not reject all norms, but rather makes claims to group "norms that are higher or more pressing" (Sykes & Matza, 1957, p. 668).

In identify these neutralizing techniques, Skyes and Matza do not suggest that any one possesses ultimate efficacy, or carries more weight in determining justice outcomes. Instead, techniques of neutralization emerge as a way to place doubt upon an opponent's arguments while allowing the wrong doer to identify with societal norms. Therefore, unlike Schopenhauer's (2004) "Art of Being Right," techniques of neutralization are not explicitly about winning debates but are discursive strategies to diminish the social opprobrium of accusations, and subsequent penalties, when the accused is aware they have violated social norms. As Sykes and Matza (1957, p. 669) claim, these techniques are only "tangential or glancing lows at a dominant normative systems rather than creating an opposing ideology." Their use allows actors to drift between value systems—wider social norms and subcultural values (Mooney, 2007)—while providing a context dependent explanation for deviant behaviour. However, a telling critique of techniques of neutralization is that they cannot explain the causes of deviance, only actors' *post-factum* explanations, rationalizations, and justifications for rule breaking (Cavanagh, Dobash, Dobash, & Lewis, 2001).

My contention is that these neutralization techniques emerge as a form of knowledge of governance to legitimate, rationalize, or deflect critiques in contentious coastal governance processes. In this milieu they are spatialized. Here, I adopt a broadly materialist interpretation of spatiality which recognises that space "is socially produced, exists in both substantial forms (concrete spatialities) and as a set of relations between individuals and groups, an 'embodiment' and medium of social life itself" (Soja, 1989, p. 120). Material spatiality then creates dialectical, recursive, and reciprocal sites for action through the interplay of social processes and geophysical space (Massey, 2005). From this perspective, spatiality is constituted through overlapping domains: political-juridical (the institutions regulating space such as planning, interwoven political systems of scale and territory), economic (space as a site for the creation of investment and profits, production and consumption), and ideological (the symbolic use of space—the coast as a cultural expression of *freedom* and *nature* or of *progress* and *development*). No single domain is determining. Within these domains, spatiality is also metaphorical, for it contains a multiplicity of potential social uses and meanings, and paradoxical, as it is produced by different knowledge forms and is consequently simultaneously knowable and unknowable (Kitchin, 2009). This kaleidoscopic spatiality creates numerous opportunities to deploy the spatialized techniques of neutralization in the governance of coastal developments.

Spatialized Techniques of Neutralization

Denial of Responsibility: Political-Juridical Structures

Actors in coastal land use governance recognize the larger political-juridical structures bearing on their actions. These can thwart, obstruct, or constrain efforts to incorporate protective measures, alter development proposals or to stop environmental protections outright. In its spatialized form, the denial of responsibility is a deferral to broader planning processes, their bureaucratic limitations, and the pro-development logics of growth coalitions. As an LGA planner laments in an interview, the scalar hierarchy of planning means "[s]tate government are the ones who have ownership of our development plan and we're the ones who try to fit it in." To neutralize complaints over coastal development, state planners who make final assessments deploy this technique. Their repertoires for action are circumscribed by external planning hierarchies, zonings, and regulations that are beyond their control:

> … we don't design the development, we don't select the location but we get a copy of the application with the design and that determines the nature of the development and the application process we follow and we make the assessment on that against the plan, and then make recommendations to accept, amend, or reject the application. (State planner)

> … the [land use zone] document might be old, deficient, very general, it may not be worded in a way that is easily understandable, it might be open to interpretation—it doesn't matter; that is what we have and that's what we have to work with. (State planner)

Here, denial for responsibility for future environmental impacts is conjured through reference to due process: Hopelessly propelled by legitimate procedure, the state planners reached a balanced decision regarding the development. A strident critique of this mode of neutralizing is that it limits accounting for exceptions. This is presented by actors less central to the making of coastal development decisions—scientists and engineers who provide expert advice:

> … [planners and bureaucrats] tick the checklist—if they get away from the checklist mentality and work towards a decision, in other words become involved in a process rather than being the police of the process, then I think they'll have a much better [outcome]. (Consultant, environmental scientist)

Moreover, advocates of environmental protection see planning assessments weighted towards valorizing the coast as economic space. Inversely, an iteration of the denial of responsibility is that the marshalling alternative forms of evidence to counter development proposals are burdensome:

> … I think economic and environmental arguments are considered differently, probably in the community as well as government, and economic arguments of 'this is going to be the benefit' are fairly quick to be accepted, whereas the environmental argument is often, I'm exaggerating here for effect, 'well prove that four different ways.' There is a different burden of proof required for environmental arguments and environmental issues than there is on economic ones—that's my personal view not a departmental view—I will make that clear! (State engineer)

Overall, this spatialized neutralization deflects environmental or future risks that cannot be managed within the existing political-juridical domain of planning assessment. This allows planners and other governance actors to claim that they are acted upon by larger structures and systemic forces. Overlapping with the denial of responsibility is the coast as an economic space.

Denial of Injury and Victim: Legitimating Economic Spatiality

Implicit in planning is the ongoing use of the coast as economic space. Materially, new developments benefit private users through land price appreciation and LGAs through rate inflation, whereas recreation hubs such as surf lifesaving clubs carry positive economic and social externalities. Hence, an urban LGA planner seeking to implement prudent coastal development recognises economic pressures (while engaging in a denial of responsibility):

> The barriers [to implementing environmental protections or blocking coastal development] are poor [meaning legally weak] state government policy for the coast. We don't have much to stop development. There is really nothing in any of the state government policies that we have to draw from to put in our development plan, stopping development. Development is always going to happen but it's a matter of mitigating the issues that go with it. That's all we can ever do. There is no policy to re-establish the [sand] dunes when a house reaches the end of its life.

In addition, the shorter term pressures for using coastal space for economic action follows the political imperatives of *growth coalition* driven economic development. As an LGA planner managing the development of a new coastal township notes:

> All the big projects are in [politically] marginal seats. [We] can't do anything as development plans are under the control of state governments. The Economic Development Board is mostly real estate developers and people who invest money.

To mitigate paradoxical norms—governance as protecting property from environmental risk; governance as enabling built development—the denial of injury and the denial of victim overlap. Both create a distinction between acts that are wrong in themselves and acts that are illegal but not immoral: no significant physical or individual harm arguments are developed. When deployed at sites where projected SLR will impact on new developments, these techniques deny future injuries and victims through an appeal to a wider, and more diffuse, social group. SLR will impact other communities so why restrict economic development at this site? For example, a LGA councillor supporting the large peri-urban township development dismisses SLR projections through reference to other communities that might be affected:

> ... but a lot of the project [township development] is above places like [an adjacent industrial and residential] Peninsula, which are actually below the levels of where most of this development is. So if we do have the problems of the rising sea levels it is going to hit a lot of other areas first!

Likewise, for a representative of a surf life-saving club redevelopment abutting the Adelaide metropolitan coast:

> ... we've done what we can to observe what the [state planner] said. One of their comments was that 3-meter waves will hit the building. If that is going to happen then the whole metropolitan coast will be in danger so you can't prepare against that. Although it might sound ridiculous to those that live along here, you simply can't prevent that if it is going to happen and the whole state would be under threat, not just us. I said, 'If you feel our building is under that sort of threat then you had better tell everybody that lives along the coast line to sell now and don't expect your kids to inherit the building, the house, on the seafront because it won't be there.' (Private developer)

Such statements could be interpreted as a denial of capability—the inevitability that SLR cannot be held back. However, this denial of victim and injury invokes a tacit socialization. Here the projected injuries of new developments become parsed through appeals to a generalized coastal community, present and future. Potential injury and victim caused by a new coastal development can be denied; if everyone else on the coast is affected, there is no specific victim or injury. There is disaster.

Therefore, the ongoing use of the coast as economic space can be legitimated by broadening the victim base, which is further diffused through reference to future uncertainty. Debates over the coast as economic space then broaden out into ideological clashes between pro- and anti-development actors, and the condemnation of condemners.

The Condemnation of the Condemners: Ideological Spatiality I

This technique is a brazen "rejection of the rejecters" (Sykes & Matza, 1957, p. 668). It calls into question the motives and behaviours of those who disapprove. This could be disapproval over developments or disapproval over pro-environmental/prudential development positions within governance settings. These techniques were widely articulated. Although descending into sophisticated name-calling, this technique echoes the metaphorical and paradoxical dimensions of the coast; its symbolic uses can be presented in different ways. Two indicative examples from pro- and anti-development MPs illustrate how ideological interests are condemned, first through reference to the uncertainty of coastal science, and second through growth coalition solidarity. The state MP managing the development process of a surf lifesaving club, whose location on a promenade sees it buffeted by storm surges, states:

> One thing I know for certain is that if I whacked half a dozen (coastal) scientists in this room we might come up with six different views, and the one thing they will all agree on is the need for more funding for research. What we often see is scientists being pitted against each other as opposed to an agreed position. The advocates for not doing anything will always say 'but we need more research.' (State MP)

Meanwhile, a state MP opposing a periurban township development on a coastal flood plain argues:

> I've made many speeches in parliament about the conflict of interest [the state government] has under the GAI—the Growth Areas Investigation. They did the major study for government for areas for growth on the outskirts of the city [including coastal zones]. They did that work for government and at the same time they represented private developers who had been buying up land on the fringes … for ages. Then, surprise, surprise they happen to recommend to government that areas that their clients owned were suitable for urban development! (State MP)

By making the self-interest within putatively transparent governance practices hyper visible, these MPs are condemning their opponents' motives. Scientists want more money, growth coalitions want to consolidate power over space. Broadly, pro-development groups can condemn by pointing to the coast's paradoxical nature—it is known and unknown—whereas antidevelopment groups condemn growth coalitions that privatize profits and socialise costs. In governance debates, they offer archetypical counterweights over the valorization of coastal space. A similar strategy appears in appeals to higher loyalties.

Appeal to Higher Loyalties: Ideological Spatiality II

In criminology literature, appeals to higher loyalties refer to gangs, brotherhoods, or immediate family whose interests are placed above the law or societal expectations. This approach then acts through reference to an alternative value system carried by subgroups. In its spatialized form, this occurs through the privileging of subgroups along paradoxical ideological scales. One iteration of this technique is to displace critiques over environmental risks through a claim to the higher loyalty of the local community as the ideologically privileged scale. Here, a pro-development LGA councillor mixes the condemnation of condemners with the high loyalties of local community:

> … because of this vocal minority (referring here to community environmental activists) drive things, the [planning] bureaucracy listens to them. But we didn't just have the vocal minority but a whole range of people, which gave us what I call 'the real people,' and not just self-appointed activists. As I say, I work with the people. Activists are vocal and dishonest. (LGA councillor)

An alternative higher loyalty is the belief in science and rational communicative action. The subgroup loyalty is to specialist expertise that is increasingly questioned in "post-truth" public and political discourse (see Kelly & McGoey, 2018). Yet, the higher loyalty to global scientific authorities can legitimate planning:

> There is public scepticism, but if you are presenting a project, like the (coastal resort) project or something like that, you include in your information that you've considered the IPCC (Intergovernmental Panel on Climate Change) guidelines and things like that. These recommendations that you're putting forward go to government departments or local council or something like that. And while there may be a lot of scepticism in the community in general, you don't get that back from government departments or local council … It's all done to get development approval on projects. (Consulting engineer)

Appeals to higher loyalties—or the other neutralizing techniques—are no guarantor of success in reaching objectives. They emerge in *on-the-ground* planning process where debates descend into a series of governance tropes or ritualistic manoeuvres to signal towards, and call to account, the actions of opponents. They are a gesture to say: "We know what you are doing and this argument/rationalisation might not change the development process, but you should be aware that we know what you are doing."

Within these examples, there is of course slippage between narratives and practices, blurred lines of authority and condemnation, and multiple, imperfect paths to legitimation within coastal planning systems. There are also evolving governance networks, ongoing tension between technical knowledge and planning systems, and the incessant economic pressures on land use, which means governance is a painful, state-managed process of negotiation. Here, the spatialized techniques of neutralization are but one component of a suite of tacit tools used in negotiations. What these techniques then intimate is a spatial meta-narrative: the neutralizing of space through reference to time.

Neutralizing Space Through Time?

Human geographers view the space-time dialectic as a core theme and organising paradigm. Nevertheless, these two concepts are not created equal in either academic or everyday narratives. As Soja (1989) argues, too often space is subsumed under time; time is active and becoming, space is dead and inert—the physical crystallization of time. In studies of global capitalism, such as Harvey's work from 1989, capital's accumulation dynamics are the annihilation of inert space (distance) by time through new technologies and institutional convergence, whereas land simultaneously provides a *spatial fix* to rounds of capital investment. Massey (2005) is more moderate in her claims, yet sees the space-time dialectic as coconstituting:

> Here the representation of space takes place through its convening into a temporal sequence. The challenge of space is addressed by the imagination of time … (modernity is one space viewed through time) … The real import of spatiality, the possibility of multiple narratives, was lost. The regulation of the world into a single trajectory, via the temporal convening of space, was, and still often is, a way of refusing to address the essential multiplicity of the spatial. It is the imposition of a single universal. (p. 71)

One implication of Massey's argument is that this modernist privileging of time over space becomes folded into spatialized techniques of neutralization. This occurs through a double hermeneutic (Giddens, 2013), that is, both academic ontology and common sense, with the actors involved in land-use governance evincing a geographical imagination. In this quasifolk knowledge of coastal governance, actors recount space as *time-indifferent*. Regardless of its unique and relational spatial histories, actors viewed coastal space primarily as physical space to be remade for immediate or future uses. A pro-development LGA councillor summarizes this time-space ethos:

> … is the world going to stop because [coastal development] happens? No. So why worry about what way the sea-level is going to do! It becomes insignificant on a day-to-day basis of what's happening in my life and what may happen in the next generation. After that we don't care enough about the following generation.

In contrast, historical development and private interests sees a coastal engineer call upon future levels of expertise and public funding to defend the coast. Time saves space:

> … you only have to do a quick calculation of the value of the waterfront property and then come to the conclusion—and these places are worth over a million dollars for every 15 m of frontage—once you've (got a government implementing a retreat from SLR strategy) buying those up what about the next row? The sea doesn't stop there. So hundreds of hectares of land behind the first row are subject to flooding if you give up on the front properties. You are not talking one or two rows of houses, you're talking about going back a couple of kilometres of flood-prone land on the coast—it is ridiculous to think we would even retreat. It doesn't even stack up economically. From a structural point of view you protect, you need to protect. We will just be following what the Dutch have always done for the last 100s of years; really you don't need to be a clairvoyant to work that one out.

Time is summoned to neutralize coastal risks in "place" by using claims to an unknown spatial future. Notwithstanding, the growing sophistication of computer-aided actuarial modelling, and the temporal logics of the money economy, planning systems cannot factor in all future spatial effects. SLR and climate change is an unknown, for it is based on projections (Whatmore, 2009); therefore, why put off what you can do today until tomorrow? There is no absolute certainty over the environmental or sociopolitical future, so this future cannot be privileged over the present. Moreover, in South Australia's context, planners and politicians do not have time to wait for economic development; action needs to be taken now. If coastal space is threatened by SLR, this was to be acted upon when equipped with yet-to-manifest levels of human ingenuity and finance, underpinned by new modes of consensual coastal management. Fixed space is neutralized by fluid time. Echoing Massey (2005):

> This kind of space of modernity, in other words, doesn't see space as emerging from interaction, nor as the sphere of multiplicity, nor as essentially open and ongoing. It is the taming of the challenge of the spatial. This is a far deeper victory of time over space than the often-referred-to deprioritisation. (p. 71)

Here, such spatialized techniques of neutralization are not just cognitive rationalizations but political tools (Fooks et al., 2013). They are a performance of power to control the responses and actions of others (Dahl, 1956). These techniques can neutralize in public domains as well by recasting the political-economic pressures of scale and place. In a modernist narrative where governance actors cannot control the past but can control the future, the privileging of time instantaneously appeals to specific communities, bearing the promise of wider social good and immediate and future material benefits, and implies coastal space is tameable regardless of future geophysical processes. Yet research shows growing belief (and experience) in the effects of climate change, and growing concern that action needs to be taken now (Giddens, 2009). Closer to the spirit of Sykes and Matza (1957), actors employing such appeals to time neutralize the concerns of wider society without questioning the legitimacy and authority of coastal land governance.

Conclusion

This analysis of the spatialized techniques of neutralization evinces a methodological problem. I conducted the interviews *postfactum*; following Bourdieu (2004), when interviewed actors create stories that are semitheoretical or seek to impress the interviewer, while presenting a particular image and identity of the participant—one that conforms to their self-image. Whether these techniques and spatial metaphors are used *in situ* governance practices or only emerge as rationalizations in interviews following the event is an area for further research.

Nevertheless, in this chapter I have approached the theme of knowledge for governance from the angle of knowledge of governance—how to articulate contests in coastal development. Bringing techniques of neutralization from the sociology of deviance to governance contexts, I have foregrounded different—and subversive—forms of knowledge. Cosituated within the formal, technical, and rational process of land development is a tacit knowledge of how to legitimate planning decisions. Actors use spatial imagery and metaphors to create neutralizing narratives: denying responsibility in the political-juridical domain; denying injury and victims to legitimate the coast as economic space; using competing ideologies of space to frame the condemning of condemners and appeals to higher loyalties; and a metanarrative privileging time over space. Such ritualistic efforts in neutralizing opponents' claims emerge not from a need to directly protect individuals or coastal space, but to deflect attention from planning arrangements and governance processes that enable—with a dull inevitability—environmentally and socially risky coastal development to occur. Hopefully, this initial dialogue between techniques of neutralization and the human geography of land governance may open new avenues for researchers in other governance settings, allowing them to explore how the techniques of neutralization circulate at the nexus of knowledge of, and knowledge for, governance.

Acknowledgments This research was funded by the CSIRO Coastal Collaboration Cluster (2010–2013). I would like to thank Johannes Glückler for the invitation to attend the *Knowledge for Governance* symposium, Heidelberg, June to July 2017, and Anna Mateja Punstein and Laura Suarsana for administrative support to do so. I would also like to thank the conference attendees for their insightful comments and critiques of an earlier version of this chapter. Thanks to the editors Johannes Glückler, Gary Herrigel, and Michael Handke and to the technical editors, for further refinement of the argument presented here.

References

Antonetti, P., & Maklan, S. (2014). Feelings that make a difference: How guilt and pride convince consumers of the effectiveness of sustainable consumption choices. *Journal of Business Ethics, 124*(1), 117–134. https://doi.org/10.1007/s10551-013-1841-9

Bevir, M. (Ed.). (2011). *The SAGE handbook of governance*. Thousand Oaks: SAGE.

Boiral, O. (2016). Accounting for the unaccountable: Biodiversity reporting and impression management. *Journal of Business Ethics, 135,* 751–768. https://doi.org/10.1007/s10551-014-2497-9

Bourdieu, P. (2004). *Science of science and reflexivity*. Cambridge: Polity.

Cavanagh, K., Dobash, R. E., Dobash, R. P., & Lewis, R. (2001). 'Remedial work': Men's strategic responses to their violence against intimate female partners. *Sociology, 35,* 695–714. https://doi.org/10.1017/S0038038501000359

Clarke, B., Stocker, L., Coffey, B., Leith, P., Harvey, N., Baldwin, C., Baxter, T., Bruekers, G., Galano, C. D., Good, M., Haward, M., Hofmeester, C., Martins De Freitas, D., Mumford, T., Nursey-Bray, M., Kriwoken, L., Shaw, J., Shaw, J., Smith, T., Thomsen, D., Wood, D., & Cannard, T. (2013). Enhancing the knowledge-governance interface: Coasts, climate and collaboration. *Ocean & Coastal Management, 86,* 88–99. https://doi.org/10.1016/j.ocecoaman.2013.02.009

Coast Protection Board. (1991). Policy on coast protection and new coastal development. Adelaide: Government of South Australia. Retrieved from http://www.environment.sa.gov.au/files/fb3b339d-50ac-4e68-aa9c-9e3900ec438d/cpb_policy_1991.pdf

Dahl, R. A. (1956). *A preface to democratic theory.* Chicago, IL: University of Chicago Press.

Elkington, J. (1999). The triple bottom line: Sustainability's accountants. In M. V. Russo (Ed.), *Environmental management: Readings and cases* (pp. 49–66). Los Angeles: SAGE.

Enticott, G. (2011). Techniques of neutralizing wildlife crime in rural England and Wales. *Journal of Rural Studies, 27,* 200–208. https://doi.org/10.1016/j.jrurstud.2011.01.005

Fischer, F. (1990). *Technocracy and the politics of expertise.* Los Angeles: SAGE.

Flyvbjerg, B. (2001). *Making social science matter: Why social inquiry fails and how it can succeed again.* Cambridge, UK: Cambridge University Press.

Fooks, G., Gilmore, A., Collin, J., Holden, C., & Lee, K. (2013). The limits of corporate social responsibility: Techniques of neutralization, stakeholder management and political CSR. *Journal of Business Ethics, 112,* 283–299. https://doi.org/10.1007/s10551-012-1250-5

Giddens, A. (2009). *The politics of climate change.* Cambridge: Polity.

Giddens, A. (2013). *New rules of sociological method: A positive critique of interpretative sociologies.* Hoboken: John Wiley & Sons.

Government of South Australia. (2014). *Development act.* Retrieved from http://www.legislation.sa.gov.au/LZ/C/A/Development%20Act%201993.aspx

Hajer, M. (2001). The need to zoom out: Understanding planning processes in a post-corporatist society. In A. Madanipour, A. Hull, & P. Healey (Eds.), *The governance of place: Space and planning processes* (pp. 178–202). Aldershot: Ashgate.

Hansmann, R., Bernasconi, P., Smieszek, T., Loukopoulos, P., & Scholz, R. W. (2006). Justifications and self-organization as determinants of recycling behavior: The case of used batteries. *Resources, Conservation & Recycling, 47,* 133–159. https://doi.org/10.1016/j.resconrec.2005.10.006

Harris, L. C., & Daunt, K. L. (2011). Deviant customer behaviour: A study of techniques of neutralization. *Journal of Marketing Management, 27,* 834–853. https://doi.org/10.1080/0267257X.2010.498149

Harvey, D. (1978). The urban process under capitalism: A framework for analysis. *International Journal of Urban and Regional Research, 2,* 101–131. https://doi.org/10.1111/j.1468-2427.1978.tb00738.x

Harvey, D. (1989). *The condition of postmodernity.* Oxford: Blackwell.

Harvey, N., & Caton, B. (2010). *Coastal management in Australia.* Adelaide, AU: University of Adelaide Press. https://doi.org/10.20851/j.ctt1sq5x5j

Johnstone, M.-L., & Tan, L. P. (2015). An exploration of environmentally-conscious consumers and the reasons why they do not buy green products. *Marketing Intelligence & Planning, 33,* 804–825. https://doi.org/10.1108/MIP-09-2013-0159

Kelly, A. H., & McGoey, L. (2018). Facts, power and global evidence: A new empire of truth. *Economy and Society, 47*(1), 1–26. https://doi.org/10.1080/03085147.2018.1457261

Kitchin, R. (2009). *Space II.* In R. Kitchin & N. Thrift (Eds.), *International encyclopedia of human geography* (pp. 268–275). New York: Elsevier.

Krippner, G. R. (2002). The elusive market: Embeddedness and the paradigm of economic sociology. *Theory and Society, 30,* 775–810. https://doi.org/10.1023/A:1013330324198

Logan, J. R., & Molotch, H. L. (2007). *Urban fortunes: The political economy of place* (20th anniversary ed.). Berkeley, CA: University of California Press.

MacLeod, G., & Goodwin, M. (1999). Space, scale and state strategy: Rethinking urban and regional governance. *Progress in Human Geography, 23,* 503–527. https://doi.org/10.1191/030913299669861026

Massey, D. B. (2005). *For space.* Los Angeles: SAGE.

McGoey, L. (2012). Strategic unknowns: Towards a sociology of ignorance. *Economy and Society, 41*(1), 1–16. https://doi.org/10.1080/03085147.2011.637330

Meyer, R. E., & Höllerer, M. A. (2016). Laying a smoke screen: Ambiguity and neutralization as strategic responses to intra-institutional complexity. *Strategic Organization, 14,* 373–406. https://doi.org/10.1177/1476127016633335

Mooney, J. (2007). Shadow values, shadow figures: Real violence. *Critical Criminology, 15,* 159–170. https://doi.org/10.1007/s10612-007-9023-7

North, D. C. (1990). *Institutions, institutional change and economic performance*. Cambridge, UK: Cambridge University Press.

Noy, C. (2008). Sampling knowledge: The hermeneutics of snowball sampling in qualitative research. *International Journal of Social Research Methodology, 11,* 327–344. https://doi.org/10.1080/13645570701401305

Nursey-Bray, M. J., Vince, J., Scott, M., Haward, M., O'Toole, K., Smith, T., Harvey, N., & Clarke, B. (2014). Science into policy? Discourse, coastal management and knowledge. *Environmental Science & Policy, 38,* 107–119. https://doi.org/10.1016/j.envsci.2013.10.010

Polanyi, K. (1957). *The great transformation: The political and economic origins of our time* (2nd ed.). Boston: Beacon.

Rhodes, R. A. W. (1997). *Understanding governance: Policy networks, governance, reflexivity and accountability*. London: Open University Press.

Schopenhauer, A. (2004). *The art of always being right: Thirty eight ways to win when you are defeated* (T. B. Saunders, Trans.). London: Gibson Square. (Original work published 1831)

Scott, M., Balaev, M., & Clarke, B. (2018). Political embeddedness in environmental contexts: The intersections of social networks and planning institutions in coastal land use. *Environmental Sociology, 4,* 286–298. https://doi.org/10.1080/23251042.2017.1374232

Scott, M., & Harvey, N. (2016). Translating science into coastal development decisions: The articulations of science and planning in South Australia. *Journal of Environmental Policy & Planning, 18*(1), 85–101. https://doi.org/10.1080/1523908X.2015.1053108

Soja, E. W. (1989). *Postmodern geographies: The reassertion of space in critical social theory*. London: Verso.

Stuart, D., & Worosz, M. R. (2012). Risk, anti-reflexivity, and ethical neutralization in industrial food processing. *Agriculture and Human Values, 29,* 287–301. https://doi.org/10.1007/s10460-011-9337-7

Sykes, G. M., & Matza, D. (1957). Techniques of neutralization: A theory of delinquency. *American Sociological Review, 22,* 664–670. https://doi.org/10.2307/2089195

Talbot, D., & Boiral, O. (2015). Strategies for climate change and impression management: A case study among Canada's large industrial emitters. *Journal of Business Ethics, 132,* 329–346. https://doi.org/10.1007/s10551-014-2322-5

Taussik, J. (2007). The opportunities of spatial planning for integrated coastal management. *Marine Policy, 31,* 611–618. https://doi.org/10.1016/j.marpol.2007.03.006

Teh, P.-L., Ahmed, P. K., & D'Arcy, J. (2015). What drives information security policy violations among banking employees? Insights from neutralization and social exchange theory. *Journal of Global Information Management, 23*(1), 44–64. https://doi.org/10.4018/jgim.2015010103

van Buuren, A. (2009). Knowledge for governance, governance of knowledge: Inclusive knowledge management in collaborative governance processes. *International Public Management Journal, 12,* 208–235. https://doi.org/10.1080/10967490902868523

Weber, M. (1978). *Economy and society: An outline of interpretive sociology* (Vol. 1). Berkeley, CA: University of California Press.

Whatmore, S. J. (2009). Mapping knowledge controversies: Science, democracy and the redistribution of expertise. *Progress in Human Geography, 33,* 587–598. https://doi.org/10.1177/0309132509339841

Yeow, P., Dean, A., & Tucker, D. (2014). Bags for life: The embedding of ethical consumerism. *Journal of Business Ethics, 125*(1), 87–99. https://doi.org/10.1007/s10551-013-1900-2

Chapter 4
The Atmosphere of Democracy: Knowledge and Political Action

Nico Stehr

Society and Nature

It is well known that in recent years, the scientific community's consensus about man-made climate change has not only become more robust but that a number of recent studies point to far more dramatic and long-lasting consequences of global warming than previously assumed. Although commonly referred to simply as "global warming," the consequences to expect are increasing average global temperatures, rising sea levels, and more frequent occurrences of extreme weather. Given the accumulation of greenhouse gases in the atmosphere, their retention time of hundreds or more years and, despite many efforts to reduce emissions, enhance resilience, and implement new technologies, the relationship between society and the climate is bound to change in novel and unpredictable ways (see Stehr & Machin, 2019).

Under the circumstances, how is it possible, many scientists now ask, that robust science-based evidence does not motivate and encourage major political action in society as a whole and change the conduct of civil society members worldwide? How is it possible that democracies in particular have done so little to effectively combat the risks of climate change and simply failed to pay attention to the dangers

In my discussion of the relation between knowledge, expertise, and democracy, I am drawing on a couple of earlier reflections, such as Stehr (2016a, 2016b). I am grateful to Michael Handke for his comprehensive and constructive review of my manuscript. I thank Scott McNall for his helpful comments.

N. Stehr (✉)
Karl Mannheim Chair for Cultural Studies, Zeppelin University, Friedrichshafen, Germany
e-mail: nico.stehr@t-online.de

J. Glückler et al. (eds.), *Knowledge for Governance*, Knowledge and Space 15,
https://doi.org/10.1007/978-3-030-47150-7_4

of global warming?[1] After all, the nature of the future present is very much predicated on decisions taken now.

Being disenchanted with the workings of democracy and blaming democracy for a host of social, economic, and political ills is not a new complaint: "Lamenting the failings of democracy is a permanent feature of democratic life, one that persists through governmental crises and successes alike" (Runciman, 2013). However, the referent of the expression "climate change" is a novel reason for a fundamental concern about the fate and future of democracy.

Blaming Democracy

Climate scientists, social scientists, and the media as well as environmental activist groups (NGOs) concerned with climate change refer to a "future present" of exceptional circumstances[2] and protest that "evolution did not design us to deal with such problems" (Jamieson, 2014, p. 61; di Paola & Jamieson, 2018).[3] Members of the same groups assert impatiently that no one is listening to the diagnosis of historically unprecedented risks and dangers.[4]

In important respects, therefore, the discourse of climate science having established the fact of anthropogenic climate change has by necessity become forward looking. The focus has shifted to how will it be possible to govern societies in the not too distant future under the massive impact of global warming. How will it be

[1] I am using the concepts of "risk" and "dangers" not as overlapping terms, but in the sense in which Niklas Luhmann (2005, p. 23) introduced them as contrasting concepts. The risks of climate change can be *attributed* human-made decisions, while humanity is exposed to the dangers of climate change. An example of risk-taking decisions related to climate change can be studied in today's State of California: People are moving into high fire-risk zones, that is, the population of California grew by 3 million between 2000 and 2010, and, "in 2017 over a quarter of the state's population lived near moderate or high-risk fire corridors. With this increase in population comes a higher possibility of a human-made wildfire. And as people move into these high-risk areas, more buildings are in harm's way: structures generally burn longer than vegetation, allowing fire more time to spread" (cf. Adolphe, 2018).

[2] The useful concept of a "future present" is Niklas Luhmann's (1976, p. 140) terminology: "If we characterize processes or activities as beginning or ending, we use a terminology which belongs to the present. If we use these expressions to refer to distant dates—for example: The Roman Empire began to fall—we refer to a past present or to a future present."

[3] An incessant amplification in the discourse of imminent threats (many may recall the 1986 SPIEGEL title with Cologne cathedral underwater) can paradoxically turn out to be supportive of the opposite virtue, namely, as a defense of the present and encouraging skepticism toward scenarios of impending dangers. This represents a psychological mechanism not unlike the everyday attitude toward weather extremes widely interpreted as an affirmation of the normal course of the climate (cf. Stehr, 1997; Stehr & Machin, 2016b, 2019).

[4] As Bill McKibben (2018), for example, notes: "Over and over we've gotten scientific wake-up calls, and over and over we've hit the snooze button. If we keep doing that, climate change will no longer be a problem, because calling something a problem implies there's still a solution."

possible to govern a future present that is anticipated to be altogether different from the societal context in which democratic systems originated and flourished in the past? In the cases which I will identify, strong opinions promoting the need to suppress political liberties in the wake of profound future environmental changes are no longer unusual, yet have not received systematic attention in social science.

In this essay, therefore, I will bring this disenchantment with democracy, especially in its currently dominant liberal version, under the spotlight. My essay is about the struggle to align politics and policy with science. I will critically probe the argument that policy makers are going to have to act, even without a broad public mandate and legitimacy. Time is very short before a future of disastrous damage becomes locked in. But rather than lamenting the inconvenience of democratic governance, it is important to reflect upon ways of enhancing democracy, not despite but *especially* in light of the massive challenges of a changing climate. Coping with major environmental challenges is best accomplished, as history shows and as I will argue, within the bounds of democratic rather authoritarian political systems. In this essay, I reframe our changing climate as an issue of political governance and not as merely an environmental or as an economic issue.

I will advance my argument in a number of steps. First, I will address the growing assertion that social science contemporary democracies face exceptional circumstances. Second, I will reflect on the classical and present-day social science discourse on the erosion of the foundations of democracy. Third, I will describe the growing sentiment of an *inconvenient democracy* among climate scientists, other scholars, NGOs, and the media. Climate scientists propose overcoming modern democracies' inability to cope with the disastrous consequences of climate change by abolishing democracy. The alternative, of course, is to strengthen democracy. Fourth, I will consider the proposed shift in role for climate scientists as policy makers. In the final section, I will examine the serious deficiencies in the assertion of contemporary society as an inconvenient democracy.

The Rise of Exceptional Circumstances

As never before, the continuity from past to future is broken in our time. Niklas Luhmann (1998, p. 67).

In the past, actors typically used war-like conditions and major disasters to justify the abolition of democratic liberties, if only temporarily. The present appeal to *exceptional circumstances* from the critics of dominant government climate policies around the world echoes this sentiment, demanding the elevation of a single sociopolitical purpose to ultimate political supremacy.[5]

[5] For a discussion of exceptionalism in political theory, critical security, and citizenship studies, see Best (2018).

With climate change, the world is confronted with a historically novel situation and future present. Climate change within historical times is locked in. Most of the scientific discourse has been devoted to establishing that anthropogenic climate change exists. Researchers have settled the issue of whether climate change is anthropogenic, and it has become clear that unless increasingly vigorous political, economic, and societal measures are implemented, the planet will continue to experience warming "greater than it has been for more than half a million years" (Nordhaus, 2013, p. 325). What scientists have not settled is a range of important questions such as the speed of global warming, or the nature of the consequences of climate change on various significant attributes of human existence and in different regions of the world.

Governing the consequences of climate change must include a time scale and anticipated societal transformations that are clearly beyond human imagination and current political institutions. Except for reference to singular historical events, such as war, revolution, economic collapse, or the struggle for national liberation, there are no large-scale human experiences within historical times to which the climate science community can appeal as it begins to reflect on a future present in which massive impacts of climate change have set in. This relates to all levels of society and its relations abroad, such as how the world makes and uses energy, the virtue of the nation state, migration patterns, the global economy, and civil societies. In such contexts, crisis conditions promote the creation of emergency powers, the delegitimation of the previous political order, the abolition of liberty and justice, and the installment of revolutionary governance. The past is by no means a foolproof guide to the future—it is, however, often the only guide we have.

Appeal is therefore made to extraordinary circumstances or a war-like footing (Lovelock, 2009; McKibben, 2016) that necessitates the suspension of freedoms and climate scientists' political ascent. As the French political scientist Pierre Rosanvallon (2013, p. 184) stresses: "The central nation state is seen as the only source of security in the face of radical risk. It is the hope that an appeal to extraordinary circumstances, that is, to a threat to the very existence of civilization if not humankind alone might be able to give capacity and … energy back to a failing or hampered [political] will." Frank Fischer (2017, p. 54) complements this in criticizing that "current political-economic efforts on part of contemporary democratic systems to deal with problems such as global warming … [are] little more than limited symbolic gestures, especially given the pressing constraints of time." The problem of global warming and its consequences does not merely pertain to contemporary democratic governance and a missing commitment of citizens to change their ambitions and behavior. Above all, a future perspective is needed (Lovelock, 2009). The future perspective imposes its own norms on the present (cf. Jonas, 1984, p. 143).[6]

[6] Hans Jonas (1984, p. 143) interrogates the Baconian idea (executed, e.g., within Marxism) of dominating nature by increasing the humanity's power over it in his search for an ethic of the technological age. Jonas designates the Baconian ideal as the source of an ethic aimed predominantly at the future and therefore imposes its norms on the present.

But how does one govern well under exceptional circumstances? This question encounters two countervailing forces: that of an *inconvenient mind*[7] and of *inconvenient social institutions*. The former relates to a public that is assumed to be "present-centric" (Skidelsky & Skidelsky, 2012, p. 130), in other words, comfortable with the status quo, and that justifies imposing one's own (superior) ideas on those of future generations citizens (because should one really need to care whether the future public cares?). The latter relates to a strong state in the form of a command society. In other terms, good governance of society based on citizen participation must be subordinated by almost any means to the defeat of the exceptional circumstances.

It is the single purpose of defeating those exceptional circumstances that legitimizes the temporal suspension of liberties (Hayek, 1944, p. 189). However, is any massive absorption of powers in the hand of the state and its representative's reversible, in the long run? And are the potential consequences of climate change the equivalent of (abrupt) war-like conditions? How can one pinpoint the onset of exceptional circumstances?

Democratic governments' deficiencies are many and far exceed the issue of climate change and its societal consequences; but is it therefore justified to reach a conclusion as disparaging as the diagnosis of an inconvenient democracy? After all, authoritarian and totalitarian governments do not have a record of environmental accomplishments; nations that have followed the path of "authoritarian modernization/environmentalism" such as China or Russia cannot claim to have a better record.[8] Nonetheless, the disenchantment with democracies continues to be advanced, and perhaps is becoming even more vocal as entrenched climate policies fail to live up to their promise.

Inconvenient Democracy

Those who assert exceptional circumstances and the concomitant promotion of the need to overcome an inconvenient democracy derive their intellectual sustenance from a range of new and classical considerations, which lead to different forms of blaming with different addressees.

---

[7] The reference to the inconvenient mind is of course a play on words rooted in the better-known metaphor "an inconvenient truth." A fairly straightforward example of an inconvenient mind in the case of climate change is to suggest that the science of climate change is much too complicated for the average citizen to comprehend. A less "neutral" version of the inconvenient mind would be to suggest that the public is intellectually incapable of grasping the idea of global warming and its consequences.

[8] As Bruce Gilley (2012, p. 287) explains, "authoritarian environmentalism" is used to refer to "an emerging theory of public policymaking in the face of severe environmental challenges. It has been discussed both as a prescriptive model of how countries should effectively respond to such challenges, and as a descriptive model of how they are likely to respond."

The Erosion of Democracy: The Classical Perspective

In the classical social science literature, many observers believe that the threat to democracy that issues from an uneven access and distribution of knowledge in societies, for example, on social inequality formation in society (see Stehr & Machin, 2016a), has radically displaced earlier, optimistic enlightenment views regarding the resilience and even the possibility of a democracy based on a general circulation of knowledge in society.[9] Numerous authors, from Max Weber to Robert Michels, have explicated these and other threats to representative democracy.

Given the unstoppable advance of bureaucracy in modern societies, Max Weber (1918/1994), for example, feared a kind of *pacifism of social impotence* of the citizenry, for in the face of a

> growing indispensability and hence increasing power of state officialdom … how can there be any guarantee that forces exist which can impose limits on the enormous, crushing power of this constantly growing stratum of society and control it effectively? How is democracy even in this restricted sense to be *at all possible*? (Weber, 1918/1994, p. 159)

Robert Michels (1915), in his classical study of the undemocratic tendencies in the social democratic party, a political organization that actually aspires to and fights for democratic goals, refers to an almost "natural" state of incompetence and immaturity of the mass of people in modern democracies. And because those of rank and file are incapable "… of looking after their own interests, it is necessary that they should have experts to attend to their affairs" (Michels, 1915, p. 93). Seldom is the rank and file willing to throw off the authority of the expert leaders and dismiss them from control.[10] Numerous of the classical concerns about the viability of democratic governance find an echo in contemporary reflections about the fragility of democracy.

---

[9] There is good reason to be skeptical of the idea that either the notion or the realities of the knowledge gap or the information overload, however defined, are genuinely new. One has only to refer to the convergence of societal diagnoses proposed, at the dawn of the last century, by thinkers such as Georg Simmel, Sigmund Freud, and Walter Benjamin, of a cultural age displaying severe overstimulation, discontinuities, and overload.

[10] Whether the disillusioned conclusion Robert Michels (1915, p. 95) draws in light of the tendencies he observes, namely that "social democracy is not democracy, but a party fighting to attain democracy" is inevitable, that is, universally applicable as a kind of iron law, is surely contestable, although many observers are prepared to concede that Michels has discovered one of the few law-like relations in social science. For more recent studies by economists, sociologists, and political scientists who take Michels's challenge about the inevitability of oligarchic tendencies in organizations on board, see Williamson (1975, 1985, 1994), Granovetter (1985), Foucault (2005), Stehr and Adolf (2018, pp. 321–324).

The Erosion of Democracy: The Modern Perspective

A deep-rooted pessimism about the psychological make-up of human beings, the temporality of human thought, the failure to mobilize individuals for the cause of effective climate policies, the inability of government given constitutional constraints to attend to long-term goals, the fragility of the political order, the influence of vested interests on the political agendas of the day, and in the case of anthropogenic climate change, the addiction to fossil fuel, as well as, last but not least, the ineffectiveness of the climate science community itself insure that their message does not fall on deaf ears.[11]

Blaming the People

Daniel Kahneman sums up the growing skepticism regarding citizen motivation when he states:

> The bottom line is that I'm extremely skeptical that we can cope with climate change. To mobilize people, this has to become an *emotional issue*. It has to have the immediacy and salience. A distant, abstract, and disputed threat just doesn't have the necessary characteristics for seriously mobilizing public opinion. (Cited in Marshall, 2014, p. 57, emphasis added)

The mass of citizens, it seems, simply cannot be won over to endorse and follow the course of scientifically based policy options. The large majority of citizens is basically inclined to act irrationally (cf. Schumpeter, 1942, pp. 262–263). The climate scientist Hans-Joachim Schellnhuber (Elger & Schwägerl, 2011, p. 29)[12] gloomily relates why climate change communication does not reach civil society: "[M]y own experience and everyday knowledge illustrate that comfort and ignorance are the biggest flaws of human character. This is a potentially deadly mix." However, to view democracy and politics in terms of the competence of the individual citizens is to argue in favor of a micro-sociology without a macro-sociology. The reference to the public perceptions of science and expert knowledge goes beyond the implicit or explicit assumption that the public has basically deficient information and knowledge, is perhaps even reactionary, and tends to respond to complexity with trepidation (cf. Gauchat & Andrews, 2018).

The apparently widely shared ability to avoid knowing what the future could bring can of course also be interpreted as a psychological "incentive" to live with

---

[11] Efforts in climate change communication are predicated on the conviction

> that if the public only knew the facts about climate change and began to understand just how serious the problem is, they would raise their voices and demand that our governments and corporations do something (Revkin, 2014).

[12] The climate scientist Hans Joachim Schellnhuber, in an interview with DER SPIEGEL (Issue 12, March 21st, 2011, p. 29) in response to the question of why science's messages do not reach society.

the knowledge about the limited knowledge on the outcome of events that are located in the future (cf. Gigerenzer & Garcia-Rettamero, 2017). Meanwhile, political scientists, who have in many ways been concerned about the voters' lack of information, have begun to stress that the democratic-political system works in spite of citizens' ignorance (Kuklinski, 1990). Or, as Petersen and Aarøe (2013, p. 289) have more recently documented, despite the widespread lack of extensive political knowledge, "citizens readily form opinions on what constitutes the best and most efficient policies."

Seymour Martin Lipset and his colleagues (1962) advance an appraisal more in support of the political virtue of knowledgeable citizens: lack of information, passivity, and lack of interest of rank-and-file members in the affairs of an organization is in the interests of the powerful and supports their capacity to perpetuate power advantages. It seems that it is not so much the volume of knowledge or information that citizens command that impacts the relation between democracy and knowledge, but rather the importance of democracy-enhancing individual and collectively shared *value-orientations*; or, as Robert Dahl (1977, p. 1) argues: It is "the ways in which we think about ourselves as a people" that support the existence and the stability of democracy. Of course, value-orientations and educational achievement are connected: "Education presumably broadens men's outlook, enables them to understand the need for norms of tolerance, restrains them from adhering to extremist and monistic doctrines, and increases their capacity to make rational electoral choices" (Lipset, 1959, p. 79).

Blaming the Political Class

In the eyes of many from the climate science community, not only citizens but also politicians are not ready to pursue policies that effectively address climate change. Climate activist, climate scientists, some politicians, and many other observers agree that the recent climate summits in Copenhagen, Cancun, Durban, and Warsaw were failures. The summits did not result in a new global agreement to cope with the emissions of greenhouse gases. The subsequent 2015 Paris Agreement, widely regarded as an historical achievement, seemingly marks a general scientific and public consensus that anthropogenic climate change is a very serious threat to human civilization and its environments. The treaty, however, is non-binding. There are no formal sanctions if a country should fail to live up to its commitments regarding the efforts in terms of mitigation, adaptation, or finance, and there is no guarantee how far reaching the Paris Agreement will be. This problem came to the fore on June 1st, 2017, when the United States, led by President Donald Trump, announced their formal withdrawal from the treaty, rejecting the scientific consensus that greenhouse gas emissions are warming the planet.

Although under the terms of the Paris Agreement the U.S. cannot formally begin the process of withdrawal until November 2019, the current administration is already embarked upon a strong anti-environmentalist agenda. In his announcement, Trump was fulfilling his campaign pledge to "end the war on coal" and his purported aim to reclaim sovereignty for the American people and put "America first." But as has been much remarked upon, shortly after his announcement in which Trump emphasized that he was elected to represent "the people of Pittsburgh and not Paris," the mayor of the state of Pittsburgh, Bill Peduto, voiced his criticism of the withdrawal and proclaimed the state's commitment to the treaty. Indeed, a number of American states and cities will continue to follow its announced climate policies, offering "a profound counter to Trump's anti-environmental crusade" (Bomberg, 2017, p. 5). What this scenario illustrates is the high degree of politicization of the issues of climate and climate change in the contemporary world.

The nature of the relation between temporality and democracy indeed justifies doubts about the effectives of democratic governance in the face of longer-term future risks and dangers of climate change. Issues of temporality refer to at least a couple of significant matters driven by distinctive but related systemic conditions of democratic governance. On the one hand, democratic governance is captivated by the *immediacy* of frequently changing events that often come and go rapidly, as much as it is affected, on the other hand, by constitutional rules of representation that prescribe relatively short frames of *temporality*. The public perception of the urgency of political issues is dynamic and relative. The attention that actors give to climate change very much depends on their perception of the importance of other political issues at any given time, especially on the perception of pressing economic issues.

Are democracy and societal institutions constrained by short-term constitutional frames and governed by principles of liberty, such as the market, capable of dealing with harms and risks to society that are located in the future? How can democracies sustain interest in a future present that is a couple of decades away and thereby escaping the typical media issue attention cycle (Downs, 1972; McDonald, 2009) of events?

There is a parallel discourse in social science to which I now turn, in which scientists express strong doubts about the "sustainability" of modern democracies. They highlight symptoms of a crisis that is not only triggered by major environmental problems but also by various structural and secular challenges faced by present-day democratic governance.

Are Democracies Dying?

The climate science community's discussions about democratic governance's inadequacies converges with assessments of the present state and future of democracy in the social sciences. It was only a few years ago that political scientists proclaimed the end to history (Fukuyama, 2018) and with it the ultimate victory of democracy. Today, political scientists—Francis Fukuyama (2018) included—are much more likely contemplating the dissolution of democracy. Even titles like "The Future of Freedom" (Zakaria, 2003), "The Retreat of Western Liberalism" (Luce, 2017),

"How Democracy Ends" (Runciman, 2018), "How Democracies Die" (Levitsky & Ziblatt, 2018), "The People vs. Democracy" (Mounk, 2018), and "Can Democracy Survive Global Capitalism? (Kuttner, 2018) give an indication of it. The dispute about climate change and climate policies plays a central role in the contemporary shift of the debate about the well-being of democracy. In response to multiple societal changes underway, the arguers concludes, democracy loses its legitimacy in the eyes of its citizens.

The conclusion of social science observers must therefore be that contemporary democracy—in many ways whether by design or at the outcome of structural economic, political and moral changes—is on its way to autocratic forms of governance. For instance, the erosion of democracy manifests itself in processes of de-politicization, the substitution of politics by techniques of management or the restriction of the public sphere, or (cf. Rosanvallon, 2006, p. 228; also Swyngedouw, 2011) "in a hollowing out of citizenship, the marketization of the public sector, the soul-destroying targets and audits that go with it, the denigration of professionalism and the professional ethic, and the erosion of public trust" (Marquand, 2004, p. 172). Democratic governance is increasingly muted by the rapid abolition of democratic principles of political equality, and even replaced by autocratic forms of governance that echo Robert Michels's (1915) century-old iron law of oligarchy.

What distinguishes the discussion about the poor health of democracy among social scientists and climate scientists is the remedy that both sides advocate. On the one hand, social scientists discuss efforts that could restore democracy, such as rebuilding "a society of similar individuals" (Rosanvallon, 2013) through the active participation of a large number of citizens that shaping the agenda of public life. On the other hand, climate scientists and other observers of global climate change disparage democratic governance's very capacity to cope effectively with the large-scale environmental problems and therefore call for a more authoritarian state and/or a state where decision making by technical experts is given weight. But then democracy is allegedly dismantling itself.

Colin Crouch (2004, p. 4), for example, describes democracy's transition to post-democracy in the following terms: "Under the conditions of a post-democracy that increasingly cedes power to business lobbies, there is little hope for an agenda of strong egalitarian policies for the redistribution of power and wealth, or for the restraint of powerful interests."

Post-democracy is also accompanied by the swift erosion and disavowal of democratic rights and values, as Richard Rorty (2004, p. 10) argues:

> At the end of this process of erosion, democracy would have been replaced by something quite different. This would probably be neither military dictatorship nor Orwellian totalitarianism, but rather a relatively benevolent despotism, imposed by what would gradually become a hereditary nomenklatura.

In some of the images of post-democracy as a state of the state, a return to aristocratic society has already been achieved. Self-appointed elites claim to carry out the wishes of the masses.[13] In short, as Pierre Rosanvallon (2006, p. 228) emphasizes, politics has been replaced, "leaving room for one sole actor on the scene: international society, uniting under the same banner the champions of the market and the prophets of the law." This marks a political development that representatives of the climate science community very much welcome.

The radical conclusion some observers draw, especially those who favor and promote the role of experts and expertise as a form of enlightened leadership, is that democracy itself is inappropriate, that the slow procedures for the implementation and management of specific, policy-relevant scientific knowledge leads to massive, unknown risks and dangers. Civilization-as-we-know-it may come to an end. Assuming it is not already too late, appropriate environmental governance must look very different. To create a globally sustainable way of life, the world immediately needs, in the words of German climate scientist Hans Joachim Schellnhuber (cf. WBGU, 2012), a "great transformation." Part, if not the core of the required great transformation appeared to be a new political regime and forms of governance. For example, as expressed by the Australian scholars David Shearman and Joseph Wayne Smith (2007, p. 12) in their book The Climate Change Challenge and the Failure of Democracy: "We need an authoritarian form of government in order to implement the scientific consensus on greenhouse gas emissions." Mark Beeson (2010, p. 289) argues in the same vein when he brings into play the notion of good authoritarianism:

> [G]iven the unprecedented and unforgiving nature of the challenges we collectively face … forms of 'good' authoritarianism, in which environmentally unsustainable forms of behavior are simply forbidden, may become not only justifiable, but essential for the survival of humanity in anything approaching a civilised form.

Another proposal is for a distinctively political role of climate scientists. In most countries, climate scientists are successful in equipping governments with the authority of the correct point of view about climate change. However, climate scientists fail to ensure that governments act on the authority of science.What is the alternative? One alternative is an exchange of leadership and the rule of the knowledgeable class. The idea to exchange political leadership is not only to put science and scientists at the center of governance, but also to depoliticize the issue of climate change (cf. Aitken, 2012; Swyngedouw, 2010).

---

[13] Hans Jonas's (1984, p. 147) sober response to such a claim is quite appropriate and worth citing in this context: "[I]f … only an elite can assume, ethically and intellectually, the kind of responsibility for the future which we have postulated—how is such an elite generated and recruited, and how is it invested with the power for its exercise?"

Enlightened Leadership?

Within the broad field of climatology and climate policy, one is able to discern growing frustration with the virtues of democracy and a mounting appeal to exceptional circumstances and the promotion of the role of scientists and experts in policy making. The impatience with democracy and the shifting understanding of the role of scientists can be observed with a change in the function of the *International Panel of Climate Change* (IPCC). IPCC no longer considers itself a scientific organization with the mandate to offer alternative policy options for political discussion and decision, but a body of experts demanding that the options for political action it identifies be rapidly realized.

Robert Stavins, the director of Harvard's Environmental Economics Program and a co-author of the IPCC Working Group 3 report, notes a

> bottom up demand which normally we always want to have and rely on in a representative democracy, is in my view unlikely to work in the case of climate change policy as it has for other environmental problems … It's going to take enlightened leadership, leaders that take the lead.[14]

The social scientist Evelyn Fox Keller (2017, p. 107) makes the strong case for an immediately effective, practical political role of climate science, given the seriousness of the problem of global warming:

> There is no escaping our dependence on experts; we have no choice but to call on those (in this case, our climate scientists) who have the necessary expertise … Furthermore, for the particular task of getting beyond our current impasse, I also suggest that climate scientists may be the only ones in a position to take the lead … [and] given the tacit contract between scientists and the state which supports them on the other, I … also argue that climate scientists are not only in a position to take the lead, but also that they are obliged to do so.

Science, Knowledge, and Democracy

The strong desire to reach specific policy outcomes spelled out by the climate science community lead many to believe that scientific knowledge is somehow immediately performative or is an immediately persuasive form of knowing. Endorsers of such a conception of knowledge privilege knowledge as a policy instrument by ignoring the limits of the power of knowledge (Prewitt, 2010; Sarewitz, 2010; Stehr, 1991). On this doubtful basis alone, it is unsurprising that climate scientists at least sympathize with the suspension of democratic process.

However, the inconvenient democracy position contains a number of obvious weaknesses that I will enumerate now in some detail. I have organized my observations into five counterarguments.

---

[14]As quoted in Andrew Revkin, "A risk analyst explains why climate change risk misperception doesn't necessarily matter," *New York Times*, April 16th, 2014.

First, and importantly, one encounters a flawed understanding of scientific knowledge and its potential role in political contexts. Scientific knowledge is neither immediately performative (knowledge equals control and represents practical reason) nor immediately persuasive (that is, knowledge convinces unencumbered). Knowledge alone does not generate a profit or score goals (cf. van Dijk, 2014). One of the fundamental flaws in the portrait of an inconvenient democracy is the failure to recognize the social character of knowledge in general and the contested and often ambivalent nature of political knowledge in particular. Recognizing the proper function of knowledge assures a premature political closure, in other words, the depoliticization of the issue of climate change and climate policies.

It is more appropriate to characterize knowledge not as "something that is so," but as a generalized *capacity to act* on the world, as a model *for* reality, or as the ability to set something in motion (Grundmann & Stehr, 2012; Stehr, 1994; Stehr & Adolf, 2018). The German term that best describes knowledge as a generalized capacity to act would be *Handlungsvermögen*. The verb *vermögen* signals "to be able to do," whereas the noun *Vermögen*, in this context, is best translated as "capacity" (rather than "fortune" or "wealth").[15] The capacity to act—the ability to put something into motion—extends to the capacity to generate "symbolic action." For example, symbolic action may involve the ability to formulate a hypothesis, carry out a ritual, find a new metaphor for an established term,[16] assess "facts," organize the literature on a topic, or defend a thesis against "new facts." The capacity to act, in other words, refers not merely to the possibility of accomplishing something in terms of a material and physical performance such as, for example, making fire or driving a car. Capacities to act also refer to *intellectual* abilities as well as the production of *meaning*, such as may be found in the detailed description of the bundle of skills that I call *knowledgeability* (cf. Stehr, 2016a). This is most likely also the reason why Norbert Elias (1984, p. 252) defines knowledge as "the social meaning of human made symbols, such as words or figures, in its *capacity as means of orientation*" (emphasis added).

Knowledge, as a generalized capacity for action, acquires an "active" role (that is, is put to work) in the course of social action only under certain circumstances, namely where social action does not follow purely stereotypical (effortless) patterns (Max Weber), or is strictly regulated in some other fashion. Under conditions of ritualized social conduct, a break in the continuity between past and future will not occur. Past and future are securely looked in through taken-for-granted sequences of events.

Niklas Luhmann's observations about the conditions for the possibility of making decisions in the first instance perhaps allows for an even broader understanding of the use of knowledge but also confirms my description of the likely usefulness of

[15] Georg Simmel (1890, p. 276), in his discussion of money as a generalized code, uses the concept *Vermögen* to describe the fact that money is more than merely a medium of exchange; his definition of money thus transcends a merely functional understanding of its social capacities.

[16] I refer in this context, for example, to Donald Schon's (1963) reflections in *Displacement of Concepts* (cf. also Haldane, 2013).

knowledge only under conditions of degrees of openness of the circumstances of action. Decision making, Luhmann (1998, p. 67) writes, "is possible only if and insofar as what will happen is uncertain."

The circumstances of action that I have in mind may also be described as actors' capacity to alter or stabilize a specific reality. However, the capacity "to get things done," to alter and affect reality, as well as the ability to intervene in a context that otherwise would change, is not symmetrical with the capacity to act (knowledge). Knowledge and control should not be symmetrical: "Foresight and control is highly fragile in reality, it can be shown that a persistent progress of knowledge neither leads necessarily to an improvement of foresight nor to an improvement of control" (Tenbruck, 1977, p. 223). One's ability to do something is dependent on one's control over the conditions of action. The lack of control over the political conditions of action is an apt description of the societal role that fits the position of climate scientists today and will continues to be the case as long as they have not appropriated political power.

Second, one of climate science critics' leading assumptions of democracy is a misunderstanding of the climate problem and a misleading framing of the policy process.[17] The result of this misunderstanding of the climate problem and of the climate policy process is a fundamental framing error, its perpetrators representing climate change as a conventional environmental "problem" that is capable of being "solved." It is neither of these.

Rather than being a *discrete* problem to be solved, climate change is better understood as a persistent condition that must be coped with and can only be partially managed more or less well. The climate issue is one part of a larger complex of such conditions encompassing population, technology, wealth disparities, public values, resource use, and so on. Hence, it is not straightforwardly an "environmental" problem either. It is axiomatically as much an energy problem, an economic development problem, or a land-use problem and may be better approached through these *multiple avenues* than as a problem of managing the behavior of the Earth's climate by changing the way that humans use energy.

This makes climate change a "wicked" problem.[18] A wicked problem is the impossibility of giving the policy issue a definitive formulation: the information needed to understand the problem is dependent upon one's idea for solving it. Furthermore, wicked problems lack a stopping rule: One cannot know whether one has a sufficient understanding to stop searching for more understanding. There is no end to causal chains in interacting open systems of which the climate is the world's prime example. Climate change policies are best embedded in comprehensive policy perspectives whose holders attack climate change indirectly, accepting, for

[17] In my critique of the dominant framing of the climate problem, I draw on our *Hartwell Paper* (Prins et al., 2010).

[18] Wicked problems are embedded in multiple social systems. Originally described by C. West Churchman (1967) and later explicated more comprehensively by Horst Rittel and Melvin Webber (1973) in the context of urban planning, wicked problems are issues that are often formulated as if they were susceptible to a simple, unilinear solution when in fact they are not.

example, that decarbonization will only be achieved successfully as a benefit contingent upon other goals that are politically attractive and pragmatic.

Third, in a related manner, proponents of the dominant political approach concentrate almost exclusively on a *single effect* that governance ought to achieve, namely a reduction of greenhouse gas emissions and perhaps necessary measures of adaptation to climate change. In doing so, they exclude other, more complex forms and conditions of action. By focusing on the goals of political action rather than its conditions, they reduce the contentious issue of climate change to scientific or technical issues. Sociopolitical issues are neglected. The politicization of climate science leads to a depolitization of climate changes. Matters relevant to the public are being permanently removed from politics (see also Jasanoff, 2012).

Equally deficient in this context is the focus on a *single approach* to attack climate change, namely a reduction of greenhouse gases, especially CO2. Those exclusively framing climate policy as directed toward a reduction of emissions ignore what Roger Pielke Jr. (2010) calls the "iron law" of climate policy. The iron law merely states that although people are often willing to pay a certain price for environmental policy goals, their willingness has its limits. That exact limit, of course, varies from place to place and household to household. The massive resistance of the "Yellow Vest" protests in France against the government's plans to implement regular fuel tax increases to fight global warming in the early winter of 2018 is a perfect example of Pielke's law. The protests of the Yellow Vest movement forced the government to cancel the tax increase. Public support for climate policies declines as a function of the impact of such policies on the household costs. A convergence of ecological and economic policies is not impossible. However, such a convergence likely tilts toward the economic part of the equation when emission reduction policies collide with economic growth or labor market policies.

Fourth, the generally pessimistic assessment of the ability of democratic governance to respond to, cope with, and control exceptional circumstances is linked, if only implicitly, to the then peculiar optimistic assessment of the potential of large-scale planning in the sense of social engineering. Planning on any scale is hardly straightforward. Not only the capacity of governments but also the general possibility to plan for the future present of societies is rather limited, perhaps absent (see Tenbruck, 1977, p. 138). Economic and social planning conceptions widely discussed in the affirmative decades ago have fallen into disrepute (see Giddens, 2009, pp. 94–100). Certain schemes to improve the human condition have failed, James Scott (1998) demonstrates case-by-case in his book *Seeing like a State*. The once active academic program of, and enthusiastic support for, futurology about desirable futures has vanished (Seefried, 2015). Modern de-centered, functionally differentiated societies preclude de-differentiated, society-wide social planning in principle (Luhmann, 1976, 1998).

Fifth, in the reasoning of the impatient critics of democracy, one notes an inappropriate fusion of nature and the nature of society. The uncertainties (related to climate) that the sciences of the natural processes claim to have eliminated and the authoritative consensus that the sciences have thereby acquired are simply transferred to the domain of societal processes. Consensus on the evidence, it is argued, should motivate a consensus on political action. What becomes desirable is a rational design of social order "commensurate with the scientific understanding of natural laws" (Scott, 1998, p. 4), for instance, a comprehensive engineering of human settlement and production. Designing society top-down is schematic and ignores the essential realities of any truly existing social order: The constitutive uncertainties, fragility, and complexity of social, political, and economic events, the difficulty of anticipating the future present are treated as minor obstacles that can be encircled as soon as possible—of course by a top-down approach—by implementing policies that the faith in scientific knowledge prescribes. This undermines the dignities, pluralities, and conflicts that are immanent features of contemporary knowledge societies.

Finally, there is the remarkable resilience of advanced capitalist democracies confronted with major "shocks" from their beginnings in the early twentieth century through one of the most turbulent modern centuries. Democracy is a more effective adaptive organism than other forms of governance (Luce, 2017, p. 87) Although the past is not necessarily a solid foundation for anticipating the future state of affairs, there is "a near-zero probability of rich democracies reverting to authoritarianism" (Iversen & Soskice, 2019; see also Przeworski & Limongi, 1997). Obviously, exceptions exist. But one cannot yet know whether the exceptional circumstances of climate change in the future present will be of such magnitude that the past indeed can be no guide to the future health of democracies.

What Is to Be Done? Enhancing Democracy?

What is good governance under exceptional circumstances? Is democratic governance effective governance? And why should a more democratic as well as egalitarian society be beneficial as the socio-political foundation for coping with extreme circumstances?

In their disenchantment with democracy, the discourse of the impatient scientists privileges hegemonic players such as world powers, states, transnational organizations, and multinational corporations. Participatory strategies are only rarely in evidence. Likewise, global mitigation has precedence over local adaptation. "Global" knowledge triumphs over "local" knowledge. However, societal trends appear to operate into the opposite direction. The ability of large societal institutions to impose their will on citizens is declining (Stehr, 2001). As a result, people mobilize around local concerns and efforts, including those of the consequences of climate change—thereby enhancing the democratic in democratic governance.

The discussion of options for future climate policies supports the impression that the same failed climate policies must remain in place and are the only correct approach; it is simply that these policies have be become more effective and "rational." It follows that international negotiations must lead to an agreement for concrete, but much broader, emission reduction targets. Only a super-Kyoto can still help. But how the noble goals of a comprehensive emission reduction can be practically and politically *enforced* remains in the fog of general declarations of intent and only sharpens scientists' political skepticism.

The still dominant line of attack to climate policy shows little evidence of success, whether at the state level or on the global scale. On the contrary, everything that actors continues to set in motion worldwide is aimed at a persistent economic growth, which prevents emissions from declining. An alternative model is needed—a model in which action under ambivalent, uncertain and unexpected circumstances can be compelled. A model whose utilizers recognize, moreover, that climate change is a wicked problem that can only be attacked indirectly and requires persistence over a longer period of time. That kind of model will only be found through revitalized rather than less democratic interaction.

Climate policy must be compatible with democracy, or the threat to civilization will be much more than just changes to the world's physical environment. Climate change demands complex solutions that require worldwide empowerment and knowledgeability of individuals, groups, and movements that labor on environmental issues. More democracy combined with political efforts to move toward a more equitable society could be the key toward sustainable climate policies. By definition, more democracy comes with greater political participation, especially among those now typically standing on the sidelines of political participation, such as the young and the economically disadvantaged strata.[19]

A more egalitarian society "would not necessarily maintain rational ecological policies, but it would be more likely to do so" (Best & Connolly, 1975, p. 59). When life chances are more equally distributed, assuring that no one can escape the benefits and costs of a resolution of a serious public problem,[20] one should expect that "the political system is very likely to generate collective responses to common dangers and burdens" (Best & Connolly, 1975, p. 59). The English political scientist David Runciman (2013, p. 316) spells out two further distinct, practical advantages of democracies over authoritarian governments faced by extraordinary circumstances: „The first is their ability to pull together when the threat becomes too big to ignore … The second is their ability to keep experimenting and adapting to the challenges they encounter.".[21]

A war-like footing, in contrast, has exactly the opposite effect. A war-like approach reduces the complexity of social and political life in as much as war "nationalizes people's life. Private activities … [are] largely shaped by collective constraints" (Rosanvallon, 2013, p. 183), as would be the case under authoritarian rule. Under modern conditions, the heightened cognitive and social abilities of ordinary citizens especially predicates successful policies and good governance on their political participation.

[19] Concrete advice on how to avoid oligarchic tendencies in organization may be found, for example, in Robert K. Merton's (1966) essay "Dilemmas of Democracies in the Voluntary Association."

[20] The systematic reduction of patterns of social inequality in modern societies enhances democratic governance and political participation (Soci, Maccagnan, & Mantovani, 2014, p. 46).

[21] Hans Jonas (1984, p. 146) advances a similar observation about systematic inability of authoritarian governments to transcend policy mistakes.

Moreover, a further denationalization of governance will assist in producing new, multiple forms of social solidarity and obligations, strengthen local/regional responses to climate change, and enhance the understanding of social interdependence. In addition, social institutions' self-sufficiency must be guaranteed and—if necessary—re-created in order to transcend boundaries, joining allegedly distinctive motives and practices of different social institutions, for example, joining economic and moral incentives and enhancing the complexity of needs.

The tendency to overestimate and overreach in assigning a crucial role to the singularity of knowledge (and information) in social conduct is evident as one considers *how much knowledge* is needed to carry a specific task, let alone how deeply and subtly one needs to know it. Curiosity about how much one needs to know also extends to the question of what one does not need to know. In the first instance, this happens to be an issue that is rarely systmatically examined. Second, the inclination is prevelant to assume that the resource of knowledge is somehow sufficient to carry a specific transaction. A more adequate conjecture would be to expect that actors carry out most decisions and actions with rather limited knowledge and information (cf. Akerlof, 1970; Smith, 2015) about future conditions of action and that they are cognizant of how little knowledge they are typically able to mobilize in many situations. The pressure to act that characterises everyday life ensures that, despite the limited knowledge and information of most actors, decisions are taken and action taken. The fact that actors are often forced to act with limited knowledge is not a constitutive deficiency of democracy. "Life cannot wait" (Durkheim, 1965, p. 479; see also Gehlen, 1988, pp. 296–297). In most social contexts, the need to act takes precedence over the need to know.

The erosion of democracy may seem "convenient" to some, such as populists, but surely is an unnecessary suppression of social complexity. Friedrich Hayek (1960, p. 25) pointed out a paradoxical development: As science advances, the observation that we should "aim at more deliberate and comprehensive control of all human activities" tends to strengthen. Hayek pessimistically adds: "It is for this reason that those intoxicated by the advance of knowledge so often become the enemies of freedom."

That democratic governance is *slow* compared, for example, to the speed at which with decisions are made in the modern economy (see Stehr & Voss, 2019) cannot be denied. In the eyes of many citizens, naturally including climate scientists, the slowness and the deliberateness of decision making generates permanent discontent. Climate scientists, with their escalating warnings about imminent risks and dangers of climate change repercussions and their communication of politicians' failures to heed these forewarnings do nothing to stem such civil restlessness. Democratic actors therefore face the major challenge of speeding up political decision making as well as enhancing opportunities for participation in democratic decision making in places such as the workplace (cf. Herzog, 2019) and the local political community.

Conclusions

> Certain kind of states, driven by utopian plans and an authoritarian disregard for the values, desires and objections of their subjects, are indeed a mortal threat to human well-being. James Scott (1998, p. 7)

As an editor of *Nature* (December 4, 2014, p. 8) editorializes: "The magnitude of … climate change is worryingly uncertain. Even more uncertain are the physical, social and economic side effects of global warming. There is every reason to believe that, by and large, they will be harmful." The central issue is no longer whether is climate change occurring. It is rather what should be done about it. Climate change is the biggest threat humanity has faced in historical times. Suspending democratic debate and decision making including extensive citizen participation in order to do what is necessary would either demand elevating experts to become decision makers or delegating power to policymakers (who happen to believe a certain group of experts). Neither the first, the technocratic or social engineering vision, nor the idea of a more authoritarian environmentalism has appeal.

I have collected and advanced arguments that speak to the need to enhance rather than abolish democracy as the best political foundation for policies suited to addressing climate change as a wicked problem. It is important to push back against simplified solutions to climate change. In debating, researching, and understanding climate and climate change, actors would do well to heed the complex interconnections of the climate system, but also the societal processes, practices, and tensions through which science, society, nature, and climate permeate, accompany, cover, and envelop each other (for such a theoretical perspective, see Stehr & Machin, 2019).

References

Adolphe, J. (2018, November 11). Why are California wildfires so bad? *The Guardian* (London). Retrieved from https://www.theguardian.com/world/ng-interactive/2018/sep/20/why-are-california-wildfires-so-bad-interactive

Aitken, M. (2012). Changing climate, changing democracy: A cautionary tale. *Environmental Politics, 21,* 211–229. https://doi.org/10.1080/09644016.2012.651899

Akerlof, G. A. (1970). The market for "lemons": Quality uncertainty and the market mechanism. *The Quarterly Journal of Economics, 84,* 488–500. https://doi.org/10.2307/1879431

Beeson, M. (2010). The coming of environmental authoritarianism. *Environmental Politics, 19,* 276–294. https://doi.org/10.1080/09644010903576918

Best, J. (2018). Technocratic exceptionalism: Monetary policy and the fear of democracy. *International Political Sociology, 12,* 328–345. https://doi.org/10.1093/ips/oly017

Best, M. H., & Connolly, W. E. (1975). Market images and corporate power: Beyond the "economics of environmental management". In K. M. Dolbeare (Ed.), *Public policy evaluation* (pp. 41–74). Thousand Oaks: SAGE.

Bomberg, E. (2017). Environmental politics in the Trump era: An early assessment. *Environmental Politics, 26,* 956–963. https://doi.org/10.1080/09644016.2017.1332543

Churchman, C. W. (1967). Guest editorial: Wicked problems. *Management Science, 14,* B141–B142.

Crouch, C. (2004). *Post-democracy*. Cambridge: Polity.

Dahl, R. A. (1977). On removing certain impediments to democracy in the United States. *Political Science Quarterly, 92,* 1–20. https://doi.org/10.2307/2148546

di Paola, M., & Jamieson, D. (2018). Climate change and the challenges to democracy. *University of Miami Law Review, 72,* 369–424. Retrieved from https://repository.law.miami.edu/umlr/vol72/iss2/5

Downs, A. (1972). Up and down with ecology: The "issue-attention cycle". *The Public Interest, 28*, 38–50.

Durkheim, E. (1965). *The elementary forms of the religious life* (J. W. Swain, Trans.). New York: Free. (Original work published 1912)

Elger, K., & Schwägerl, C. (2011, March 21). Diktatur des Jetzt [Dictatorship of the now]. *Der Spiegel*. Retrieved from https://www.spiegel.de/spiegel/print/d-77531589.html

Elias, N. (1984). Knowledge and power: An interview by Peter Ludes. In N. Stehr & V. Meja (Eds.), *Society and knowledge*: *Contemporary perspectives on the sociology of knowledge* (pp. 251–292). New Brunswick: Transaction.

Fischer, F. (2017). *Climate crisis and the democratic prospect: Participatory governance in sustainable communities*. Oxford, UK: Oxford University Press. https://doi.org/10.1093/oso/9780199594917.001.0001

Foucault, M. (2005). Die Maschen der Macht [The meshes of power]. In M. Foucault (Ed.), *Schriften in vier Bänden: Dits et Ecrits I–IV* (Vol. 4, pp. 224–244). Frankfurt: Suhrkamp.

Fukuyama, F. (2018). *Identity: Contemporary identity politics and the struggle for recognition*. London: Profile.

Gauchat, G., & Andrews, K. T. (2018). The cultural-cognitive mapping of scientific professions. *American Sociological Review, 83,* 567–595. https://doi.org/10.1177/0003122418773353

Gehlen, A. (1988). *Man*: *His nature and place in the world*. New York, NY: Columbia University Press. (Original work published 1904)

Giddens, A. (2009). *The politics of climate change*. Cambridge: Polity.

Gigerenzer, G., & Garcia-Retamero, R. (2017). Cassandra's regret: The psychology of not wanting to know. *Psychological Review, 124,* 179–196. https://doi.org/10.1037/rev0000055

Gilley, B. (2012). Authoritarian environmentalism and China's response to climate change. *Environmental Politics, 21,* 287–307. https://doi.org/10.1080/09644016.2012.651904

Granovetter, M. (1985). Economic action and social structure: The problem of embeddedness. *American Journal of Sociology, 91,* 481–510. https://doi.org/10.1086/228311

Grundmann, R., & Stehr, N. (2012). *The power of scientific knowledge*: *From research to public policy*. Cambridge, UK: Cambridge University Press. https://doi.org/10.1017/CBO9781139137003

Haldane, A. (2013). Rethinking the financial network. In S. A. Jansen, E. Schröter, & N. Stehr (Eds.), *Fragile Stabilität—stabile Fragilität* (pp. 243–278). Wiesbaden: Springer VS. https://doi.org/10.1007/978-3-658-02248-8

Hayek, F. A. (1944). *The road to serfdom*. London: Routledge. https://doi.org/10.4324/9780203991718

Hayek, F. A. (1960). *The constitution of liberty*. London: Routledge. https://doi.org/10.4324/9781315832081

Herzog, L. (2019). *Die Rettung der Arbeit* [Saving the work]. Munich: Hanser Berlin.

Iversen, T., & Soskice, D. (2019). *Democracy and prosperity: Reinventing capitalism through a turbulent century*. Princeton, NJ: Princeton University Press.

Jasanoff, S. (2012). *Science and public reason*. London: Routledge. https://doi.org/10.4324/9780203113820

Jamieson, D. (2014). *Reason in a dark time: Why the struggle against climate change failed—and what it means for our future*. New York, NY: Oxford University Press. https://doi.org/10.1093/acprof:oso/9780199337668.001.0001

Jonas, H. (1984). The imperative of responsibility: In search of an ethics for the technological age. *Human Studies, 11,* 419–429.

Keller, E. F. (2017). Climate science, truth and democracy. *Studies in History and Philosophy of Biological and Biomedical Sciences, 64,* 106–122. https://doi.org/10.1016/j.shpsc.2017.06.006

Kuklinski, J. H. (1990). Information and the study of politics. In J. A. Ferejohn & J. H. Kuklinski (Eds.), *Information and democratic processes* (pp. 391–395). Urbana, IL: University of Illinois Press.

Kuttner, R. (2018). The crash that failed. *The New York Review of Books, 65*(18). Retrieved from https://www.nybooks.com/articles/2018/11/22/financial-crash-that-failed/?

Levitsky, S., & Ziblatt, D. (2018). *How democracies die: What history tells us about our future.* New York: Crown.

Lipset, S. M. (1959). Some social requisites of democracy: Economic development and political legitimacy. *American Political Science Review, 53,* 69–105. Retrieved from http://www.jstor.org/stable/1951731

Lovelock, J. (2009). *The vanishing face of Gaia: A final warning.* New York: Basic.

Luce, E. (2017). *The retreat of western liberalism.* New York: Little, Brown and Company.

Luhmann, N. (1976). The future cannot begin: Temporal structure in modern society. *Social Research, 43,* 130–152. Retrieved from https://www.jstor.org/stable/40970217

Luhmann, N. (1998). *Observations on modernity*. Stanford, CA: Stanford University Press.

Luhmann, N. (2005). *Risk: A sociological theory* (with a new introduction by Nico Stehr and Gotthard Bechmann). London: Aldine Transaction.

Marquand, D. (2004). *Decline of the public: The hollowing out of citizenship.* Cambridge: Polity.

Marshall, G. (2014). *Don't even think about it: Why our brains are wired to ignore climate change.* New York: Bloomsbury.

Martin, S. L., Trow, M., & Coleman, J. S. (1956/1962). *Union democracy: The internal politics of the international typographical union.* New York: Doubleday & Company.

McDonald, S. (2009). Changing climate, changing minds: Applying the literature on media effects, public opinion, and the issue-attention cycle to increase public understanding of climate change. *International Journal of Sustainable Communities, 4,* 45–63.

McKibben, B. (2016, August 15). A world at war: We're under attack by a powerful enemy—and our only hope it to mobilize like we did in WWII. *The New Republic* (New York). Retrieved from https://newrepublic.com/article/135684/declare-war-climate-change-mobilize-wwii

McKibben, B. (2018, October 25). A very grim forecast. *New York Review.* Retrieved from https://www.nybooks.com/articles/2018/11/22/global-warming-very-grim-forecast/

Merton, R. K. (1966). Dilemmas of democracy in the voluntary associations. *The American Journal of Nursing, 66,* 1055–1061. https://doi.org/10.1097/00000446-196605000-00039

Michels, R. (1915). *Political parties: A sociological study of the oligarchical tendencies of modern democracy* (E. Paul & C. Paul, Trans.). New York: Hearst's International Library.

Mounk, Y. (2018). *The people vs. democracy: Why our freedom is in danger & how to save it.* Cambridge, MA: Harvard University Press.

Nordhaus, W. (2013). *The climate casino: Risk, uncertainty, and economics for a warming world.* New Haven, CT: Yale University Press.

Petersen, M. B., & Aarøe, L. (2013). Politics in the mind's eye: Imagination as a link between social and political cognition. *American Political Science Review, 107,* 275–293. https://doi.org/10.1017/S0003055413000026

Pielke, R. (2010). *The climate fix: What scientists and politicians won't tell you about global warming.* New York: Basic.

Prewitt, K. (2010). Introduction: Limits to knowledge? No easy answer. *Social Research, 77,* 901–904. Retrieved from https://www.jstor.org/stable/40972298

Prins, G., Galiana, I., Green, C., Grundmann, R., Hulme, M., Korhola, A., Laird, F., Nordhaus, T., Pielke, R., Rayner, S., Sarewitz, D., Shellenberger, M., Stehr, N., & Tezuka, H. (2010). *The Hartwell paper: A new direction for climate policy after the crash of 2009.* London, UK: London School of Economics and Political Science.

Przeworski, A., & Limongi, F. (1997). Modernization: Theories and facts. *World Politics, 49,* 155–183. https://doi.org/10.1353/wp.1997.0004

Revkin, A. C. (2014, April 16). A risk analyst explains why climate change risk misperception doesn't necessarily matter. *The New York Times*. Retrieved from https://dotearth.blogs.nytimes.com/2014/04/16/a-risk-analyst-explains-why-climate-change-risk-misperception-doesnt-necessarily-matter/?_php=true&_type=

Rittel, H. W. J., & Webber, M. M. (1973). Dilemmas in a general theory of planning. *Policy Sciences, 4,* 155–169. https://doi.org/10.1007/BF01405730

Rosanvallon, P. (2006). *Democracy: Past and future.* New York, NY: Columbia University Press.

Rosanvallon, P. (2013). *The society of equals*. Cambridge, MA: Harvard University Press.

Rorty, R. (2004). Post-democracy: Richard Rorty on anti-terrorism and the national security state. *London Review of Books, 26*(7), 10–11. Retrieved from https://www.lrb.co.uk/the-paper/v26/n07/richard-rorty/post-democracy

Runciman, D. (2013). *The confidence trap: A history of democracy in crisis from World War I to the present.* Princeton, NJ: Princeton University Press.

Runciman, D. (2018). *How democracy ends*. London: Profile.

Sarewitz, D. (2010). Normal science and the limits on knowledge: What we seek to know, what we choose not to know, what we don't bother knowing. *Social Research, 77,* 997–1010. Retrieved from https://www.jstor.org/stable/40972304

Schön, D. A. (1963). *Invention and the evolution of ideas*. London: Tavistock.

Schumpeter, J. A. (1942). *Capitalism, socialism and democracy*. New York: Harper & Brothers.

Scott, J. C. (1998). *Seeing like a state: How certain schemes to improve the human condition have failed.* New Haven, CT: Yale University Press.

Seefried, E. (2015). *Zukünfte: Aufstieg und Krise der Zukunftsforschung 1945–1980* [Futures: The Rise and Crisis of Futurology 1945–1980]. Berlin: De Gruyter Oldenbourg.

Shearman, D., & Wayne, J. (2007). *The climate challenge and the failure of democracy*. Westport: Praeger.

Simmel, G. (1890). *Über sociale Differenzierung: Sociologische und psychologische Untersuchungen* [On social differentiation: Sociological and psychological studies]. Leipzig: Duncker & Humblot.

Skidelsky, R. J. A., & Skidelsky, E. (2012). *How much is enough? Money and the good life.* New York: Other.

Soci, A., Maccagnan, A., & Mantovani, D. (2014). *Does inequality harm democracy? An empirical investigation on the UK.* Paper presented at the Sixth International Scientific Conference on Economic and Social Development and Third Eastern European ESD Conference: Business Continuity, Vienna. Paper retrieved from https://bib.irb.hr/datoteka/759841.Book_of_Proceedings_esd_Vienna_2014.pdf#page=54

Smith, C. W. (2015). *What the market teaches us: Limitations of knowing and tactics for doing.* Oxford, UK: Oxford University Press. https://doi.org/10.1093/acprof:oso/9780198745112.001.0001

Stehr, N. (1991). The power of scientific knowledge—and its limits. *Canadian Review of Sociology and Anthropology, 28,* 460–482. https://doi.org/10.1111/j.1755-618X.1991.tb00165.x

Stehr, N. (1994). *Knowledge societies*. London: SAGE.

Stehr, N. (1997). Trust and climate. *Climate Research, 8,* 163–169. https://doi.org/10.3354/cr008163

Stehr, N. (2001). *The fragility of modern societies: Knowledge & risk in the information age.* London: SAGE.

Stehr, N. (2015). Climate policy: Democracy is not an inconvenience. *Nature, 525,* 449–450. https://doi.org/10.1038/525449a

Stehr, N. (2016a). *Information, power, and democracy: Liberty is a daughter of knowledge.* Cambridge, UK: Cambridge University Press. https://doi.org/10.1017/CBO9781316343159

Stehr, N. (2016b). Exceptional circumstances: Does climate change trump democracy? *Issues in Science and Technology, 32,* 37–44.

Stehr, N., & Adolf, M. (2018). *Ist Wissen Macht? Wissen als gesellschaftliche Tatsache* [Is knowledge power? Knowledge as social fact]. Weilerswist-Metternich: Velbrück Wissenschaft. https://doi.org/10.5771/9783845298948

Stehr, N., & Machin, A. (2016a). Inequality in modern societies: Causes, consequences and challenges. In N. Stehr & A. Machin (Eds.), *Understanding inequality: Social costs and benefits* (pp. 3–36). Wiesbaden: Springer VS. https://doi.org/10.1007/978-3-658-11663-7

Stehr, N., & Machin, A. (2016b). Trusting the climate: Catastrophe vs. stability. *Society, 53,* 573–580. https://doi.org/10.1007/s12115-016-0069-6

Stehr, N., & Machin, A. (2019). *Society and climate: Transformations and challenges.* Singapore: World Scientific.

Stehr, N., & Voss, D. (2019). *Money: A theory of modern society.* New York: Routledge. https://doi.org/10.4324/9780429331565

Swyngedouw, E. (2010). Apocalypse forever? Post-political populism and the spectre of climate change. *Theory, Culture and Society, 27,* 213–232. https://doi.org/10.1177/0263276409358728

Swyngedouw, E. (2011). Interrogating post-democratization: Reclaiming egalitarian political spaces. *Political Geography, 30,* 370–380. https://doi.org/10.1016/j.polgeo.2011.08.001

Tenbruck, F. H. (1977). Grenzen der staatlichen Planung [Limits of national planning]. In W. Hennis, P. G. Kielmansegg, & U. Matz (Eds.), *Regierbarkeit: Studien zu ihrer Problematisierung* (Vol. 1, pp. 134–149). Stuttgart: Klett-Cotta.

van Dijk, T. A. (2014). *Discourse and knowledge: A sociocognitive approach.* Cambridge, UK: Cambridge University Press. https://doi.org/10.1017/CBO9781107775404

Williamson, O. E. (1975). *Markets and hierarchies: Analysis and antitrust implications.* New York: Free.

Williamson, O. E. (1985). *The economic institutions of capitalism: Firms, markets, relational contracting.* New York: Free.

Williamson, O. E. (1994). Transaction cost economics and organization theory. In N. J. Smelser & R. Swedberg (Eds.), *The handbook of economic sociology* (pp. 77–107). Princeton, NJ: Princeton University Press.

WBGU. (2012). Welt im Wandel—Gesellschaftsvertrag für eine Große Transformation [World in transition—Society contract for a Great Transformation]. Retrieved from https://www.wbgu.de/de/publikationen/publikation/welt-im-wandel-gesellschaftsvertrag-fuer-eine-grosse-transformation

Zakaria, F. (2003). *The future of freedom: Illiberal democracy at home and abroad.* New York: Norton.

Chapter 5
Risk Governance: From Knowledge to Regulatory Action

Ortwin Renn

Governance Requirements for Complex Risks

In today's world of globalized trade, travel, and communication, an ever-larger number of risks have a transboundary impact, crossing national and regional frontiers: Large-scale electricity blackouts, chemical accidents, and risks related to emerging technologies have all affected various parts of the world. Even these risks seem limited, however, when compared to those that affect our living conditions globally. A highly topical example is that of cyberattacks. Other examples include pandemics, global energy security, the financial collapse, and the impacts of climate change.

The International Risk Governance Council (IRGC) has developed a framework for risk governance designed to assist societies in generating the necessary inter- and transdisciplinary knowledge to address and respond to such global risks (IRGC, 2005). To this end, the IRGC's framework maps out a structured approach that guides its user through the process of investigating global risk issues and designing appropriate governance strategies. The designers of this approach combine scientific evidence with economic considerations as well as social concerns and societal values and, thus, ensure that any risk-related decision draws on the broadest possible view of risk. They also state the case for an effective engagement of all relevant stakeholders. The idea is that governance comprises more than government: It includes all the actors and institutions that play a role in assessing, managing, communicating, and regulating risks. The IRGC framework is inspired by the concept of adaptive institutional learning (Armitage, Marschke, & Plummer, 2008). Such a learning process is based on both solid knowledge about risk reduction measures as well as flexible responses with feedback incorporation in complex situations. The role of risk knowledge in such a process is to provide interdisciplinary, inclusive, and integrative

O. Renn (✉)
Institute for Advanced Sustainability Studies, Potsdam, Germany
e-mail: ortwin.renn@iass-potsdam.de

J. Glückler et al. (eds.), *Knowledge for Governance*, Knowledge and Space 15,
https://doi.org/10.1007/978-3-030-47150-7_5

expertise for the various actors involved (Rosa, Renn, & McCright, 2013, pp. 99, 196): "Our risk knowledge … must be traced to an amalgam of actors and institutions, as well as to the outcomes of exercising individual reflexivity in terms of making intuitive sense of conflicting knowledge claims and evaluation criteria" (p. 197).

The IRGC framework has been tested for efficacy and practicability—for example, can the framework help ensure that all relevant issues and questions are being addressed, and does it support the development of appropriate risk governance strategies (IRGC, 2005)? Researchers conduct tests in the form of short case studies, applying the framework to different risks, including those related to genetically modified organisms, stem cells, nature-based tourism, and the European gas infrastructure (all case studies are described in detail in Renn & Walker, 2008a). The results from these tests have given input to several revisions to the framework (Renn & Klinke, 2014).

The framework offers two major innovations to the risk field: the inclusion of the societal context and a new categorization of risk-related knowledge (Renn, 2008).

Inclusion of the Societal Context: In addition to the generic elements of risk assessment, risk management, and risk communication, the framework adds two other phases to the risk governance cycle: preassessment and evaluation. The preassessment phase highlights the risk topic's framing and boundaries, for example what kind of risks and consequences to include in an analysis of genetically modified organisms (GMOs). Should analysts focus only on health and environmental impacts, or should they also consider risks of economic concentration or the creation of dependencies on large suppliers of GMOs? Risk evaluation is a process by which to determine the acceptability of a given risk (or risk education strategy). This is the place where plural values, multiple evaluation criteria, and social preferences play a large role in defining what acceptability means to whom. Linking the social and cultural context with risk evaluation, the framework reflects the important role of stakeholder involvement and the need for resolving risk-risk trade-offs.

Categorization of Risk-Related Knowledge: The framework also proposes a categorization of risk that is based on the different states of knowledge about each particular risk, distinguishing between *simple*, *complex*, *uncertain*, and *ambiguous* risk problems. The characterization of a particular risk depends on the degree of difficulty of establishing the cause-effect relationship between a risk agent and its potential consequences, the reliability of this relationship, and the degree of controversy with regard to both what a risk actually means for those affected and the values to be applied when judging whether or not something needs to be done about it. Examples of each risk category include, respectively, known health risks such as those related to smoking, the failure risk of interconnected technical systems such as the electricity transmission grid, atrocities such as those resulting from the changed nature and scale of international terrorism, and the long-term effects and ethical acceptability of controversial technologies such as nanotechnologies. For each category, the researcher then derives a strategy for risk assessment, risk management, as well as the level and form of stakeholder participation, supported by proposals for appropriate methods and tools.

In the following sections, I will first introduce the characterization of risk knowledge according to the three components *complexity*, *uncertainty*, and *ambiguity*. This opens the path for explaining the various phases of the IRGC risk governance framework and its further development by Klinke and Renn (2012). I will then conclude by addressing the issue of stakeholder involvement and public participation, a major element of inclusive governance.

Three Characteristics of Risk Knowledge

Risk governance faces specific challenges raised by three risk characteristics that result from a lack of knowledge and/or competing knowledge claims about the risk problem: complexity, scientific uncertainty, and sociopolitical ambiguity (Klinke & Renn, 2010, 2018; Renn, Klinke, & van Asselt, 2011).

Complexity

Complexity refers to the difficulty of identifying and quantifying causal links between a multitude of potential candidates and specific adverse effects. A crucial aspect here concerns the applicability of probabilistic risk assessment techniques. If the chain of events between a cause and an effect follows a linear relationship (as, e.g., in car accidents, or when a building collapses due to a hurricane), simple statistical models are sufficient to calculate the probabilities of harm. Such simple relationships may still be associated with high uncertainty, for example, if only few data pieces are available or the effect is stochastic by its own nature (e.g., an earthquake). If the relationship between cause and effects becomes more complex, more sophisticated models of probabilistic inferences are required (Renn & Walker, 2008a). The nature of this difficulty may be traced back to interactive effects among these candidates (synergisms and antagonisms, positive and negative feedback loops), long delay periods between cause and effect, interindividual variation, intervening variables, and others. It is precisely these complexities that make sophisticated scientific investigations necessary, because the cause-effect relationship is neither obvious nor directly observable. Complexity requires sensitivity to both nonlinear transitions and scale (on different levels). Examples of highly complex risk include nested chemical facilities that may threaten nearby settlements, synergistic effects of potentially toxic substances in urban air, the failure risk of large interconnected infrastructures such as water and electricity grits, and the risks that critical loads pose to sensitive ecosystems within human settlements.

Scientific Uncertainty

Scientific uncertainty may result from unresolved complexity, in particular if the cause-effect models show large confidence intervals. It relates to the limitedness or even absence of scientific proof for a causal or functional relationship that makes it difficult to exactly assess the probability and possible outcomes of undesired effects (cf. Filar & Haurie, 2009). In the context of risk assessment, it is essential to acknowledge that human knowledge is always incomplete and selective, and, thus, contingent upon uncertain assumptions, assertions, and predictions (Funtowicz & Ravetz, 1992; Renn, 2008, pp. 75–77). It is obvious that the modeled probability distributions within a numerical relational system can only represent an approximation of the empirical relational system that helps elucidate and predict uncertain events. It therefore seems prudent to include additional aspects of uncertainty (van Asselt, 2000, pp. 93–138). Uncertainty may be linked to lack of reliable data, to imprecision in the analytical model, in the statistical treatment of the use of inductive statistical tools, or in the interpretation of ambiguous results (Funtowicz & Ravetz, 2008). Examples of high uncertainty include many natural disasters, such as earthquakes, possible health effects of air-borne pollutants below the threshold of statistical significance, acts of violence—such as terrorism and sabotage—and long-term effects of high social mobility on personal wellbeing and social cohesion.

Sociopolitical Ambiguity

While more and better data and information may reduce scientific uncertainty, more knowledge does not necessarily lessen ambiguity. Ambiguity thus indicates a situation of ambivalence in which different and sometimes divergent streams of thinking and interpretation about the same risk phenomena and their circumstances are apparent (cf. Zahariadis, 2003). Renn and Klinke (2015) distinguish between interpretative and normative ambiguity, which both relate to divergent or contested perspectives on the justification, severity, or wider "meanings" associated with a given threat.

Interpretative ambiguity denotes the variability of (legitimate) interpretations based on identical observations or data assessments results, for example an adverse or nonadverse effect. Variability of interpretation, however, is not restricted to expert dissent. Laypeople's perception of risk often differs from expert judgments because it is related to qualitative risk characteristics such as familiarity, personal or institutional control, or assignment of blame. Moreover, in contemporary pluralist societies diversity of risk perspectives within and between social groups is generally fostered by divergent value preferences, variations in interests, and very few, if any, universally applicable moral principles. This is all the more true if risk problems are complex and uncertain.

This leads to *normative ambiguity*, which alludes to different concepts of what can be regarded as tolerable, referring to aspects such as ethics, quality of life parameters, or distribution of risks and benefits. A condition of ambiguity emerges where the problem lies in agreeing on the appropriate values, priorities, assumptions, or boundaries to be applied to the definition of possible outcomes. Dealing with ambiguities requires governance approaches that emphasize mutual learning across different academic and practical communities as well as promote the cocreation of joint knowledge and practical applications. Examples for high interpretative ambiguity include exposure to low dose radiation (ionizing and non ionizing), low concentrations of genotoxic substances, food supplements, and—in the social domain—the gentrification of urban quarters or the loss of social cohesion in a disaster-prone community. Normative ambiguities can be associated, for example, with passive smoking, restricted mobility regimes in highly congested cities (such as congestion pricing), zoning laws for hazard-prone areas, or busing of schoolchildren from different social classes.

Most risks are characterized by a mixture of complexity, uncertainty, and ambiguity. Passive smoking may be a good example of low complexity and uncertainty, but high ambiguity. Nuclear energy may be a good candidate for high complexity and high ambiguity, but relatively little uncertainty. The use of IT in smart urban environments could be cited as an example for high complexity, uncertainty, and ambiguity.

Adaptive and Integrative Capacity of Risk Governance

The ability of risk governance institutions to cope with complex, uncertain, and ambiguous consequences and implications has become a central concern to scientists and practitioners alike. Adaptive and integrative governance on risk can be broadly understood as the ability of politicians and society to collectively design and implement a systematic approach to organizational and policy learning in institutional settings that are conducive to resolving complexity, uncertainty, and ambiguity in various risk arenas.[1]

This dynamic governance process is characterized by continuous and gradual learning and adjustment. Adaptive and integrative capacity in risk governance processes encompasses a broad array of structural and procedural mechanisms by which politics and society can handle collectively relevant risk problems. The main task is to collect robust knowledge about potential risk management measures by integrating systematic, experiential, and tacit knowledge (Renn, 2010) and by initiating a well designed but flexible learning process by which systematic collection of feedback and responses inform the adaptive processes of adjusting to new situations, surprises, or unforeseen events (Kerzner, 2017, pp. 613–620). In practical

[1] To the definition and understanding of adaptive capacity, see, for example, Webster, Gasser, Young, & Choucri (2008).

terms, adaptive and integrative capacity is the ability to design and incorporate the necessary steps in a risk governance process that allow risk managers to reduce, mitigate, or control the occurrence of harmful outcomes resulting from collectively relevant risk problems in an effective, efficient, and fair manner (cf. Brooks & Adger, 2004).

In 2005, the International Risk Governance Council proposed a process model of risk governance based the authors' work (IRGC, 2005; Renn, 2008; Renn & Walker, 2008a). With this framework, its designers structure the risk governance process in four phases: preassessment, appraisal, characterization/evaluation, and risk management. They conceptualized communication and stakeholder involvement as constant companions to all four phases of the risk governance cycle. Based on this framework and informed by many comments on the original framework (i.e., the edited volume by Renn & Walker, 2008b), Klinke and Renn (2012) modified the original IRGC proposal. The new framework the two authors suggested consists of the following steps: preestimation, interdisciplinary risk estimation, risk characterization, risk evaluation, and risk management, all related to risk governance institutions' abilities and capacities to use resources effectively (see Fig. 5.1).

Appropriate resources include institutional and financial means as well as social capital (e.g., strong institutional mechanisms and configurations, transparent decision making, allocation of decision making authority, formal and informal networks that promote collective risk handling, education), technical resources (e.g., databases, computer soft- and hardware, etc.), and human resources (e.g., skills, knowledge, expertise, epistemic communities, etc.). Therefore, the adequate involvement of experts, stakeholders, and the public in the risk governance process is a crucial dimension to produce and convey adaptive and integrative capacity in risk governance institutions.

Preestimation

A systematic reviewer of the preestimation stages would begin with *screening* as an exploration of a large array of actions and problems, searching for those with a specific risk-related feature. It is important to explore what major political and societal actors such as governments, companies, epistemic communities (e.g., the community of risk analysis specialists, associations for toxicology or epidemiology, or communities for disaster management), nongovernmental organizations, and the general public identify as risks and what types of problems they label as problems associated with risk and uncertainty. This is called *framing* and it specifies how society and politics rely on schemes of selection and interpretation to understand and respond to those phenomena that are socially constructed as relevant risk topics (Kahneman & Tversky, 2009; Reese, Gandy Jr., & Grant, 2001). Interpretations of risk experience depend on the frames of reference. The process of framing corresponds with a multiactor and multiobjective governance structure, since governmental authorities (national, supranational, and international agencies), risk and

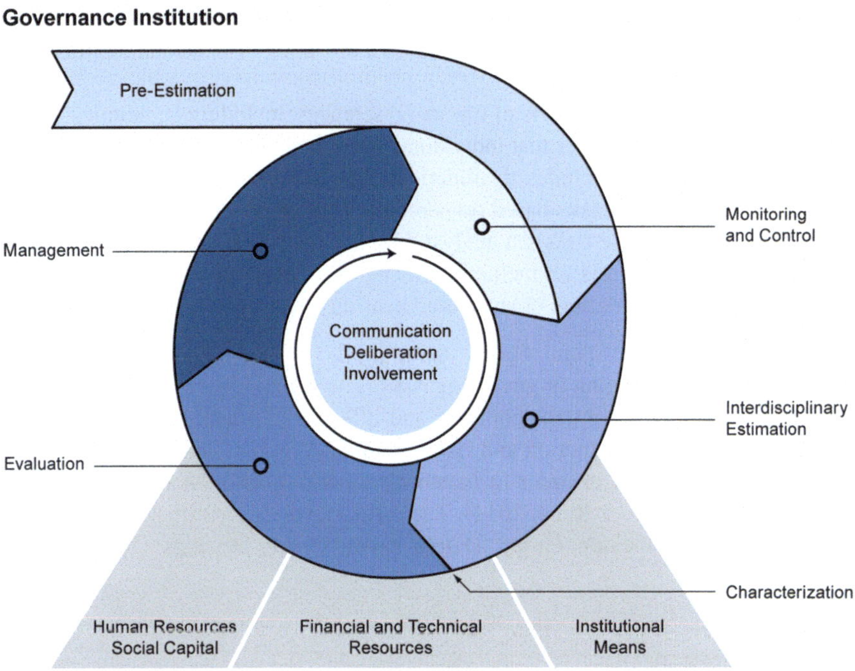

Fig. 5.1 Adaptive and integrative risk governance model. Adapted from Klinke and Renn (2012, p. 276). Copyright 2012 by the Journal of Risk Research. Adapted with permission of the Journal of Risk Research. (The adaptive and integrative risk governance model is based on a modification and refinement of the IRGC framework (IRGC, 2005, 2017))

opportunity producers (e.g., industry), those affected by risks and opportunities (e.g., consumer organizations, environmental groups), and interested bystanders (e.g., the media or an intellectual elite) are all involved and often in conflict about the appropriate frame to conceptualize the problem. What counts as risk may vary among these actor groups. Whether an overlapping consensus evolves about what requires consideration as a relevant risk depends on the legitimacy of the selection rule (Renn & Klinke, 2014).

Interdisciplinary Risk Estimation

Interdisciplinary risk estimation requires the cooperation of all disciplines that are necessary to generate a common understanding of all risk consequences (physical, monetary, social, cultural). The estimation process is comprised of two stages (cf. IRGC, 2005; Renn & Walker, 2008a):

1. *Risk assessment:* Experts of the natural and technical sciences produce the best estimate of the physical harm that a risk source may induce. Such harm could be

the collapse of buildings; discontinuation of central services to residents such as water, electricity, or information; breakdown of traffic; or inadequate infrastructural support.

2. *Concern assessment:* Experts of the social sciences, including economics, identify and analyze the issues that individuals or society as a whole link to a certain risk. This portfolio includes dysfunctional social services, risks of economic subsistence, but also risks based on perceptions of crime or insecurity. To identify and explore these risks, an analyst may use the repertoire of the social sciences such as survey methods, focus groups, econometric analysis, macroeconomic modeling, or structured hearings with stakeholders.

The second step in risk estimation is including the concerns and expectations of those involved in managing or governing risks. The main idea is here to collect the necessary knowledge from stakeholders and affected citizens about their preferences in terms of risk reduction and risk handling. Although analysts often forget this step, it is essential in order to match physical risk assessments with human perception (van Asselt & Renn, 2011). The instruments to perform such a concern assessment might include Group Delphi processes or hearings (Renn, 2008, pp. 336–337.).

Risk Evaluation

Actors in the risk governance process heavily dispute how best to classify a given risk and justify an evaluation about its societal acceptability or tolerability (see Fig. 5.2). In many approaches, they rank and prioritize risks based on a combination of probability (how likely is it that the risk will occur) and impact (the consequences should this take place). In the so-called traffic light model, analysts locate risks in the diagram of probability versus expected consequences and identify three areas: green, amber, and red (Renn, 2008, pp. 149–154.).

A risk falls into the green area if its occurrence is highly unlikely and its impact negligible. No further formal intervention is necessary in this case. Analysts view a risk is as tolerable when serious impacts might occur occasionally (amber area). The benefits are worth the risk, but risk reduction measures are necessary. Finally, they view a risk as intolerable when the occurrence of catastrophic impacts is most likely (red area). The risk's possible negative consequences are so catastrophic that they cannot be tolerated, despite the potential benefits.

Drawing the lines between *acceptable* (green area), *tolerable* (amber area), and *intolerable* (red area) is one of the most controversial tasks in the risk governance process. The UK Health and Safety Executive developed a procedure for chemical risks based on risk-risk comparisons (Löfstedt, 1997). Some Swiss cantons such as Basle County experimented with Round Tables as a means to reach consensus on drawing the two demarcation lines, whereby participants in the Round Table represented industry, administrators, county officials, environmentalists, and

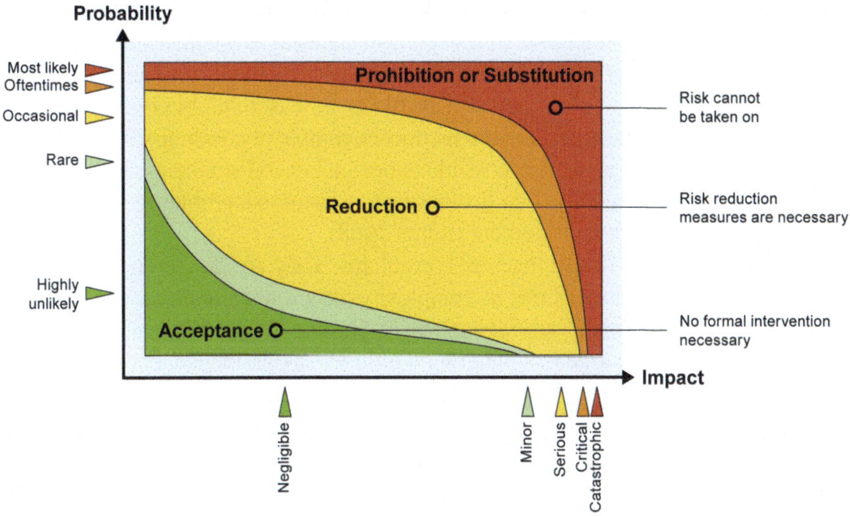

Fig. 5.2 Risk areas: intolerable (red), tolerable (amber), acceptable (green). Slightly modified version from the original illustration published in IRGC (2005, p. 37). Copyright 2005 by the International Risk Governance Council. Adapted with permission of the International Risk Governance Council

neighborhood groups. The Round Table was facilitated by a professional mediator charged with reaching a consensus between the various groups. Although such a consensus was difficult to achieve, all parties ultimately agreed to a solution by which the demarcation lines were only determined for a limited time, with the clear understanding that standards would be tightened if more risk reduction measures were to become available and further developed (RISKO, 2000).

Irrespective of the means selected to support this task, the judgment on acceptability or tolerability is contingent on making use of a variety of different knowledge sources. One needs to include the data and insights resulting from the risk assessment activity and additional data from the concern assessment.

Risk Management

Risk management analysts begin by reviewing all relevant data and information generated in the previous steps of interdisciplinary risk estimation, characterization, and risk evaluation. The systematic analysis of risk management options focuses on still tolerable risks (amber area) and those where tolerability is disputed (light green and orange transition zones). The other cases (green and red area) are fairly easy to deal with. Intolerable risks demand prevention and prohibition strategies as a means of replacing the hazardous activity with another activity leading to identical or similar benefits. The management of acceptable risks is left to private actors (civil society and economy). They may initiate additional and voluntary risk reduction measures or seek insurance to cover possible but rather minor or negligible losses. If risks are classified as tolerable, or if experts disagree as to whether they are in the transition zones of tolerability, public risk managers must design and implement actions that make these risks either acceptable or at least tolerable by introducing reduction strategies. Based on the distinction in complexity, scientific uncertainty, and sociopolitical ambiguity, it is possible to design general strategies for risk management that can be applied to four distinct categories of risk problems, thus simplifying the process of risk management (Renn, 2008).

The first category refers to linear risk problems: They are characterized by their low scores across the dimensions of complexity, uncertainty, and ambiguity. They can be addressed by *linear risk management* because they are normally easy to assess and quantify. Routine risk handling within risk assessment agencies and regulatory institutions is appropriate for this category, since the risk problems are well known, sufficient knowledge of key parameters is available, and there are no major controversies about causes and effects or conflicting values. The management includes risk-benefit analysis, risk-risk comparisons, and other instruments of balancing pros and cons.

If risks are ranked high on complexity but rather low on both uncertainty (i.e., the complexity can be widely resolved by adequate scientific models) and ambiguity, they require the systematic involvement and deliberation of experts who represent the relevant epistemic communities that produce the most accurate estimate of these complex relationships. It does not make much sense to integrate public concerns, perceptions, or any other social aspects for resolving complexity unless specific knowledge from the concern assessment helps to untangle complexity. Complex risk problems therefore demand *risk-informed management,* which scientists and experts can offer by applying methods of expanded risk assessment, determining quantitative safety goals, consistently using cost-effectiveness methods, and monitoring and evaluating outcomes.

Risk problems that are characterized by high uncertainty but low ambiguity require *precaution-based management.* Because sufficient scientific certainty is currently either unavailable or unattainable, expanded knowledge acquisition may help to reduce uncertainty and, thus, to revert the risk problem back to first stage of handling complexity. If, however, uncertainty cannot be reduced by additional knowledge, risk management should foster and enhance precautionary and resilience-building strategies and decrease vulnerabilities in order to avoid irreversible effects. Appropriate instruments include containment, diversification, monitoring, and substitution. Because the focal point here is to find the adequate and fair balance between being overly cautious and overly reckless, a reflective processing involving stakeholders is necessary to ponder concerns, economic budgeting, and social evaluations.

Finally, if risk problems are ranked high on ambiguity (regardless of whether they are low or high on uncertainty), *discourse-based management* is required, a process that demands participative processing. This includes the need to involve major stakeholders as well as the affected public. The goals of risk management are to produce a collective understanding among all stakeholders and concerned members of the public on interpretative ambiguity or to find legitimate procedures of justifying collectively binding decisions on acceptability and tolerability. It is important to achieve a consensus or compromise between those who believe that the risk is worth taking (perhaps because of self-interest) and those who believe that the pending consequences do not justify the potential benefits of the risky activity or technology.

Risk Communication

All four phases must be accompanied by intensive risk communication efforts. Communication should not be limited to sharing information but must include an effort to create both a common understanding of the problems and challenges as well as a joint agreement on the most acceptable risk reduction solutions. Such a concept of communication requires a transdisciplinary approach to problem solving that involves the strong participation of all relevant stakeholders in the creation of knowledge and risk reduction options and a mutual learning process in which all actors share their knowledge and insights (Hirsch-Hadorn et al., 2008; Newig, Kochskämper, Challies, & Jager, 2016). In this understanding, communication should already have begun during the preestimation phase. It should convey the basic concepts and what these concepts entail in terms of opportunities and risks. Analysts can arrange feedback channels on the internet as to evaluate the responses of stakeholders and affected citizens. During the risk estimation phase, the communication process should emphasize the process by which the research and planning team conducts the risk assessments. The main goal here is to promote trust in the risk-handling authorities (Löfstedt, 2005).

It might be helpful to ask stakeholders and citizens for additional knowledge that public officials may not possess. More input from the public is to be encouraged during the evaluation phase. First of all, the process of how tradeoffs are assigned and justified must be made transparent to all stakeholders as well as the general public. Furthermore, depending on the degree of ambiguity, it might be useful to have procedures in place that systematically collect feedback and concerns with respect to the planned urban risk management measures. During the management phase, it is essential to familiarize all affected persons with the chosen or deliberated risk reduction measures, in particular those that rely on the cooperation of the affected public (such as evacuation or sheltering plans). Instruments for making risk reduction plans known to the public are open meetings, brochures, websites, TV shows, and other popular forms of information transfer (Earle & Cvetkovich, 1994).

Although risk communication implies a stronger role for risk professionals to provide information to the public rather than vice versa, it should be regarded as a mutual learning process. Concerns, perceptions, and experiential knowledge of the targeted audience(s) should thus guide risk professionals in their selection of topics and subjects: It is not the task of the communicators to decide what people need to know, but to respond to the questions of what people want to know (this is normally referred to as the "right to know" concept). The step from risk communication to stakeholder and public involvement is only gradual and should be seriously considered any time that risk communication addresses issues of major concerns and contesting claims.

Inclusive Governance: The Need for an Effective Involvement of Experts, Stakeholders, and Civil Society

The effectiveness and legitimacy of the risk governance process depends on the management agencies' capacity to resolve complexity, characterize uncertainty, and handle ambiguity by means of communication and deliberation. In the following, I introduce a particular procedural mechanism of communication and deliberation to address each of the specific challenges raised by complexity, scientific uncertainty, and sociopolitical ambiguity. I illustrate the various steps of involvement of larger stakeholder groups in Fig. 5.3.

Instrumental Processing Involving Governmental Actors (Linear Mode)

Dealing with linear risk issues, which are associated with low scores of complexity, scientific uncertainty, and sociopolitical ambiguity, requires hardly any changes to conventional public policymaking. The data and information of such linear (routine) risk problems are provided by statistical analysis, law or statutory requirements determine the general and specific objectives, and the role of public policy is to ensure that all necessary measures of safety and control are implemented and enforced. The aim is to find the most cost-effective method for a desired regulation level. If necessary, deliberators may include stakeholders, as they have information and knowhow that may provide useful hints for increased efficiency.

Epistemic Processing Involving Experts and Stakeholders (Complex Mode)

Resolving complex risk problems requires dialogue and deliberation among experts and representatives of stakeholder groups with special knowledge and experience. Involving members of various epistemic communities who demonstrate expertise and competence is the most promising step for producing more reliable and valid judgements about the complex nature of a given risk. Epistemic discourse is the instrument for discussing the conclusiveness and validity of cause-effect chains relying on available probative facts, uncertain knowledge, and experience that can

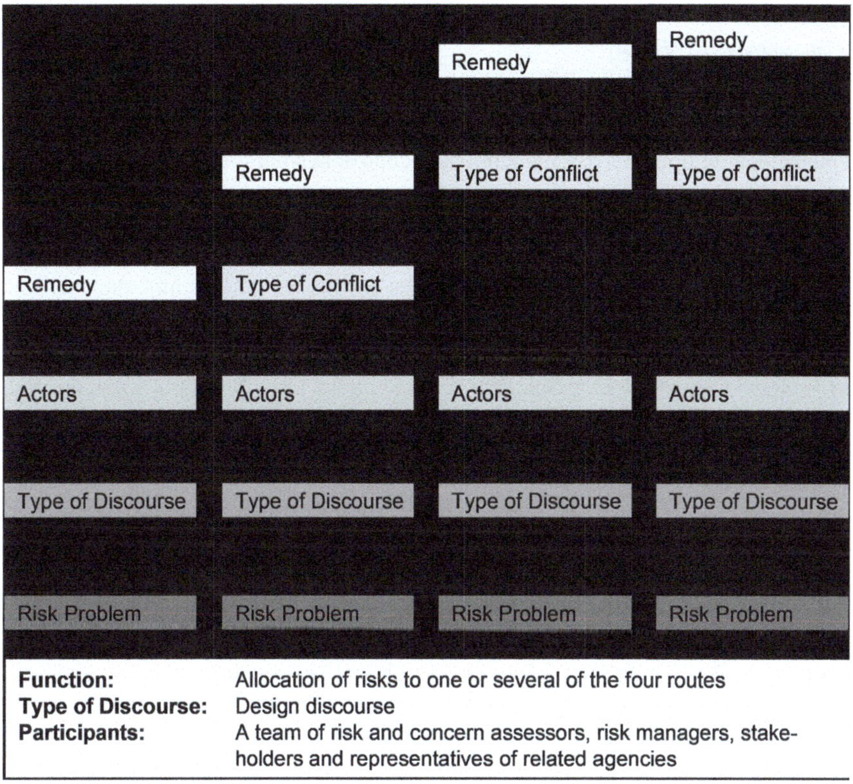

Fig. 5.3 The risk management escalator. Modified version from the original illustration published in IRGC (2005, p. 53). Copyright 2005 by the International Risk Governance Council. Adapted with permission of the International Risk Governance Council

be tested for empirical traceability and consistency. The objective of such a deliberation is to find the most cogent description and explanation of the phenomenological complexity in question as well as a clarification of dissenting views (i.e., by addressing the question of which environmental and socioeconomic impacts specific community action plans are expected to produce). The deliberation among experts might generate a profile of the complexity of the given risk issue on selected intersubjectively chosen criteria. It may also reveal more uncertainty and ambiguity hidden in the case than the initial appraisers had anticipated. It is advisable to include natural as well as social scientists in the epistemic discourse, in order to anticipate potential problems with risk perception. If this were done, fewer unsuspected controversies would occur.

Reflective Processing Involving Stakeholders (Uncertainty Mode)

Characterizing and evaluating risks as well as developing and selecting appropriate management options for risk reduction and control in situations of high uncertainty pose particular challenges. How can risk managers characterize and evaluate the severity of a risk problem when the potential damage and its probability are unknown or highly uncertain? Scientific input is therefore only the first in a series of steps during a more sophisticated evaluation process. It is crucial to compile the data and information relevant to the different types of uncertainties to inform the process of risk characterization. The risk characterization's outcome provides the foundation for a broader deliberative arena, which out to include not only policy makers and scientists, but also directly affected stakeholders and public interest groups including representatives of the affected public, in order to discuss and ponder the "right" balances and trade-offs between potential over- and under-protection. This reflective involvement of stakeholders and interest groups is aimed at finding a consensus on the extra margin of safety that potential victims would be willing to tolerate and potential beneficiaries of the risk would be willing to invest in order to avoid potentially critical and catastrophic consequences. The reflective involvement of policy makers, scientists, stakeholders, and public interest groups can be accomplished by a spectrum of different forms, such as negotiated rule making, mediation, round table or open forum, or advisory committee (cf. Beierle & Cayford, 2002; Rowe & Frewer, 2000; Stoll-Kleemann & Welp, 2006).

Participative Processing Involving the Public (Ambiguity Mode)

If risk problems are associated with high ambiguity, it is not enough to demonstrate that risk regulation addresses the issues of public concerns. In these cases, the evaluation process must also be open to public input and new forms of deliberation. This begins with revisiting the question of proper framing. Is the issue really one of risk, or of lifestyle or future vision? Often, both the benefits and risks are contested. The debate about smart cities may illustrate the point that observers may be concerned not only about technical risks of network failures or privacy issues being violated by information transfer, but also about the acceptability of the desired goal to reduce choices for individuals by means of paternalistic design of choice situations (Kahneman, 2012; Thaler & Sunstein, 2008). Thus, the controversy is often much broader than simple risk management. The aim here is to find an overlapping consensus on the dimensions of ambiguity that must be addressed in comparing risks and benefits and balancing pros and cons. High ambiguity would require the most inclusive strategy for involvement, because both directly affected and indirectly affected groups should have an opportunity to contribute to this debate. Resolving ambiguities in risk debates necessitates the public's participatory involvement to openly discuss competing arguments, beliefs, and values. The set of possible forms to involve the public includes citizen panels or juries, citizen forums, consensus conferences, public advisory committees, and similar approaches (cf. Abels, 2007; Beierle & Cayford, 2002; Hagendijk & Irwin, 2006).

Wider Governance Issues

When considering the wider environment of risk handling in modern societies, many classes of influential factors come into play. I can mention only a few here. For example, the distinction between government and governance I introduced in the first section of this document can be helpful in describing and analyzing cases of risk handling in different countries and contexts (Zürn, 2000). In addition, analysts must address the interplay between economic, political, scientific, and civil society actors when looking beyond just governmental or corporate actions. Furthermore, looking at organizational capacity opens a new set of wider risk governance issues that relate to the interplay between the governing actors and their capability to fulfill their role in the risk governance process.

In Fig. 5.4, I present external influencing factors that I cannot place within the risk framework itself, and have additionally placed illustrations for each level within this figure. I have selected four cases: listeria, gas transportation, acrylamide, and genetically modified organisms (for each case, see Renn & Walker, 2008b). The listeria case concerns the risk of nonpasteurized milk, which is traditionally used by local cheese manufacturers, specifically in France and Mexico. This risk can be resolved completely within the core risk governance framework, as it is well manageable within the four phases outlined above. The case of gas transportation from Russia to Western countries involves additional aspects such as the risk of political dependence on Russia or the possibility of terrorist attacks on the pipelines (Vatansever, 2017). Managing the risk of gas pipelines requires governing institutions to wield specific skills, assets, and strategies that go beyond risk assessment and management. The case of acrylamide is an example of how strongly an issue can depend on the cooperation of different societal actors. Acrylamide is a natural carcinogen that has been found in baked food items such as French-fried potatoes and crispbread. Food regulators, producers, NGOs, and various science communities have extensively negotiated to define the risks and adopt appropriate risk reduction measures (Bonneck, 2017). Finally, the case of genetically modified organisms (GMOs) underlines how the social and political culture influence the debate about the role of GMOs have to play in the future.

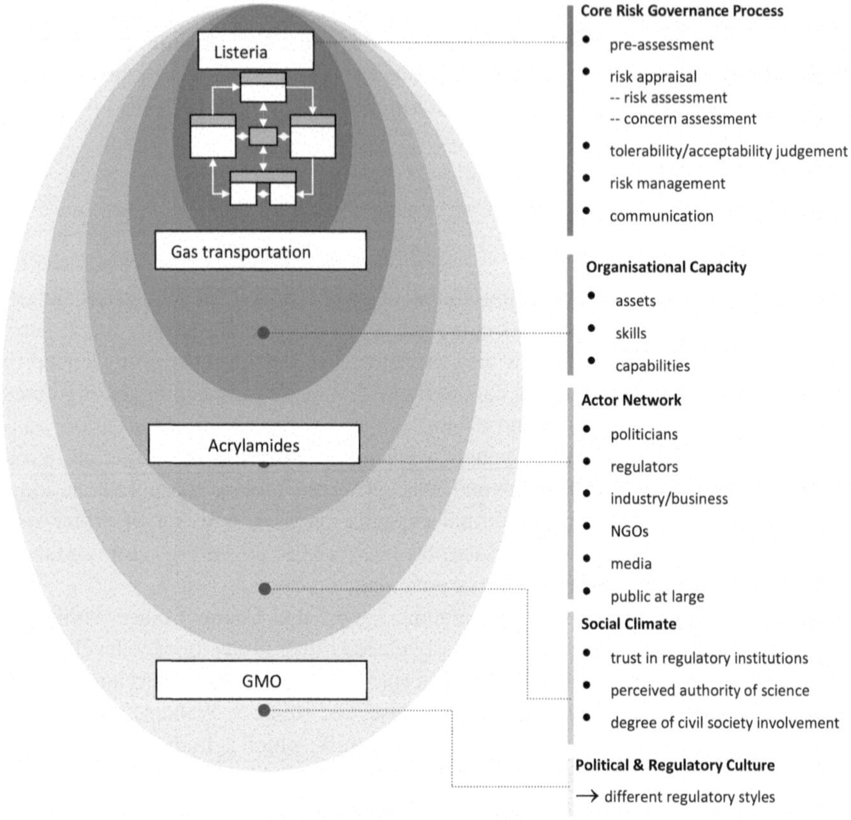

Fig. 5.4 Wider governance issues. Source: Design by author

Conclusion

One of the main functions of a comprehensive risk governance framework is to assist risk or concern assessors and managers in exploring and handling risks and to promote effective and fair approaches for improving, and enhancing the visibility of, the present risk governance processes. With the framework I present here, I aim to offer guidance and advice on how to approach the complexities, uncertainties, and ambiguities of risk issues and to promote a wider understanding of their interconnectedness and transgressional nature, particularly in relation to newly emerging systemic risks. To this end, the framework integrates different sources of knowledge that include scientific, experiential, anecdotal, and tacit aspects and includes effective and appropriate engagement of stakeholders—not least to ensure that both risk appraisal and risk management strategies command the widest possible acceptance and support.

I have designed the framework, on the one hand, to include enough flexibility to allow its users to do justice to the wide diversity of risk governance structures and, on the other hand, to provide sufficient clarity, consistency, and unambiguous orientation across a range of different risk issues and countries. I do not intend the framework to serve as a recipe or a checklist that can guarantee that analysts have considered all relevant aspects when analyzing a risk and its governance process and structures. However, by building into conventional risk analysis and management such "soft" issues as societal values, concerns, and perceptions of risk, and by taking into account the interactions between the various actors involved in the process, the risk governance framework can contribute to the development of more inclusive and effective risk governance strategies and the enhancement of decision making under uncertainty.

References

Abels, G. (2007). Citizen involvement in public policy-making: Does it improve democratic legitimacy and accountability? The case of pTA. *Interdisciplinary Information Sciences, 13*(1), 103–116. https://doi.org/10.4036/iis.2007.103

Armitage, D., Marschke, M., & Plummer, R. (2008). Adaptive co-management and the paradox of learning. *Global Environmental Change, 18*(1), 86–98. https://doi.org/10.1016/j.gloenvcha.2007.07.002

Beierle, T. C., & Cayford, J. (2002). Democracy in practice: Public participation in environmental decisions. *Perspectives on Politics, 1,* 181–182. https://doi.org/10.1017/S1537592703450159

Bonneck, S. (2017). *Gesellschaftlicher Umgang mit Risiken: Fallstudie zur Steuerungsleistung gesellschaftlicher Institutionen in Bezug auf Bewältigung von Unsicherheit (Risk Governance)* [Societal handling of risk: Case study on the governance of societal institutions for managing uncertainty] (Doctoral dissertation). University of Stuttgart, Germany. https://doi.org/10.18419/opus-9139

Brooks, N., & Adger, W. N. (2004). Assessing and enhancing adaptive capacity. In B. Lim & E. Spanger-Siegfried (Eds.), *Adaptation policy frameworks for climate change: Developing strategies, policies and measures* (pp. 165–181). Cambridge, UK: Cambridge University Press.

Earle, T. C., & Cvetkovich, G. (1994). Risk communication: The social construction of meaning and trust. In B. Brehmer & N. E. Sahlin (Eds.), *Future risks and risk management* (pp. 141–181). Amsterdam: Kluwer.

Filar, J. A., & Haurie, A. (2009). *Uncertainty and environmental decision making: A handbook of research and best practice*. New York: Springer US.

Funtowicz, S. O., & Ravetz, J. R. (1992). Three types of risk assessment and the emergence of post-normal science. In S. Krimsky & D. Golding (Eds.), *Social theories of risk* (pp. 251–273). Westport: Praeger.

Funtowicz, S. O., & Ravetz, J. R. (2008). Values and uncertainties. In G. Hirsch-Hadorn, H. Hoffmann-Riem, S. Biber-Klemm, W. Grossenbacher-Mansuy, D. Joye, C. Pohl, U. Wiesmann, & E. Zemp (Eds.), *Handbook of transdisciplinary research* (pp. 361–368). Dordrecht: Springer.

Hagendijk, R., & Irwin, A. (2006). Public deliberation and governance: Engaging with science and technology in contemporary Europe. *Minerva, 44,* 167–184. https://doi.org/10.1007/s11024-006-0012-x

Hirsch-Hadorn, G., Biber-Klemm, S., Grossenbacher-Mansuy, W., Hoffmann-Riem, H., Joye, D., Pohl, C., Wiesmann, U., & Zemp, E. (2008). The emergence of transdisciplinarity as a form of research. In G. Hirsch-Hadorn, H. Hoffmann-Riem, S. Biber-Klemm, W. Grossenbacher-Mansuy, D. Joye, C. Pohl, U. Wiesmann, & E. Zemp (Eds.), *Handbook of transdisciplinary research* (pp. 19–42). Dordrecht: Springer.

IRGC (International Risk Governance Council). (2005). *Risk Governance—Towards an Integrative Approach* (White Paper No. 1, with an Annex by P. Graham). Geneva: IRGC. Retrieved from https://irgc.org/wp-content/uploads/2018/09/IRGC_WP_No_1_Risk_Governance__reprinted_version_3.pdf

IRGC (International Risk Governance Council). (2017). *Introduction to the IRGC risk governance framework*. Revised version. Lausanne: EPFL International Risk Governance Center. https://doi.org/10.5075/epfl-irgc-233739

Kahneman, D. (2012). *Thinking, fast and slow.* New York: Penguin.

Kahneman, D., & Tversky, A. (2009). *Choices, values, and frames*. New York, NY: Cambridge University Press.

Kerzner, H. (2017). Project management: A systems approach to planning, scheduling, and controlling. Hoboken: John Wiley & Sons.

Klinke, A., & Renn, O. (2010). Risk governance: Contemporary and future challenges. In J. Eriksson, M. Gilek, & C. Rudén (Eds.), *Regulating chemical risks: European and global challenges* (pp. 9–28). Dordrecht: Springer.

Klinke, A., & Renn, O. (2012). Adaptive and integrative governance on risk and uncertainty. *Journal of Risk Research, 15*(3), 273–292. https://doi.org/10.1080/13669877.2011.636838

Klinke, A., & Renn, O. (2018). Distributed responsibility in risk governance. In P. A. Wilderer, O. Renn, M. Grambow, M. Molls, & K. Mainzer (Eds.), *Sustainable risk management* (pp. 19–31). Cham: Springer.

Löfstedt, R. E. (1997). *Risk evaluation in the United Kingdom: Legal requirements, conceptual foundations, and practical experiences with special emphasis on energy systems* (Discussion paper No. 92). Stuttgart: Center of Technology Assessment in Baden-Württemberg.

Löfstedt, R. E. (2005). *Risk management in post-trust societies.* London: Palgrave Macmillan. https://doi.org/10.1057/9780230503946

Newig, J., Kochskämper, E., Challies, E., & Jager, N. W. (2016). Exploring governance learning: How policymakers draw on evidence, experience and intuition in designing participatory flood risk planning. *Environmental Science & Policy, 55,* 353–360. https://doi.org/10.1016/j.envsci.2015.07.020

Reese, S. D., Gandy Jr., O. H., & Grant, A. E. (2001). *Framing public life: Perspectives on media and our understanding of the social world.* Mahwah: Lawrence Erlbaum Associates. https://doi.org/10.4324/9781410605689

Renn, O. (2008). *Risk governance: Coping with uncertainty in a complex world.* London: Earthscan. https://doi.org/10.4324/9781849772440

Renn, O. (2010). The contribution of different types of knowledge towards understanding, sharing and communicating risk concepts. *Catalan Journal of Communication & Cultural Studies, 2*(2), 177–195. https://doi.org/10.1386/cjcs.2.2.177_1

Renn, O., & Klinke, A. (2014). Risk governance: Application to urban planning. *A|Z ITU Journal of the Faculty of Architecture, 11*(1), 5–19. Retrieved from http://www.azitujournal.com/jvi.aspx?pdir=itujfa&plng=eng&un=ITUJFA-66588

Renn, O., & Klinke, A. (2015). Complexity, uncertainty and ambiguity in inclusive risk governance. In T. J. Andersen (Ed.), *The Routledge companion to strategic risk management* (pp. 13–30). London: Routledge.

Renn, O., Klinke, A., & van Asselt, M (2011). Coping with complexity, uncertainty and ambiguity in risk governance: A synthesis. *AMBIO, 40,* 231–246. https://doi.org/10.1007/s13280-010-0134-0

Renn, O., & Walker, K. (2008a). Lessons learned: A re-assessment of the IRGC framework on risk governance. In O. Renn & K. Walker (Eds.), *The IRGC risk governance framework: Concepts and practice* (pp. 331–367). Dordrecht: Springer.

Renn, O., & Walker, K. (2008b). *The IRGC risk governance framework: Concepts and practice*. Dordrecht: Springer.

RISKO. (2000). Mitteilungen für Kommission für Risikobewertung des Kantons Basel-Stadt: Seit 10 Jahren beurteilt die RISKO die Tragbarkeit von Risiken [Notes of the Commission for Risk Assessment of the Cantone of Basle City: For 10 years the Commission RISKO assesses the tolerability of risk]. *Bulletin of the City of Basle, 3*, 2–3.

Rosa, E. A., Renn, O., & McCright, A. M. (2013). *The risk society revisited: Social theory and governance*. Philadelphia, PA: Temple University Press.

Rowe, G., & Frewer, L. J. (2000). Public participation methods: A framework for evaluation. *Science, Technology, & Human Values, 25*, 3–29. https://doi.org/10.1177/016224390002500101

Stoll-Kleemann, S., & Welp, M. (2006). *Stakeholder dialogues in natural resources management: Theory and practice*. Heidelberg: Springer. https://doi.org/10.1007/978-3-540-36917-2

Thaler, R. H., & Sunstein, C. R. (2008). *Nudge: Improving decisions about health, wealth and happiness*. New Haven: Penguin.

van Asselt, M. (2000). *Perspectives on uncertainty and risk: The PRIMA approach to decision support*. Boston: Kluwer.

van Asselt, M., & Renn, O. (2011). Risk governance. *Journal of Risk Research, 14*, 431–449. https://doi.org/10.1080/13669877.2011.553730

Vatansever, A. (2017). Is Russia building too many pipelines? Explaining Russia's oil and gas export strategy. *Energy Policy, 108*, 1–11. https://doi.org/10.1016/j.enpol.2017.05.038

Webster, D. G., Gasser, L., Young, O. R., & Choucri, N. (2008). *Adaptive governance: The dynamics of Atlantic fisheries management*. Cambridge, MA: MIT Press.

Zahariadis, N. (2003). *Ambiguity and choice in public policy: Political decision making in modern democracies*. Washington, D.C.: Georgetown University Press.

Zürn, M. (2000). Democratic governance beyond the nation-state: The EU and other international institutions. *European Journal of International Relations, 6*, 183–221. https://doi.org/10.1177/1354066100006002002

Chapter 6
Knowledge and Governance: Can Systemic Risk in Financial Markets Be Managed? The Case of the Euro Crisis

Stephen Bell and Andrew Hindmoor

In *Masters of the Universe, Slaves of the Market*, Bell and Hindmoor (2015) portray a highly "structuralist" account of the 2008 Global Financial Crisis (GFC) that originated in the Wall Street and London markets. Such an account implies that structural influences, here defined as material forces in financial markets, are important not only as a result of action by relevant agents, but also as subsequent influences of agents' options and behavior, especially in driving forced asset fire-sales and credit freezes. A structural account of this kind can be contrasted with an institutional account, in which an agent's behavior is shaped by institutional factors—primarily rules, norms, or operating procedures in formal or informal organizational or institutional settings. Hence, structures and institutions are somewhat similar in that they both shape behavior, but the sources of incentives or constraints (emanating from either structures or institutions) are different in each case.

Prior to and during the 2008 crisis, bankers in the key New York and London markets faced institutional pressures from within their own organizations for high returns, linked also to remuneration packages and professional prestige. Bankers were also influenced by wider structural forces, such as growing competitive pressures for high short-term returns in financial markets. The favored strategy to achieve such returns was highly leveraged trading in mortgage-backed financial assets, all of which finally ended in calamity. Other structural dynamics were also at work. One was the growth of "systemic risk." Here, agents (largely unknowingly)

S. Bell (✉)
School of Political Science and International Studies, University of Queensland, Brisbane, QLD, Australia
e-mail: stephen.bell@uq.edu.au

A. Hindmoor
Department of Politics and International Relations, University of Sheffield, Sheffield, UK
e-mail: a.hindmoor@sheffield.ac.uk

J. Glückler et al. (eds.), *Knowledge for Governance*, Knowledge and Space 15,
https://doi.org/10.1007/978-3-030-47150-7_6

had constructed a fragile system of highly complex independencies between financial institutions that enhanced the prospect of contagion effects in markets that would rapidly escalate into panic and the domino-like collapse of global credit markets and many major banks and financial institutions. This was essentially a structural characteristic of the system. Systemic risk of this kind gradually built up during the 2000s, eventually "crystallizing" into a full-blown crisis of panic-selling and a global credit crash in the wake of the Lehman Brothers' collapse in late 2008. All this was driven, most proximately, by the behavior of bankers who were structurally ensnared in a context of their own making that left them little choice but to ultimately pursue collectively destructive behaviors: a malign form of structural power, mutually exerted.

In recent decades, social scientists participating in the long-running debates on agency and structure have tended to give a greater focus to agency. Yet the above account of the GFC is grimly structuralist, suggesting little room for agency or at least for agent choice (see also Bell & Hindmoor, 2018a, b). This raises the question of whether such situations of structural exigency can be prevented or managed. The question is important given the prevalence and costs of financial crises in recent decades.

A key question is then whether it might be possible to shape or manage banker behavior in order to try and avoid the panicked herding and fleeing behavior, asset fire sales, and withholding of credit that are typical during the crystallization of systemic risk amidst a full-blown financial crisis. Our response is tentatively affirmative, based on the experience of the Euro crisis, whose actors have thus far avoided both the crystallization of systemic risk and a banking crisis, despite the ongoing fragility of the large European banks (Bell & Hindmoor, 2018a, b). We argue that the key to such an outcome hinges on the relationship between agency, knowledge, and governance arrangements. In particular, we argue from this case that if agents are knowledgeable and aware of the potential for catastrophe, they may illicit institutional responses and governance arrangements that may serve to build resolution strategies. The aim here is to stabilize market sentiment, thus helping to avoid the (unwanted) crystallization of systemic risk via the mutual exertion of structural power by key agents in financial markets that proved so destructive during the GFC.

In this chapter, we therefore focus on the Euro crisis, which is, thus far, quite unlike the GFC. The latter saw bankers and key agents stumble blindly into disaster, with little knowledge or forewarning of the complex chain of events that awaited them (Bell, 2017). By contrast, key agents of the Euro crisis (bankers, regulators, and policy makers) have proven far more aware of the potential for the crystallization of systemic risk and a banking crisis. As we argue more fully below, this kind of knowledge has been central, and cognizant agents have proved capable of using institutions and governance strategies to (thus far) forestall a potential debt and banking collapse in Europe.

We use the resources of political science to explore these issues, unpacking the concepts of systemic risk and structural power in more detail. We then examine the relevant governance and historical institutionalist literature in political science to find useful tools and concepts to help us probe the knowledge, relational, and contextual issues in question in order to better understand if or how systemic risk in financial markets might be managed.

Authors of governance literature have focused on the role of both the state and non-state actors in forging governance partnerships to develop solutions to complex collective action problems. As we shall lay out below, private actors and state-supported networks prevented the crystallization of systemic risk amidst the Euro crisis and thereby played an important role in managing systemic risk, essentially by arranging a form of collective action amongst financial interests, supported by states, that helped contain the crisis. As we shall argue, however, the governance literature contains only a limited account of agency and agential knowledge. It is here that the historical institutionalist literature is better able to analyze how agents use ideas in institutional settings to forge change or reform strategies; in this case with a focus on how relevant agents dealt with the unfolding Euro crisis. We will begin, however, by unpacking concepts such as systemic risk and structural power in financial markets.

Systemic Risk and Structural Power in the GFC

In the run-up to the GFC, bankers and financiers produced a very complex asset-and-debt structure that was fragile in the extreme and difficult to understand and ultimately control. Market actors thus produced systemic risk, leading to a structural context of fragility that was marked by complex and highly leveraged securities trading and myriad other intricate and often opaque interdependencies in the financial system (Bell & Hindmoor, 2015). How should scholars theorize this interaction between agents and structure? Archer (2000, p. 465) argues that "structures only exert an effect when mediated through the activities of people. Structures are only ever relational emergents and never reified entities existing without social interaction." How, then, did agents actualize and mediate such structural effects?

In the context of the GFC, agents were unaware of the full complexity and fragility of the system they had created (Bell, 2017). Yet they became increasingly aware of at least some of the relevant dangers when the US mortgage market saw a downturn from mid-2007, threatening the value of mortgage-backed assets that formed the basis of what would later be known as the *toxic* securities trading at the center of the crisis. As the crisis mounted through 2008, bankers and financiers sought more funding whilst selling assets, which saw further falls in asset values and growing concerns about debt exposures and risk. The panic really set in when Lehman Brothers collapsed in September 2008, leading to a global freeze in credit markets and further bank runs and failures. In this structural context, panic and herding fuelled the liquidity crisis and greatly increased the scale of the overall financial crisis. Agents thus *actualized* the structural impacts of the context they had established. This was not a matter of structure *over* agency (cf. Kim & Sharman, 2014), but one of the mutual constitution of structures and agents.

The crystallization of systemic risk thus involved a structural power context in which bankers exerted mutually generated power over other bankers, forcing them into unwanted fire sales and the withholding of credit, which essentially defined the crisis. "Bankers and financiers were thus caught in a series of severe collective action problems stemming from an institutional and structural environment they had helped create and eventually could not control" (Bell & Hindmoor, 2015, p. 69). Concepts of power are central here because agents were forced by other agents to act against their will. Power is often thought of as a resource that is deployed or used by agents, usually in a strategic manner. But there is another category of power, in which agents mutually exert power over one another, through the way they interact in a structured context. In the case we are examining, bankers were subject to a form of power that they themselves had created and exerted collectively, though unwillingly. They did not wish to withhold credit or engage in asset fire sales, but were forced to do so by the structural pressures and incentives they confronted. This was not about power being exercised by those in a structurally privileged position (Lindblom, 1977). Nor was it about exerting power through controlling discourse (Foucault, 1979). Instead, this was about agents exerting power over each other in a structured context that brought on collective ruin. Agents thus produced large structural effects that they did not anticipate and could not ultimately control. Prior to the crisis, bankers and financiers thought the institutions and relationships they had created were built on sound risk management and rational contracting. But as Bell and Hindmoor (2015, pp. 70–71) argue:

> Prevailing ideas and assumptions concealed the true nature of the structural dynamics confronting agents. In this sense, ideas and structures proved to be dangerously congruent. Only as the crisis was breaking did bankers come to realize what they had created. They ceased being 'true believers' in financial markets at precisely the moment that it became too late to escape.

Solutions?

Because the crystallization of systemic risk stemming from mutually exerted forms of structural power arises from the behavior of bankers and financiers in financial markets, any solution to these problems must ultimately involve modifying the behavior of these actors. To avoid the crystallization of systemic risk, they must attempt to collectively mold or shape their behavior to avoid such an outcome. The relevant behavior here is shaped by actors' knowledge, ideas, and motives, as well as by the governance, institutional, and structural terrains in which they operate. We explore these factors below, starting with the contribution of authors of governance literature.

The Governance Literature

In recent decades, the authors of the burgeoning governance literature have highlighted the interactions of public and private actors in the governance of public affairs. Much of this literature has a *society-centred* perspective, in the sense that it emphasizes relatively horizontal forms of governance networks, which are said to have marginalized government (Bell & Hindmoor, 2009). As Sørensen and Torfing (2007, p. 3) put it: "[T]he sovereign state … is losing its grip and is being replaced by new ideas about pluricentric government based on interdependence, negotiation, and trust." Hence, this literature's authors have allegedly shifted from "government to governance," involving interactions between a wide range of actors in formal and informal "self-organising networks." For Stoker (1998, p. 17), "the essence of governance is its focus on governing mechanisms which do not rest on recourse to the authority and sanctions of government." Sørensen and Torfing (2007) similarly emphasise the centrality of "non-hierarchical forms of governance" (p. 3), the "absence of top-down authority" (p. 44), and the "role of horizontal networks of organised interests" (p. 3). Bevir and Rhodes (2003, pp. 55–56) argue that "networks are the defining characteristics of governance," and offer a "coordinating mechanism notably different from markets and hierarchies." In this account, key dynamics in politics, such as hierarchy, power, and conflict, tend to recede, to be replaced by more horizontal forms of negotiation, networking, mutual dependence, reciprocity, and trust relations.

Bell and Hindmoor (2009) argue that this approach, although useful in highlighting multiple actors in governance, downplays the role of the state and of hierarchy that are typically found in politics and in governance practices. Indeed, even when governments *choose* to govern in alternative ways, in using markets for example, governments and state agencies typically remain important players in establishing and operating the agendas and rules for such strategies, in sanctioning the role of key players, and in providing resources and support. Indeed, the relational aspects of governance can often be seen as a way of *strengthening* state capacity. As Andersen (2004, p. 7) argues:

> Many researchers have claimed that the restructuring of governance is a general retreat of government and the state … yet there is no reason to assume that the rise of governance necessarily leads to a decline of government … the main reason for the rise in state capacity through restructuring is … the fact that the state is now able to influence hitherto non-governmental spheres of social life through partnerships, in other words, an enlargement of state competencies.

In this view, posing a choice between society-centred and more state-centric approaches to governance is misleading because both sets of dynamics are often involved. Moreover, this approach sees governance as an extension of traditional forms of public policy, with the state as a key actor but utilizing a wider variety of governing strategies and actors, often involving non-governmental actors, including business, unions, associations, NGOs, or communities.

The governance literature's strength, then, is its focus on the role of both the state and non-state actors in forging governance partnerships aimed at working out solutions to complex collective action problems. As we shall see below, in preventing the crystallization of systemic risk amidst the Euro crisis, the role of private actors and networks working in tandem with the state has been of central importance.

Although useful, the governance literature in general and the society-centred governance literature in particular have several limitations. First, their authors fail to adequately deal with questions of knowledge and agency. At the extreme, there is a highly interpretivist account within the society-centred literature that focusses almost exclusively on agents and their ideas (e.g., Bevir & Rhodes, 2003). This, however, fails to adequately account for the dialectical interaction between agents and wider institutional or indeed structural contexts in which agents operate, largely because researchers view such contexts as the artefacts of an agent's interpretation. State-centric governance accounts harbor almost the opposite problem, as their authors focus on the dynamics of state-society relationships but spend little time dealing with detailed questions of agency. Admittedly, the authors of certain works in this approach have discussed how states use persuasion as a governance strategy, leveraging the ideas and cognition of relevant actors in reshaping behavior (Bell & Hindmoor, 2009; Bell, Hindmoor, & Mols, 2010). Yet even this approach's proponents does not delve deeply enough into how agents themselves actually operate in ideational terrains and appraise and respond to the knowledge and information they confront.

Second, scholars advocating the society-centred version of governance literature, in particular, with its emphasis on horizontal networks, largely ignore the role of hierarchy, not only in relation to the role of the state, but also within societal networks themselves. As we illustrate in the case below, the collective action responses that European bankers and financiers were able to achieve were orchestrated not by horizontal networks but by organized hierarchies within such networks, centred, in particular, around the associational role of the Institute of International Finance (IIF), the international bank lobbying organisation, which represented the major global banks and financial institutions and which worked in tandem with relevant state actors.

These gaps in relation to agency and hierarchical organisation within the governance literature are significant. The agency issue is especially important because a key issue that emerges from both the GFC and the Euro crisis is that key agents' knowledge, ideas, and perceptions crucially shape their role and actions in financial markets (Bell, 2017). Knowledge and ideational factors matter in relation to whether systemic risk is perceived and whether it eventually crystalizes, and they also matter in forging governance strategies and responses. Researchers must therefore know how agents think and respond in a cognitive and ideational sense to the situations they confront.

The Institutionalist Literature

Arguably, contemporary institutionalist scholars are better able to flesh out the interactions between the cognitive and ideational realm of agency and the institutional and structural terrains in which agents operate.

Agents do not operate in a vacuum, but shape and are shaped by their institutional and wider structural contexts. As Karl Marx once famously observed, "men make their own history, but they do not make it as they please; they do not make it under self-selected circumstances, but under circumstances existing, already given, and transmitted from the past" (quoted in Tucker, 1978, p. 575). Institutional contexts are important in this respect. As Scharpf (1997, pp. 41–42) argues, "once we know the institutional setting of interaction, we know a good deal about the actors involved, about their options, and about their perceptions and preferences." Institutions are primarily about the rules and norms (formal or informal) that shape actor behavior. Institutions matter because they shape actor identities, interpretations, and preferences, the norm and rule-based scope of agents' discretion, and the resources and opportunities available to agents within organizations or institutions. As Farrell (2018, p. 26) puts it: "Institutions are not historical constants; rather, they are themselves the product of human agency, and as humans enact institutions, they correspondingly transform them."

In political science, one of the main versions of institutional theory is historical institutionalism (HI). There has been a problem, however, because proponents of various strands of institutional theory, including strands of HI, have tended to emphasize highly constraining notions of institutions. Prominent theorists such as North (1990, p. 3) define institutions as "the humanly devised *constraints* that shape human interaction" (our emphasis). This is a *sticky* form of institutional theory. It has a limited account of agency and is better at explaining institutional continuity than change. Blyth (1997, p. 230) is among many critics who argue that institutional theorists view institutions as largely "constraining rather than enabling political action." Weyland (2008, p. 281) similarly argues that "institutionalism has emphasised inertia and persistence," offering a *static* view of institutional life. Schmidt (2008, p. 314) also sees established theorists as "subordinating agency to structure," whilst Crouch and Keune (2005, p. 83) argue that "institutional configurations are often presented as a straitjacket from which endogenous actors cannot escape." We are sympathetic to such criticisms and wary of subscribing fully to overly *sticky* versions of HI theory (Bell, 2011). For example, the sweeping institutional changes that constituted the revolution in banking institutions and practices with the rise of highly leveraged trading during the 1990s and 2000s in the core financial markets of the US and UK suggest there is something wrong with such accounts (Bell & Hindmoor, 2015). Thelen (2004) similarly finds a pattern of agent-driven institutional change in the German vocational training system, whereby incremental changes led to more profound changes over time.

In recent years, scholars have made a number of revisions to HI. The authors of more flexible accounts within HI have shifted to a more agency-centred, "post-determinist" (Crouch, 2007) analysis, recognizing that institutions are subject to endogenous, agency-driven change and that dialectical interactions between agents and institutions are central to institutional life and in driving institutional change (see Bell, 2011; Bell & Feng, 2014; Campbell, 2004; Steinmo & Thelen, 1992). A major step in this direction was Thelen's work with various colleagues, pointing to sources of agency-based discretion as a basis for *incremental* institutional innovation and change (Mahoney & Thelen, 2010; Streeck & Thelen, 2005; Thelen, 2004). Proponents of this approach recognize that agents operate with varying degrees of initiative and discretion, in a context in which institutions are both constraining and enabling (Bell, 2011). Hence, agents both shape and are shaped by institutions:

> Institutional and/or structural environments can exert potential, though always agency-actualised effects, by imposing costs or benefits on agents, by shaping actor interpretations and preferences, the scope of bounded discretion, and the resources and opportunities available to actors. (Bell, 2011, p. 892)

Glückler and Lenz (2016, p. 257) add that agency is manifest when "legitimate mutual expectations" about rules and behavior help reinforce stable patterns of interaction within institutions.

In recent work, constructivist institutionalists also argue that agents can actively interpret institutional rules and norms, again creating at least some room for agency (Blyth, 2002; Hay, 2007; Schmidt, 2010). Actors use ideas and typically rely on agreed understandings to interpret and navigate such institutional terrains. Importantly, however, institutions and structures are also "distinct strata of reality" that are not simply reducible to the actors that inhabit them. Bell (2011) has therefore cautioned that ideational accounts must ground agents squarely and dialectically within institutional and wider settings.

Knowledge and Ideas

As noted, institutions and structures only exert an effect when mediated through the activities of people (Archer, 2000). This suggests that people's ideas, knowledge, and basic behavioral biases shape how they interact with institutions and wider structural forces. Hence, an "agency-based HI approach can easily integrate constructivist notions of interpretive agency and give full recognition to the fact that ideas, knowledge, language and inter-subjective discursive processes provide the crucial building blocks for establishing meaning and understanding and thus of purposeful action in politics and institutional life" (Bell, 2011, p. 893).

Amidst the GFC, for example, most of the participants in the financial system were not simply responding to skewed incentive structures such as highly competitive pressures for profits or bank remuneration schemes that rewarded risk taking. They were also on the whole "true believers." The assumption made within many banks was that their trading activity and leverage were largely risk-free. Bank traders, bank CEOs, regulators, investors, and even many politicians were "boundedly rational," operating on incomplete information, myopia, group think, herding, and over-optimism (Bell, 2017). This led them to discount or neglect inconvenient or complex information as well as warning signals. Such ideas and motives mattered because in an uncertain environment the assumptions key actors made about how markets work, how other actors would behave, and how governments would respond, shaped their perceptions and actions (Bell, 2017; Hindmoor & McConnell, 2013).

A further important and related finding from behavioral studies comes from prospect theory (Kahneman & Tversky, 1979; Tversky & Kahneman, 1992). Its proponents argue that agents subjectively define value in terms of gains or losses from a given (often current) reference point rather than in terms of final gains or overall wealth positions. A finding of relevance to systemic risk issues is that individuals thus tend to be loss averse; they will worry about downside risks, and will take bigger risks (compared to securing a gain) to avoid losses, which helps explain market panics and asset fire sales.

Overall, the literatures above provide key elements with which to analyze agent behavior within institutional, structural, and governance settings in financial markets that feature the potential for systemic risk. From the above, it is clear that researchers must be attentive to the role of agency, as shaped by the knowledge and cognitive and ideational drivers of behavior and how agents use them to understand and react to the situations in which they find themselves. From the institutionalist literature, we also need to factor in the way in which agents are shaped and in turn shape the institutional and structural contexts in which they operate. Important here also is the notion that agents are not only pressured by institutional and structural contexts, but that they also have some scope to shape and manage these contexts. Finally, the authors of the governance literature emphasize the possibility that agents might band together in hierarchical networks capable of dealing with collective action and governance challenges and that, more often than not, the state is an important part of such arrangements. In the next section, we probe the basic elements of the Euro crisis and show how the explanatory elements outlined above can help reveal how relevant institutionally-situated agents used knowledge and ideas to ascertain the nature of the systemic risks they confronted and then forged collective action solutions that thus far have prevented the crystallization of systemic risk.

The Euro Crisis

One key institutional context pertinent to the unfolding Euro crisis was the establishment of the European Central Bank (ECB), a body charged with overseeing European monetary integration. The introduction of a common currency in 1999 was a further key institutional development. These institutional contexts shaped the behavior of major banks and debt markets and effectively meant that the markets treated all members of the Euro area in broadly similar ways. Crucially, this form of monetary integration and the ideas held by lenders meant that the less productive economies on Europe's periphery—the likes of Portugal, Spain, Ireland, and especially Greece—were able to borrow freely at lower interest rates than might have been the case otherwise. These countries subsequently piled on debt to fund what would turn out to be real estate bubbles in Portugal, Spain, and Ireland, as well as a sovereign debt crisis in Greece. Lenders viewed these economies as part of the EU monetary system, assuming that Irish or Greek debt, and so forth, would be treated similarly to German debt. Lenders also assumed that member states and the EU would support the major banks if troubles developed.

If the peripheral economies had their own national currencies, financial market agents might have imposed a degree of discipline on borrowing. But because the countries in question were members of the Euro, external market pressures were not effective in bringing about change in economies that would normally have been judged to be "living beyond their means," potentially facing a market-driven currency depreciation. For example, before Greece joined the European Monetary Union, its large and rising public debts would have probably initiated rising interest rates and/or falling exchange rates as markets reacted to the rise in perceived risk. This would have helped to stem debt increases. However, the EMU system largely eliminated such market constraints on debt, and the EU's own policy and administrative monitoring regarding debt and fiscal balances also clearly failed.

The periphery's various and growing private and sovereign debt problems were exacerbated by the GFC from 2008 onwards and would become a key structural problem for the Eurozone. In Ireland, for example, debt helped fuel the "growth miracle" that developed into a massive property bubble, which eventually collapsed, exposing Irish and European banks. Indeed, the huge burden of public and private debt at the center of the Euro crisis was largely held by German, French and UK banks. The German banks, for example, were "structurally hugely vulnerable to crisis," according to Thompson (2015, p. 856), whilst the French banks' exposure was even larger. It is estimated that the French, German, and UK banks' combined exposure to peripheral European debt at the height of the crisis was as high as two trillion US dollars (Kalaitzake, 2017, p. 396; Thompson, 2015, p. 857). Any default on such would place enormous pressure on these banks, risking a bank-run and banking insolvency, in turn forcing national governments to try and bail the banks out. This posed a major systemic risk to the European banking system, not only because of the debt exposures and highly fragile nature of the large Euro banks in question (Bell & Hindmoor, 2018a, b), but also because of these banks' sheer scale relative to the national GDPs involved. This posed the question of whether bailouts would even be fiscally feasible for the relevant governments and authorities. The scale and uncertainties surrounding these potential problems thus constituted an acute context of systemic risk and raised wider institutional questions about the design of the Euro system and its capacities for crisis management and adjustment. For the debt-laden peripheral economies the problem was that they were locked into a relatively inflexible institutional system that was never designed for such crises and that, for example, foreclosed currency depreciation as an adjustment mechanism.

Preventing the Crystallization of Systemic Risk Amidst the Euro Crisis

As noted above, agents tend to be loss averse and will often take drastic action in the face of impending losses and instability. In this context, the challenge was to try and prop up European debt markets and prevent a run on the major European banks. In other words, the challenge was to prevent the crystallization of systemic risk whereby financial market actors could potentially bring on collective ruin by exerting a form of structural power, mutually exercised, thus presenting market actors with a severe and urgent collective action problem.

In the literature on structural power, researchers often treat the structural dimensions of such power separately from so-called instrumental dimensions of business power, with the latter based on overt business activism, organization, and lobbying. But these two forms of power should not for the most part be analytically separated and can in fact interact. Indeed, structural power need not be deterministic or automatic and can be mediated through agency, ideas, and collective organization (Bell, 2012).

Kalaitzake (2017) uses this framework to analyze the response to the Greek crisis and the way in which European and international bankers, and especially the leaders of the IIF, understood the challenges and risks they confronted and used ideas, knowledge, and experience regarding the dynamics of previous banking crises and debt resolution strategies in a range of developing countries to help chart a way forward in this new European situation. The IIF thus emerged as an important enabling institution that was knowledgeable, expert, and well-connected with European leaders and officials. Over the course of the crisis, the IIF was able to frame the key issues, articulate clear response strategies, and use associative means to help organize collective action responses. The IIF also worked closely with European state leaders and officials in the European Commission, the European Central Bank, and the IMF, and was valued by the authorities as a knowledgeable and organizationally capable partner. All of these actors had a key knowledge advantage compared to those involved in the GFC, who were essentially overwhelmed by a more complex and completely unexpected chain of events, starting with the collapse of mortgage-backed assets and ending with the meltdown in global wholesale funding markets (Bell, 2017). In contrast, European actors were confronted with a more conventional and more clearly understood debt and potential bank run type of crisis. As noted, the IIF leaders were well versed in such crises, and this was in contrast to the level of experience and knowledge held by many European leaders and officials who "had little grasp of the technical issues involved" (quoted in Kalaitzake, 2017, p. 399). Above all, it was well understood that it was essential to uphold market confidence and prevent panic and contagion and that the only way to do this was to organize responses that would stabilize the debt situation and above all convince relevant market actors that the situation was in hand. According to one private sector participant, the European authorities thus recognized the IIF "as a valuable platform to coordinate policy objectives with the majority of bondholders as a unified block [and] foster stronger policy communication to financial markets more broadly, allowing officials to better manage policy expectations and market reactions" (Kalaitzake, 2017, p. 399).

In the case of the Greek sovereign debt crisis, tough medicine in the form of fiscal restraint and other austerity measures were imposed as the condition for liquidity assistance from the European authorities and the IMF. As the crisis worsened in May 2010, a 110bn Euro bailout was announced to facilitate Greek debt servicing. This was largely aimed at buying time and reassuring markets that the situation was in hand. Policymakers initially directed their efforts at avoiding a debt write down and a private sector debt haircut in order to avoid market panic and the potential for contagion in other stressed markets in the Euro periphery. As Kalaitzake (2017) notes, it was initially feared that "a creditor write-down in Greece would trigger a 'Lehman-type event' resulting in bank runs and rising borrowing costs." The Euro leaders also developed an emergency bailout or lending fund—the European Financial Stability Facility. ECB President Mario Draghi's famous promise in 2012 to do "whatever it takes" to resolve the crisis was also intended to both reassure skittish financial markets and to justify buying "unlimited" quantities of sovereign bonds. As the IMF (2013, p. 28) argued, the imposition of pain on Greece but not initially on creditors "provided a window for private creditors to reduce exposures and shift debt into official hands." This process saw almost 100 billion Euros of Greek debt pass from the private to the state sector between the initial bailout and a further debt restructuring deal announced in 2012, the latter prompted by the continuing instability of the Greek debt situation. Having bought some breathing space with the first bailout, the IMF began to insist that any further official assistance would need to be supported by a private sector bail-in or haircut in which creditors would be exposed to losses. In this context, the IIF worked to organize a collective banking response and forged agreement with Euro leaders that by March 2012 was focused on a second Greek bailout, though this time with a substantial degree of private sector bail-in, amounting to an over 50% write down for bond holders. Yet policymakers sweetened the bail-in with generous offsets, including swaps for certain amounts of Greek debt for official bonds of various maturities. Overall, these arrangements benefited the banks and private sector creditors by avoiding a disorderly default, and reduced the risk of panic and contagion, the threat of more coercive government measures to restructure debt markets, and market exposure to the Greek crisis—all based on essentially voluntary, collective private sector responses and organization, backed and supported by the EU authorities. In contrast to Woll (2014), who argues that private financial sector disorganization forced governments into more generous banks bailouts in some countries after the GFC, this case illustrates the advantages of private financial sector knowledge and organization in cases of sovereign debt crises.

This episode shows that systemic risk in financial markets can be managed even in the face of a potential *Lehman-type event*, but only if market and state actors are able to perceive the looming threat and act in an organized manner with sufficient institutional back-up and resources to avoid the crystallisation of systemic risk. Ideas, financial expertise, experience, the willingness and capacity to act collectively, all supported by appropriate institutions and public-private partnerships and effective governance strategies were all involved in this case. The key contrast here, compared to what happened in the case of the GFC, is that bankers and state leaders in the Euro situation were able to clearly understand and perceive the potential for the crystallisation of systemic risk and act to avert it.

Finally, it is true that the bailouts and actions taken to assuage market actors exposed the states and authorities in question to moral hazard—a form of reassurance and an outcome that is likely to reduce risk perceptions and embolden market actors going forward. This situation always puts states and the authorities in a bind, but the reality is that financial markets are now so large that systemic financial collapse cannot be countenanced. This essentially structural market shift now means that the potential crisis-induced collateral damage to wider economics and even to the fate of nations is now so great in most cases of large, complex, and interconnected financial markets that concerns about moral hazard now take second place to the need to avoid a financial meltdown.

Conclusion

Knowledge, collective capacity, and governance can clearly matter in understanding and managing complex human interactions amidst financial markets that are structurally prone to systemic risk and its crystallization. This occurred in a dramatic and damaging way during the 2008 GFC, but thus far at least the Euro crisis has not morphed into a Euro banking crisis, although many Euro banks remain fragile and vulnerable (Bell & Hindmoor, 2018a). As we have demonstrated, at the crisis's peak around 2010, the Euro authorities and states managed to avoid the crystallization of systemic risk through collective action, with the help of a knowledgeable and organized private sector.

We have argued that various strands of research and theory in political science offer useful tools for understanding such dynamics and outcomes. The authors of governance literature point to the importance of the state and the orchestration of public-private cooperation in meeting governance challenges. But we have argued that institutional analysis also offers a way of locating agents in relevant institutional and structural contexts, through tracing dialectical relations of mutually shaping interactions over time. This approach also offers a way of bringing in deeper insights about agents' cognitive and ideational processes in shaping the way agents use knowledge to help appraise and react to situations and in building institutional and collective responses to risk environments. Future researchers must bring together these elements in wider studies of how knowledge and governance strategies have been deployed, studying other cases of financial market dynamics or in broader settings.

References

Andersen, H. (2004). Governance and regime politics in Copenhagen (pp. 1–17). Eurez Lecture: Vol. 16. University of Copenhagen, Department of Geography.

Archer, M. S. (2000). For structure: Its reality, properties and powers: A reply to Antony King. *Sociological Review, 48,* 464–472. https://doi.org/10.1111/1467-954X.00226

Bell, S. (2011). Do we really need a constructivist institutionalism to explain institutional change? *British Journal of Political Science, 41,* 883–906. https://doi.org/10.1017/S0007123411000147

Bell, S. (2012). The power of ideas: The ideational shaping of the structural power of business. *International Studies Quarterly, 56,* 661–673. https://doi.org/10.1111/j.1468-2478.2012.00743.x

Bell, S. (2017). Historical institutionalism and new dimensions of agency: Bankers, institutions and the 2008 crisis. *Political Studies, 65,* 724–739. https://doi.org/10.1177/0032321716675884

Bell, S., & Feng, H. (2014). How proximate and 'meta-institutional' contexts shape institutional change: Explaining the rise of the People's Bank of China. *Political Studies, 62,* 197–215. https://doi.org/10.1111/1467-9248.12005

Bell, S., & Hindmoor, A. (2009). *Rethinking governance: The centrality of the state in modern societies.* New York, NY: Cambridge University Press. https://doi.org/10.1017/CBO9780511814617

Bell, S., & Hindmoor, A. (2015). *Masters of the universe, slaves of the market.* Cambridge, MA: Harvard University Press.

Bell, S., & Hindmoor, A. (2018a). Are the major global banks now safer? Structural continuities and change in banking and finance since the 2008 crisis. *Review of International Political Economy, 25,* 1–27. https://doi.org/10.1080/09692290.2017.1414070

Bell, S., & Hindmoor, A. (2018b). *Structural forces can come with instruction sheets and blindside agents: Structure and agency in financial markets in Europe and the United States* (Unpublished paper). University of Queensland, St.Lucia.

Bell, S., Hindmoor, A., & Mols, F. (2010). Persuasion as governance: A state centric-relational perspective. *Public Administration, 88,* 851–870. https://doi.org/10.1111/j.1467-9299.2010.01838.x

Bevir, M., & Rhodes, R. A. W. (2003). *Interpreting British governance.* London: Routledge. https://doi.org/10.4324/9780203135914

Blyth, M. (1997). Any more bright ideas: The ideational turn in comparative political economy. *Comparative Politics, 29,* 229–250. https://doi.org/10.2307/422082

Blyth, M. (2002). *Great transformations: Economic ideas and institutional change in the twentieth century.* New York, NY: Cambridge University Press. https://doi.org/10.1017/CBO9781139087230

Campbell, J. L. (2004). *Institutional change and globalization.* Princeton, NJ: Princeton University Press.

Crouch, C. (2007). How to do post-determinist institutional analysis. *Socio-Economic Review, 5,* 527–567. https://doi.org/10.1093/ser/mwm001

Crouch, C., & Keune, M. (2005). Changing dominant practice: Making use of institutional diversity in Hungary and the United Kingdom. In W. Streeck & K. Thelen (Eds.), *Beyond continuity: Institutional change in advanced political economies* (pp. 83–102). Oxford, UK: Oxford University Press.

Farrell, H. (2018). The shared challenges of institutional theories: Rational choice, historical institutionalism, and sociological institutionalism. In J. Glückler, R. Suddaby, & R. Lenz (Eds.), *Knowledge and institutions* (pp. 23–44). Knowledge and Space: Vol. 13. Cham: Springer. https://doi.org/10.1007/978-3-319-75328-7_2

Foucault, M. (1979). *Discipline and punish: The birth of the prison.* New York: Pantheon.

Glückler, J., & Lenz, R. (2016). How institutions moderate the effectiveness of regional policy: A framework and research agenda. *Investigaciones Regionales, 36,* 255–277.

Hay, C. (2007). Constructivist institutionalism. In R. A. W. Rhodes, S. A. Binder, & B. A. Rockman (Eds.), *The Oxford handbook of political institutions* (pp. 56–74). Oxford, UK: Oxford University Press. https://doi.org/10.1093/oxfordhb/9780199548460.001.0001

Hindmoor, A., & McConnell, A. (2013). Why didn't they see it coming? Warning signs, acceptable risks and the global financial crisis. *Political Studies, 61,* 543–560. https://doi.org/10.1111/j.1467-9248.2012.00986.x

IMF (International Monetary Fund). (2013). *Global financial stability report: October 2013.* Washington, DC: International Monetary Fund.

Kahneman, D., & Tversky, A. (1979). Prospect theory: An analysis of decision under risk. *Econometrica, 47,* 263–291. https://doi.org/10.2307/1914185

Kalaitzake, M. (2017). The political power of finance: The Institute for International Finance and the Greek debt crisis. *Politics & Society, 45,* 389–413. https://doi.org/10.1177/0032329217707969

Kim, H. J., & Sharman, J. C. (2014). Accounts and accountability: Corruption, human rights and individual accountability norms. *International Organization, 68,* 418–448. https://doi.org/10.1017/S0020818313000428

Lindblom, C. E. (1977). *Politics and markets: The world's political-economic systems.* New York: Basic.

Mahoney, J., & Thelen, K. (2010). A theory of gradual institutional change. In J. Mahoney & K. Thelen (Eds.), *Explaining institutional change: Ambiguity, agency and power* (pp. 1–37). New York, NY: Cambridge University Press. https://doi.org/10.1017/CBO9780511806414

North, D. C. (1990). *Institutions, institutional change and economic performance.* New York, NY: Cambridge University Press.

Scharpf, F. W. (1997). *Games real actors play: Actor-centred institutionalism in policy research.* Boulder: Westview. https://doi.org/10.4324/9780429500275

Schmidt, V. A. (2010). Taking ideas and discourse seriously: Explaining change through discursive institutionalism as the fourth "new institutionalism". *European Political Science Review, 2,* 1–25. https://doi.org/10.1017/S175577390999021X

Schmidt, V. A. (2008). Discursive institutionalism: The explanatory power of ideas and discourse. *Annual Review of Political Science, 11,* 303–326. https://doi.org/10.1146/annurev.polisci.11.060606.135342

Sørensen, E., & Torfing, J. (2007). Introduction: Governance network research: Towards a second generation. In E. Sørensen & J. Torfing (Eds.), *Theories of democratic network governance* (pp. 1–24). Basingstoke: Palgrave Macmillan. https://doi.org/10.1057/9780230625006_1

Steinmo, S., & Thelen, K. (1992). Historical institutionalism in comparative perspective. In S. Steinmo, K. Thelen, & F. Longstreth (Eds.), *Structuring politics: Historical institutionalism in comparative analysis* (pp. 1–31). Cambridge, UK: Cambridge University Press.

Stoker, G. (1998). Governance as theory: Five propositions. *International Social Science Journal, 68,* 15–24. https://doi.org/10.1111/issj.12189

Streeck, W., & Thelen, K. (2005). *Beyond continuity: Institutional change in advanced political economies.* Oxford, UK: Oxford University Press.

Thelen, K. (2004). *How institutions evolve: The political economy of skills in Germany, Britain, the United States and Japan.* Cambridge, UK: Cambridge University Press. https://doi.org/10.1017/CBO9780511790997

Thompson, H. (2015). Germany and the Euro-Zone crisis: The European reformation of the European banking crisis and the future of the Euro. *New Political Economy, 20,* 851–870. https://doi.org/10.1080/13563467.2015.1041476

Tucker, R. C. (1978). *The Marx-Engels reader.* New York: Norton.

Tversky, A., & Kahneman, D. (1992). Advances in prospect theory: Cumulative representation of uncertainty. *Journal of Risk and Uncertainty, 5,* 297–323. https://doi.org/10.1007/BF00122574

Weyland, K. (2008). Toward a new theory of institutional change. *World Politics, 60,* 281–314. https://doi.org/10.1353/wp.0.0013

Woll, C. (2014). *The power of collective inaction: Bank bailouts in comparison.* Ithaca, NY: Cornell University Press. https://doi.org/10.7591/cornell/9780801452352.003.0003

Part II
How Knowledge Drives the Effectiveness of Governance

Chapter 7
Explaining Subnational Governance: The Role of Governors' Codified and Uncodified Knowledge

Claudia N. Avellaneda, Ricardo Andrés Bello-Gómez, and Johabed G. Olvera

Scholars, donors, and practitioners worldwide search for the tools and means to improve governance. Throughout this search, stakeholders recognize the importance of bringing actors together from public, private, and non-profit sectors for problem solving. The cross-sector effort is expected to lead to effective governance, which in turn should result in improved management, effective implementation of instruments, improved service delivery, and higher outcomes. "Governance comprises the legal, social, political, economic, environmental, and administrative arrangements put in place to ensure the intended outcomes for stakeholders are defined and achieved"[1] (CIPFA & IFAC, 2013, p. 8). In Latin America, for example, decentralization has been one of the adopted governance arrangements. By increasing autonomy and assigning more responsibilities to subnational governments, decentralization is expected to improve service delivery and peoples' lives. Despite the generalized increase in the role of subnational governments, considerable performance variation exists across regions, leading us to question what factors explain subnational governments' performance.

Although the terms of subnational governments are very broad, encompassing second-level (state or province) and third-level government (municipalities) (Herrera Gutierrez, 2015), we here refer to the second level of government as states/provinces/departments/regions. In explaining governance performance,

[1] Based on this definition, governments are a component of the variety of arrangements needed for public action to happen. So, in this paper, when referring to governance, we are talking about the different arrangements to pursue national goals and when using the term government, we refer to the organizations and branches/levels that comprise the public sector.

C. N. Avellaneda (✉) · R. A. Bello-Gómez · J. G. Olvera
O'Neill School of Public and Environmental Affairs, Indiana University Bloomington, Bloomington, IN, USA
e-mail: cavellan@indiana.edu; rabellog@indiana.edu; jolverae@indiana.edu

J. Glückler et al. (eds.), *Knowledge for Governance*, Knowledge and Space 15, https://doi.org/10.1007/978-3-030-47150-7_7

scholars have stressed the role of political factors, such as partisanship[2] (Brollo & Nannicini, 2010; Cox & McCubbins, 1986; Solé-Ollé & Sorribas-Navarro, 2008), party ideology (Strøm, 1990; Wittman, 1990), electoral competitiveness (Holbrook & van Dunk, 1993; Key, 1949), and electoral cycle (Ames, 1987; Buchanan & Tullock, 1962; Nordhaus, 1975). Others explain performance as a function of resources (Sharkansky, 1967), oversight mechanisms (Blair, 2000), intergovernmental/interorganizational networks (Agranoff & McGuire, 1998, 2003), population size and nature (Durant & Legge, 1993), and organizational capacity (Collins & Gerber, 2006; Hall, 2008; Terman & Feiock, 2015). Finally, another branch of research identifies collaborative arrangements as key drivers of governance performance (Meier & O'Toole, 2001; Page, 2008).

Without denying their explanatory power, we find the above factors neglect the potential role of government CEOs' knowledge. In this second level of governments, the CEO/manager of the state/province/department is a directly elected governor or an appointed intendent (as in Chile), who performs both political and administrative functions and who enjoys managerial autonomy and discretion. Given his/her significant autonomy and discretionary power, his/her knowledge influences decision-making, and, in turn, governance performance. A top manager's individual traits are thus expected to contribute to her decision-making in setting strategic and tactic goals, selecting middle-level managers, rearranging organizational structure, risk taking, adopting innovation, networking with other levels of government, and/or adopting a participatory process.

In this study, we consequently specifically seek to assess the effect of governors' knowledge (codified and uncodified) on performance at the subnational level. In doing so, we rely on data derived from 32 Mexican states and 32 Colombian departments/provinces during the 1995–2010 and 2004–2013 periods, respectively. We assess subnational performance in terms of two objective indicators—enrollment in secondary education and infant mortality rate (IMR)—that are mainly the responsibility of both Mexican states and Colombian departments.

Identifying the drivers of governance performance at the subnational level is relevant for several reasons. In some countries, half the national budget is currently allocated to subnational governments (Herrera Gutierrez, 2015). Moreover, as delivery of health, education, and development programs happens at the subnational level, long-term development is a function of governance performance at this level. Finally, subnational performance is vital for strengthening democracy, transparency, and citizens' trust of government. Moreover, most researchers studying subnational governance performance focus on the United States and/or the United Kingdom. However, both countries enjoy highly developed, post-industrial economies, as well as fully democratic systems. Therefore, it is unknown whether the identified drivers of performance in these developed settings apply to

[2] In political science, partisan and partisanship refer to strong adherence and loyalty to a political party or group (see, e.g., Allan & Scruggs, 2004, or Bartels, 2000).

underdeveloped and new democracies, such as Mexico and Colombia, where existing institutions might generate different incentives.

In the following section, we present the theoretical underpinnings that link knowledge to governance performance. In the chapter we then offer the rationale for alternative explanations, which we will also test in the empirical analysis. In the next section, we provide a brief background of the Mexican and Colombian estates, as well as a description of the data collection and variable operationalization. Afterward, we present the empirical analysis, interpret and discuss the results, identify some of the limitations, and develop our conclusions.

Explaining Governance Performance

As mentioned above, several factors seem to contribute to governance performance. Although no clear drivers seem to explain variation in performance, at risk of simplifying, researchers have paid considerable attention to seven drivers: resources, market competition, accountability, organizational structure, political/environmental/internal context (O'Toole & Meier, 2015), collaborative arrangements (Meier & O'Toole, 2001; Page, 2008), and management (Boyne, 2003; Lynn, 2003; Meier & O'Toole, 2002). According to O'Toole and Meier, (1999), management (i) provides organizational stability and structure, (ii) coordinates achievement of organizational goals, (iii) exploits opportunities, and (iv) buffers the organization from external shocks.

For some, public management makes the difference between the success and failure of policy implementation (Avellaneda, 2009; Lynn, 1987; Meier & O'Toole 2002; O'Toole & Meier, 1999). Under this view, qualified management—the "management-quality" hypothesis—facilitates program success, contributing to overall organizational performance. The notion is that managers are expected to rely on structure, craft, and institutions (Lynn, 2003, p. 2) to direct routine activities in purposeful organizations. However, managerial influence works through different causal pathways (Meier & O'Toole, 2002), as management influences performance through multiple mechanisms: goals, material and human resources, regulation, representation, workforce diversity, organizational strategy, and leadership. Nevertheless, "it is increasingly clear that (individual) managers can improve program effectiveness, sometimes in substantial ways" (Boyne, Meier, O'Toole, & Walker, 2005, p. 634). For it is "[a]t the higher levels of the bureaucracy and among the elected officials, for example, that important decisions on what services to deliver or how to deliver them are made that limit a street-level bureaucrat's ability to affect service distributions" (Meier, Stewart, & England, 1991, p. 158, see also Avellaneda, 2009). Or as Lynn (1987, p. 103) posits, "[i]n its most concrete and observable sense, the activity of government agencies is the product of the behavior of identifiable individuals who occupy responsible positions." However, how do individual managers/leaders contribute to governance performance?

Governors' Codified Knowledge and Subnational Governance

Although education is codified knowledge, experience is uncodified knowledge. The mechanisms by which codified knowledge contributes to performance and governance can be explained at the macro- and micro-level. At the macro-level, Adam Smith highlighted the role of education as a contributor to the process of the production of assets and services (Sen, 1997). Specifically, technical formation and learning contribute to production (Sen, 1997, p. 70). Codified knowledge, that is education, is part of human capital. Solow's (1956) theory of economic growth stresses the importance of human capital on a country's growth and development (see also Besley, Montalvo, & Reynal-Querol, 2012; Jones & Olken, 2005). Since Solow (1956), many others have highlighted the role of human capital in production (Lucas, 1988; Mankiw, Romer, & Weil, 1992), encouraging empirical research that reports a positive relationship between school years and growth rate (Bassanini & Scarpetta, 2001).

At the micro-level, codified knowledge also contributes to governance performance through several mechanisms. Education brings recognition, leads to better choices and empowers leaders to argue and communicate strategies. Hence, "[c]ognitive resource theory assumes that more intelligent and knowledgeable leaders make better plans and decisions than do those with less ability and knowledge" (Fiedler, 1986, p. 533). Knowledgeable leaders are expected to be more innovative and more direct in communicating plans, decisions, and strategies. As Dearborn and Simon (1958) posited, "[w]hen presented with the same problem, executives with different functional areas defined the problem largely in terms of the activities and goals of their own areas." Therefore,

H1: The higher a governor's codified knowledge, the higher the state/province performance.

Governors' Uncodified Knowledge and Subnational Governance

Besides the codified, scientific, and technocratic knowledge, the uncodified, intuitive, and artistic knowledge of managers also influences performance (Lynn, 1996, pp. 112–113). Hence, "[k]nowledge is, by no means, the only cognitive resource expected to influence a leader's performance" (Avellaneda, 2009, p. 289). Leaders acquire scientific knowledge at the university and/or workshop level. On the contrary, uncodified, intuitive knowledge is obtained through mentorship and job experience (Avellaneda, 2009; Lynn, 1996). This is considered learning though experience and practice (Arrow, 1962). According to Fiedler (1986, p. 32), experience likely affects a leader's performance in several ways: "by (a) providing useful and job-related knowledge, (b) enhancing the ability to cope with stressful conditions, and (c) engendering a feeling of greater self-confidence and control of the leadership situation." That is, experience helps governors anticipate technical and administrative obstacles and allows them to search for previous adopted strategies to overcome them. Experience should also help governors cope with task difficulty.

According to the resource-based view, experience and expertise are rare, valuable, inimitable, and non-substitutable organizational resources, which contribute to its competitive advantage (Penrose, 1959; Rumelt, 1984; Wernerfelt, 1984). Riccucci (1995), Doig and Hargrove (1990), and Avellaneda (2009, 2012, 2016) have examined the experience-performance relationship and find that political leaders' experience positively influences their organizational effectiveness. Others, however, report no empirical support for the experience-performance relationship (Fernandez, 2005; Fiedler, 1966; Fiedler & Chemers, 1968). The use of different indicators for qualitative and quantitative dimensions of experience may explain the inconsistent results (Olvera & Avellaneda, 2019). Nevertheless, the uncodified knowledge-performance relaxations deserve to be tested in underdeveloped settings. Therefore,

H2: Governors' uncodified knowledge (experience) is positively correlated with state performance.

Alternative Explanations of Subnational Governance Performance

Governance performance may also be influenced by elected officials' ideological alignment with other elected officials at higher or equal levels of government. Researchers have mainly tested this partisan alignment hypothesis in studies to explain grant approval. Proponents of this hypothesis specifically posit that due to risk aversion, grants tend to be allocated to co-partisan jurisdictions (Cox & McCubbins, 1986). There is some empirical evidence supporting this claim. Brollo and Nannicini (2010) find that Brazilian municipalities in which the mayor is affiliated with the coalition of the president received 36 to 43 percent larger transfers than non-aligned municipalities in the final 2 years of the mayoral term. Likewise, Solé-Ollé and Sorribas-Navarro (2008) find that, in Spanish municipalities, grants to co-partisans led to some political support, but grants to opposition parties did not bring in more votes. However, a number of studies also support a contrary hypothesis that grants tend to be directed toward jurisdictions with a high number of swing voters as a means of winning support (Dahlberg & Johansson, 2002; Dixit & Londregan, 1998; Johansson, 2003; Lindbeck & Weibull, 1987; Mejía Guinand, Botero, & Rodriguez Raga, 2008). The plausibility of both hypotheses calls for further testing at the subnational level in understudied contexts. Therefore,

H3: States whose governors are ideologically aligned with other elected officials at the state and national level tend to have higher performance.

Besides partisanship, government ideology also has been suggested as a driver of governance performance. Downs (1957) suggested that party competition takes place along a left-right ideological spectrum, suggesting that political parties are policy seekers, rather than just vote seekers (Strøm, 1990; Wittman, 1990). Although some question the validity of this one-dimensional scale, researchers have used left-right continuum to test several theoretical propositions. Regarding social policies, for example, the debate centers on whether or not parties of the left spend more money than parties of the right (Blais, Blake, & Dion, 1993; Solano, 1983; Swank, 2002). Greater social spending, in turn, may enhance program coverage. Blais et al. (1993) and Swank (2002) found that parties make a difference, whereas Solano (1983) and Avellaneda (2009) report no party ideology effect at all. The potential explanatory power at the state level in developing settings justifies testing party ideology. Here, the expectation is that states led by governors affiliated with rightist parties tend to expend less, resulting in lower performance.

Finally, others link program implementation and performance to electoral competitiveness (Holbrook & van Dunk, 1993; Key, 1949). Proponents of the electoral competitiveness hypothesis suggest that when elections are tight, candidates and incumbents tend to provide more services in order to gain political support from many segments (Key, 1949). Party competition hypothesis has received some support (Holbrook & van Dunk, 1993), but others report no support or little impact (Dye, 1966). This inconclusiveness calls for additional tests of the competitiveness hypothesis at the subnational level in new democracies.

We will test the generated hypothesis and expectations in a data set derived from the Mexican states and Colombian departments/provinces. In the next section, therefore, we provide a brief background of these subnational governments.

Case Analysis: Mexican States and Colombian Departments

Since the adoption of decentralization in the 80s, Mexican states and Colombian departments have expanded their fiscal, political, and administrative autonomy. Despite having the same responsibilities, considerable variation exists across Mexican states and Colombian departments in terms of several indicators and dimensions of performance. Understanding the drivers of this variation is central to this study, and we suggest that governors' knowledge explains this subnational variation. Beside cross-state variation, both Mexican and Colombian states have experienced violence. Nevertheless, Mexico is a federal system, but Colombia is a unitary system. In addition, their party systems also exhibit variations. Mexico has a highly institutionalized three-party (PRI, PAN, and PRD) system versus Colombia's multi-party system (11 parties represented in the 2018 Senate elections). Although the common features serve as controls, differences justify testing our propositions to understand their impact on governance performance at the subnational level.

Mexican States

We test our hypotheses using data derived from the 32 states in the Estados Unidos Mexicanos (United Mexican States) and the 32 departments of Colombia (Fig. 7.1). In this section, therefore, we provide a background on both Mexican and Colombian subnational governments. The Constitution of 1917 formally established Mexico as a federal republic. However, the country was highly centralized until 1980. From the early 1980s to late 1990s, the federal government initiated a process of administrative, political, and fiscal decentralization, delegating many governmental responsibilities, including health and education services, to states (Rowland & Ramírez, 2001). Consequently, Mexican states have attained a high degree of political, administrative, and budgetary autonomy (Falleti, 2010).

State autonomy also has increased governors' power (Falleti, 2010; Modoux, 2006), enabling governors to block federal policies (Rodríguez, 2003) and control decisions for municipalities' resource allocation (Alvarado, 1996). This significant expansion of responsibility has inspired numerous studies addressing the role Mexican governors play in federal and local politics (e.g., Figueras Zanabria, 2009;

Fig. 7.1 Mexican States. Source: Design by Volker Schniepp, Department of Geography, Heidelberg

Hernández Rodríguez, 2008; Langston, 2010; Modoux, 2006; Montero Bagatella, 2014; Morales y Gómez & Salazar Medina, 2009; Oikion Solano, 2012; Rodríguez, 2003). However, no systematic study has examined the effects of governors' characteristics on state performance.

Mexican governors are democratically elected for a six-year term, with no re-election allowed. They head the executive branch, perform political and administrative functions, and are accountable to state legislatures. State legislatures are unicameral and populated by directly and indirectly elected representatives, who serve three-year terms. Consequently, each governor serves alongside two legislative bodies. According to the *Ley General de Educación*, the federal government defines the general principles and goals of Mexico's education policy. However, states are responsible for designing and implementing necessary programs to ensure federal education goals are reached (DOF, 2013). For instance, the governor through the *State Government Plan* determines strategies to achieve education enrollment levels established by the federal government.

Mexican states have two kinds of revenues: (1) their own revenues, and (2) federal transfers. Their own revenues are comprised of state taxes (e.g., vehicle-ownership tax, purchase or sale of used cars, lotteries, etc.), social security fees, provision of public services to individuals (e.g., expedition of driver's licenses), public works that differentially benefit particular individuals, sale of state-owned real estate, and any other revenue derived from the execution of the state's faculties (e.g., traffic ticket fines). Around 8% of states' revenues come from these sources (Ramírez-Cedillo & Lopez-Herrera, 2016). Federal transfers resulted from the *Ley de Coordinación Fiscal* (Law of Fiscal Coordination) enacted in 1978 (Ley de Coordinación Fiscal, 1978). Under this law, states agreed to yield some of their tributary faculties to the federal government in exchange for a share of federal taxes. The two types of federal transfers are *participaciones* (participations) and *aportaciones* (contributions). *Participaciones* are determined according to a formula incorporating these elements: (1) level of tax collection, (2) population, and (3) compensations to less advantaged states. States can spend money from this source at their discretion. For the average Mexican state, *participaciones* account for 31% of revenues (Ramírez-Cedillo & Lopez-Herrera, 2016). *Aportaciones* are earmarked funds dedicated to education, health, social infrastructure, municipalities strengthening, etc. On average, this type of federal transfer represents 52% of states' revenues (Ramírez-Cedillo & Lopez-Herrera, 2016).

Governors can implement different strategies to increase revenues from any source. For example, to increase its own revenues, the Mexico City government installed cameras across the city to detect drivers going over the speed limit and fined the speeders. Governors also might apply actions to improve tax collection. For example, a state government may require a tax payment for an individual to use a vehicle. In particular, Mexico City's government may request vehicle owners to pay vehicle ownership taxes to obtain environmental verification needed to travel across the city (Reglamento de Tránsito de la Ciudad de México, 2016).

In terms of health care, states manage primary-care hospitals and are responsible for nutrition, epidemiology, maternity care, and visual and hearing health.

Decentralization reforms made states responsible for providing medical attention to most uninsured populations (Martinez Fritscher & Rodriguez Zamora, 2016). Some argue decentralization of health services allowed governors to use administrative and technical posts as political currency (Gonzalez-Block, Leyva, Zapata, Loewe, & Alagon, 1989). To expand health and education services, governors may seek to increase states' education and health budgets assigned by the federal government. A special federal commission reviews budget requests related to education and health issues from different actors, including governors, mayors, ministries, and associations. Consequently, lobbying strategies and political networks are essential assets for governors seeking resources for education and health.

Colombian Departments

Colombia is a unitary but decentralized republic (Const. 1991, art. 1). According to the 1991 Constitution (art. 311), Colombia's entire territory has been divided into 32 departments and the capital district of Bogotá.[3] Colombian departments coordinate and mediate between the national state and the municipalities, which are the main service providers and fundamental territorial authorities (Fig. 7.2). In the last two decades, Colombia has implemented changes over its territorial structure in order to advance fiscal, political, and administrative decentralization (Falleti, 2005).

Fiscal decentralization started with the institution of the "Situado Fiscal," the first attempt of intergovernmental transfers in 1968. The 1991 Constitution strengthened the transfers system and vested the territorial entities with the responsibility of providing social services, such as education, health care, and basic sanitation. These changes furthered administrative decentralization. In 2001, Congress adopted the Participations General System (SGP), which modified the transfer allocation method to achieve fiscal sustainability and improve equality (Bello & Espitia, 2011).

The first mayoral and gubernatorial elections took place in 1998 and 1992, respectively, giving way to political decentralization. Since then, mayors and governors were first elected for a three-year period, increasing to a four-year period on Jan. 1, 2004. An elective body, called the Department Assembly (Const. 1991, art. 299), oversees the subnational executives, approves state budgets, determines size and structure of the departmental government, and creates or suppresses municipalities, among other functions (Const. 1991, art. 300).

Departments raise revenue mainly from three categories of sources: taxes, transfers, and royalties, representing around 85% of their total revenue from 2000 to 2012.[4] During the same period, transfers' share of departments' revenue has fluctu-

[3] Even though other districts exist and enjoy greater autonomy than municipalities, they relate to specific issues (tourism, cultural affairs) and for most purposes, they are part of the respective departments.

[4] The remainder corresponds to capital projects co-financed by the national government, and other non-tax revenues.

Fig. 7.2 Colombian departments. Source: Design by Volker Schniepp, Department of Geography, Heidelberg

ated between 45.7% and 56.8%, taxes have ranged between 23.2% and 27.5%, and royalties have shifted between 10.0% and 16.4%.[5] Levies on beer, liquor, and tobacco constitute most of the tax revenue collected by the departments, even though the central government determines the respective tax rate. Royalties are directly related to the exploitation and transportation of natural resources, such as oil, gas, metals, and minerals. Nevertheless, the Participations General System (SGP) determines the way the central government transfers resources to each department, district, and municipality. The system allocates a total amount indexed to inflation, the variation of the national current revenue, and the target population for the provision of public services. The resources assigned are earmarked to the provision of education (58.5%), health care (24.5%), basic sanitation (17%), and

[5] Percentages calculated from Departamento Nacional de Planeación (2013).

other minor purposes (4%) (Barón & Meisel, 2003; Bello & Espitia, 2011; Ley 715, 2001).

Resources are assigned to departments for their administration and distribution, according to municipalities' needs. When municipalities obtain certifications to administer their own education and health resources, transfers bypass the department and go directly to the municipality (Bello & Espitia, 2011). To obtain the National Education Ministry's certification, municipalities must demonstrate they satisfy certain conditions of administrative capacity (Ministerio de Educación Nacional, 2004). In 2002, 46 municipalities were certified, and in 2008, 16 more were certified, bringing the total to 62 localities certified to directly provide secondary education without state involvement. After excluding the certified localities, Colombian departments are still in charge of providing secondary education to about 42% of the potential target population in Colombia (see Table 7.2). The education ministry considers children between 11 and 16 years old as the potential target population for secondary education (Ministerio de Educación Nacional, 2014). By 2013, more than 1.1 million Colombian teenagers still did not have access to secondary education. In Fig. 7.3, we summarize the process of education provision in Colombia, highlighting the role of the departments.

In the case of the provision of health, municipalities can be certified by either the department or the Public Health Ministry (Ley 60, 1993). However, departments still exercise a monitoring role and can subject a municipality to a performance management regime if the municipality fails to maintain a certain level of service

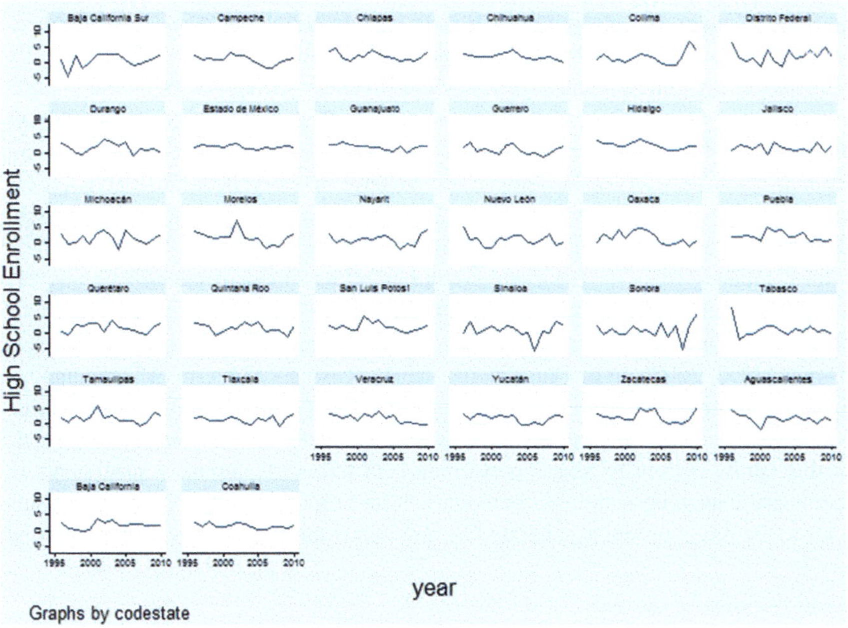

Fig. 7.3 High school enrollment for Mexican States (1995–2010). Source: Design by authors

provision and capacity (Ley 60, 1993). Indeed, between 2002 and 2013, departments recentralized health provisions in 92 municipalities.[6] In other words, departments influence the achievement of public health outcomes in their territory either by direct administration of the health system or by close monitoring of decentralized municipalities.

Data and Variable Operationalization

The unit of analysis in this study is state-year. Based on data availability, we built two panel data sets—one with information from 32 Mexican states over a 16-year period (1995–2010)—and another panel for 32 Colombian departments from 2004 to 2013. The period for the Mexican states includes three gubernatorial administrations for most states (1994–2000, 2000–2006, and the first 4 years of the 2006–2012 administration), but some states have shifted electoral cycles. The period for the Colombian department covers three gubernatorial administrations (2004–2007, 2008–2011, and the first 2 years of the 2012–2015 administration).

Assessing Governance Performance

Objectively assessing performance has become central in the public management literature. Walker, Boyne, and Brewer (2010) summarized the models used to measure performance. According to them, academics and practitioners have basically followed either the *economy-efficiency-effectiveness (3Es) model* or the *inputs-outputs-outcomes (IOO) model.* Proponents of both models try to capture performance as the ability of the organization to (i) acquire resources from the environment (*system-resource approach*), and (ii) to achieve the organizational goals (*goal approach*).[7] However, as Walker et al. (2010) recognize, these models lack insight about the organization's internal stakeholders, as well as several "responsiveness values" (Rainey, 2014, p. 105), such as human rights and accountability.

Another perspective to assess performance focuses on the different dimensions of organizational performance that aim to complement the 3Es and IOO models (Boyne, 2002, 2003). In this case, performance is assessed in terms of accountability, effectiveness, efficiency, democratic outcomes (representation, participation, etc.), equity, justice, responsiveness, quantity, and quality of outputs and outcomes. In this study, we opt to assess performance in terms of outputs—enrollments in secondary education—and one outcome—infant mortality rate. Although outputs refer to the direct products generated by an organization, outcomes denote the final

[6] From original data collected by Instituto Geográfico Agustín Codazzi (2018).

[7] See Rainey (2014) for more detail on the alternative approaches to organizational performance.

effect a product has on society (Boyne, 2003). For instance, in reducing infant mortality rate (the final policy goal, i.e., outcome), a governor might promote children vaccination (output) across her state. Although no single indicator is perfect, these two indicators are objective and fully measure a state's performance, as they are implemented at the state/provincial level. Assessing performance in terms of school enrollment and infant mortality rate seems appropriate because Mexico and Colombia face a considerable gap between potential and actual enrollees in secondary education and considerable variations in IMR within the countries.

For Mexico, we obtained high school enrollment data through the Secretaría de Educación Pública (Federal Ministry of Education). This rate measures the percentage of eligible children who enroll in high school in a given state in a particular year. Infant mortality rates are defined as deaths in the first year of life per 1000 live births. We obtained data to measure this variable from Instituto Nacional de Estadística y Geografia (INEGI), the National Center for Statistics and Geographical Data.

The Panel Municipal del CEDE (Acevedo & Bornacelly, 2014) collected a wide range of socioeconomic variables for all municipalities in Colombia. We aggregated the number of high-school enrollees in non-certified municipalities for each department, thus obtaining a total enrollment figure at the department level. We also calculated the total number of children between 11 and 16 years old in such non-certified municipalities, as a measure of the total targeted population for secondary education in each department. The high-school enrollment rate is the ratio between the number of enrollees and the targeted population in each given year. The Panel Municipal del CEDE (Acevedo & Bornacelly, 2014) also reports figures of infant mortality at the municipal level. By aggregating these figures at the department level and determining a ratio of the total number of births in a given year, we obtain infant mortality rates for each department in a given year (Figs. 7.4, 7.5 and 7.6).

In order to test H1, we obtained governors' codified knowledge via background and biographical information of the governors as subnational chief executives. Besides collecting information on their age and place of birth, we originally calculated the number of years of formal education. However, given the large number of governors with pre-graduate education in both countries, we opted to create a dummy variable receiving "1" if the governor has postgraduate education and, otherwise, "0," to test the effect of governors' codified knowledge on state performance. As noted in Fig. 7.7, only 2.63% of Mexican governors have less than a college education, yet 10.31% of Colombian governors do not have college education. In addition, most Mexican governors obtained college degrees (66.67%), but only 29.06% of Colombian governors have college degrees. On the other hand, the majority of Colombian governors (60.62%) have some kind of postgraduate degree (specialty, master or Ph.D.), yet only 30.7% of Mexican governors attained postgraduate degrees.

To test the effect of governors' uncodified knowledge (H2), we also collected information on the years of experience in both the public and private sectors. Moreover, we classified governors' experience at the local, state/department and national levels. As we can see in Fig. 7.8, the distributions of public sector

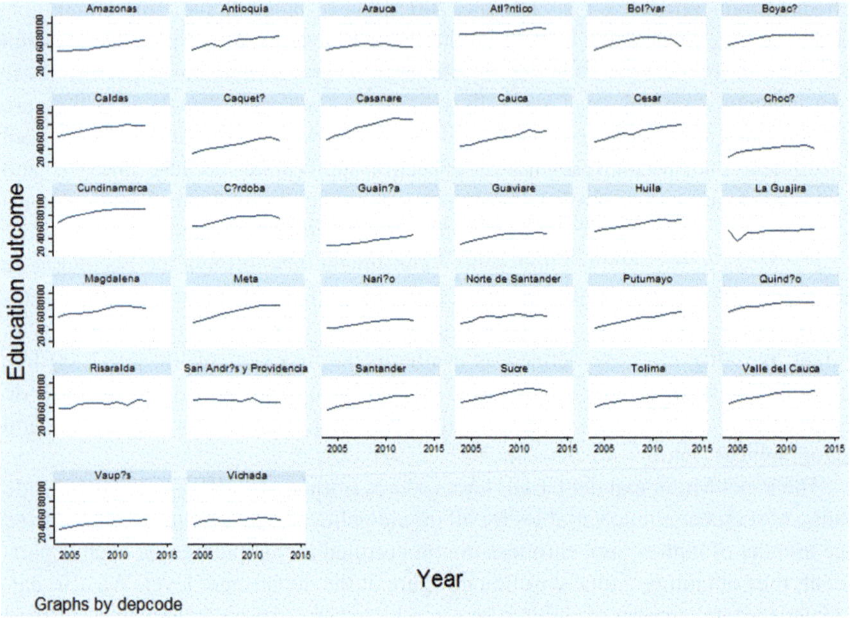

Fig. 7.4 High school enrollment for Colombian departments (2004–2013). Source: Design by authors

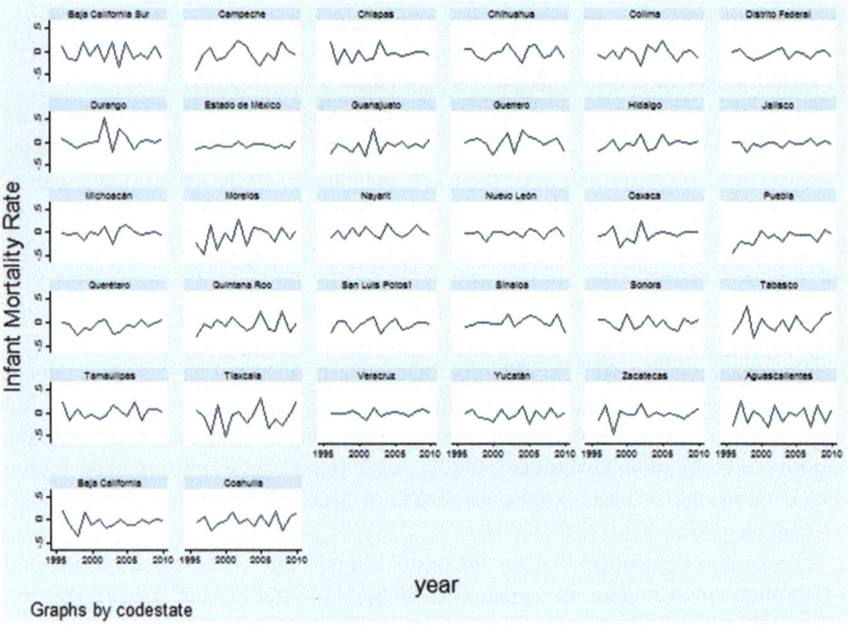

Fig. 7.5 Infant mortality rates for Mexican States (1995–2010). Source: Design by authors

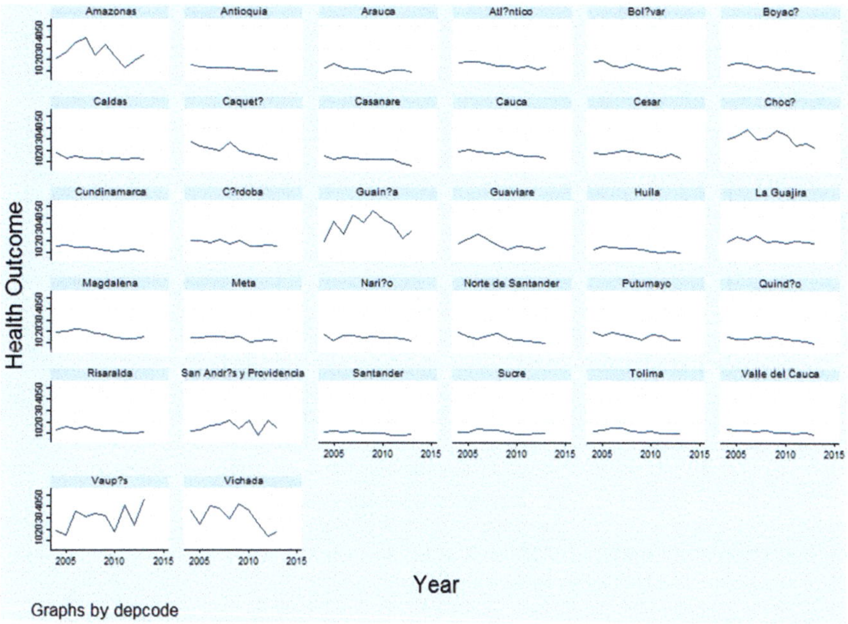

Fig. 7.6 Infant mortality rate for Colombian departments (2004–2013). Source: Design by authors

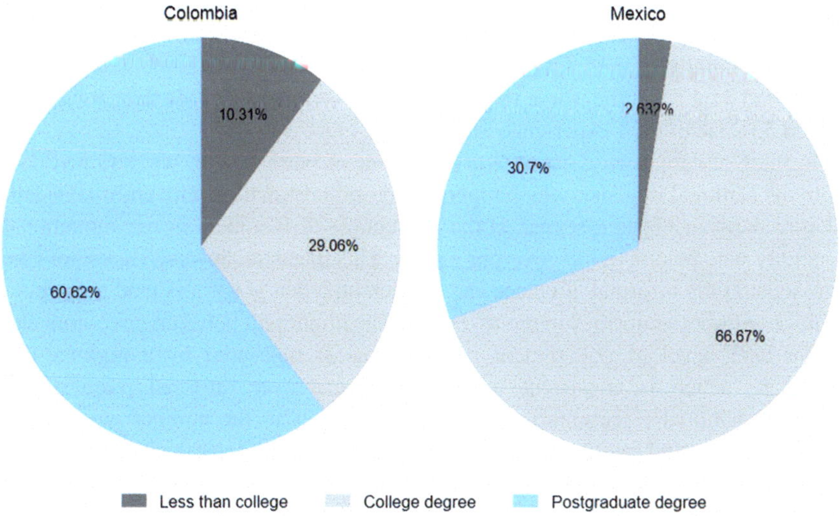

Fig. 7.7 Codified knowledge of Mexican and Colombian Governors, by level of education. Source: Design by authors

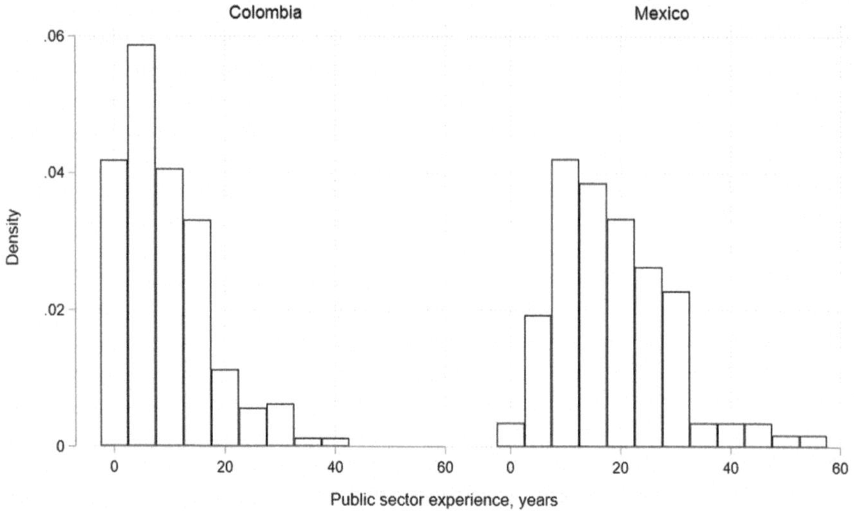

Fig. 7.8 Uncodified knowledge of Mexican and Colombian Governors: by total years of public sector experience. Source: Design by authors

experience in both Mexico and Colombia are negatively skewed. In particular, most Colombian governors have less than 20 years of experience, and most Mexican governors have less than 40 years of public sector experience. Of years spent in the public sector, Colombian governors dedicated 32.78% and Mexican governors dedicated 10.96% to local governments. At the state/department level, Colombian governors dedicated 35.8% of their time, and Mexican governors invested 35.79% of their time. Finally, of the total time spent in the public sector, Colombian governors committed 31.42% of their time to the national government, but Mexican governors gained 53.27% of their experience at that level (Fig. 7.9).

In the Colombian case, to test for the effect of partisanship, we obtained data from the National Registry, which reports electoral information. Governor-state legislature partisanship is assessed as the percentage of members in the subnational assembly who belong to the governor's party. This measure, however, does not take into account the informal alliances and partnerships that governors tend to build to achieve working majorities in the assembly. The alignment between governors and the national level of government in Colombia is measured by a regional-ties approach, which is consistent with this concern over informal partnerships. Governor-national representatives' alignment equates to the number of senators, representatives, and cabinet members who are native to the department. Meanwhile, to measure governor-ministry of health/education alignment, we created a dummy variable receiving "1" if the respective ministry was native to the department; otherwise it is "0." For Mexico, governor-national representatives' partisanship equates to the percentage of both senators and representatives ideologically aligned to the governor's party. Finally, to measure governor-ministry of health/education

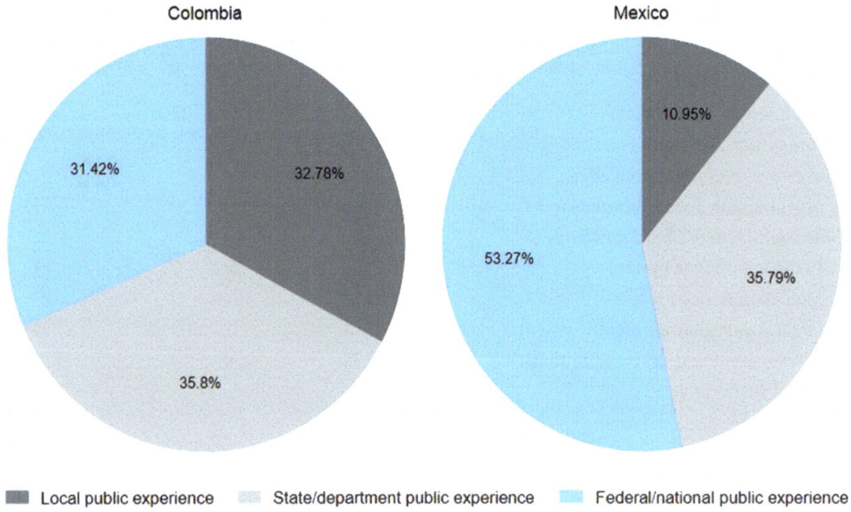

Fig. 7.9 Uncodified knowledge of Mexican and Colombian Governors by level of public experience. Source: Design by authors

partisanship, we created a dummy variable receiving "1" if the governor and respective ministry belonged to the same party; otherwise it is "0."

In Mexico's case, given its three-party system, we created two dummy variables to assess governors' party ideology: PRI and PRD, with PAN indicating the excluded category. The PRI category receives value of "1" if the governor belongs to the PRI; otherwise, it is "0." Likewise, the PRD category receives a value of "1" if the governor belongs to the PRD; otherwise, it is "0." These two categories will be compared to the excluded category, PAN. In Colombia's case, given its multi-party system, it becomes difficult to classify all small parties within the left-right continuum. Historically, Colombia maintained a two-party system. But in the last two decades, dissidents from the traditional parties have created many small and new parties, numbering 59 in the Congressional elections held in March 2006. Therefore, we created a dummy variable for the conservative category and compared it with the base category, in which we group liberal and leftist parties. For both countries, we assessed margin of victory in the gubernatorial as the difference in percentage points between the winner and the runner-up.

For both countries, we also controlled for other economic, fiscal, and sociodemographic factors. For the Colombian case, we specifically controlled for GDP per capita, and state royalties and total state revenues, both of which are reported per capita to make these measures comparable across units. We also controlled for the percentage of rural population per state. Given Colombia's long-lasting armed conflict, we included a proxy of violence as the number of internally displaced people per 1000 inhabitants. These data were obtained by aggregating municipal figures from the Panel Municipal del CEDE (2014).

Table 7.1 Descriptive statistics for Mexican States (1995–2010)

	Mean	Std. Dev.	Min.	Max.
Independent variables				
High school enrollment	52.52	11.70	22.40	99.90
Infant mortality rate	12.70	4.55	3.00	33.18
Governors' codified knowledge				
Postgraduate degree (dummy)	0.26	0.44	0	1
Governors' uncodified knowledge				
Local experience (years)	2.06	2.56	0	11
State experience (years)	6.62	5.68	0	27
Federal experience (years)	9.84	9.03	0	38
Private sector experience (dummy)	0.35	0.48	0	1
Political factors				
Partisanship				
Governor-State Deputies (%)	0.44	0.10	0.05	0.81
Governor-Federal Deputies (%)	0.39	0.15	0.10	0.62
Governor-Minister of Health (dummy)	0.10	0.30	0	1
Governor-Minister of Education (dummy)	0.36	0.48	0	1
Party ideology				
PAN (dummy)	0.23	0.42	0	1
PRD (dummy)	0.13	0.34	0	1
PRI (dummy)	0.64	0.48	0	1
Margin of electoral victory (%)	17.88	17.57	0.53	81.32
Controls				
Substitute Governors	0.05	0.22	0	1
Homicides/capita (absolute number)	2.64	2.29	0.31	24.31
State revenues/capita (in millions)	6.58	3.85	0.54	20.20
Rural population (absolute number)	761,426.60	693,358.30	22.85	2,976,060

Note. Source: Design by authors

For the Mexican case, we control for four factors. First, we created a dummy variable, receiving "1" if the governor was a substitute, rather than an elected official. Given the Mexican experience with violence derived from drug wars, we control for the rate of homicides per capita. We obtained these data from INEGI. Finally, we control for state income per capita and percentage of rural population, data we also obtained from INEGI. In Tables 7.1 and 7.2, we provide descriptive statistics for Mexico and Colombia, respectively.

Results

We used a two-way fixed effects model to elicit the effect of our independent variables on secondary education provision and infant mortality rate (IMR). This method allows us to control for inherent, unobservable differences among the states/

Table 7.2 Descriptive statistics for Colombian departments (2004–2013)

Variable	Mean	Std. Dev.	Min	Max
Dependent variables				
High school enrollment (%)	64.83	14.50	27.66	91.82
Infant mortality rate	16.90	7.40	6.91	47.29
Governors' codified knowledge				
Postgraduate degree (=1)	0.61	0.49	0.00	1.00
Governors' uncodified knowledge				
Local experience (years)	3.24	4.06	0.00	20.00
State experience (years)	3.50	4.83	0.00	25.00
National experience (years)	3.11	5.06	0.00	24.00
Private experience (years)	2.75	6.92	0.00	40.00
Political factors				
Partisanship				
Governor-State Deputies (%)	24.12	17.33	0.00	87.50
Governor-National Representatives (%)	7.49	6.65	2.00	36.00
Minister of Education (=1)	0.03	0.16	0.00	1.00
Minister of Health (=1)	0.03	0.17	0.00	1.00
Ideology				
Conservative Governor (=1)	0.14	0.35	0.00	1.00
Margin of victory (%)	15.41	15.84	0.05	81.09
Controls				
Displaced people (per 1000)	8.76	8.28	0.31	51.18
Revenue/capita (millions of COP)	0.72	0.55	0.19	3.09
GDP/capita (millions of COP)	7.16	4.33	2.24	25.18
Rural population density (per sq. km)	31.76	79.41	0.33	480.39

Note. Source: Design by authors

departments, as well as for general shocks over time. The standard errors are robust and clustered at the state/department level. The Variance Inflation Factor (VIF) suggests that multi-collinearity is not an issue.

In Table 7.3, we report estimations for our two measures of performance in the 32 Mexican states from 1995–2010. Observations number 512 in each model. In explaining high school enrollment (model 1), postgraduate degree (codified knowledge) does not reach statistical significance. From our measures of uncodified knowledge, only private sector experience is statistically significant at the 0.10 level and in the expected direction. States whose governors come to office with previous private sector experience tend to exhibit 1.47 percentage points less in high school enrollment than states whose governors do not have that experience. In model 1, none of our measures of partisan alignment reaches statistical significance. Of our control variables, homicides and state income per capita are statistically significant at the 0.10 and 0.05 level, respectively.

Model 2 of Table 7.3 shows results for our health outcome. According to these results, the governor's codified knowledge does not explain IMR. Governors' previous experiences in local, state, and federal government achieve statistically

Table 7.3 Explaining state performance in Mexican States (1995–2010)

	(1). School Enrollment	(2) IMR
Governors' codified knowledge'		
Postgraduate degree (dummy)	−0.800	−0.016
	(0.532)	(0.040)
Governors' uncodified knowledge'		
Local experience (years)	0.076	−0.011*
	(0.155)	(0.006)
State experience (years)	0.074	−0.010**
	(0.055)	(0.003)
Federal experience (years)	−0.045	0.006**
	(0.039)	(0.003)
Private sector experience (dummy)	−1.470*	0.110**
	(0.735)	(0.037)
Political factors: Partisanship		
Governor-State Deputies (%)	−0.477	0.196
	(2.844)	(0.148)
Governor-Federal Deputies (%)	2.744	−0.260*
	(2.395)	(0.142)
Governor-Minister of Health (dummy)		−0.098*
		(0.050)
Governor-Minister of Education (dummy)	−0.871	
	(0.755)	
Party ideology		
PRD	−0.620	0.134*
	(1.572)	(0.076)
PRI	−0.809	0.149**
	(1.049)	(0.043)
Margin of electoral victory	−0.014	0.002
	(0.023)	(0.002)
Controls		
Substitute Governor	1.095	0.033
	(1.052)	(0.062)
Homicides/capita (lg)	−1.658*	0.095**
	(0.837)	(0.033)
State revenues/capita	−0.967**	0.015
	(0.299)	(0.010)
Rural population (lg)	−0.068	0.015*
	(0.124)	(0.008)
State and year fixed effects	Yes	Yes
Constant	45.330***	1.127***
	(3.239)	(0.186)
N	512	512

Note. Source: Design by authors
*p< .10; **p< .05; ***p<.001

significant effects on IMR at the 0.10 and 0.05 levels, respectively. Holding all else constant, experience at the local and state government decreases IMR. One additional year of experience at the local level decreases IMR by 0.011, and one additional year of experience at the state level decreases IMR by 0.010. In contrast, one additional year of experience at the federal level increases IMR by 0.006. Likewise, private sector experience is associated with worse health outcomes. Specifically, states whose governors took office with private sector experience have an IMR higher by 0.11 IMR on average than states whose governors arrived without private sector experience. Governors' partisanship with federal deputies exhibits statistical significance at the 0.10 level. A 1% increase of federal deputies aligned with governors' party decreases IMR in 0.26. Similarly, partisan alignment with the minister of health at the federal level decreases IMR in 0.098 and this coefficient is statistically significant at the 0.10 level. Another political factor that affects our health outcome is political ideology. Compared to PAN (conservative) governors, states with governors representing PRD (left) and PRI (center) ideologies have on average higher levels of IMR. Of our controls, homicides per capita and rural populations are both associated with higher levels of IMR, as their coefficients are positive and statistically significant at the 0.05 and 0.10 levels, respectively.

In addition, Table 7.4 reports estimations for our two measures of performance in the 32 Colombian departments during 2004–2013. Observations number 273 in each model. In explaining high school enrollment (model 1), neither governors' codified nor uncodified knowledge (postgraduate) reach statistical significance. Moreover, none of the political factors serves to explain variation in high school enrollment. Two control factors, GDP/capita and rural population, do reach statistical significance with the expected direction. Specifically, states with higher GDP/capita and lower rural population exhibit higher enrollment in high school, holding everything else constant.

In explaining the IMR health outcome for Colombian departments, according to Table 7.4, model 2, governors' codified knowledge fails to reach statistical significance. From our measures of uncodified knowledge, only experience at the state level is statistically significant at the 0.10 level and in the expected direction. That is, states whose governors come to office with previous state-level experience tend to exhibit 0.07 percentage points more in high school enrollment than states whose governors come to office without state experience. None of the political factors explains variation in high school enrollment in Colombian departments. Finally, none of our control measures reaches statistical significance. Of our control variables, homicides and state income per capita are statistically significant at the 0.10 and 0.05 level, respectively.

Table 7.4 Explaining subnational performance in Colombian Departments (2004–2013)

	(1) High School Enrollment	(2) IMR
Governors' codified knowledge		
Postgraduate degree (=1)	0.745	0.0466
	(0.757)	(0.486)
Governors' uncodified knowledge		
Local experience (years)	−0.00973	−0.0218
	(0.0962)	(0.0799)
State experience (years)	0.121	0.0757*
	(0.0768)	(0.0411)
National experience (years)	−0.0222	0.0109
	(0.0640)	(0.0318)
Private experience (years)	0.0121	0.0553
	(0.0394)	(0.0444)
Political factors – Partisanship		
Governor-State Deputies (%)	−0.0247	−0.0267
	(0.0291)	(0.0246)
Governor-Nat. Representativas	0.0444	−0.00637
	(0.290)	(0.118)
Minister of Education (=1)	−1.557	
	(2.466)	
Minister of Health (=1)		−0.601
		(1.236)
Political factors – Ideology		
Conservative Governor (=1)	1.469	0.203
	(1.470)	(0.654)
Electoral competitiveness		
Margin of victory (%)	0.000896	0.0312
	(0.0281)	(0.0261)
Controls		
Displaced people (lg)	0.397	−0.198
	(0.388)	(0.826)
Revenue/capita (millions COP)	1.404	−2.447
	(1.269)	(2.607)
GDP/capita (millions COP)	0.570**	0.140
	(0.248)	(0.198)
Rural population (lg)	−24.03*	−2.333
	(12.03)	(12.03)
State and year fixed effects	Yes	Yes
Constant	101.2***	24.33
	(27.24)	(25.71)
N	273	273

Note. Source: Design by authors
*p<.1; **p<.05; ***p<.001

Discussions and Conclusions

This chapter seeks to explain governance performance at the subnational level in terms of educational output (high school enrollment) and a health outcome (IMR). In doing so, we suggest states CEOs' codified and uncodified knowledge contribute to state/province performance. We test the explanatory power of governors' knowledge against political factors, such as partisanship, electoral competitiveness, and government ideology, while controlling for other state-level factors. The suggested explanations tested are two data sets derived from the 32 Mexican states and the 32 Colombian departments.

Results are inconsistent across both countries and the two indicators of state/department performance. For instance, although few factors seem to explain high school enrollment across both countries, some differences are still worth mentioning. In the Mexican case, governors' prior experience in the private sector is negatively correlated to high school enrollment. On the other hand, in the Colombian case, neither governors' education nor their experience type is statistically correlated to high school enrollment. The lack of statistical significance suggests other factors, not taken into account in this study, may explain state/department variation in high school enrollment.

In explaining the IMR health outcome, results show large inconsistencies across both countries. Specifically, although IMR is positively correlated with governors' federal and private-sector experience in the Mexican case, none of these drivers is statistically correlated with IMR in the Colombian case. Likewise, although governors' local and state experience seems to be negatively correlated with high school enrollment in the Mexican case, governors' state experience is positively correlated with IMR in the Colombian case. This inconsistency in results calls for caution when interpreting results. Although the operationalization of variables is consistent across both countries, intra-country variation in terms of aggregation of data and other standard mechanisms may still be an issue.

The results seem to provide more support for the role of political factors in explaining state/department performance, but only for the Mexican case. Hence, scholars have emphasized the power of partisanship, especially in settings where the distribution of resources is contingent on political ties. Our results suggest that political factors play no role in explaining high school enrollment or IMR in the Colombian case. However, in the Mexican case, as expected, governors' partisanship with federal deputies and with the minister of health tends to reduce IMR. Moreover, party government also serves to explain state performance, but only in the Mexican case. States whose governors are ideologically affiliated with the PRD and PRI tend to report higher IMR, compared to governors affiliated to the PAN, considered a right-wing party. In the Colombian case, conservative governors do not perform statistically differently from governors affiliated to other parties in terms of education or health. Given the lack of party discipline and the multi-party system, party ideology in Colombia does not seem to be a good predictor of governance performance.

Our study has limitations. First, more indicators of performance are needed to fully assess the role of governors' knowledge on performance. Implementation of different policies/programs require different skills, knowledge, and experience. Therefore, future studies should assess state performance in other policy areas. Secondly, our study disregards the role of organization capacity, in terms of human resources, on state/department performance. Data unavailability impedes us from testing the explanatory power of organizational capacity. Finally, our study is limited to two countries. Although our study is one of the first comparative studies in governance performance, future research should replicate this study across other countries in the region.

The two-country study presented here conveys three broad lessons. First, what drives performance in one country may not hold the same explanatory power in another country. Hence, what serves to explain performance in Mexican states differs from the reasons boosting performance in Colombian departments. This finding highlights the importance of refraining from extrapolating conclusions to different contexts. Second, what boosts performance in one policy area may not do so in another policy area. Although governors' uncodified knowledge (experience) does explain health performance in Mexican states, leaders' uncodified knowledge fails to improve educational outputs. Third, although political factors (e.g., partisanship and party ideology) help explain performance in Mexican states, demographic and socioeconomic factors (GDP and rural population) do so in the Colombian departments.

In sum, with this study we provide one of the few comparative tests of the role of chief executives' knowledge on governance in an understudied region. We are cautious about generalizing results based on our findings. What works in a country in a particular policy area may not work in another country in the same policy area. Indeed, additional research should explore the contingences to governors' traits-performance relationship. Although governors are the decision-makers at the subnational level, implementation of their decisions is outside of their control, for it takes place outside of their realm. In such case, characteristics of administrative personnel and street-level bureaucrats may moderate the performance-governors' codified and uncodified knowledge relationship. In addition, given the considerable differences between Colombian and Mexican governors' education attainments and experiences, future studies should explore whether party system, electoral rules, and/or campaign rules contribute to explain the type of politicians' traits in a particular country. The above research is needed to gain a better understanding of the workings of subnational governments, for they are important actors in service delivery in regions where performance improvement is desperately needed.

References

Acevedo, K., & Bornacelly, I. (2014). *Panel municipal del CEDE* (No. 012223) [Municipal panel of CEDE]. Bogotá: Universidad de Los Andes, Facultad de Economía, CEDE. Retrieved from https://economia.uniandes.edu.co/components/com_booklibrary/ebooks/dcede2014-26.pdf

Agranoff, R., & McGuire, M. (1998). Multinetwork management: Collaboration and the hollow state in local economic policy. *Journal of Public Administration Research and Theory, 8,* 67–91. https://doi.org/10.1093/oxfordjournals.jpart.a024374

Agranoff, R., & McGuire, M. (2003). *Collaborative public management: New strategies for local governments*. Washington, D.C.: Georgetown University Press.

Allan, J. P., & Scruggs, L. (2004). Political partisanship and welfare state reform in advanced industrial societies. *American Journal of Political Science, 48,* 496–512. https://doi.org/10.1111/j.0092-5853.2004.00083.x

Alvarado, A. (1996). Los gobernadores y el federalismo mexicano [The governors and Mexican federalism]. *Revista Mexicana de Sociología, 58*(3), 39–71. https://doi.org/10.2307/3540874

Ames, B. (1987). *Political survival: Politicians and public policy in Latin America*. Berkeley, CA: University of California Press.

Arrow, K. J. (1962). The economic implications of learning by doing. *The Review of Economic Studies, 29* (3), 155–173. https://doi.org/10.2307/2295952

Avellaneda, C. N. (2009). Municipal performance: Does mayoral quality matter? *Journal of Public Administration Research and Theory, 19,* 285–312. https://doi.org/10.1093/jopart/mun001

Avellaneda, C. N. (2012). Do politics or mayors' demographics matter for municipal revenue expansion? *Public Management Review, 14,* 1061–1086. https://doi.org/10.1080/14719037.2012.662442

Avellaneda, C. N. (2016). Government performance and chief executives' intangible assets: Motives, networking, and/or capacity? *Public Management Review, 18,* 918–947. https://doi.org/10.1080/14719037.2015.1051574

Barón, J. D., & Meisel, A. (2003). La descentralización y las disparidades económicas regionales en Colombia en la década de 1990 [The decentralization and regional economic inequalities en Colombia in the 90s]. *Coyuntura Económica, 9,* 105–141. Retrieved from https://www.repository.fedesarrollo.org.co/handle/11445/935

Bartels, L. M. (2000). Partisanship and voting behavior, 1952–1996. *American Journal of Political Science, 44,* 35–50. https://doi.org/10.2307/2669291

Bassanini, A., & Scarpetta, S. (2001). Does human capital matter for growth in OECD countries? Evidence from pooled mean-group estimates (OECD Economics Department Working Papers No. 282). Paris: OECD. https://doi.org/10.1787/424300244276

Bello, R., & Espitia, J. (2011). Distribución regional de las transferencias intergubernamentales en Colombia 1994–2009 [Regional distribución of intergubernamental transferencias in Colombia, 1994–2009]. *Documentos y aportes en administración pública y gestión estatal, 11,* 7–50. https://doi.org/10.14409/da.v1i16.1263

Besley, T., Montalvo, J. G., & Reynal-Querol, M. (2012). Do educated leaders matter? *The Economic Journal, 121,* 205–227. https://doi.org/10.1111/j.1468-0297.2011.02448.x

Blair, H. (2000). Participation and accountability at the periphery: Democratic local governance in six countries. *World Development, 28,* 21–39. https://doi.org/10.1016/S0305-750X(99)00109-6

Blais, A., Blake, D., & Dion, S. (1993). Do parties make a difference? Parties and the size of government in liberal democracies. *American Journal of Political Science, 37,* 40–62. https://doi.org/10.2307/2111523

Boyne, G. A. (2002). Theme: Local government: Concepts and indicators of local authority performance: An evaluation of the statutory frameworks in England and Wales. *Public Money and Management, 22*(2), 17–24. https://doi.org/10.1111/1467-9302.00303

Boyne, G. A. (2003). Sources of public service improvement: A critical review and research agenda. *Journal of Public Administration Research and Theory, 13,* 367–394. https://doi.org/10.1093/jopart/mug027

Boyne, G. A., Meier, K. J., O'Toole, L. J. Jr., & Walker, R. M. (2005). Where next? Research directions on performance in public organizations. *Journal of Public Administration Research and Theory, 15,* 633–639. https://doi.org/10.1093/jopart/mui037

Brollo, F., & Nannicini, T. (2010). Tying your enemy's hands in close races: The politics of federal transfers in Brazil (IGIER Working Papers No. 358). Milan, Italy: Bocconi University.

Buchanan, J. M., & Tullock, G. (1962). *The calculus of consent: Logical foundations of constitutional democracy.* Ann Arbor, MI: University of Michigan Press.

CIPFA (Chartered Institute of Public Finance and Accountancy), & IFAC (International Federation of Accountants). (2013). *International framework: Good governance in the public sector.* Retrieved from https://www.cipfa.org/policy-and-guidance/standards/international-framework-good-governance-in-the-public-sector

Collins, B. K., & Gerber, B. J. (2006). Redistributive policy and devolution: Is state administration a road block (grant) to equitable access to federal funds? *Journal of Public Administration Research and Theory, 16,* 613–632. https://doi.org/10.1093/jopart/muj010

Constitución Política de Colombia [Political Constitution of Colombia]. (1991). Retrieved from http://pdba.georgetown.edu/Constitutions/Colombia/colombia91.pdf

Cox, G. W., & McCubbins, M. D. (1986). Electoral politics as a redistributive game. *The Journal of Politics, 48,* 370–389. https://doi.org/10.2307/2131098

Dahlberg, M., & Johansson, E. (2002). On the vote-purchasing behavior of incumbent governments. *American Political Science Review, 96,* 27–47. https://doi.org/10.1017/S0003055402004215

Dearborn, D. C., & Simon, H. A. (1958). Selective perceptions: A note on the departmental identifications of executives. *Sociometry, 21,* 140–144. https://doi.org/10.2307/2785898

Departamento Nacional de Planeación. (2013). Ejecuciones Presupuestales Departamentales 2000–2012 (Versión 4 de septiembre de 2013). Retrieved from http://www.dnp.gov.co

Diario Oficial de la Federación (2013). Ley General de Educacion [General Education Law]. Mexico.

Dixit, A., & Londregan, J. (1998). Ideology, tactics, and efficiency in redistributive politics. *The Quarterly Journal of Economics, 113,* 497–529. https://doi.org/10.1162/003355398555667

Doig, J. W., & Hargrove, E. C. (1990). *Leadership and innovation: Entrepreneurs in government.* Baltimore, MD: Johns Hopkins University Press.

Downs, A. (1957). *An economic theory of democracy*. New York: Harper.

Durant, R. F., & Legge, J. S. (1993). Policy design, social regulation and theory building: Lessons from the traffic safety policy arena. *Political Research Quarterly, 46,* 641–656. https://doi.org/10.1177/106591299304600310

Dye, T. R. (1966). *Politics, economics and the public.* Chicago: Rand McNally.

Falleti, T. G. (2005). A sequential theory of decentralization: Latin American cases in comparative perspective. *American Political Science Review, 99,* 327–346. https://doi.org/10.1017/S0003055405051695

Falleti, T. G. (2010). *Decentralization and subnational politics in Latin America*. New York, NY: Cambridge University Press.

Fernandez, S. (2005). Developing and testing an integrative framework of public sector leadership: Evidence from the public education arena. *Journal of Public Administration Research and Theory, 15*(2), 197–217. https://doi.org/10.1093/jopart/mui014

Fiedler, F. E. (1966). The effect of leadership and cultural heterogeneity on group performance: A test of the contingency model. *Journal of Experimental Social Psychology, 2,* 237–264. https://doi.org/10.1016/0022-1031(66)90082-5

Fiedler, F. E. (1986). The contribution of cognitive resources and leader behavior to organizational performance. *Journal of Applied Social Psychology, 16,* 532–548.

Fiedler, F. E., & Chemers, M. M. (1968). *Group performance under experienced and inexperienced leaders: A validation experiment* (Technical report No. 69). Retrieved from https://pdfs.semanticscholar.org/b1e5/866121d4ada612e4de1dc1ecc21b5359c6a6.pdf

Figueras Zanabria, V. M. (2009). Apertura política y descentralización fiscal en México: el papel de los gobernadores [Political openness and fiscal decentralization in Mexico: The role of state governors]. *Nuevo Mundo Mundos Nuevos*. https://doi.org/10.4000/nuevomundo.55760

Gonzalez-Block, M., Leyva, R., Zapata, O., Loewe, R., & Alagon, J. (1989). Health services decentralization in Mexico: Formulation, implementation and results of policy. *Health Policy and Planning, 4,* 301–315. https://doi.org/10.1093/heapol/4.4.301
Hall, J. L. (2008). Assessing local capacity for federal grant-getting. *The American Review of Public Administration, 38,* 463–479. https://doi.org/10.1177/0275074007311385
Hernández Rodríguez, R. (2008). *El centro dividido: La nueva autonomía de los gobernadores* [The divided center: the new autonomy of governors]. Mexico City, Mexico: Colegio de México.
Herrera Gutierrez, A. (2015). What are we talking about when we talk about "subnational" governments? [Blog post]. Retrieved from https://blogs.worldbank.org/latinamerica/what-are-we-talking-about-when-we-talk-about-subnational-governments
Holbrook, T. M., & van Dunk, E. (1993). Electoral competition in the American states. *American Political Science Review, 87,* 955–962. https://doi.org/10.2307/2938827
Instituto Geográfico Agustín Codazzi. (2018). *Sistema de información geográfica para la planeación y el ordenamiento territorial* [System of geographic information for planning and territorial zoning]. Retrieved from https://sigot.igac.gov.co/
Johansson, E. (2003). Intergovernmental grants as a tactical instrument: Empirical evidence from Swedish municipalities. *Journal of Public Economics, 87,* 883–915. https://doi.org/10.1016/S0047-2727(01)00148-7
Jones, B. F., & Olken, B. A. (2005). Do leaders matter? National leadership and growth since World War II. *Quarterly Journal of Economics, 120,* 835–864. https://doi.org/10.1093/qje/120.3.835
Key, V. O. (1949). *Southern politics in state and nation*. New York: Alfred A. Knopf.
Langston, J. (2010). Governors and "their" deputies: New legislative principals in Mexico. *Legislative Studies Quarterly, 35,* 235–258. https://doi.org/10.3162/036298010791170132
Ley 60 [Law No. 60]. (1993, August 12). Colombia. Retrieved from https://www.mineducacion.gov.co/1759/articles-86098_archivo_pdf.pdf
Ley 715 [Law No. 715]. (2001, December 21). Colombia. Retrieved from https://www.mineducacion.gov.co/1621/articles-86098_archivo_pdf.pdf
Ley de Coordinación Fiscal [Fiscal Coordination Law]. (1978, December 27). Mexico. Retrieved from http://www.diputados.gob.mx/LeyesBiblio/ref/lcf/LCF_orig_27dic78_ima.pdf
Lindbeck, A., & Weibull, J. W. (1987). Balanced-budget redistribution as the outcome of political competition. *Public Choice, 52,* 273–297. https://doi.org/10.1007/BF00116710
Lucas, R. E. (1988). On the mechanics of economic development. *Journal of Monetary Economics, 22,* 3–42. https://doi.org/10.1016/0304-3932(88)90168-7
Lynn, L. E. (1987). *Managing public policy*. Boston: Little, Brown and Co.
Lynn, L. E. (1996). *Public management as art, science, and profession*. London: Chatham House.
Lynn, L. E. (2003). Public management. In B. Guy Peters & J. Pierre (Eds.), *Handbook of public administration* (pp. 14–24). Thousand Oaks: SAGE.
Mankiw, N., Romer, D., & Weil, D. N. (1992). A contribution to the empirics of economic growth. *The Quarterly Journal of Economics, 107,* 407–437. https://doi.org/10.2307/2118477
Martinez Fritscher, A., & Rodriguez Zamora, C. (2016). An evaluation of the 1997 expenditure decentralization reform in Mexico: The case of the health sector. *Public Finance Review, 44,* 563–588. https://doi.org/10.1177/1091142115571578
Meier K. J., & O'Toole, L. J. Jr. (2001). Managerial strategies and behaviour in networks: A model with evidence from U.S. public education. *Journal of Public Administration Research and Theory, 11,* 271–294. https://doi.org/10.1093/oxfordjournals.jpart.a003503
Meier, K. J., & O'Toole, L. J. Jr. (2002). Public management and organizational performance: The effect of managerial quality. *Journal of Policy Analysis and Management, 21,* 629–643. https://doi.org/10.1002/pam.10078
Meier, K. J., Stewart, J., & England, R. E. (1991). The politics of bureaucratic discretion: Educational access as an urban service. *American Journal of Political Science, 35,* 155–177. https://doi.org/10.2307/2111442

Mejía Guinand, L. B., Botero, F., & Rodriguez Raga, J. C. (2008). Pavimentando con votos? Apropiación presupuestal para proyectos de infraestructura vial en Colombia, 2002–2006 [Paving roads with votes? Assigning budget for infraestructural project in Colombia, 2002–2006]. *Colombia Internacional, 68,* 14–42.

Ministerio de Educación Nacional. (2004). Guía No. 9 (Guía Certificación de Municipios Menores de Cien Mil Habitantes) [Certifitication of municipalites with less than one hundred thousand population]. Retrieved from https://www.mineducacion.gov.co/1759/articles-81012_archivo_pdf.pdf

Ministerio de Educación Nacional. (2014). Sistema Nacional de Indicadores Educativos para los Niveles de Preescolar, Básica y Media en Colombia [National Educational Indicators for Basic, Media and Pre-school Education]. Retrieved from https://www.mineducacion.gov.co/1621/articles-329021_archivo_pdf_indicadores_educativos_enero_2014.pdf

Modoux, M. (2006). Geografía de la gobernanza: ¿la alternancia partidaria como factor de consolidación del poder de los gobernadores en el escenario nacional mexicano? [Geography of governance: the partisan shift as a power-consolidation factor of governors in the Mexican national cotext]. *Foro Internacional, 46,* 513–532. Retrieved from https://forointernacional.colmex.mx/index.php/fi/article/view/1833

Montero Bagatella, J.C. (2014). Equilibrios de poder en las entidades federativas. Estudio longitudinal de las legislaturas locales en México en el periodo 2000–2014 [Political balances in Mexican states: Longitudinal research on local congresses in the period 2000–2014]. *Revista Mexicana de Análisis Político y Administración Pública, 3,* 125–154. Retrieved from http://www.remap.ugto.mx/index.php/remap/article/view/91

Morales y Gómez, J. M., & Salazar Medina, J. (2009). La administración pública en los estados y reflexiones sobre el federalismo [Public administration at the states and reflections on federalism]. *Convergencia, Revista de Ciencias Sociales, 49,* 229–250. Retrieved from http://www.scielo.org.mx/scielo.php?script=sci_arttext&pid=S1405-14352009000100009

Nordhaus, W. D. (1975). The political business cycle. *The Review of Economic Studies, 42*(2), 169–190. https://doi.org/10.2307/2296528

Oikion Solano, V. (2012). Gobernadores: Elecciones y poder local en el estado de Hidalgo, México, 1869–1975 [Governors, elections, and local power in the State of Hidalgo]. *Tzintzun, 56,* 233–240. Retrieved from http://www.scielo.org.mx/scielo.php?script=sci_arttext&pid=S0188-28722012000200009

Olvera, J. G., & Avellaneda, C. N. (2019). Subnational government performance: Testing multiple dimensions of governors' experience. *Public Administration Review, 79,* 383–398. https://doi.org/10.1111/puar.13035

O'Toole, L. J. Jr., & Meier, K. J. (1999). Modeling the impact of public management: Implications of structural context. *Journal of Public Administration Research and Theory, 9,* 505–526. https://doi.org/10.1093/oxfordjournals.jpart.a024421

O'Toole, L. J. Jr., & Meier, K. J. (2015). Public management, context, and performance: In quest of a more general theory. *Journal of Public Administration Research and Theory, 25,* 237–256. https://doi.org/10.1093/jopart/muu011

Page, S. (2008). Managing for results across agencies: Building collaborative capacities in human services. In L. Blomgren Bingham & R. O'Leary (Eds.), *Big ideas in collaborative public management* (pp. 138–161). New York: M. E. Sharpe.

Penrose, E. (1959). *The theory of the growth of the firm.* New York: John Wiley and Sons.

Rainey, H. G. (2014). *Understanding and managing public organizations.* New York: John Wiley & Sons.

Ramírez-Cedillo, E., & Lopez Herrera, F. (2016). La relación de los ingresos y gastos en el nivel subnacional en México: 1989–2014 [The relationship between income and revenue at the subnational level in Mexico: 1989–2014]. *Paradigma Económico, 8*(2), 111–138. Retrieved from https://paradigmaeconomico.uaemex.mx/article/view/4805

Reglamento de Tránsito de la Ciudad de México. (2016). [Transit Regulation for Mexico City]. Retrieved from http://data.ssp.cdmx.gob.mx/reglamentodetransito/

Riccucci, N. (1995). *Unsung heroes: Federal execucrats making a difference*. Washington, D.C.: Georgetown University Press.

Rodríguez, R. H. (2003). Cambio político y renovación institucional: Las gubernaturas en México [Political change and institutional renovation: The governorships in Mexico]. *Foro Internacional, 43*(4), 789–800. Retrieved from https://forointernacional.colmex.mx/index.php/fi/article/view/1699

Rowland, A., & Ramírez, E. (2001). La descentralización y los gobiernos subnacionales en México: una introducción [The decentralization and the subnational governments in Mexico: An introduction] (CIDE Número 93). Retrieved from http://aleph.academica.mx/jspui/bitstream/56789/3636/1/DOCT2064572.PDF

Rumelt, R. (1984). Towards a strategic theory of the firm. In R. Lamb (Ed.), *Competitive strategic management* (pp. 556–570). Englewood Cliffs: Prentice-Hall.

Sen, A. (1997). Inequality, unemployment and contemporary Europe. *International Labour Review, 136*(2), 155–171. Retrieved from http://ilo.org/public/english/revue/download/pdf/sen.pdf

Sharkansky, I. (1967). Government expenditures and public services in the American States. *American Political Science Review, 61*, 1066–1077. https://doi.org/10.2307/1953407

Solano, P. L. (1983). Institutional explanations of public expenditures among high-income democracies. *Public Finance, 38*, 440–458.

Solé-Ollé, A., & Sorribas-Navarro, P. (2008). The effects of partisan alignment on the allocation of intergovernmental transfers: Differences-in-differences estimates for Spain. *Journal of Public Economics, 92*, 2302–2319. https://doi.org/10.1016/j.jpubeco.2007.06.014

Solow, R. M. (1956). A contribution to the theory of economic growth. *The Quarterly Journal of Economics, 70*, 65–94. https://doi.org/10.2307/1884513

Strøm, K. (1990). *Minority government and majority rule*. Cambridge, UK: Cambridge University Press.

Swank, D. (2002). *Global capital, political institutions, and policy change in developed welfare states*. Cambridge, UK: Cambridge University Press.

Terman, J. N., & Feiock, R. C. (2015). Improving outcomes in fiscal federalism: Local political leadership and administrative capacity. *Journal of Public Administration Research and Theory, 25*, 1059–1080. https://doi.org/10.1093/jopart/muu027

Walker, R. M., Boyne, G. A., & Brewer, G. A. (2010). Introduction. In R. M. Walker, G. A. Boyne, & G. A. Brewer (Eds.), *Public management and performance: Research directions* (pp. 1–33). New York, NY: Cambridge University Press.

Wernerfelt, B. (1984). A resource-based view of the firm. *Strategic Management Journal, 5*(2), 171–180. https://doi.org/10.1002/smj.4250050207

Wittman, D. (1990). Spatial strategies when candidates have policy preferences. In J. M. Enelow & M. J. Hinich (Eds.), *Advances in the spatial theory of voting* (pp. 66–98). New York, NY: Cambridge University Press.

Chapter 8
The (De-)Contextualization of Geographical Knowledge in Forest Fire Risk Management in Chile as a Challenge for Governance

Michael Handke

With their annual reports, large and integrated forestry companies and insurance companies in Chile signal specific knowledge and a sophisticated language on the value of risk to the financial markets (Kalthoff, 2005). In the course of this, they have developed their own way of risk communication, which decouples risk from its geographical context. This is remarkable in that forestry activities are initially deeply rooted into physical space and socially embedded in regional communities.

In January 2017, Chile experienced the worst forestry calamity in recent history. More than 120 simultaneous wildfires in the O'Higgins, Maule, and Biobio regions affected nearly 467,000 hectares of native forest and tree plantations (CONAF, 2017). While the real ecological and economic damage caused by the fires was significant and had far-reaching social consequences for many people, the associated financial losses on the timber markets proved to be manageable. Economic players spread it among themselves on several shoulders. Empresas CMPC S.A., for example, a multinational holding company of Chilean origin, a paper manufacturer, and also the country's second largest forestry company, announced that the plantation property affected by the fires reached approximately 19,000 hectares, equivalent to US$73 million of economic damage. However, CMPC also assured that the timber supply for plant operations remained unaffected and that the financial stability of the company was not at risk (CMPC, 2018). In retrospect, insurance companies with whom CMPC had signed insurance policies compensated up to US$17 million of the damage. The Chilean insurance sector reported similar outcomes. HDI Seguros S. A., for example, the fourth-largest insurance company in Chile and an innovator in forest fire insurance policies, recorded losses due to the catastrophe amounting to US$20.2 million, of which 17.2 million (or 85%) were still covered by reinsurance contracts (HDI, 2018). Statements like these make clear that from an economic point of view it is possible to decouple the physical dimension of the risk

M. Handke (✉)
Department of Geography, Heidelberg University, Heidelberg, Germany
e-mail: handke@uni-heidelberg.de

J. Glückler et al. (eds.), *Knowledge for Governance*, Knowledge and Space 15,
https://doi.org/10.1007/978-3-030-47150-7_8

(i.e., the occurrence of forest fire events in space) from its economic (i.e., the losses of forested areas) and financial dimension (i.e., profit warnings).

The risk management of the Chilean forest companies combines hierarchical orders with abstract market mechanisms: On the one hand, forestry companies reduce the probability of wildfire occurrence and the magnitude of losses with modern standards of tree-plantation management that include, among other things, the maintenance of watchtowers and the employment of fire brigades. With their sophisticated forest logistics, they are able to extract timber resources even from burnt trees. Hierarchical orders are executed in space trough corporate routines (see also Perrow, 1986, 1972; or Becker & Knudsen, 2005). On the other hand, they apply a market solution of risk management as they purchase insurance policies and other services from the financial markets that allow them to individual hedge the risk of economic losses (Chichilnisky & Heal, 1998). Insurance transforms risk into a commercial good. The insured pays an insurance premium to compensate the insurer for assuming the risk. In other words: The insured gives up part of his profits and turns incalculable uncertainties into calculable stability (Dean, Doyle, & Ericson, 2003; Freeman & Kunreuther, 1997). Insurance markets as an instrument of risk management to reduce the economic vulnerability of social actors have gained importance in Chile in recent years (Cifuentes, Desormeaux, & González, 2002; Loewe, Corti, Ruiz, & Lobo, 2017).[1]

The strategies and practices of dealing with forest fire risks in Chile are a proper object of conceptual and empirical research (Úbeda & Sarricolea, 2016)—especially from an economic geographical perspective. When aggregated figures that primarily contain economic indicators express the magnitudes of risks decoupled from their geographical context, it is no longer easy to understand how wildfires occurred in first place and what additional socio-economic consequences they might have for the people living in forestry regions. Quantification is useful for the economic control of risks, but it restricts knowledge about risk (Beck, 1992; Luhmann, 1991/1993; Viscusi & Magat, 1987). A society that relies mainly on economic risk management practices consequently loses its ability to respond appropriately to changing causes and consequences of risk (Rosa, Renn, & McCright, 2014).

The devastating forest fires in Chile in early 2017 have certainly increased the society's overall sensitivity to this kind of risk scenario. They uncovered the vulnerability of various stakeholders in society and their powerlessness in making real decisions about risks. While the risk management of forestry and insurance companies mainly takes place behind closed doors, whereby annual balance sheets signal that forest fires are economically controllable, other affected stakeholders are struggling to deal with the consequences. This not only raises the question of how a society should deal with risk in appropriate ways; simultaneously, there is a growing interest in opening the black box of the companies' internal risk management systems to screen them for unintended negative external effects (Bottaro, Roco, Pettenella, Micheletti, & Vanhulst, 2018; van Dam, 2006).

[1] As Chile is frequently hit by natural disasters, the country's insurance industry is required by law to make use of reinsurance. This regulation aims at both stimulating the national insurance market and making it more predictable and financially stable.

The extent to which markets and hierarchical types of risk management (and the corresponding risk knowledge) are suitable for improving forest fire risk prevention in the Chilean society as a whole or, generally speaking, for increasing society's resilience to risk, which is the concern of the broader concept of risk governance, remains an open question that motivates the following analysis. How can actors utilize hierarchical and market approaches to risk management jointly in the process of risk governance?

Risk management and risk governance are not the same. The differences lie not only in the perspective of the involved actors (individual versus collective risk handling) or in the time horizon underlying the risk practices (short-term versus long-term orientation), but above all in the production and usage of sophisticated risk knowledge: Risks arise in knowledge, and therefore in knowledge they can be reduced, enlarged or simply eliminated from consciousness (Beck, 1992). A perspective of risk knowledge, therefore, not only promises to clarify the differences between management and governance but also the interrelations between different risk management practices.

By revealing the strengths and weaknesses in the interaction between hierarchical and market forms of forest fire risk management in Chile and focusing on the epistemological challenges related to the geographical (de-)contextualization of risks, in this paper I contribute to a better understanding of the societal benefits and challenges of explicitly regional risk governance approaches. Those approaches, however, have yet to be put in practice in Chile.

I proceed in four parts. Following this introduction, I use the first section to explain and justify the relevance of the research problem concerning the challenges of different knowledge perspectives in dealing with risks. I make an explicit distinction between risk management and a risk governance perspective. In the second section, I analyze the different characteristics of forest fires in Chile. Applying the risk governance approach of Rosa et al. (2014, see also Chap. 5 by Renn), I argue that the relationships between wildfires' causes and effects are epistemologically complex, ambiguous, and uncertain. I make it clear that a deliberate spatiotemporal view is needed to understand these relationships and to be able to react to evolving risk situations. In the third section, I then contrast these findings with an analysis of the management practice of forestry and insurance companies in dealing with wildfire risks in Chile. Drawing on semi-structured interviews with forest owners and executives of insurance companies conducted between 2014 and 2019, I examine how hierarchical and market-based forms of risk management complement each other and, in parallel, limit the use of risk knowledge. It becomes clear that where specific knowledge is lacking or too expensive to produce, actors adopt risk avoidance strategies rather than investing in collective learning processes. In the fourth section I interpret these risk management practices as a decontextualization of risk and risk knowledge that stands in the way of a more collective and regional approach of forest fire risk governance. The paper concludes with a call for a broader and explicitly geographical perspective of wildfire risk governance in Chile.

The Challenges of Governing Economic Uncertainties

Researchers use the notion of risk to describe the contrast between reality and possibility (Hacking, 1990) and refer to a future that is visible only through the eyes of the present. People who take risks know that their decisions are accompanied by several possible consequences, but only when the risk actually turns into losses are they aware of its true characteristics and consequences. In the meantime, actors use discursive risk knowledge to shape (and strategically manipulate) the ideas and perceptions of risk that circulate in society (González-Hidalgo & Zografos, 2017; Martin, Martin, & Kent, 2009). Risks are real phenomena that are simultaneously socially constructed and discursively amplified. Actors must therefore grasp, assess, and manage the use of sophisticated and interdisciplinary knowledge (Rosa et al., 2014), which also places high demands on knowledge for governance (Glückler, Rehner, & Handke, 2019).

From an economic point of view, the future's unpredictability is nothing to be intimidated by. Risks are part of entrepreneurial ventures and promise above-average returns. Whoever succeeds in controlling risks better than others will be prosperous in the market (Clark, 2018; Knight, 1921). With this line of argument, it is easy to ignore that individual decisions about risk in the economy most often coincide with external effects for other social stakeholder. Forest fire risks are no exception. Certain tree-plantation management practices, for example, which are associated with varying degrees of accident probability, can trigger forest fire events, which, under certain climatic conditions, can quickly spread from their location of origin to neighboring areas and even endanger human settlements as they burn through the landscape (Bottaro et al., 2018). Many may suffer losses, even if they did not originally take any decisions on risk. While risk managers can easily justify economic losses due to risk in retrospect—by simply referring to the limited knowledge of risk that was available at the time of the decision and assure the best possible precautionary handling of it (Luhmann, 1991/1993)—other social stakeholders have greater difficulties in explaining and enforcing their positions. Social responsibilities in risk taking often remain unclarified.

Incomplete knowledge about risks, unintended side effects, and other "unknown unknowns" (Beck, 2006, p. 335), which can be summarized under the term *systemic risk*, are fundamental challenges for risk governance (Rosa et al., 2014). Ultimately, the way knowledge of risk is communicated determines the success or failure of management and governance practices. Yet, what exactly are the differences between them?

In general, *risk management* aims at objectively defining probable outcomes of decision making in order to reduce uncertainty to a list of probable events (following Beckert, 2016). More specifically, organizational risk management guides and legitimizes decision-making processes at the management level of a corporation and helps coordinating the available resources in pursuit of strategic objectives (Lundqvist, 2015; Soin & Collier, 2013). Risk from the point of view of an enterprise is technically regarded as a cost factor, where the probabilities of harmful events' occurrences are offset against the value of expected losses (Knight, 1921).

Risk managers calculate risks by making use of historic and quantifiable data. Nowadays, sophisticated risk models allow risk management to be based on individual decisions made by experts who aim at maximizing profit or—to put it differently—reducing the company's vulnerability. Risk management is short- to medium-term in nature depending on the forecasting capability of the available risk models.

The perspective *of risk governance* broadens the scope of the actors involved in risk management and their relations towards each other. From a societal perspective, governance in general terms serves to coordinate the collective actions of legally independent stakeholders toward the achievement of consensual goals. It extends beyond the scope of a single authority and requires negotiation between vested interests (Glückler et al., 2019). Risk governance, in particular, covers processes that lead to collectively binding decisions and the establishment of legitimized risk-management standards and practices that help to regulate, reduce, or control collective problems of risk (Crouch & Keune, 2012; Renn, 2008; van Asselt & Renn, 2011). In this sense, researchers of risk governance analyze the institutional structures, power constellations between various stakeholders, and political processes in society. Governance is a collective learning process with a long-term time horizon. It focuses on context-specific collective solutions aimed at increasing the resilience of societal stakeholders (Young, 2010).

The challenges of risk governance are clearly associated with the complexity, ambiguity, and/or uncertainty of knowledge about risk (Rosa et al., 2014). Only rarely can actors comprehensively describe and calculate risk via linear correlations. Complexity, ambiguity, and uncertainty are attributes of risks that—if empirically distinguishable—imply different governance mechanisms. Therefore, risk governance uncovers and negotiates dissimilar risk interpretations and creates new knowledge to deal with inconclusive and unknown risk correlations.

How can risk management and risk governance practices beneficially complement each other in the context of forest fire risk? Perhaps they stand in each other's way because of their incommensurable handling of knowledge about risk? In the following I draw attention to these unanswered questions and apply them to the empirical case of Chile's risk-laden forestry sector.

Methodology

In order to address the aforementioned questions, I process the findings from several empirical studies on risk in the Chilean forestry and timber industry. These studies were conducted between 2014 and 2018 as part of different seminar courses in the M.Sc. Governance of Risk and Resources Master's program at the Heidelberg Center for Latin America. The courses contrasted theoretical insights into risk governance with the practical efforts of economic actors in dealing with real risk phenomena. From the outset, the empirical investigations aimed at recording the context-related risk perceptions of different actors exposed to wildfires and analyzing their influence on joint efforts in dealing with them regionally. Interestingly, the many surveyed actors rarely perceived forest fires in Chile as a collective risk. Individual efforts to minimize economic losses predominate.

Since wildfires cannot be assessed by deterministic or probabilistic hypotheses alone, a triangulation of different methods of qualitative social research proved to be appropriate for the collection of the empirical data (Flick, 2018): (i) First and foremost, I based this contribution on an extensive evaluation of scientific literature on forest fire risks in Chile. In this way, I was able on the one hand to shed light on the fundamental interdisciplinary complexity of the relationships between causes and effects in wildfires. On the other hand, I could get familiar with the way in which scientific knowledge on wildfires is reduced and generalized in contemporary risk models. (ii) I supplemented the literature review with a content analysis of discursive arguments on the causes and effects of forest fires, as they are discussed in Chilean trade journals such as *Lignum*, *Revista Mundo Forestal*, and *Revista CIFOR*, or in the general press.[2] I included more than 250 newspaper articles and reports on forest fires in Chile that occurred between 2008 and 2018 in my analysis. This not only provided additional insights into the region-specific particularities of forest fire risks, but also confirmed the wide spread of strongly generalized explanations of the phenomenon of forest fires. Both indicate the discursive use of knowledge on risk (van Dijk, 2014). (iii) Thirdly, I have included an analysis the official Chilean forest fire statistics (CONAF, 2018) as well as an evaluation of the annual reports of the large Chilean forestry companies in the present study to highlight the regional diversity and temporal variability of risk. The statistics clearly reveal the quantitative extent of individual catastrophic wildfire events, which vary greatly from year to year and from region to region. They also differentiate between different affected parties (owners of natural forests versus owners of plantation forest of different size). (iv) Finally, I based this contribution on 25 expert interviews with risk managers from insurance companies in Santiago de Chile and economic actors at risk in the Chilean forest regions (primarily in the Maule and Biobio regions). I conducted the interviews as open, guideline-based interviews.

Precisely because different economic actors perceive forest fire risk differently—depending on their contextual experiences—and evaluate and communicate it differently—depending on the use of risk-calculating methods—it is necessary to survey these risk experts and their interactions with each other as an additional object of investigation. Contextually differentiated risk knowledge influences the risk behavior of these actors in many ways (Müller-Mahn, Everts, & Stephan, 2018). Ultimately, it determines the possibilities to develop a collective view on risk as the result of governance efforts.

[2] These include national newspapers such as www.elmercurio.com; www.latercera.com; www.elmostrador.cl; www.cnnchile.com, but also regional media such as www.diarioelcentro.cl; www.redmaule.com; www.diarioconcepcion.cl; www.biobiochile.cl. Only online articles from these newspapers were analyzed. Critical online news portals such as www.terram.cl; www.mapuexpress.org; www.laizquierdadiario.cl were also included in the research.

Qualitative research does not end with a comparison of the collected data with the aim of explaining similarities between the units of research. Rather, it uncovers phenomena and associated variants for which case-specific explanations have to be developed (Crang, 2002). With my qualitative research approach, I was able to uncover contradictions and decode the relationship between realities and interpretations of risk (see also Eisenhardt, 1989). Contradictions in knowledge and practices are manifold when it comes to forest fire risks in Chile.

Complexity, Ambiguity, Uncertainty? Forest Fire Risks in Chile

Wildfires are typical for Mediterranean climate zones that are characterized by mild and rainy springs that are followed by dry summers. They are also a natural phenomenon in several regions in south-central Chile (McWethy et al., 2018). Wildfires are eminently spatial: They have an unmistakably definable place of origin from which they extend by geophysical laws to nearby spaces. However, nowadays they are mainly man-made phenomena. In many cases, they are caused by carelessness, accidents, or even arbitrary arson (O'Flanagan, 1997). In this sense, wildfires represent negative externalities of the expansion of human activities.

In Fig. 8.1, I provide an overview of the frequency of wildfire events in Chile since the 1990s. On average, almost 6,000 forest fires occur in the course of a year and around ten hectares of forested land are lost per event. Two thirds of the affected areas are natural landscapes such as natural forests, shrubs, and grasslands. However, plantation forests are also affected, and this trend is rising (Julio, 2014). Most recently, a drought phase that lasted several years led to the biggest forest fires in Chilean history (González, Gómez-González, Lara, Garreaud, & Díaz-Hormazábal, 2018). The overall dimension of destructiveness of the latest "firestorm" (Gobierno de Chile, 2017) in terms of burned areas stands out in Fig. 8.1.

To some extent, researchers can assess the risk of wildfires with linear correlations that refer to physical-geographical conditions (Castillo, Molina-Martínez, Rodríguez y Silva, & Julio, 2013). In the jargon of forestry authorities, who monitor the wildfire risk, one speaks for example of *the rule of 30-30-30*. The rule says that temperatures above 30 °C, gusts of wind of the order of 30 km/h or more, and a relative humidity of less than 30% raises the risk of wildfires significantly. It is undisputed that under extreme weather conditions wildfires are hard to control and can quickly expand into neighboring spaces. On the basis of these findings, authorities then may publish timely risk warnings to raise public awareness of the potential hazard. Topography has another direct effect on the speed at which forest fires spread. The steeper the slope, the greater the flames' inclination, which increases heat development on the ground and allows biomass to burn faster and more intensively. This knowledge is of practical importance in firefighting (Vélez, 2009).

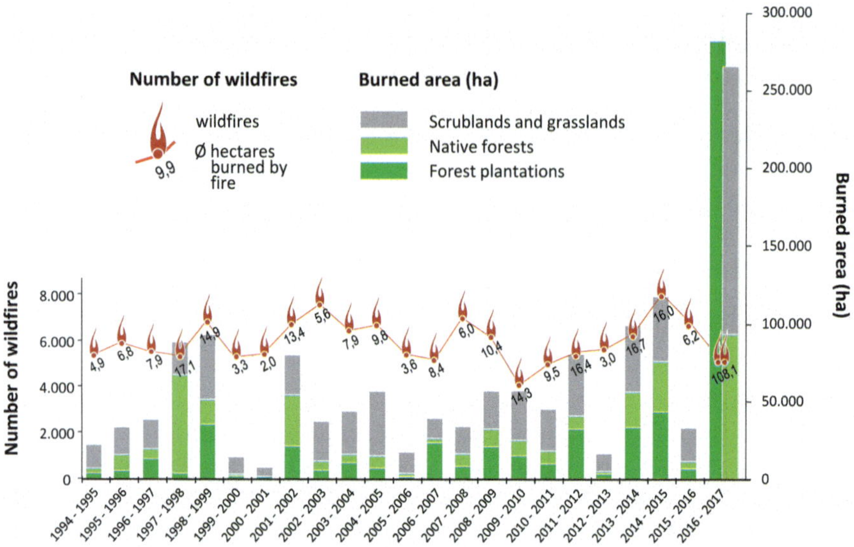

Fig. 8.1 The evolution of forest fire events and damage in Chile from 1994–2017. Source: Design by author. Data: Own calculations based on CONAF (2018)

However, natural factors alone cannot fully explain the extent of forest fires in Chile (Gómez-González et al., 2019). They may explain the size and spread of the flames and therefore the damage dimension of the risk. The occurrence and causes of fires, however, usually have an anthropogenic background. Risk analyses, for example, reveal the highest probabilities of forest fires occurring near settlements and along road networks (Martínez, Martínez, & Martín, 2004). There, man is reckless in treating nature. Land use forms and the compositions of tree species in Chile's economically exploited forests plantations influence on the dimension of forest fire risk as well. This relationship, however, is anything but conclusively clarified (see, e.g., the controversial debate published in Mundo Forestal by Goméz-González & Espósito, 2017). To be able to capture the phenomena of forest fires in Chile, researchers must consider both physical geographical and anthropogenic contextual factors jointly. Interdisciplinary research is indispensable, which, however, leads to challenges of complexity, ambiguity, and uncertainty of knowledge about risk (Rosa et al., 2014).

It is very common in the risk management literature to describe risks as complex, ambiguous, and uncertain in order to differentiate them epistemologically (Brugnach & Ingram, 2012; Cabantous, 2007; Ericson & Doyle, 2004; Ilin & Varga, 2015; Johansen & Rausand, 2015; Knight, 1921; Müller-Mahn et al., 2018; Perrow, 1986, 1972). In short, the three characteristics of risk can be distinguished as follows (see also Rosa et al., 2014): *Complexity* characterizes a condition where it is difficult to identify and quantify exactly the potential causal relationships between aspects of risk and possible adverse effects of decision making on risk. *Ambiguity* corresponds to the phenomenon that various actors know risk differently, resulting in a variability of legitimate risk interpretations with respect to observations and evaluations of identical data. *Uncertainty* refers to unknown knowledge about certain causal relationships. That includes not knowing what one does not know.

One way of dealing with these challenges is through regulation. Standards and norms help actors reduce complexity, clarify ambiguities, and improve the reliability of mutual expectations (North, 1990). In Chile, for example, in 2015 the National Institute for Standardization published the norm NCh3380 that aimed at uniformly measuring wildfire risks (INN, 2015). The norm establishes a terminology and classification of risks of forest fires in plantations, for which it defines and delimits spatially different levels of risks, based on the evaluation of probabilities of occurrence and the different impacts caused by wildfires. The norm consolidates interdisciplinary knowledge and is considered "[...] applicable to any existing forest plantation in the country, regardless of the species it contains, size, location, among others" (INN, 2015, p. 2). This, however, raises a new challenge. Although the norm covers geographically correlated aspects of risk and makes them comparable, it misleads one to believe that the resulting risk models can be applied uniformly, that is independently of time and space. Such an approach in particular would mask and decontextualize the uncertain and ambiguous circumstances of the risk. Standardized models always entail the challenge of being able to capture evolving or systemic risks (MacKenzie, 2011). Well, what exactly leads to complexity, ambiguity and uncertainty in relation to forest fire risks in Chile and how to deal with it?

Complexity: Multidirectional Correlations Between Physical and Anthropogenic Factors

It is an obvious choice to start with the Chilean forestry sector with its specific ways of organizing economic activities in space (Clapp, 1995; Gatica, 2012) to illustrate how complex links between anthropogenic and natural factors of wildfire risks are created. Forestry companies, in first place, decide the composition of tree species in their forest plantations. They prefer exotic species such as eucalyptus or pine, as these are adapted to dry climatic conditions, grow rapidly, and are therefore economically very profitable.[3] However, as both species have a higher water consumption than native trees, they can intensify droughts on a microgeographical scale and fuel the effects of fires (Little, Lara, McPhee, & Urrutia, 2009). At the same time, however, industrial measures of plantation management significantly reduce the magnitude of forest fire risk. In their quest for higher yields, forestry companies remove "interfering biomass" from forest plantations and thus control an important fire accelerator. By investing in watchtowers that they strategically place within their territories, they create and maintain an interconnected monitoring system that

[3] Currently, pines account for 58% of all plantation area in Chile, while eucalyptus represent 36% (CONAF, 2018).

allows for wildfires being detected in time (Tapia & Castillo, 2014). The companies' own fire brigades can quickly reach out to a fire's source through laid forest paths. If the probability of the occurrence of fires remains constant, plantation management measures reduce the damage dimension of risk—although this relationship is not linear. Many risk-increasing and risk-reducing factors cannot be measured simultaneously in an accurate way. This is how complexity is created, and—by the way—a methodological problem for many forecasting risk models (Castillo et al., 2013). Normative assumptions on the weightings of individual factors must be chosen for these models to work.

Table 8.1 summarizes regional forestry activities in Chile and illustrates the complexity of the relationships between physical and anthropogenic factors of wildfire risks. In Chile, the probability of forest fire occurrence should actually decrease in the direction of the southern forest regions, as these are characterized by climatic conditions with lower average temperatures over the course of the year and generally record higher precipitation (Altamirano, Salas, Yaitul, Smith-Ramirez, & Ávila, 2013). A glance at Table 8.1, however, reveals that forest fires in the south occur more frequently in absolute terms and measured in terms of available forest areas than in the northern forestry regions. It is also a fact that in the regions of Bio Bio and Araucanía, forest fires occur much more frequently in plantation forests than in natural forests. Simultaneously, however, plantation fires are significantly less destructive than fires that affect native forest.

The search for explanations of these findings leads to the question of who invests in firefighting in the first place and who decides which sources of fire are extinguished first when several fires occur simultaneously. Since firefighting in Chile is largely subject to a market in which private companies offer their services—and in which the economic value of the timber resources often determines the availability of fire brigades—it is understandable that wildfires in Chile provoke critical interpretations of the country's neoliberal forestry model (Barton & Román, 2012; Reyes

Table 8.1 Forestry activities and forest fires in Chile

	O'Higgins (VI)	Maule (VII)	Bio Bio (VIII)	Araucania (IX)
Forested areas 2014 [hectares]	586,615	833,227	1,695,082	1,447,635
Forest plantations 2014 [hectares]	127,306	448,513	926,530	483,482
Annual fire events (annual average 2003–2016) [absolute number]	222	437	2,499	1,006
Forest fires per 1000 hectares	0.38	0.52	1.47	0.69
[%] of events affecting plantations	1.4	15.5	59.1	56.0
Annual damage (all forests) (annual average 2003–2016) [hectares]	7,927	7,001	13,487	5,996
Damage to native forests [hectares by fire]	35.1	17.7	11.1	9.3
Damage to plantations [hectares by fire]	38.6	17.4	1.9	3.3

Note. Personal elaboration with statistical data from CONAF (2018)

& Nelson, 2014). In relation to these interpretations, the ambiguous characteristics of risk come into play.

Ambiguity: The Coexistence of Several Equally Logical Explanations for Forest Fires in Chile

"Risks only gain influence in the social world to the extent that they are communicated" (Renn, 2008, p. 57). Therefore, it is through *discursive knowledge* (van Dijk, 2014) that risks are socially constructed. Those who have experienced flames that directly affected their *livelihoods* describe and explain wildfire risks differently than those who only look at them from a distance. Geographical knowledge of complex risk relationships gives profound explanatory substance to these discourses. However, with geographically founded, context-dependent explanations actors can quickly confront the distanced with supposed ambiguity.

In Chile, many different discursive explanations circulate about the origins and the hazardous effectiveness of forest fires. They certainly feed on the heterogeneity of the society's risk knowledge. The ambiguity of risk discourses, on the one hand, reflects the perceived threats by different social actors (Mermoz, Kitzberger, & Veblen, 2005), and on the other hand, is due to linguistic difficulties in articulating risk. Those who have experienced risk at first hand, often describe their experiences in an opportunistic way, for example, to maximize the influence on distanced political decision makers who they want to win over to their cause (Farré, 2005). This may lead to interpretations related to forest fires that are artificially exaggerated and intentionally false. Ambiguity can thus be understood as a function of conflict over preferences, intersubjectivity of discourse, and uncertainty over the technical aspects of risk communication (Hanson & Kysar, 1999; Johansen & Rausand, 2015). The consequential ambiguity of the discourses does not refer so much to an uncertain future as to the uncertainty related to past and present experiences.

Across all Chilean regions, the main causes of wildfires are accidents and the carelessness of man (58%). However, almost one in three fires is caused by delinquency, in other words, intentionally caused ignitions, while 15% of the causes of fires remain in the dark. It is above all the unknown and intentional causes of wildfires that are reported and speculated on in detail in the Chilean press (Millones, 2017; see also the review of Aylwin, 2017).

Based on the statistics by the Chilean National Forest Cooperation (CONAF, 2018), Table 8.2 reveals a regionally unequal distribution of the causes of the wildfires in Chile—especially related to fires that are intentionally started. This is also strongly reflected in the causal explanations one gets when asking experts and affected parties about the background of the arsons in Chile:

1. Some argue that wildfires are caused by forest owners themselves in order to circumvent existing land-planning regulations (Caviedes, 2017). In the proximity of urban areas, for example, landowners speculate that burnt areas may be

Table 8.2 Causes of forest fires in Chile (Annual average 2003–2016)

	O'Higgins (VI)	Maule (VII)	Bio Bio (VIII)	Araucania (IX)
Annual fire events (annual average 2003–2016) [absolute number]	222	437	2,499	1,006
Causes: Accidents [%]	67.5	69.3	27.7	37.6
Causes: Agricultural and forestry activities [%]	14.3	15.5	8.4	12.0
Causes: Arson [%]	14.4	11.0	35.7	46.5
Unknown causes [%]	3.8	4.2	28.2	3.9

Note. Personal elaboration with statistical data from CONAF (2018)

designated as urban developing land that awaits construction permits. This is a reality in many parts of the world (Salvati & Ranalli, 2015). In Chile, these causes are only narratively discussed in public, since hasty accusations can have legal consequences.

2. Intentional forest fires are also interpreted as a deliberate strategy of forestry expansion. Such interpretations, in particular, arise in communities where landowners subsequently reforest burned natural forests as forest plantations (Gerber, 2011). In Chile, since the new "law on native forest recovery" (CONAF, 2008) came into force in 2008, forest owners are prohibited from converting natural forests into forest plantations with exotic tree species. However, after a fire event, reforestation is allowed, for example, in the form of plantations with endemic species. In this way, the natural forests are made accessible and transformed for industrial exploitation (Moreno del Valle, 2015).
3. It is further argued that forest owners themselves burn their forests in order to benefit from forest fire insurances that they have purchased beforehand (Focacci, 2017). This argument in its simple form might seem plausible but most often turns out to be misleading and requires detailed and contextualized explanations: Insurers are generally very aware of the problem of the so-called *moral hazard* (Stiglitz, 1983), in other words, the possible deliberately caused burn of an insured forest. Therefore, insurance companies design contracts in such a way that the insured always assumes part of the risk (Agroseguro, 2018). High deductibles, for example, lower incentives to carry out fraud. Those incentives, however, can be quite different for forest owners who are on the verge of ruin, for example, in cases where an insect calamity significantly reduced the value of a standing forest so that reforestation becomes necessary. Was this the case in 2017 in certain parts of Chile's forest regions (see, e.g., the case-specific fact check in Mapuexpress, 2017)? As long as the contractual conditions that forest companies negotiate with the insurance sector are not publicly transparent, discursive and ambiguous speculation about this kind of motivation for arson will last.
4. In some cases, intentional fires in Chile are the product of deep social discontent. A widespread public opinion is that specifically in the Araucanía region, members of the Mapuche people intentionally set fires to protest against socioeconomic

disparities and political exclusion. Wildfires seem to have become the symbol of their conflict with the Chilean state (Montalba-Navarro & Carrasco, 2003; Rojas & Miranda, 2015). It also seems that wildfires are directly targeted against large forestry companies that mark land uses with their tree plantations and dictate the regions' main economic activities (van Holt, Binford, Portier, & Vergara, 2016). As late as after colonization by the Spanish, with the country's nationalization, Southern Chile experienced a capitalist appropriation and valorization of land that created vulnerable people and is still ongoing today (Latorre & Rojas, 2016). Though the accusations are made very quickly, the circumstances of wildfires in the Araucanía region are not at all clear. Some of them turned out to have been caused by the private sector and even by public officials in order to blame the Mapuches for the crime and thus keep the conflict over land usage alive in a discursive manner (Seguel, 2018; Sepúlveda, 2013). In this way, fires serve to justify forest expansion (retrospectively) (González-Hidalgo & Zografos, 2017).

The coexistence of these different discourses confronts political and economic actors with decision-making problems. They obscure the true motives of the actors behind the arsons, which leads to the uncertain characteristics of forest fire risk in Chile. For the insurance sector in particular, arson represents an uncertainty, in other words, risk that is impossible to calculate.

Uncertainty: Hidden Self-Reinforcing Social Amplification of Forest Fire Risk

Unknown and uncertain risks are induced by scientific uncertainty. While in the case of unknown risks researchers are aware that they do not know enough, for example, to predict the frequency and severity of catastrophic events (Chichilnisky & Heal, 1998), uncertain risks remain hidden from the society's radar for a long time (Beck, 1992; O'Malley, 2004). This is partly due to the fact that the causes of modern risks—but also their effects—are no longer confined to one place. Often there is also no legally identifiable entity as the perpetrator of risk. Risk decisions can have a long latency period before they materialize. For political actors, this is the main source of uncertainty: "[I]t is precisely unknown unknowns which provoke far-reaching conflicts over the definition and construction of political rules and responsibilities with the aim of preventing the worst" (Beck, 2006, p. 335).

Researchers also refer to uncertainty in risk to address processes of systemic, self-reinforcing risk amplification (Kaufman & Scott, 2003). Uncertain risks are systemic if they trigger unnoticed chain reactions. Accumulated negative externalities can trigger devastating effects. In the following, I will focus on three examples of systemic aspects of uncertainties related to wildfire risk in Chile. They raise awareness of the fact that depending on the local and regional context, the occurrence of a forest fire in Chile can "ignite" subsequent fires and reinforce risk. The first example is certainly simplified, but it serves as a textbook example to illustrate the underlying systemic effect empirically. In the second example I refer once again to the discussion of the ambivalent symbolic attributions of arson that I introduced in the previous chapter. In the third example I summarize a content-analytical interpretation of interview statements from several empirical field studies in the Maule region conducted between 2015 and 2019.

1. In individual and isolated cases in Chile, it turned out that wildfires were deliberately caused by young people "[...] to experience once again the spectacular use of airplanes or helicopters in the firefighting process" (interview with forest firefighters in Constitucion/Maule, 2015). This problem seems particularly relevant in the transition zone between urban and forested areas. The legitimate concern of this type of wildfire causes is also reflected in rulebooks that guide fire brigades (Vélez, 2009). Pragmatic rules stipulate that helicopter missions in the event of fire should be restricted to areas outside urban areas. This may be understood as aimed at preventing the experienced spectacles of past fires from causing new fires in the future.
2. Another systemic aspect of risk is evidenced in the strategic appropriation of the symbolic effects that fire and flames can create (Segovia, Basulto, & Zambrano, 2018). As a social imaginary, forest fires acquire effectiveness for different groups in the Chilean society—even if they live far away from forested areas. Fires are lit repeatedly due to their medial efficacy and the sociocritical imaginations originating from them, and as was demonstrated in the previous chapter, these imaginations are ambiguous in Chile. For some, they represent the struggle of the vulnerable against the neoliberal alliance between the state and the industrialized forestry business. For others, they gain importance in justifying that the State takes sides and imposes even harsher control over *resource peripheries* (see also Chap. 16 by Hayter and Clapp). Different groups expect individual advantages from wildfires. The risk transcends the boundaries of systems when its effects extend from the forest landscape to the economic and political spheres (see also Beck, 2006).
3. In the Maule region, where one might not expect it at first because there has been no open conflict over land use so far, forest fires have recently created a systemically heated atmosphere that has the potential to further increase the inherent risks. The explanations behind this case are as follows: Competition in the regional timber market in Maule is fierce. Within a radius of 30km of Constitución, about 60 small and medium-sized family owned sawmills compete with each other for access to raw timber resources. Some work as contractual suppliers for the large forestry companies in the region and thereby gain privileged access to their clients' timber resources. Others have emancipated themselves from the big players in search of their own customers. In the course of time, they also have acquired their own forest property, which, however, is not sufficient to supply them with raw materials all year round. They experience that their expansion opportunities are limited by the supply of regional timber that turns out to be scarce and sensitive in price—a situation that has worsened since 2017

(Hechavarria, 2018). This is an important first detail in understanding the regional conflict and the subliminal systemic risk of forest fires that accumulates in the Maule region.

Speculation and mutual accusations on the causes of recent wildfires spread in the region: Large forest owners, for example, claim that their smaller neighbors act as *free-riders* in the face of forest fire risk. In other words, "[...] they do not buy insurance and [instead] rely opportunistically on the fact that their forest property is observed and cared for by the large company and their fire brigades" (Interview conducted with the manager of a medium-sized sawmill in San Ramón, March 10, 2016). This expectation would—according to the interviewee—reduce smaller landowners' incentives to actively monitor their plantations. From the point of view of large forest owners, this increases the risk of a forest fire in their neighborhood and therefore "[...] through geographical expansions, the risk of damage to our plantations is uncontrollably increased" (Interview conducted with the operational risk manager of a large forest company in Constitution, March 11, 2016). Implicitly, this deep concern of the large company gives rise to the carelessness of the small neighbors in the first place.

According to this logic, it is understandable that large forest owners in Chile have incentives to expand their forest ownership over a large area and in a coherent manner: The less mixed the regional mosaic of forest properties, the more controlled the risk will be (Vergara-Díaz, Sandoval-Vásquez, & Herrera-Machuca, 2017). In response to the discourses of the large companies that accuse small landowners of being opportunistic, these, in contrast, accuse the big players of not adequately protecting the property of their neighbors in the event of a forest fire. More precisely, in interviews with medium-sized forest owners in Maule in 2016, large forestry companies were accused of directing fire fronts towards the property of small forest owners to avoid major damage to their own forest plantations. Still others claim that extensively insured forest areas stand in the way of the real efforts of large forestry companies to fight the flames. Of course, it is not possible within the framework of this contribution to resolve the absolute truth behind these testimonies or to confront it with the technical aspects of firefighting. It should be noted, however, that firefighting follows military hierarchical command structures and, depending on the situation, includes the right to subordinate private property, which is otherwise very strongly protected in Chile. At least the narratives presented above give an idea of the subliminal conflict between small and large forestry actors in the Maule region. The fear of pyromaniac acts in the region in response to economic repression and exclusion is growing for years (see also the statements of economic actors quoted in Saavedra, 2017).[4]

[4] In more recent interviews with forest owners in the Maule region, implications of the arsonists of the 2017 forest fires were collectively avoided. "We don't want to make any false accusations, and in particular we want to hinder the discourses from Chile's southern forest regions to reach the Maule region. We want to avoid the false interpretation that forest fires have become an inflicted result of the forest industry itself," one sawmill owner openly admitted in an interview early in 2019.

Systemic risks are uncertain until they materialize and therefore cannot be calculated ex ante (Ilin & Varga, 2015). To address them, a readjustment of risk perception is needed. In the first case, this was done, as described, by new firefighting rules that limit the use of actually effective helicopters to specific areas. In the second case, however, the systemic effect of risk remains out of reach for risk managers in forestry. At best, they have the option of relying on risk-avoidance strategies—that is to distance themselves from the neoliberal model of forestry—, which in this case could mean the end of their business model; and even that does not guarantee that other actors will not continually try to exploit the symbolic effect of forest fires. The interviews quoted for the third case show that regional economic actors at least develop a feeling for the accumulating systemic risk.

As should have become clear from the above, the epistemological challenges of the complexity, ambiguity, and uncertainty of forest fire risk arise from the fact that all three characteristics are closely intertwined and lead to an obscuring of risk relationships. Against this background, it is understandable that risk managers usually aim at simplifying and separating the individual attributes of risk. This enables them to transform ambiguity or uncertainty into complexity that they can then structure and handle by making use of technical risk models (Ericson & Doyle, 2004). In the following chapter, I explain how this works in practice. While it becomes clear that risk management is not the same as risk governance, the chapter reveals the complementary aspects of different management practices and explores their possible uses in explicitly regional risk governance approaches.

The Complementarity of Risk Management Practices

Advocates of theoretical insights from the New Institutional Economics (Furubotn & Richter, 2005), which are widely applied in financial risk management practices, proclaim markets and hierarchies as in a sense ideally opposed governance forms for the coordination of economic interaction. The characteristics of a transaction as well as the degree of asymmetrically distributed information between economic actors determine whether either hierarchical orders through company routines or market-based pricing processes under competitive conditions provide transaction-cost-efficient coordination. Interestingly, hierarchical and market-based forms of risk management are not necessarily opposed to each other, but rather mutually supportive. The case of risk-management practices in Chile's forestry sector illustrates that a hierarchically organized risk management even can be the prerequisite for the emergence of and access to insurance market solutions.

To fully understand the complementary logic behind these practices, it is important to note that forestry companies and insurers only slightly differ in their basic approach to risk and uncertainty. Both actors are able and willing to handle risk that they can calculate by themselves, and both reject uncertainty in risk management. Then, however: What can a market for forest fire insurances look like if the preferred and undesirable risks of both players are more or less the same? Of course, it depends on the small differences in dealing with particular characteristics and aspects of risks. Additionally, one has to look at the uncertainty avoidance strategies of both actors to understand the complementary character of Chilean forest fire risk management practices.

For insurance companies, calculated risks undisputedly represent a core business. Insurers are able to diversify them with the help of the *law of large numbers*, which states that the empirical reality (temporal and spatial) of risk can be aggregated into mean values (Chichilnisky & Heal, 1998). Forestry companies take a positive view on risks as well. Correctly managed, risks promise above-average returns and competitiveness advantages (Lundqvist, 2015). In contrast to insurers, however, forestry companies have to find ways to handle causes and effects of forest fires individually and in context-specific ways.

Uncertainty, that is incalculable risk, is something that neither actor is looking for. Forestry companies and insurers alike are trying to externalize or completely avoid risks that are rare (and at the same time very destructive), highly specific, and uncertain. Large forestry companies in particular, as I showed in the previous chapter with the example of systemic forest fire risks in Maule, are afraid of risks that "are taken" independently and opportunistically by others and that "endanger" the success of their own businesses in the sense of a negative externality (Luhmann, 1991/1993). Insurance companies, in turn, do everything to avoid moral hazard behavior of their clients, which can even lead to the situation that there is no market supply for insurances at all (Hellwig, 1983; Stiglitz, 1983). This line of argumentation fits with the transaction-cost approach, whose advocates stipulate purely hierarchical control for dealing with uncertain risks (see also Knight, 1921). How does the insurance market emerge when market participants share a common desire to waive or avoid uncertainty? The answer has to do with diverse risk management practices of standardization and categorization of risk in order to reduce its complexity, ambiguity, and uncertainty characteristics.

Risk Management in Chilean Forestry

In Chile's industrialized forestry sector, forest plantations are managed in a way that minimizes both the likelihood of a forest fire and the possibility of uncontrolled expansion. The homogenization of contextual conditions, the simplification of risk relationships, and the decoupling of certain risk elements go hand in hand and influence each other reciprocally: (i) The simplification of risk begins with the focus on certain causes of forest fires. In particular, the main cause of accidents can be minimized through employee training and clear routine instructions. Routines are instrumental for the implementation of plantation management practices that aim at creating industrial economies of scale. Only a few specialists are needed to monitor the compliance of the routines. (ii) The decoupling of forest fires' causes and effects is achieved through targeted infrastructure investments, such as the formation of firebreaks that slow down or stop fires that have broken out. Often, firebreaks are strategically created in the course of regular clear cuts. In this case, plantation management practices and infrastructure investments coincide. Furthermore, forest roads are a strategic investment that serves risk management purposes as they allow fire brigades quick access. (iii) Finally, the homogenization of the physical space complements the decoupling and simplification of risk relationships. The large forestry companies in Chile aim at a consolidated and coherent forest property. Over the past 40 years, they have achieved this largely through continuous acquisitions and forestation of tree plantations. Only few other private forest properties still separate the plantations of the large companies. Connected forested areas lead to economies of scale in forestry logistics. They can also be better monitored with fire protection watchtowers, which in turn further contribute to the decoupling of risks as fires can be quickly detected once they have broken out.

All these plantation management practices are based on combining centralized and decentralized risk management logics. In addition, multiple spatial references are evident: Complexity reduction via routines, for example, is planned and coordinated hierarchically top-down from the companies' headquarters (see also Perrow, 1986). The site-specific implementation and monitoring of the routines, however, takes place in decentralized way. The watchtowers in the forest regions are an important node in the risk communication network. In case of a forest fire, their occupants communicate to neighboring units and the company headquarters alike so that responses can strategically be elaborated in a timely manner. In the event of a wildfire, military-like chains of command are activated and take control. Firefighters are coordinated centrally but can decide locally in order to be able to react quickly to changing conditions (Arnaldos, Navalón, Pastor, Planas, & Zárate, 2004). With this combination of centralized and decentralized risk-management routines, actors in forestry enterprises are in a good position to react to ambiguous risk signals from forest regions and actively shape local practices of handling risk. However, through locally adapted action, they are also coresponsible for creating ambiguity in the interpretation of wildfire risks in Chile on a national scale.

Risk Management in Insurance Companies

Insurance companies implement risk management practices that rely on standardization and categorization based on quantification and mathematical procedures (Dean et al., 2003; Jarzabkowski, Bednarek, & Spee, 2015). Since quantification disconnects existing data from local narratives or general stories that stabilized the meaning of risk in the first place (Müller-Mahn et al., 2018), insurance companies must create their own meanings of risk. Prices in insurance markets, for example, are such a standard for risk quantification that also provides meaning for economic operations (Hayek, 1945; Kessler, 2015). Prices communicate the magnitude of the risk to third parties such as policyholders. Risk maps provide a similar communication frame (Dransch, Rotzoll, & Poser, 2010). Maps not only can be used to locate the origins of historical forest fires events and illustrate their propagation in space.

Loss statistics incorporated into the maps also provide insurers with a spatial overview of the risk's frequency and extent. Risk maps, therefore, serve marketing purposes. Insurance companies use them to signal to their clients why insurance policies in so-called high-risk areas contain a high insurance premium. Risk maps give meaning to prices. This kind of risk communication, however, first requires a sophisticated valuation process (Aspers, 2009). Risk valuation applied in insurance companies' risk management practices is mainly based on mathematical models and correlation logics. Actors using these models are able to capture the complexity of individual aspects and conditions that cause or amplify forest fires and break them down into their individual components by means of multiple (and even spatial) regressions (Castro & Chuvieco, 1998). Spatially differentiated risk models allow insurers to simulate their values at risk.

The better the insurance company manages to break down the complexity of wildfires into modeled chains of linear causes and effects, the better it can design and offer different kinds of insurance contracts. On the one hand, insurance companies are always free to decide whether and where to offer an insurance policy to a customer. This means that they can exclude areas in which forest fires occur very frequently and for unknown reasons from accessing insurance. The market is simply rationed geographically (Hellwig, 1983). On the other hand, insurance companies have a strong contract design tool at their disposal: *self-selection* (Furubotn & Richter, 2005). By offering customers alternative contract designs that differ, for example, in the amount of the deductible in the event of a loss, insurers obtain detailed information about a customer's risk exposure. By accepting certain contracts and rejecting others, the insured reveals his risk attitude and self-assesses his exposure to risk. Self-selection is part of a bilateral negotiation process. In negotiations, the customers themselves offer to fulfill certain conditions in order to lower the prices for insurance. For example, the better the forestry company itself controls the risk through simplification, homogenization and decoupling its effects, the cheaper the insurance premium offered.

Mutually Complementary Risk Management Practices and Risk Avoidance Strategies

The hierarchical and market-oriented risk management practices of forestry companies and the Chilean insurance sector complement each other at least in two ways: On the one hand, this can be seen from the fact that forestry management practices stimulate innovation in insurers' services. Without economies of scale from industrialized plantation management, the market for forest fire insurance in Chile would have been too small, too transaction-intensive, and possibly nonexistent (see also

Chichilnisky & Heal, 1998).[5] On the other hand, complementarity most obviously arises from how insurers influence the forestry management standards of forestry companies through self-selection and contract design. Insurers make it clear how important it is for them that hierarchically supervised forestry workers regularly thin out the plantations (= simplification of risk correlations), that the insured plantations comprise a minimum size (= homogenization of risk), and that watchtowers and firebreaks cover areas at risk (= decoupling of risk correlations). These requirements are either specified directly in the insurance contracts, or the forestry companies are indirectly given incentives to fulfill them through promised premium discounts.

Self-reinforcing effects of hierarchical and market forms of risk management also derive from the uncertainty avoidance strategies of the actors. Forestry companies, for example, want to get rid of the unlikely but in principle possible scenario of a total loss of their assets in the event of a regional wildfire catastrophe. Although they view this total loss scenario as unacceptable, it is also incalculable and therefore incorporates the characteristics of uncertainty. Large forestry companies avoid this uncertainty by transferring it to the insurance sector. They buy insurance policies that include high deductibles. As a result, despite having invested in insurance, the forestry companies will cover many minor losses by themselves. This practice is confirmed by the operational risk manager of a large Chilean forestry company: "Even in the disaster year 2017, the compensation payments of our insurers were below the expenses of the annual insurance premiums. All our contracts included high deductibles" (Personal interview conducted in 2018). The manager claimed that the firm's shareholders explicitly wished to insure only the uncertainty of a total loss.

What forestry companies consider an incalculable and uncertain risk, caused by third parties and therefore understood as an external hazard, is, in the eyes of the insurer, complex and controllable. Insurance companies are able to absorb the uncertainty of a total loss of one of their customers because they can diversify it and convert it into a calculable and statistically low risk of a total loss across all their customers. However, insurers also retain much of their control over risk explicitly through their own uncertainty avoidance strategies: They avoid uncertainty by rationing the market. This strategy is geographically oriented in two ways: On the one hand, as already mentioned, they exclude insurance in municipalities with high potential for political tension, in other words, in territories where arson is a frequent cause of forest fires or where causes are simply unknown. On the other hand, they also exclude specific forest areas on a small scale that they declare to be uninsurable. In avoiding so called *cluster risks* and *risks of geographical contagion*, insurers deny market access to forest owners located in the immediate geographical proximity of already insured forest property. During a 2016 interview, a Chilean insurance company's sales executive referred to a map on the wall of his office, on

[5] This is illustrated, for example, by the fact that there are currently no insurance policies available for economically used natural forests.

which the sites of the currently insured forest areas in his responsibility were marked with flags. These markings gave the sales executive an indication of the surrounding areas where he could (or should) not offer any further forest fire insurance. "I avoid selling here because a single forest fire could otherwise affect several of my customers at the same time," he explained. In a market like Chile, which is characterized by a limited number of insurers (Loewe et al., 2017), this leads to the exclusion of forest owners willing to buy insurances. Rationing financial products is typical for oligopolistic insurance markets (Hellwig, 1983). However, for some clients of the insurance companies, exclusion may seem arbitrary. Even when exclusion is based on the principle of *who comes first*, for many it seems related to an exclusive relationship between the insurers and the large Chilean forestry companies. In some ways, geographically uneven access to insurance in Chile represents a negative externality of insurance-based risk management practice and can be interpreted as the result of the decontextualization of risk and risk knowledge, which I will explain in more detail in the following chapter.

The Decontextualization of Risk and Risk Knowledge

Risk managers tend to analyze different risks in isolation (MacKenzie, 2011). In more general terms, risk-exposed actors epistemologically grasp and handle the causes and effects of risks as if they were separated from each other (Rosa et al., 2014). Therefore, also conceptually it makes perfect sense to separate the material dimension of potential physical losses from the discursive dimension of how people originally perceive, communicate and socially construct risk. However, in the course of these management and academic practices, the knowledge of risks is easily decontextualized, also in a spatial sense (November, 2008).

Forestry and insurance companies in Chile design and work with economic models for decision making that are considered (and have been proven) to be useful for the handling of risk. In doing so, they clearly decontextualize the risk of forest fires in a spatial sense. Forestry enterprises, for example, decontextualize risk and risk knowledge in the course of their homogenization and standardization strategies in managing forested land. They homogenize space by purchasing adjacent areas, which they then use for reforestation, and they standardize space by anchoring the same risk management routines in their plantations. Actors using routines not only standardize risk knowledge, they also decouple risk from the unit of space, in the course of which plantation forests are transformed into assets that can be valued uniformly and according to economies of scale. A forest area managed as a monoculture decontextualizes specific site conditions. It ignores the fact that site-adapted, native tree species might be superior to exotic tree species in terms of reducing the local risk of forest fires.

The decontextualization of knowledge related to forest fire risks can also be seen in the fact that large forestry companies even succeed in acquiring insurance for their plantations in politically unstable communities, in other words, in communities in which insurance companies are actually planning a rationing strategy for the market. Large forestry companies generally own and manage forest areas spread over several forest regions in Chile, and because these are homogenized and managed in a standardized and comparable manner, they are able to acquire a package of insurance policies from the insurance companies that covers several separated plantations at once. Forest fire risks in high-risk areas are then contractually offset by risks in less endangered areas. The exact knowledge of the circumstances of the risk in the package becomes blurred and decontextualized. It no longer appears relevant due to the standardized price for the packaged risk.

By purchasing an insurance package with spatially dispersed forest ownership, also the insurer gains several advantages. Not only does a package of insurances enable the application of the law of large numbers; the insurer is also given an opportunity to spread the risks spatially, which reduces his risk of total failure. His accumulated overall risk is no longer determined by the conditions at a given site. This reduces his interest in surveying context-specific knowledge about the exact causes and effects of individual forest fires. The same is true for a high number of deductibles that policyholders accept in individual insurance contracts. Finally, the decontextualization of risk increases its quantifiability, which enables the insurer to translate it into the language of the financial markets, which in turn opens up opportunities for the insurer to resell parts of the risk to reinsurance companies.

Despite all these obvious advantages for corporate risk-management practice, the widespread practice of decontextualizing risk and knowledge on forest fire risks in Chile, coupled with reduced incentives to generate new knowledge, for example about the causes and effects of forest fires, can prove to be a step backwards in governing risk from the point of view of a resilient society. Uncertainties and ambiguities require a broadening of the risk debate and should include as many stakeholders in the evaluation process as reasonable: "Participants should be asked to find a consensus of the extra margin of safety in which they would be willing to invest in exchange for avoiding potentially catastrophic consequences" (Rosa et al., 2014, p. 144). Risk assessments that are based on economic considerations alone are accompanied by disparities. If the initial conditions for decision making on risk are not equally distributed among societal actors they open the window for self-reinforcing processes of systemic risk accumulation.

Conclusion

Forest fire risks in Chile have different spatiotemporal origins and consequences. They prove to be epistemologically complex, ambiguous, and uncertain depending on the socioeconomic and sociopolitical context of their appearance. Different physical and anthropological causes and effects of forest fires cannot be clearly attributed (= complexity). Unexplained causes give rise to various logical explanations, which run counter to each other (=ambiguity) and hamper collective efforts to deal with the risk. In many cases, actors also demonstrate a lack of sensitivity to their own risk unawareness (=uncertainty). In sum, the knowledge about the causes and effects of wildfires in Chile is heterogeneous and geographically dispersed among societal actors. Empirically it becomes evident that wildfire risks in Chile are socially constructed. They depend on how findings from natural and social science are interpreted and being implemented in standards of risk management practices. These practices also communicate risk, as in the present case via prices and access to insurance markets.

Focusing on economic-geographical explanations, in this study I have not only made it clear that forest fires are a profoundly geographical risk, but also that they represent a fundamentally collective challenge for Chilean forest regions. A single fire—regardless of its origin—can affect and destroy the forest property of neighboring actors. Accordingly, neighborhood relationships, in other words, cooperative and collective approaches to risk management practices, could have been expected. However, it turned out that forestry enterprises prefer to organize risk management on an individual basis, applying hierarchical company routines. In parallel, they buy insurance policies to transfer part of the risk to the financial markets. Their relationships with insurance companies are based on bilaterally negotiated contract designs and market prices for risk.

Forestry companies and insurance companies in Chile seem to have found a complementary way of dealing with risk by sharing jointly created technical risk knowledge as part of their routine risk-management practices. Both choose a management approach that reduces complexity, decouples risk, and decontextualizes risk knowledge to make forest fire risks calculable and manageable in mutually beneficial ways. They agree, on the one hand, on price mechanisms for insurance contracts as a standardized language, which has the consequence that detailed geographical knowledge on forest fire risks in Chile is explicitly decontextualized and reduced. On the other hand, they follow complementary risk-avoidance strategies whenever risks are characterized as ambiguous or uncertain.

The catastrophic forest fires in 2017 revealed the limits of standardized risk management practices. The law of large numbers and therefore the calculability of risk, in general terms, becomes less effective when large and unlikely events occur or when the causes of the risks are systemically interdependent. The latter is the worst case for the insurance business. Systemic interdependencies of risks can hide behind characteristics of ambiguity or uncertainty. In addition, there is a great danger that if actors only follow their own dominant risk discourses and ignore observations and interpretations that are due to different conditions of the spatiotemporal context of risk, the dynamic changes in the relationships between environmental and anthropogenic risk factors of forest fires in Chile remain invisible to decision makers. The catastrophic extent of the forest fires of 2017 can certainly be attributed first and foremost to the extreme climatic conditions. However, there are also numerous indications that the fires were ignited by arson across a broad front. Unfortunately, their exact background remains unknown to this day (Saavedra, 2017).

This calls for a coordinated effort of risk management that gives regional knowledge an explicit edge and is designed as a long-term learning process. Risk management primarily covers individual views of risk. In contrast, risk governance offers a holistic approach that assigns the social responsibilities of risk takers and also considers negative externalities for other societal stakeholders in spatiotemporal variation. Of course, this does not mean that risk management has no value at all. The risk governance concept of the International Risk Governance Council, an international think tank that aims at improving the understanding and assessment of risk and the ambiguities involved, integrates risk management into a larger process of risk governance. Risk communication, which includes the transmission of risk data, but also the transfer of sophisticated risk knowledge forms the connecting link in the governance process, which is circular and reflexive by nature (Renn, 2008, p. 374). Risk governance practices take the attributes of ambiguity and uncertainty in risk knowledge into account and open debates on risk that invite many societal stakeholders to take part in risk evaluation and risk assessment processes. In particular, risk governance is about transparency, which enables collective learning for a more resilient risk society.

Researchers must take the complexity, ambiguity, and uncertainty of knowledge on forest fire risks in Chile seriously. These attributes originate from social science perspectives on risk. Empirically, one cannot always separate them from each other, which is why they cannot have the objective of structuring risk governance uniformly. However, their transparent application in risk management or risk governance has consequences. The comparative clarification of the attributes communicates to society how risk is defined and legitimized in the future (see Bustos, Lukas, Stamm, & Torre, 2019, for a similar argument related to regional crisis management in Chile). A lack of societal participation and acceptance of normative settings can lead to adverse reaction and resistance (see, e.g., the case study on forest fires in California by Simon & Dooling, 2013).

Since forest fires are immanently spatial, the regional level promises to be a suitable scale for initializing the organization of related risk governance processes. At the regional level, changing interrelationships in risk formation can be observed and interpreted early on (Müller-Mahn et al., 2018). However, as I have shown with my statements and empirical analyses in this contribution, forest fire risks can be technically detached from the spatial dimension, especially in terms of their economic effects. Decontextualization is what makes it possible to deal with risk financially in first place. In Chile, it gave rise to a reciprocal complementary relationship between hierarchy and market forms of risk management. Are these complementarities valuable for risk governance approaches as well? Some of the knowledge and guidance on certain risk management practices on the regional scale come directly from the insurance sector. Insurers utilize contract design and negotiations to offer direct and indirect incentives to decouple wildfire risks and learn how to manage them in a routinely manner. However, the extent to which different forms of risk management practices complement each other in a regional context of risk governance and lead to a higher level of local-global knowledge that strengthens the resilience of the society in the long term remains a pending issue that requires additional research efforts beyond the example of forest fires in Chile.

References

Agroseguro. (2018). Seguro de explotaciones forestales: Condiciones [Forestry insurance: Conditions]. Retrieved from https://agroseguro.es/fileadmin/propietario/Productos/FORESTALES/319%20INCENDIOS%20FORESTALES/PLAN_2018/CES-319-18-1.0.pdf

Altamirano, A., Salas, C., Yaitul, V., Smith-Ramirez, C., & Ávila, A. (2013). Influencia de la heterogeneidad del paisaje en la ocurrencia de incendios forestales en Chile Central [Influence of landscape heterogeneity on the occurrence of forest fires in Central Chile]. *Revista de Geografía Norte Grande, 55,* 157–170. https://doi.org/10.4067/S0718-34022013000200011

Arnaldos, J., Navalón, X., Pastor, E., Planas, E., & Zárate, L. (2004). Manual de ingeniería básica para la prevención y extinción de incendios forestales [Engineering manual for the prevention and extinction of forest-fires]. Madrid: Mundi-Prensa.

Aspers, P. (2009). Knowledge and valuation in markets. *Theory & Society, 38,* 111–131. https://doi.org/10.1007/s11186-008-9078-9

Aylwin, J. (2017, February 7). Incendios y modelo forestal: Una estrecha relación [Wildfires and the forestry model: A strong relationship] [Opinion]. *The Clinic* (Santiago de Chile). Retrieved from https://www.theclinic.cl/2017/02/07/columna-jose-aylwin-incendios-modelo-forestal-una-estrecha-relacion/

Barton, J. R., & Román, Á. (2012). Social movement strategies for articulating claims for socio-ecological justice: Glocal asymmetries in the Chilean forestry sector. *Globalizations, 9,* 869–885. https://doi.org/10.1080/14747731.2012.739346

Beck, U. (1992). *Risk society: Towards a new modernity.* London: SAGE. (Original work published 1986)

Beck, U. (2006). Living in the world risk society. *Economy and Society, 35,* 329–345. https://doi.org/10.1080/03085140600844902

Becker, M. C., & Knudsen, T. (2005). The role of routines in reducing pervasive uncertainty. *Journal of Business Research, 58,* 746–757. https://doi.org/10.1016/j.jbusres.2003.10.003

Beckert, J. (2016). *Imagined futures: Fictional expectations and capitalist dynamics.* Cambridge, MA: Harvard University Press.

Bottaro, G., Roco, L., Pettenella, D., Micheletti, S., & Vanhulst, J. (2018). Forest plantations' externalities: An application of the analytic hierarchy process to non-industrial forest owners in central Chile. *Forests, 9,* 141–162. https://doi.org/10.3390/f9030141

Brugnach, M., & Ingram, H. (2012). Ambiguity: The challenge of knowing and deciding together. *Environmental Science and Policy, 15*(1), 60–71. https://doi.org/10.1016/j.envsci.2011.10.005

Bustos, B., Lukas, M., Stamm, C., & Torre, A. (2019). Neoliberalismo y gobernanza territorial: Propuestas y reflexiones a partir del caso de Chile [Neoliberalism and territorial governance: Proposals and reflections based on the case of Chile]. *Revista de Geografía Norte Grande, 73,* 161–183. https://doi.org/10.4067/S0718-34022019000200161

Cabantous, L. (2007). Ambiguity aversion in the field of insurance: Insurers' attitude to imprecise and conflicting probability estimates. *Theory and Decision, 62,* 219–240. https://doi.org/10.1007/s11238-006-9015-1

Castillo, M. E., Molina-Martínez, J. R., Rodríguez y Silva, F., & Julio, G. H. (2013). A territorial fire vulnerability model for Mediterranean ecosystems in South America. *Ecological Informatics, 13,* 106–113. https://doi.org/10.1016/j.ecoinf.2012.06.004

Castro, R., & Chuvieco, E. (1998). Modeling forest fire danger from geographic information systems. *Geocarto International, 13,* 15–23. https://doi.org/10.1080/10106049809354624

Caviedes, J. A. (2017). *Construyendo sobre cenizas: ¿Son utilizados los incendios forestales como una herramienta informal para la expansión urbana de Chile central?* [Building on ashes: Are forest fires used as an informal tool for urban expansion in central Chile?] (master's thesis). Pontificia Universidad Católica de Chile, Santiago de Chile. Retrieved from http://estudiosurbanos.uc.cl/images/tesis/2018/MHM_JonathanCaviedes.pdf

Chichilnisky, G., & Heal, G. M. (1998). Managing unknown risks: The future of global reinsurance. *The Journal of Portfolio Management, 24,* 85–91. https://doi.org/10.2139/ssrn.1375656

Cifuentes, R., Desormeaux, J., & González, C. (2002). Capital markets in Chile: From financial repression to financial deepening. *BIS Papers, 11,* 86–102. Retrieved from https://ideas.repec.org/h/bis/bisbpc/11-06.html

Clapp, R. A. (1995). Creating competitive advantage: Forest policy as industrial policy in Chile. *Economic Geography, 71,* 273–296. https://doi.org/10.2307/144312

Clark, G. (2018). Learning-by-doing and knowledge management in financial markets. *Journal of Economic Geography, 18,* 271–292. https://doi.org/10.1093/jeg/lby005

CMPC [Compañía Manufacturera de Papeles y Cartones]. (2018). Integrated Report CMPC 2017. Retrieved from https://s23.q4cdn.com/927837516/files/doc_financials/annual/Integrated-Report-2017.pdf

CONAF [Corporación Nacional Forestal]. (2008). Ley sobre recuperación del bosque nativo y fomento forestal y reglamentos [The Chilean law of recuperation of native forest and forestry promotion]. Retrieved from http://www.conaf.cl/wp-content/files_mf/1368741650LibroLey_Bosque_NativoReglamentos.pdf

CONAF [Corporación Nacional Forestal]. (2017). Descripción y efectos 'tormenta de fuego': 18 de enero al 5 de febrero de 2017, regiones O'Higgins, el Maule y Biobío [Characteristics and effects of the fire-storm 2017]. Retrieved from http://www.conaf.cl/tormenta_de_fuego-2017/DESCRIPCION-Y-EFECTOS-TORMENTA-DE-FUEGO-18-ENERO-AL-5-FEBRERO-2017.pdf

CONAF [Corporación Nacional Forestal]. (2018). Statistics published on the website. Retrieved from http://www.conaf.cl/centro-de-documentacion/

Crang, M. (2002). Qualitative methods: The new orthodoxy? *Progress in Human Geography, 26,* 647–655. https://doi.org/10.1191/0309132502ph392pr

Crouch, C., & Keune, M. (2012). The governance of economic uncertainty: Beyond the 'new social risks' analysis. In G. Bonoli & D. Natali (Eds.), *The politics of the new welfare state* (pp. 45–67). Oxford, UK: Oxford University Press. https://doi.org/10.1093/acprof:oso/9780199645244.003.0003

Dean, B., Doyle, A., & Ericson, D. (2003). *Insurance as governance*. Toronto, Canada: University of Toronto Press. https://doi.org/10.3138/9781442676220

Dransch, D., Rotzoll, H., & Poser, K. (2010). The contribution of maps to the challenges of risk communication to the public. *International Journal of Digital Earth, 3,* 292–311. https://doi.org/10.1080/17538941003774668

Eisenhardt, K. M. (1989). Building theories from case study research. *The Academy of Management Review, 14,* 532–550. https://doi.org/10.2307/258557

Ericson, R. V., & Doyle, A. (2004). *Uncertain business: Risk, insurance, and the limits of knowledge*. Toronto, Canada: University of Toronto Press.

Farré, J. (2005). Comunicación de riesgo y espirales del miedo [Risk communication and spirals of fear]. *Comunicación y Sociedad, 3,* 95–119.

Flick, U. (2018). *An Introduction to qualitative research*. London: SAGE.

Focacci, G. (2017, July 4). *Investigan relación de incendios forestales con plaga de avispas y pago de seguros* [Investigation of the relationship between forest fires, wasp infestation and insurance payments]. *biobiochile.cl* (Chile). Retrieved from https://www.biobiochile.cl/noticias/nacional/chile/2017/07/04/investigan-relacion-de-incendios-forestales-con-plaga-de-avispas-y-pago-de-seguros.shtml

Freeman, P. K., & Kunreuther, H. (1997). *Managing environmental risk through insurance*. Studies in Risk and Uncertainty: Vol. 9. Dordrecht: Springer. https://doi.org/10.1007/978-94-011-5360-7

Furubotn, E. G., & Richter, R. (2005). *Institutions and economic theory: The contribution of the new institutional economics* (2nd ed.). Ann Arbor, MI: The University of Michigan Press. https://doi.org/10.3998/mpub.6715

Gatica, F. (2012). Prácticas empresariales con potencial de vinculación territorial: El caso del sector forestal chileno [Business practices embedded in territories: The case of the Chilean forestry sector]. *Revista de Geografía Norte Grande, 52,* 127–143. https://doi.org/10.4067/S0718-34022012000200008

Gerber, J.-F. (2011). Conflicts over industrial tree plantations in the South: Who, how and why? *Global Environmental Change, 21,* 165–176. https://doi.org/10.1016/j.gloenvcha.2010.09.005

Glückler, J., Rehner, J., & Handke, M. (2019). Gobernanza, redes y territorio: Editorial [Governance, networks and territory: Editorial]. *Revista de Geografía Norte Grande, 74,* 5–20. https://doi.org/10.4067/S0718-34022019000300005

Gobierno de Chile. (2017). *Chile y la "tormenta de fuego": Informe incendios forestales—Enero-Febrero de 2017* [Chile and the "firestorm": Report on forest fires-January-February 2017]. Santiago de Chile: Ministerio del Interior y Seguridad Pública. Retrieved from http://biblioteca.digital.gob.cl/bitstream/handle/123456789/64/2017-03-29%20Chile%20y%20la%20Tormenta%20de%20Fuego%20-%20final%20-%20para%20distribucion%20limitada.pdf?sequence=1&isAllowed=y

Goméz-González, S., & Espósito, R. C. (2017). Debate sobre las causas de los incendios forestales en Chile [Debate on the causes of forest fires in Chile]. *Revista Mundo Forestal, 32,* 16–29. Retrieved from http://cifag.cl/wp-content/uploads/2017/08/RM32_baja-1.pdf

Gómez-González, S., González, M. E., Paula, S., Díaz-Hormazábal, I., Lara, A., & Delgado-Baquerizo, M. (2019). Temperature and agriculture are largely associated with fire activity in Central Chile across different temporal periods. *Forest Ecology and Management, 433,* 535–543. https://doi.org/10.1016/j.foreco.2018.11.041

González, M. E., Gómez-González, S., Lara, A., Garreaud, R., & Díaz-Hormazábal, I. (2018). The 2010–2015 megadrought and its influence on the fire regime in central and south-central Chile. *Ecosphere, 9*(8), Article e02300. https://doi.org/10.1002/ecs2.2300

González-Hidalgo, M., & Zografos, C. (2017). How sovereignty claims and "negative" emotions influence the process of subject-making: Evidence from a case of conflict over tree plantations from Southern Chile. *Geoforum, 78,* 61–73. https://doi.org/10.1016/j.geoforum.2016.11.012

Hacking, I. (1990). *The taming of chance.* Ideas in Context: Vol. 17. Cambridge, UK: Cambridge University Press. https://doi.org/10.1017/CBO9780511819766

Hanson, J. D., & Kysar, D. A. (1999). Taking behavioralism seriously: The problem of market manipulation. *New York University Law Review, 74,* 632–749.

Hayek, F. A. (1945). The use of knowledge in society. *The American Economic Review, 35,* 519–530. Retrieved from www.jstor.org/stable/1809376

Hayter, R., & Clapp, A. (2020). The remapping of forest governance: From shareholder to stakeholder. In J. Glückler, G. Herrigel, & M. Handke (Eds.), *Knowledge for governance* (pp. 375–395). Knowledge and Space: Vol. 15. Cham: Springer. https://doi.org/10.1007/978-3-030-47150-7_16.

HDI [Haftpflichtverband der Deutschen Industrie]. (2018). Solvency and financial condition report 2017. Retrieved from https://www.talanx.com/~/media/Files/T/Talanx/reports-and-presentations/2018/SFCR/2017_hdi_gruppe_sfcr_en.pdf

Hechavarria, M. (2018, May 17). Avisoran crisis económica en la industria maderera de la región [Warnings of economic crisis in the region's timber industry]. *El Centro* (Talca), p. 2. Retrieved from https://issuu.com/diarioelcentro/docs/diario_17-05-2018_36b8c13b9e1203/2

Hellwig, M. F. (1983). Moral hazard and monopolistically competitive insurance markets. *The Geneva Papers on Risk and Insurance—Issues and Practice, 8,* 44–71. https://doi.org/10.1057/gpp.1983.7

Ilin, T., & Varga, L. (2015). The uncertainty of systemic risk. *Risk Management, 17,* 240–275. https://doi.org/10.1057/rm.2015.15

INN [Instituto Nacional de Normalización]. (2015). *Riesgo y peligro de incendios forestales en plantaciones: Terminología y clasificación* [Risk and danger of forest fires in plantations: Terminology and classification]. (NCh3380).
Jarzabkowski, P., Bednarek, R., & Spee, P. (2015). *Making markets for acts of god: The practice of risk trading in the global reinsurance industry.* Oxford, UK: Oxford University Press. https://doi.org/10.1093/acprof:oso/9780199664764.001.0001
Johansen, I. L., & Rausand, M. (2015). Ambiguity in risk assessment. *Safety Science, 80,* 243–251. https://doi.org/10.1016/j.ssci.2015.07.028
Julio, G. (2014). Tendencias del problema de los incendios forestales en Chile [Tendencies of the wildfire problem in Chile]. *Revista Mundo Forestal, 26,* 11–16. Retrieved from http://cifag.cl/wp-content/uploads/2014/09/26.pdf
Kalthoff, H. (2005). Practices of calculation: Economic representations and risk management. *Theory, Culture & Society, 22,* 69–97. https://doi.org/10.1177/0263276405051666
Kaufman, G. G., & Scott, K. E. (2003). What is systemic risk, and do bank regulators retard or contribute to it? *The Independent Review, 7,* 371–391. Retrieved from https://law.stanford.edu/publications/what-is-systemic-risk-and-do-bank-regulators-retard-or-contribute-to-it/
Kessler, O. (2015). What price culture? Calculation, commensuration, contingency, and authority in financial practices. In B. Jessop, B. Young, & C. Scherrer (Eds.), *Financial cultures and crisis dynamics* (pp. 85–105). Routledge Frontiers of Political Economy: Vol. 189. Abingdon: Routledge. https://doi.org/10.4324/9781315773476
Knight, F. H. (1921). *Risk, uncertainty and profit.* Boston: Houghton.
Latorre, J. I., & Rojas, N. (2016). El conflicto forestal en territorio mapuche hoy [The forest conflict in Mapuche territory today]. *Ecología Política, 51,* 84–87.
Little, C., Lara, A., McPhee, J., & Urrutia, R. (2009). Revealing the impact of forest exotic plantations on water yield in large scale watersheds in South–Central Chile. *Journal of Hydrology, 374,* 162–170. https://doi.org/10.1016/j.jhydrol.2009.06.011
Loewe, V., Corti, D., Ruiz, J. M., & Lobo, F. (2017). La evaluación del riesgo de incendio en plantaciones forestales para mejorar el acceso de la pyme a seguros contra incendios [Fire risk assessment in forest plantations to improve SME access to fire insurance]. Paper presented at the 7th Congreso Forestal Español, Cáceres, Extremadura. Retrieved from http://secforestales.org/publicaciones/index.php/congresos_forestales/article/view/19479/19191
Luhmann, N. (1993). *Risk: A sociological theory.* Berlin: Walter de Gruyter. (Original work published 1991)
Lundqvist, S. A. (2015). Why firms implement risk governance: Stepping beyond traditional risk management to enterprise risk management. *Journal of Accounting and Public Policy, 34,* 441–466. https://doi.org/10.1016/j.jaccpubpol.2015.05.002
MacKenzie, D. (2011). The credit crisis as a problem in the sociology of knowledge. *American Journal of Sociology, 116,* 1778–1841. https://doi.org/10.1086/659639
Mapuexpress (2017, February 4). Revuelo caso incendios e "intencionalidad": Zonas devastadas estaban infestadas por plagas en plantaciones forestales [Fire and "intentionality": Devastated areas were infested by plagues in forest plantations]. *Mapuexpress* (Chile). Retrieved from https://www.mapuexpress.org/2017/02/04/revuelo-caso-incendios-e-intencionalidad-zonas-devastadas-estaban-infestadas-por-plagas-en-plantaciones-forestales/
Martin, W. E., Martin, I. M., & Kent, B. (2009). The role of risk perceptions in the risk mitigation process: The case of wildfire in high-risk communities. *Journal of Environmental Management, 91,* 489–498. https://doi.org/10.1016/j.jenvman.2009.09.007
Martínez, J., Martínez, J., & Martín, P. (2004). El factor humano en los incendios forestales: Análisis de factores socio-económicos relacionados con la incidencia de incendios forestales en España [The human factor in forest fires: Analysis of socio-economic factors related to the incidence of forest fires in Spain]. In E. Chuvieco & M. d. P. Martín (Eds.), *Nuevas tecnologías para la estimación del riesgo de incendios forestales* (pp. 101–142). Madrid: Consejo Superior de Investigaciones Científicas, Instituto de Economía y Geografía.

McWethy, D. B., Pauchard, A., García, R. A., Holz, A., González, M. E., Veblen, T. T., Stahl, J., & Currey, B. (2018). Landscape drivers of recent fire activity (2001–2017) in south-central Chile. *PLOS ONE, 13,* e0201195. https://doi.org/10.1371/journal.pone.0205287

Mermoz, M., Kitzberger, T., & Veblen, T. T. (2005). Landscape influences on occurrence and spread of wildfires in Patagonian forests and shrublands. *Ecology, 86,* 2705–2715. https://doi.org/10.1890/04-1850

Millones, M. (2017, August 3). ¿Quiénes incendiaron Chile? [Who set Chile on fire?][Opinion]. *El Mostrador* (Santiago de Chile). Retrieved from https://www.elmostrador.cl/noticias/opinion/2017/08/03/quienes-incendiaron-chile/

Montalba-Navarro, R., & Carrasco, N. (2003). Modelo forestal chileno y conflicto indígena ¿ecologismo cultural mapuche? [The Chilean forestry model and indigenous conflict: Mapuche cultural ecology?]. *Ecología Política, 26,* 63–78. Retrieved from https://dialnet.unirioja.es/servlet/articulo?codigo=1255882

Moreno del Valle, F. (2015). Orientaciones y contradicciones en la Ley de Bosque Nativo [Orientations and contradictions in the Native Forest Law]. *Justicia Ambiental, 7,* 163–185. Retrieved from https://cl.boell.org/sites/default/files/libro_fima_interior_y_tapas.pdf

Müller-Mahn, D., Everts, J., & Stephan, C. (2018). Riskscapes revisited: Exploring the relationship between risk, space and practice. *Erdkunde, 72,* 197–213. https://doi.org/10.3112/erdkunde.2018.02.09

North, D. (1990). *Institutions, institutional change and economic performance*. Cambridge, UK: Cambridge University Press.

November, V. (2008). Commentary: Spatiality of risk. *Environment and Planning A, 40,* 1523–1527. https://doi.org/10.1068/a4194

O'Flanagan, P. (1997). ¿Incendiarismo o ecocidio en la Iberia Atlántica? La transformación del paisaje a causa del fuego: Reflejo de los problemas socioeconómicos [Incendiarism or ecocide in Atlantic Iberia? The transformation of the landscape by fire: A reflection of the socio-economic problems]. *Polígonos, 7,* 77–96. https://doi.org/10.18002/pol.v0i7.1050

O'Malley, P. (2004). *Risk, uncertainty and government*. London: The GlassHouse.

Perrow, C. (1986). *Complex organizations: A critical essay* (3rd ed.). New York: Random House. (Original work published 1972)

Renn, O. (2008). *Risk governance: Coping with uncertainty in a complex world*. London: Routledge. https://doi.org/10.4324/9781849772440

Renn, O. (2020). Risk governance: From knowledge to regulatory action. In J. Glückler, G. Herrigel, & M. Handke (Eds.), *Knowledge for governance* (pp. 93–111). Knowledge and Space: Vol. 15. Cham: Springer. https://doi.org/10.1007/978-3-030-47150-7_5

Reyes, R., & Nelson, H. (2014). A tale of two forests: Why forests and forest conflicts are both growing in Chile. *International Forestry Review, 16,* 379–388. https://doi.org/10.1505/146554814813484121

Rojas, N., & Miranda, O. (2015). Dinámica sociopolítica del conflicto y la violencia en territorio mapuche: Particularidades históricas de un nuevo ciclo en las relaciones contenciosas [Socio-political dynamics of the conflict and violence in Mapuche territory]. *Revista de Sociología, 30,* 33–69. https://doi.org/10.5354/0719-529X.2015.46411

Rosa, E. A., Renn, O., & McCright, A. M. (2014). *The risk society revisited: Social theory and governance*. Philadelphia, PA: Temple University Press.

Saavedra, O. (2017, January 29). *Productores de madera: "Tenemos temores fundados ... Nos enfrentamos al terrorismo"* [Timber producers: "We have well-founded fears ... We are facing terrorism"]. *El Mercurio* (Santiago de Chile). Retrieved from http://www.economiaynegocios.cl/noticias/noticias.asp?id=331220

Salvati, L., & Ranalli, F. (2015). "Land of fires": Urban growth, economic crisis, and forest fires in Attica, Greece. *Geographical Research, 53,* 68–80. https://doi.org/10.1111/1745-5871.12093

Segovia, P., Basulto, O., & Zambrano, P. (2018). Imaginarios sociales y representaciones: Su aplicación a análisis discursivos en tres ámbitos diferentes [Social imaginaries and representations: Its application to discursive analysis in three different fields]. *EMPIRIA: Revista de Metodología de Ciencias Sociales, 41,* 79–102. https://doi.org/10.5944/empiria.41.2018.22605

Seguel, A. (2018). Chile: Mega incendios forestales, crímenes empresariales e impunidad [Chile: Mega forest fires, corporate crime and impunity]. *World Rainforest Movement Bulletin, 238.* Retrieved from https://wrm.org.uy/es/articulos-del-boletin-wrm/seccion1/chile-mega-incendios-forestales-crimenes-empresariales-e-impunidad/

Sepúlveda, J. (2013, January 4). La historia de los autoatentados incendiarios de forestal Mininco en 1999 [The story of the self-inflicted arsonists of Mininco in 1999]. *Gamba* (Santiago de Chile). Retrieved from http://www.gamba.cl/2013/01/la-historia-de-los-autoatentados-incendiarios-de-forestal-mininco-en-1999/

Simon, G. L., & Dooling, S. (2013). Flame and fortune in California: The material and political dimensions of vulnerability. *Global Environmental Change, 23,* 1410–1423. https://doi.org/10.1016/j.gloenvcha.2013.08.008

Soin, K., & Collier, P. (2013). Risk and risk management in management accounting and control. *Management Accounting Research, 24,* 82–87. https://doi.org/10.1016/j.mar.2013.04.003

Stiglitz, J. E. (1983). Risk, incentives and insurance: The pure theory of moral hazard. *The Geneva Papers on Risk and Insurance, 8,* 4–33. https://doi.org/10.1057/gpp.1983.2

Tapia, G., & Castillo, M. (2014). Propuesta de diseño de un sistema de torres de detección de incendios forestales: Aplicación a la región metropolitana de Chile central [Proposal for the design of a forest fire detection tower system: Application to the metropolitan region of central Chile]. *Bosque, 35,* 399–412. https://doi.org/10.4067/S0717-92002014000300014

Úbeda, X., & Sarricolea, P. (2016). Wildfires in Chile: A review. *Global and Planetary Change, 146,* 152–161. https://doi.org/10.1016/j.gloplacha.2016.10.004

van Asselt, M., & Renn, O. (2011). Risk governance. *Journal of Risk Research, 14,* 431–449. https://doi.org/10.1080/13669877.2011.553730

van Dam, C. (2006). Empresas forestales y comunidades rurales en el centro-sur de Chile: Externalidades sociales de un modelo 'exitoso' [Forestry companies and rural communities in south-central Chile: social externalities of a 'successful' model]. *Debate Agrario, 40–41,* 225–243.

van Dijk, T. A. (2014). *Discourse and knowledge.* Cambridge, UK: Cambridge University Press. https://doi.org/10.1017/CBO9781107775404

van Holt, T., Binford, M. W., Portier, K. M., & Vergara, R. (2016). A stand of trees does not a forest make: Tree plantations and forest transitions. *Land Use Policy, 56,* 147–157. https://doi.org/10.1016/j.landusepol.2016.04.015

Vélez, R. (2009). *La defensa contra incendios forestales: Fundamentos y experiencias* [Protection against forest fires: Fundamentals and experiences] (2nd ed.). Madrid: McGraw-Hill.

Vergara-Díaz, G., Sandoval-Vásquez, V. A., & Herrera-Machuca, M. A. (2017). Spatial distribution of forest plantations in southern Chile, an area with a pulp mill. *Revista Chapingo Serie Ciencias Forestales y del Ambiente, 23,* 121–135. https://doi.org/10.5154/r.rchscfa.2015.09.045

Viscusi, W. K., & Magat, W. A. (1987). *Learning about risk: Consumer and worker responses to hazard information.* Cambridge, MA: Harvard University Press.

Young, O. R. (2010). Institutional dynamics: Resilience, vulnerability and adaptation in environmental and resource regimes. *Global Environmental Change, 20,* 378–385. https://doi.org/10.1016/j.gloenvcha.2009.10.001

Chapter 9
Carbon Markets, Values, and Modes of Governance

Janelle Knox-Hayes, Jarrod Hayes, and Erik-Logan Hughes

Market Governance at the Interface of Competing Logics and Modes of Organization

Market governance of climate change is situated at the interface of two competing logics: universalistic governance predicated on technocratic norms, and the particularities of politics embedded in local cultures. Here, *local* refers to subglobal political units that are internally constructed as political entities. The issue of how global-level ideational frameworks are translated into local contexts has formed an important part of discussions on the green economy and environmental governance. A particular focus within these debates is whether and the extent to which green economy concepts and practices constitute a post-political project. The term "post-political" here references the idea that political problems surrounding environmental futures and resource allocation are removed from political discourse and recast in technical language (Garsten & Jacobsson, 2007; see also Chap. 4 by Stehr).

To this end, markets represent a universalistic or technocratic logic whose proponents contend that global problems can be solved with the application of one-size-fits all solutions derived from science and economics (Bailey & Wilson, 2009). Interwoven into this logic is the idea that markets can be made to work more effectively for environmental and social equity through the recalibration of economic valuations aimed at making environmental investments more desirable not just on moral grounds or as responses to practical problems, but also because they offer profitable investment options (Newell & Paterson, 2010). This logic is reductionist,

J. Knox-Hayes (✉) · E.-L. Hughes
Department of Urban Studies and Planning, Massachusetts Institute of Technology, Cambridge, MA, USA
e-mail: jankh@mit.edu

J. Hayes
Department of Political Science, University of Massachusetts, Lowell, MA, USA
e-mail: Jarrod_Hayes@uml.edu

J. Glückler et al. (eds.), *Knowledge for Governance*, Knowledge and Space 15,
https://doi.org/10.1007/978-3-030-47150-7_9

operates at scale, and is built from codified knowledge that exists independent of the particularities of location.

Treating climate change as a problem of internalizing economic externalities through the pricing of greenhouse gases gives authority to a particular set of economic principles. The technocratic prescriptions for resolving climate change rely on metrics to measure the effects of climate change, establishing quantitative baselines, pricing emissions that exceed quantity limits, and building a market-based governance institution to control emissions quantities through price. The role of the Intergovernmental Panel on Climate Change (IPCC) and United Nations Framework Convention on Climate Change (UNFCCC) is to build and disseminate the universalistic solution. Carbon markets will be more successful to the extent that they operate on a global scale, so proponents craft the architecture to achieve fluidity and permeate other spheres of economic activity. In theory, a global market would also allow emissions reductions to be achieved in the places of least cost by the actors who can most afford to pay for emissions reductions.

The universalistic assumptions underlying global climate policy are similar to the high-modernist logics in centrally planned state projects (Scott, 1998). These projects often fail because they are based on presumptions about the effectiveness of state coordination and do not take into account local or practical knowledge (*metis*) that arises from everyday life. Proponents of Community Based Conservation similarly argue that centralized conservation planning, by failing to engage local knowledge, sets development and conservation programs up for disaster (Goldman, 2003). The application of technocratic approaches to climate governance highlights both the universalistic assumptions about market financialization (cf. Hayek, 1945) that underlie the post-political consensus and how those holding them fail to appreciate the ways scale and place impact on political and economic behavior. The Kyoto Protocol's requirement for commoditization, standardization of measurement, and homogenization matches well with high-modernist ideology.

The Kyoto Protocol's authors set out the trade of certified emissions reductions as a principal policy response to climate change. The Protocol's framework thus positions commodification through the expansion of carbon trading systems as central to resolving the problem of climate governance. As institutions, carbon markets can be expected to transfer and develop with a degree of isomorphism. Indeed, to some extent the hope of building global governance through interconnection of regional schemes relies on the compatibility of markets in each jurisdiction. However, by focusing on market mechanisms, the Protocol's authors marginalize other policies (taxation, command-and-control, technology transfer) that may be more effective in some political economies (Knox-Hayes, 2012). Although governments have mobilized a primarily techno-economic fix to address climate change, this is not necessarily indicative of the solutions desired by various societal stakeholders (Owens & Cowell, 2011).

Furthermore, carbon markets must be translated from the global scale into the particularities of each operating jurisdiction. Here the universalistic logic encounters different sociopolitical and cultural governance logics and values operating on the ground. Variations in political-economic cultural contexts—understood as socially established norms, rules, and expectations that define how social actors operate and interact (Katzenstein, 1996, p. 6)—play a key role in shaping how societies respond to global imperatives. Culture influences the conduct of economic interactions within specific societies as well as the role of economic processes in policy and society. As the markets are constructed, everything from the nature of the legislation that is developed, to the organizations used to operationalize the markets, to the way in which various polities respond to the idea of market-based governance, are affected. The dynamics on the ground affect not only the shape that the markets take, but also the ways in which they perform.

The variability in market form causes problems for efforts to establish climate governance arrangements that rely on universalistic assumptions about socioeconomic systems. Cultures are not monolithic, of course, and vary as the scale and location of analysis shift, and care is needed in drawing conclusions about the influence, causal or otherwise, of culture. Nonetheless, there is value in looking at the influence of culture at different scales as a means by which to draw attention to differences and the reasons for these differences. Moreover, by examining links between culture and climate governance we recognize that climate policy is fundamentally political in nature.

Another important point of analytical concern is the issue of scale. Subnational governments, nongovernment organizations, corporations and government agencies in "hybrid" (combining state and non-state actors), and transnational environmental governance networks play an increasingly important role in climate governance (Andonova, Betsill, & Bulkeley, 2009; Betsill & Bulkeley, 2007; Hoffmann, 2011). One of the effects of this dispersal of governance is a resistance to efforts to establish universalistic environmental governance arrangements as regional/local governments, businesses, and communities reinterpret governance concepts through the particular lenses of their beliefs, traditions, circumstances, and dilemmas (Krueger & Gibbs, 2010). The creation of standards and agreements through which carbon management occurs can similarly vary (Ocampo, 2011). As a consequence, the institutional landscape of carbon governance is highly variegated across initiatives, actors, and countries. The translation of international commitments into action remains reliant on and imbedded within territorially bound politics (While, Jonas, & Gibbs, 2010). Thus, this plurality of approaches can in part be seen as a response by the various actors involved in promoting and implementing the carbon governance to integrate—and potentially challenge—neoliberal capitalist attempts to fit environmental problems within prevailing political-economic paradigms (Pattberg, 2007; Redclift, 2012).

The organizational terrain of policy further complicates the carbon governance landscape. Specifically, political economic culture interacts with modes of network governance. Provan and Kenis (2008) provide a useful typology for mapping networks: participant-governed networks (shared governance), lead organization-governed networks (lead organization), and network administrative organizations (NAOs). These network governance types are informed by their position along two dimensions. The first addresses centralization. For networks governed internally by participant organizations, those that exhibit a centralized broker are classified as lead organizations. Those in which governance is decentralized across participants are classified as shared governance. The second axis of classification deals specifically with centralized modes of governance. Here, the distinction lies between those that have a centralized broker that is internal to the network (lead organization) versus those in which the broker is an organization external to the network (network administrative organization/NAO).

Taken together, logics of governance and modes of governance represent two important axes along which climate policy can be mapped and assessed. In this chapter, we seek to assess how policy intersects with these axes and in the process provide a broad-based qualitative and quantitative assessment of how geographically specific sociocultural factors shape intersubjective understandings of markets in general and carbon markets in particular. To do so, we adopt a cross-national perspective, examining and evaluating the intersubjective meanings of carbon market formation drawn from interview data of market makers across the United States, Australia, China, the EU, Japan, and South Korea. In the next section, we address the role of technocratic norms, and in so doing provide an extensive overview of our methodology.

Technocratic Norms and Political Context

Methods of Analysis

Drawing on 245 interviews with actors from various institutions involved in climate policy and market development, including professional service firms, legal firms, and regulatory agencies in the EU, US, Australia, South Korea, China (including Hong Kong), and Japan, we utilize techniques from grounded theory to explore perspectives on market-based climate governance. The interviews were semi-structured to guide the conversations while allowing respondents to address topics they considered significant (Clark, 1998) and lasted between 45 min and 2 h. Individuals were asked about the nature of their firms, practices, network relationships, perspectives on climate policy and market-based governance, and the importance of various types of expertise in developing climate policy and markets. We then triangulated the responses with each other and with relevant policy and organizational documents.

We coded interviews to generate insights into the relationship between technocratic norms and politics in climate policy, and to identify attributes of market-based governance in each geographic region. We deployed two coding techniques. First, we treated the interview transcripts as data and analyzed them in a coding pattern from raw text to first-order concepts, and then to analytic categories (Eisenhardt & Graebner, 2007). Specifically, we coded two questions to generate insights into how market-based governance is perceived in each region: What are the advantages or disadvantages of using markets to govern climate change and what are the opportunities and challenges of the creation of carbon markets? We used our analysis of these questions to gauge positive and negative associations of market-based climate governance, as well as to identify governance concepts that are positively associated with market-based governance in the form of opportunities and challenges.

Second, following Gioia (1998), we treated interlocutors as "knowledgeable agents," people who know what they are trying to do and can explain their thoughts, intentions, and actions. This grounds the study in accounts of the informants' experience (Gioia, Corley, & Hamilton, 2013). The knowledge provided by key informants informs our understanding of the ways in which each region has negotiated tensions between technocratic governance norms and place-specific politics. We have used anonymized quotations to support key observations and combine them with relevant policy, organizational, and press material where appropriate.

Coded Concepts of Market-Based Governance

Through the coding of the interview data, we identified seven prominent perceptions of the advantages and opportunities of carbon emissions markets (efficiency, technology investment and innovation, global scale, flexibility to participate, political viability, and capacity to reduce emissions) and six prominent disadvantages and challenges (reliance on governance cohesion, political uncertainty, technical complexity, time to translate, prevalence of economics, and intangibility).

Advantages and opportunities

The concept of *governance efficiency* refers to the fact that markets are a more efficient form of governance. Respondents often identify efficiency with the ability to resolve climate change using economic techniques that would not require significant behavioral changes, particularly by bringing about technological development. Similarly, the concept of *efficacy* is described as providing transparency for government and industries to make production decisions through the carbon price. There was an associated belief that the price is useful in identifying and distributing the "real" value of low-carbon production.

Some participants identified the benefits of markets as being associated with the *flexibility* they give various actors to participate in the governance of greenhouse gas emissions. Unlike taxes and command and control types of regulation (such as efficiency standards or renewable energy portfolios), the market generates a profit incentive and allows a range of organizations from banks to professional service firms to participate. Many individuals also expressed the belief that the market is more flexible as a governance mechanism in that it allows actors to purse a variety of strategies to reduce emissions rather than having the government mandate a set course of action. For these reasons and others, participants also identified *political viability* as an advantage of emissions markets. In many jurisdictions there is less political resistance to emissions trading than to other forms of governance. Some interlocutors argue that a market is the only palatable solution for industry. This is a popular perspective in the United States as well as in China, where there is considerable resistance to the idea of taxes.

The emphasis on *investment and innovation* reflects a belief that through the creation of a price signal carbon markets effectively distribute revenue to the "best" developers of technology. Respondents additionally identify the ability to *operate on a global scale* as a market advantage, suggesting that carbon markets can provide liquidity, and the potential for offshoring (of finance) and offsetting (of financial credits) in ways other governance mechanisms cannot. This argument primarily rests on the idea that markets are transnational and do not respect national boundaries, which stands in contrast to other approaches like command and control or taxation where the generative force of action lies firmly within states. Also note the emphasis of scale and how it is seen as generating ease and affordability. Finally, several respondents (thought fewer than might be expected) identify the ability to *reduce emissions* as an advantage of market-based governance. Not surprisingly, they view the pathway to emissions reduction through the lens of technological development rather than behavioral change. Participants express confidence that a price for carbon will give renewable energy technologies an advantage.

Disadvantages and challenges

Participants identify six primary disadvantages and challenges of market-based governance. Although markets have the ability to operate at the international level, participants identified *reliance on governance cohesion*, or the idea that markets need to operate according to common rules in order to be effective, as an associated disadvantage. As witnessed by the challenges in achieving binding targets under the UNFCCC, this is a major challenge for climate governance. Additionally, participants identify domestic *political uncertainty* as a major challenge, because carbon markets are derived from and reliant upon politics for their existence. Here the belief seems to be that carbon markets are efficient except for political contestation, which is thought to destabilize them.

In addition, participants identify the *technical complexity* required for markets to operate effectively. In addition to the basic rules and regulations of the capping of emissions and distribution of allowances, to be effective the markets require a system of standards and infrastructure to measure and monitor emissions as well as enforcement mechanisms for the actors under the cap. There is considerable technical complexity required to build such a system, and unless the policy details are correct, there is strong potential for the system to fail. Many interlocutors point to the early phases of the EU Emission Trading System (ETS) as an example of the technical complexity required to build an emissions trading system as well as what happens (e.g., price collapse) when the technical details are incorrect.

Associated with this concept is the idea that it takes *time to translate* markets into new jurisdictions and that they are a slower form of governance. The concept is particularly raised by participants in Asia, with recognition that to be effective, the markets not only have to be built to operate according to a technically high capacity, but that they must also be translated to fit different sociopolitical and cultural norms in the jurisdictions where they are being built.

One of the drawbacks identified is that markets give *prevalence of economics* over the environment and often achieve economic gains without environmental benefits. In particular, markets treat climate change as though it were purely an economic problem, with economic concerns. Many actors join the markets not to reduce emissions but to pursue the profit incentive. This concept is associated with concern that markets do not reduce emissions because of the inability to *price the intangible*. Here there is recognition that the absence of carbon dioxide is an intangible commodity, and considerable skepticism over the concept of additionality of offsetting (the requirement that allowances and offsets produce emissions additional to what otherwise would have occurred). The difficulty in pricing carbon dioxide and associated negative and positive externalities generates doubt as to whether emissions reductions are real.

We have summarized these concepts with exemplary quotes in Tables 9.1a and 9.1b (Knox-Hayes, 2016, p. 53). Quotes in the first section of the table exemplify universalistic conceptions of markets as solutions to climate change. The second section embodies counter-conceptions that underline the political nature of markets and doubts about their universality. Importantly, many of the comments in the second section of the table stress the role of scale and cultural difference in shaping how universal, technocratic norms are interpreted and applied, such as the difficulties of developing common global rules for carbon markets. The two sets of observations in turn provide a basis for synthesizing these cultural variances.

Comparisons Between East and West

We identified the coded concepts as common responses to the questions of market advantages/disadvantages and opportunities/challenges. Given that there were 245 responses to the data recorded, it was possible to do statistical analysis of the data to look for macro trends across the regions. In Table 9.2 we provide a count summary of the identified concepts according to location, dividing the region of residence into six categories: the United States, the European Union, Australia, China, Japan, and Korea (Knox-Hayes, 2016, p. 59). It should be noted that to maintain statistically comparable sample sizes, respondents from Hong Kong have been grouped into the China category. Hong Kong serves as one of the financial centers for the trade of carbon out of China, and these interlocutors were asked questions specific to the developing Chinese carbon markets. London was likewise coded as part of the EU. It is one of the financial centers of the EU ETS and respondents were asked questions specific to the EU ETS. Table 9.2 shows that neither conceptions of

Table 9.1a Summary of positive concepts of market governance from coded interview data

Concept	Manifestation in the data	Illustrative quote
Advantages and opportunities		
Governance efficiency	Markets are a more efficient form of governance. It is possible to have minimal impact on the operation of business and industry and still achieve environmental objectives because markets are efficient. Efficiency is often associated in the data with the belief in a carbon price to solve climate change by driving investment into clean technology. Market efficiency suggests that technical solutions rather than behavioral change will solve climate change. The market is perceived to be the ultimate driver of technical efficiency.	Cap and trade makes sure the things we take for granted, lifestyle, transportation, etc. are not affected. With the cap and trade approach the biggest benefit is the ability to resolve the problem while maintaining business as usual! Cap and trade helps lubricate energy transaction. With the cap and trade approach you drive up the price of coal and support renewables. You could burn coal 50 years from now with the price today. There is a huge shift in renewable and other areas. The market adjusts; that is the beauty of it. (Managing Director, Market Technology Provider, New York)
Efficacy of a carbon price	The carbon price provides the transparency for government and industries to make efficient production decisions. Unlike the price generated by a tax or command and control mechanism, a market price represents the "real" value of reducing emissions. Markets, and particularly market actors, provide the transparency and liquidity in the market to transmit the price as widely as possible. The efficacy of the carbon price is often identified in in conjunction with issues of scaling up market activities.	The market provides a strong advantage of price transparency. There is liquidity in the market, which advantages everyone. They can do activities to manage their actions etc. That is why the banks are in this market, to create transparency and liquidity. (Environmental Products Team Leader, Multinational Petroleum Firm, London)
Technology investment and innovation	Through the creation of a price signal carbon markets effectively distribute revenue to the best developers of new technology. This is often described as a normative and objective process that allows the "best," "brightest," or most efficient to win. The perspective fits with the notion that markets provide individual justice in contrast to the social justice provided by governments.	That's where you get your innovation and your development and your new ideas. If you incentivize the private community—What you are effectively doing with a market-based mechanism is you're saying that the smartest person wins. And that's what drives the way to innovation. Now that might be big corporates, it might be small one-man bands; it might be the next Google in the environmental world. You're just incentivizing the smartest person at the table to be rewarded. And that's all the market-based mechanism does. (Managing Director, Brokerage, Melbourne)

(continued)

Table 9.1a (continued)

Concept	Manifestation in the data	Illustrative quote
Ability to operate on global scale	Two issues stand out under global scale. The first is the perception that markets are capable of functioning on a global scale, where other systems of governance cannot. Global scale is often associated with liquidity and greater market capitalization. These are assumed to be positive attributes of markets. The other issue of scale is the ability to dislocate emissions reduction activities through offshoring and offsetting, presumably to places where least-cost reductions can be made.	It's part of the solution, and it's the only mechanism we have that can mobilize resources and allocate them at scale, at speed. (CEO, Carbon Aggregator, Sydney) The cost structure for emissions reduction in domestic are actually quite high that there may well be better opportunities for emissions reduction offshore in the initial years … A tonne of carbon is a tonne of carbon. Whether the reduction is here or international, it shouldn't matter. (CEO, Business Council, Melbourne)
Flexibility to participate	The markets allow more individuals to participate because they are not specific only to the entities who have obligations under the cap. The markets generate a profit incentive and allow a range of organizations to participate in both the construction and operation of the market. Wide-scale participation is thought to be important to generating shared belief in the operation of markets, as well as making the markets function efficiently. The markets are additionally seen as a more flexible mode of governance; giving actors the flexibility to structure emissions reductions with the means they see best.	We appreciate cost reducing aspects of markets including flexibility and the opportunity to bring other players in. They link well and are adaptable systems that bring the cost effective emissions as well. Markets are valuable for reductions. For example with the US the market with offsets, unregulated international players can participate via offsets. It provides funds for others to get involved. (Senior Research Associate, Environmental Think Tank, Washington, DC) I think it's my personal opinion but the market approach has more advantage than recommending control, or carbon tax, or other options. That's what the research results said and that's what I think too. The company would have more flexibility to comply with carbon regulation if that's an Emission Trading Scheme rather than recommending control, or carbon tax. (Head of Climate Change Division, Accountancy, Seoul)

(continued)

Table 9.1a (continued)

Concept	Manifestation in the data	Illustrative quote
Political viability	Compared to other options like taxes and command and control regulation (energy portfolios, emissions standards, etc.), markets are more politically viable. This is in part because they operate with a profit incentive. Even if the results are theoretically comparable to those under a tax, individuals feel less penalized by a market system and that flexibility gives them greater tenure over the system. The approach also allows governments to popularize other associated green initiatives.	Low-carbon, green-growth was the national agenda of the former government of President Myung-bak Lee. ETS and Carbon-tax were considered as regulation tools for GHG reduction, but because the ETS was more symbolic, representative and popular than carbon-tax, in the political aspect, they decided to introduce an ETS in 2012. The main reason was it could be a significant signal that all business entities must be part in developing green technology or patents for GHG reduction for new green market from K-ETS. (Climate Change Leader, Government Ministry, Seoul)
		There are a couple of advantages of carbon trading schemes. They're probably harder to implement, but they're not as much viewed as a tax. It's also an issue like in the US, because we don't like taxes. Instead of taxing, you do a carbon-trading scheme. Then I guess there's also the possibility of linkages you know with other carbon trading schemes. I mean that's definitely a big trend that we're trying to go towards. (Climate and Energy Policy Director, NGO, Beijing)
Capacity to reduce emissions	Respondents express a belief that carbon markets reduce emissions. Faith in markets to reduce emissions is often attached to a technology pathway. Markets create a carbon price, which should drive investment in clean energy technology and generate emissions reductions. Associated with this perspective is often the belief that climate change is simply a problem of failing to price externalities.	As a more widespread approach, the market approach drives down cost to society, and allows the private sector to seek out the least cost methods of reducing emissions. It allows and focuses the economy on finding the least cost emissions reductions and creates investments into new technologies. We are reducing emissions by putting a price on carbon capture as an externality. The market decides the price. If we had a 100 per tonne price, we wouldn't have a climate change problem. (Director, Carbon Asset Developer, New York)

Note. Reprinted from Knox-Hayes (2016, p. 53). Copyright 2016 by Oxford University Press. Reprinted with permission

Table 9.1b Summary of negative concepts of market governance from coded interview data

Concept	Manifestation in the data	Illustrative quote
Disadvantages and challenges		
Reliance on governance cohesion	To be effective markets need to operate according to common rules and standards. Many participants identify the construction of a cohesive system of governance (at various scales) as one of the major challenges to the operation of market-based governance of climate change. Here universal logics of efficiency, efficacy and global scale compete with the messy reality of implementing policy on the ground. Even if guided by a coherent universalist perspective, policy is transformed as it is moves from region to region, leading to differences in outcome that can restrict the cohesion of markets.	I love carbon pricing and I love the notion of a budget and I love the notion of letting the free market determine the lowest cost way of doing it. Where we've been naive was that thinking the world would unite on a common set of rules. That was extraordinarily brave, and we have never done that on anything of any kind, ever. Why would this be different? We've built a system that was bound to fail from day one, which was a united global, common language. And if you just pull back for a moment, look at free trade and currency and everything else; they are geocentric, idiosyncratic, negotiated, debated. So why would carbon be any different? (Managing Director, Investment Firm, Melbourne)
Challenge of political uncertainty	Political uncertainty is associated with the fact that the markets are derived from and reliant upon politics for their existence. This is seen to be problematic both at the international scale, and in domestic politics, particularly in Australia. Respondents often express a belief that markets are efficient except for the political instability that underlines them. The efficacy of a carbon price, signaling low-carbon investment, only holds so long as the markets also have a long-term stable policy platform to underlie them.	The biggest problem in Australia with these market mechanisms is really the political uncertainty. Now we have a carbon price but you know of course today's carbon price is not what effects investments decisions it's the expected carbon price over the lifetime of an investment. And so, if half of your industry experts think the carbon price will be dead and buried in three years' time, and the other half thinks it will actually keep rising, then you've really got a problem for decision-making in investments, and that's what ultimately matters. (Professor of Economics, Sydney)

(continued)

Table 9.1b (continued)

Concept	Manifestation in the data	Illustrative quote
Technical complexity	To be effective, markets require considerable technical complexity to operate. Everything from the systems and technologies of monitoring, verifying, and reporting emissions to the rules and standards of how emissions markets operate must be built. There is considerable technical complexity and interlocutors express doubt whether the markets will function unless the technical details are correct. This is particularly a concern expressed among interlocutors in Asia. Many point to the example of the first phases of the EU ETS where lacking technical specificity (over allocation of credits for example) led to a price collapse. Additionally, from a bureaucratic standpoint, there is a sense that markets have a single correct form, and must be engineered to a correct standard in order to function effectively. This is at odds with the complexity and diversity of the types of industries that operate under a cap and require different standards and methodologies of reduction.	The disadvantages are that it is very difficult to have a fair cap setting. That is the difficult part, very difficult, the allocation of allowances and emission are theoretically impossible, so there is room for lobbying, huge room for lobbying, which actually happened in the EU and there is plenty of evidence that such fair allocation of allowances is impossible. So it is a government failure. It is a perfect system if allocation is done fairly. However, it is in reality impossible to allocate fair distribution of our caps, of our allowances, so therefore, it doesn't happen today. (Manager Market Mechanism Group, Government Think Tank, Tokyo) The ability to speculate on price is not actually achieving reductions …. We are in the process of learning how we measure, how we verify the amount of carbon footprint and how we more importantly, how we estimate the capacity or capability of companies to reduce greenhouse gases. That is critical to calculate the allowance …. If you are a government, how can you put allowance to each company that has different instruments, different processes, and different businesses? (Managing Director, Consultancy Seoul)

(continued)

Table 9.1b (continued)

Concept	Manifestation in the data	Illustrative quote
Economics takes precedence over environment	Climate change is treated as predominantly an economic problem. The solution of using markets to govern greenhouse gas emissions also creates an economic logic for solving the problem. While the profit motive brings participants into the markets, there is concern that the markets accomplish only economic and not environmental gains. Also present are concerns that economic logics of market governance take precedence over environmental concerns and benefits. The predominance of economics can be at odds with the need to generate material changes in order to reduce greenhouse gas emissions, and to transform energy supplies.	I feel like there's a lot of enthusiasm for carbon trading, but not necessarily as much concern about whether the actual emissions reduction with normal impact would be there, and how do you ensure the integrity of it. Carbon trading, like any kind of stock trading bond trading, it's a profit incentive. And I think people are really looking at carbon trading as a profit, as a business. But I think before you get to that point, you need to go into these things about the institution and data, and all that. (Climate and Energy Policy Director, NGO, Beijing) We should focus on the market, of course, but we should focus on good market, or technological market. We are not focused on, I don't know, cap and trade system, with its financing, we don't think so. In order to reduce the CO_2 emissions, we have to improve the energy efficiency or we have to reduce the carbon intensity of the energy. Then, in order to realize these two things we need renewed technologies. Technologies are really the key to saving this planet. [Cap and trade] system will not be an incentive to make research and development because [companies] are just buying credits. It seems they do not have any material impact or any real impact on emissions reductions. (Vice Chairman, Industry Association, Tokyo)

(continued)

Table 9.1b (continued)

Concept	Manifestation in the data	Illustrative quote
Time to translate	It takes time to build a market, and it particularly takes time to translate a market institution across socio-political boundaries. Many interlocutors, particularly in Asia, identified the time to operate as a drawback of the market system. They expressed the belief that to be effective the markets, while initially adapted from the model of the EU ETS, would have to be translated to take on different "cultural" norms and expectations. As part of this process, it is important that companies acculturate to the idea of market governance. Such a process takes more time than simply implementing a tax or even setting up command and control legislation.	The advantage and the disadvantage of this are because the emission trading is a totally new idea for China. So, we need a long time to develop this carbon market, to develop all the institutions that will be involved in this market, also the regulations and the regulators, or the governance of this market. It's like the stock market. The first challenge will be the registration of the permits of the market. We define the emission rights as a commodity rather than financial instruments due to the fact that in China the market is restricted, regulated by the financial security commission. That's one of the challenges we are facing at this moment. The second one is how to develop this new carbon market in China. We have to get the companies to accept the idea. We also need to develop the stakeholder engagement. (CEO, Exchange, Shenzhen)
Inability to price the intangible	There are challenges both in pricing carbon dioxide because it is an intangible asset, as well as inadequately pricing the associated positive and negative externalities attached to the carbon. The intangibility of externalities creates doubts as to whether or not emissions reductions are real. Associated with this concern is recognition that the value markets trade is a reduction in the range of social values that give meaning to environmental assets.	The disadvantages, I think, are some of the things that we haven't priced correctly over time—Ever—Are the hardest things to measure. And it's very, very difficult to include or to put a price on something that's difficult to measure .... For example, there might be local employment benefits if it's an indigenous project in the middle of Australia. There might be water quality or biodiversity benefits, which are hugely valuable in and of themselves but aren't—At the moment—Priced into the carbon market. So, we and other organizations around the world are currently trying to figure out how would you attract a price premium or place a value on these additional co-benefits. That's a market-based mechanism, and it's incredibly difficult and controversial. (Senior Associate, Consultancy, Melbourne)

Note. Reprinted from Knox-Hayes (2016, p. 53). Copyright 2016 by Oxford University Press. Reprinted with permission

Table 9.2 Count summary of governance concepts by region

	Advantages and opportunities						Disadvantages and challenges						
	Governance efficiency	Efficacy of a caron price	Flexibility	Political viability	Technology investment and innovation	Ability for global scale	Capacity to reduce emissions	Reliance on governance cohesion	Challenge of political uncertainty	Technical complexity	Time to translate	Prevalence of economics	Inability to price the intangible
US	23	23	4	2	14	22	15	17	11	1	1	1	8
Australia	25	21	0	2	7	3	1	6	11	2	0	1	6
EU	9	14	1	1	6	17	3	23	4	1	0	2	1
						213						**96**	
China	3	10	5	5	6	2	1	2	3	8	12	10	1
Japan	5	3	4	0	1	0	0	1	1	9	8	15	4
South Korea	2	1	3	2	2	2	0	0	0	9	7	2	1
						57						**93**	
Total	**270**							**189**					

Note. Reprinted from Knox-Hayes (2016, p. 59). Copyright 2016 by Oxford University Press. Reprinted with permission cultural difference nor conceptions of the technocratic/political nature of markets are universally held, thus highlighting the fundamentally contingent nature of climate policy in different national contexts despite the underlying norms that inform it at the global level.

In total, the data contain 270 identifications (some participants' responses are coded for more than one concept) of positive constructs of market-based governance and 189 identifications of negative constructs. Additionally, over half of the respondents (139 out of 245) directly identify efficiency and or the efficacy of a carbon price as advantages of market-based climate governance. The overwhelming majority of positive responses regarding the utility of carbon markets give some validity to the notion of a post-political consensus around market-based climate governance. Some interlocutors voiced support for markets even when recognizing the difficulties involved in pricing environmental externalities:

> [We] believe in a market-based instrument, a carbon price. That said, markets have a history of creating externalities and mispricing things that should have a price. So, I mean, creating a market to fix a problem that was caused by the market is ironic and I guess we understand that. But, we would say, yes, it's positive. I mean, how else can you possibly address environmental issues that are otherwise not a part of our daily economic and social life? (Senior Associate, Consultancy, Melbourne)

Surprisingly, however, only twenty respondents (8.1%) directly suggested that carbon markets have the positive benefit of reducing emissions. Many of these respondents suggested that the way in which a carbon price has a positive impact is by generating technological innovation and investment.

When the respondents are categorized as either Anglo/European (US, EU, Australia) or Asian (China, Japan, South Korea), differences in perspective are even more surprising. The respondents from Anglo/European countries produced 213 identifications of advantages and opportunities of market-based governance of climate change and only ninety-six disadvantages and challenges (a positive-negative ratio greater than 2:1). In general, respondents from these regions are much more likely to have a positive perspective of how the markets operate. In contrast, individuals from Asian countries are much more likely to have a skeptical perspective of the use of market mechanisms to govern climate change. Individuals from China, Japan, and South Korea generated fifty-seven identifications of advantages of market-based governance, in comparison with ninety-three identifications of disadvantages (2:3 positive-negative ratio).

We explore the reasons for the skepticism in Asia in greater detail below. In short, however, the differences between Anglo/European and Asian countries reflect underlying differences in beliefs regarding the role that markets play in governance, the role of different institutions and sources of authority in generating policy, as well as different aspirations and values. To better distinguish the nature of these differences, we statistically tested the significance of differences in responses across regions (Table 9.3).

Table 9.3 Correlation analyses of market constructs by region of residence

	Advantages and opportunities							Disadvantages and challenges					
	Governance efficiency	Efficacy of a caron price	Flexibility	Political viability	Technology investment and innovation	Ability for global scale	Capacity to reduce emissions	Reliance on governance cohesion	Challenge of political uncertainty	Technical complexity	Time to translate	Prevalence of economics	Inability to price the intangible
US						0.15**	0.27***			−0.22***	−0.21***	−0.23***	
Australia	0.33***	0.22***	−0.12*			−0.12**			0.20***		−0.15**	−0.13**	
EU						0.28***		0.43***		−0.12*	−0.15**		
China	−0.23***			0.12*		−0.18***	−0.22*	−0.19***			0.20***	0.11*	−0.11*
Japan		−0.15**			−0.12*	−0.16***	−0.16*	−0.16**		0.20***	0.18***	0.41***	
Korea			0.18***	0.14**						0.46***	0.36***		

Note. *Significant at the 90% Confidence Interval, **Significant at the 95% Confidence Interval, ***Significant at the 99% Confidence Interval. Reprinted from Knox-Hayes (2016, p. 60). Copyright 2016 by Oxford University Press. Reprinted with permission

Statistical Analysis of Regional and Occupations Difference

We used logistic regression and correlation analysis to evaluate differences across the regions (Table 9.3) as well as across industry sectors (Table 9.4). As with region of residence, we coded occupational background to maintain statistically comparable sample sizes. We distinguished five categories: financial services (banks, brokerages, hedge funds, exchanges), professional services (accountancies, legal firms, news wire, consulting), industry (energy companies, manufacturing companies, project developers, industry associations), government (regulatory agencies, policymakers, staffers, legislators), and NGOs (environmental advocacy, think tanks, academics).

We report statistically significant correlation coefficients in Table 9.3 (Knox-Hayes, 2016, p. 60) for region of residence tests and in Table 9.4 (Knox-Hayes, 2016, p. 61) for occupational sector analysis, and we tested responses from each category against all others. The results can be interpreted as follows: In Table 9.3, the positive 0.33 under efficiency for Australia signifies that compared to respondents in all other geographies, individuals from Australia are 33% more likely to identify governance efficiency as an advantage of market mechanisms to govern carbon emissions. The triple asterisk beside 0.33*** signifies that the difference is statistically significant at the 0.99% confidence interval. In contrast, the negative 0.23 in China's row under efficiency indicates that respondents from China are 23% less likely than respondents in all other geographies to identify governance efficiency as an advantage of market mechanisms.

Respondents in the United States are more likely to identify global scale and the ability to reduce emissions as advantages or opportunities of using emissions markets to govern climate change. In general, participants in the United States have a more positive perspective about the potential of market-based governance. There is strong belief in the utility of market-based governance in the US, which has clear linkages to the perspective that the markets can succeed in reducing emissions. The focus on global operation, particularly among individuals from the financial services industry, relates to logic that markets operate best at a global scale. Thus, carbon markets can be effective at a global level, and in the process would also displace some of the burden of reducing emissions from the United States, which is one of the largest emitters of greenhouse gases. Respondents from the United States are less likely to identify technical complexity, time to translate, and prevalence of economics as disadvantages of emissions markets. This fits with the generally neoliberal perspective and lack of concern over economic or incentive-based regulatory mechanisms.

Australian respondents are more likely to positively identify governance efficiency and efficiency of a carbon price as advantages of market-based governance. This reflects the strong emphasis on technical competence in Australian governance. Across all sectors, respondents from Australia demonstrated a high degree of technical (particularly economic) competence. By contrast, Australians are less likely to identify flexibility and global scale as advantages of market-based governance. In

Table 9.4 Correlation analyses of market governance constructs by occupational sector

Construct	Advantages and opportunities						Disadvantages and challenges						
	Governance efficiency	Efficacy of a caron price	Flexibility	Political viability	Technology investment and innovation	Ability for global scale	Capacity to reduce emissions	Reliance on governance cohesion	Challenge of political uncertainty	Technical complexity	Time to translate	Prevalence of economics	Inability to price the intangible
Financial services		0.15**				0.16***	0.13*			−0.12**			
Professional service							−0.10*			0.11*			
Industry		−0.10*											
Government									0.12*				
NGOs								−0.10*					

Note. *Significant at the 90% Confidence Interval, **Significant at the 95% Confidence Interval, ***Significant at the 99% Confidence Interval. Reprinted from Knox-Hayes (2016, p. 61). Copyright 2016 by Oxford University Press. Reprinted with permission general, the focus of Australians seems to be much more domestic, even if there is recognition that efforts to address climate change come from an obligation to be a good global citizen. With respect to disadvantages, Australians were more likely to identify the challenges of political uncertainty. This perhaps reflects the tumultuous political experience Australians have had with emissions trading. Finally, Australians are less likely to express concern over the time it takes to translate markets into local context, and less likely to identify the disadvantage of the fact that markets privilege economics over the environment (interview respondents across occupational sectors in Australia demonstrated a sophisticated understanding of, and comfort with, the application economics).

Respondents from the EU were much more likely to identify global scale as an advantage of market-based governance with a correlation coefficient of 0.28. Many respondents expressed the idea that climate change is a unifying policy because it cannot be addressed by any nation-state acting on its own. As such, it helps consolidate policy authority at the EU level. Furthermore, the EU has taken global leadership on the issue of climate change, which is a source of pride amongst policymakers and market makers. EU responses were also strongly correlated with the disadvantage of carbon markets requiring governance cohesion, with a correlation of 0.43. In line with the global focus, EU respondents acknowledge that to be effective markets must be international and interconnected, which requires a strong degree of governance cohesion that does not yet exist in the international system. The EU remains one of the strongest proponents for a binding agreement under the UNFCCC. Finally, respondents from the EU are less likely to identify technical complexity and time to translate as disadvantages of market-based governance. This aligns with the fact that the EU ETS was one of the first established systems as well as the fact that the EU, and particularly the Commission, has strong technical capacity.

Reflecting a divergence in values vis-à-vis other regions, Chinese respondents are less likely to identify efficiency, global scale, and the ability to reduce greenhouse gas emissions as advantages of market-based systems. The only advantage Chinese respondents are more likely to identify is political viability. Many interlocutors suggested that, despite the fact that the Chinese government uses many command and control measures for regulation, they are also concerned about minimizing unrest among industry and society. Markets are seen as a more politically viable approach to climate change than taxes or command and control. Chinese respondents are less likely to indicate governance cohesion as a negative aspect of market-based governance. Chinese emissions markets are in some ways more flexible because they are being tried in different regions with an experimental approach. This generates a more local focus.

The Chinese are also less likely to indicate the inability to price the intangible as a negative aspect of market-based governance. This may reflect the fact that China was the largest host country of Clean Development Mechanism (CDM) projects, and most concerns about pricing the intangible relate to offsetting in the CDM. Several respondents identified the large number of operating CDM projects as a reason the government should use markets rather than taxes or command and control to address climate change. Finally, respondents from China are more likely to identify the time to translate and the preference for economics over the environment as disadvantages of market-based governance. The respondents were concerned that markets be better tailored to cultural expectations once they are scaled to the national level and begin to have greater effect on industry. They also expressed concern that the markets would be focused more on economic benefit (i.e., profit-taking), for although the environment is increasingly a concern in China, development is still the greatest priority.

In Japan, respondents express a far more negative perspective over emissions markets than in any other region. Japanese respondents are less likely to identify efficiency, technology innovation and investment, global scale, and capacity to reduce emissions as advantages or opportunities, and more likely to identify technical complexity, time to translate, and economic prevalence over the environment as disadvantages of the markets. The Japanese place strong emphasis on the need for technological development to resolve climate change, but express considerable skepticism that carbon markets work efficiently and effectively to generate environmental impact. There is also a recognition that Japan is culturally different from other regions like the EU, where markets have been trialed, and that it would take considerable time to develop a market approach that would work in Japan. To some extent, this is the solution the Japanese have identified with their Bilateral Offset Credit Mechanism/Joint Crediting Mechanism (BOCM/JCM). Finally, the responses from Japan are negatively correlated with concern over governance cohesion, which reflects the fact that there is a much greater domestic than global focus. For example, the Joint Crediting Mechanism is designed to allow each host country that partners with Japan to independently tailor the program to their needs.

Korean respondents are more likely to identify flexibility and political viability as positive advantages of market-based governance and more likely to identify technical complexity and time to translate as disadvantages. Several interlocutors suggested that the government tends to be focused on command and control, so markets would generate a welcome change. Additionally, as in Australia there is strong technical competence in Korea, where the ministries are very influential, and therefore a high degree of concern for the creation of technical policy. Korean respondents are more likely to see technical complexity as a challenge, in part because the carbon markets are managed by the ministries, particularly the Ministry of Environment and Ministry of Trade, Energy, and Industry, and treated as technical devices. Respondents also expressed more concern with the time it would take to translate and operationalize the markets.

With respect to the occupational sector (Table 9.4), there are only a few statistically significant differences among the groups. Unsurprisingly, individuals from financial services are more likely to identify efficiency, global scale, and reducing greenhouse gas emissions as advantages or opportunities of market-based governance and less likely to identify technical complexity as a disadvantage of market-based governance. Members of the financial services are among the strongest proponents of market-based governance and are comfortable with the technical requirements of structuring carbon as a commodity. In contrast, individuals from the professional services are less likely to believe markets have the capacity to reduce emissions, and more likely to see technical complexity as a disadvantage. These firms also stand to gain from servicing emissions markets, but in contrast to the financial services industry they benefit from the markets regardless of whether or not the markets are successful because their revenue is generated from service fees. For example, legal firms generate revenue from writing contracts for carbon credits, structuring products, and litigating disputes regardless of whether or not the price of emissions is high. They are consequently more familiar with the technical complexity required for markets to operate efficiently, and more skeptical of the ability of markets to reduce emissions.

Individuals from industry are less likely to see price efficacy as an advantage of emissions markets. Even if the markets are a more flexible means of governance, they still place a burden on the industry sectors subject to a cap. Individuals from government are more likely to recognize political uncertainty as a disadvantage of emissions markets. These individuals operate most closely and are most familiar with the political tensions involved in markets. Whether or not respondents were in favor of the markets, they recognized that they operate only if they have political support. The markets require a consistent and stable regulatory framework to be effective. However, as in the case of Australia, the regulatory framework is subject to change every time there is an election. The political turnover creates market uncertainty. Finally, NGOs are unsurprisingly less likely to identify reducing emissions as an advantage of emissions trading. NGOs tended to be more concerned with the environmental impacts of governance and more skeptical of market-based approaches.

In sum, statistically significant regional variations around the idea of market-based climate governance belie the apparent consensus in the Anglo/European space regarding markets as a basis for managing climate change. European respondents emphasize global-scale coordination and governance beyond the nation-state. US respondents hold a stronger belief in the efficacy of markets as a means of reducing emissions, and Australian respondents demonstrate a more internal focus and a stronger emphasis on technocratic norms. Australian's associate market-based governance with fairness but recognize the limitations.

In the East there is much greater skepticism in general about the use of market mechanisms for governance. The respondents who favor markets tend to demonstrate different priorities. The perspective from respondents in China is perhaps at odds with the government rationale for pursuing market-based governance. In China, markets are associated with modernity and China's international standing relative to other great powers, but more importantly as the path to ecological modernization through which economic growth will be connected with environmental preservation. Nevertheless, market participants express skepticism about where the markets come from, the technical complexity involved in creating them and how they will relate to a Chinese institutional setting of governance.

In South Korea, there is a similar association with development and international standing through the construction of markets. To the extent that they provide flexibility, the actors who have considerable expertise (particularly within the ministries) see the markets as an advantage. Nevertheless, there is also recognition of the complexity involved in building effective carbon markets, and concern over the time it will take to properly operationalize them. Japan is the country most skeptical of emissions trading, with individuals across occupational sectors voicing strong concerns. Some of the most vocal critics come from within the industrial and financial service sectors. The strong negative associations of market governance, particularly regarding the technical complexity of the markets and prevalence of profit over environmental impact, are derived from cultural norms and values that differ from those of the Anglo/European context.

The Role of Politics in Shaping Political Norms

The differences in how markets are conceptualized between regions highlight the importance of intersubjective coherence in perspective within regions. The idea of market-based governance resonates in different ways in different geographies, but there is also a surprising degree of coherence in concepts within and even between places. In countries like Australia and the United States, efficiency is a major goal. In Europe, the ideas of economic opportunity and global leadership predominate in discussion of markets. The term "modernization" is common in both China and South Korea. In Japan, respondents associated their skepticism of financial markets by describing emissions markets with the term "money game." Although these differences in conception and in the language used to describe markets are interesting in and of themselves, what is particularly interesting is the way markets permeate the social conscious and resonate with their own language in different places. This suggests that scholars and policymakers cannot speak of "the market" because processes of syncretism driven by local sociopolitical factors produce multiple permutations of "the market," each with greater or lesser degrees of divergence from the abstracted ideal. This has potentially significant ramifications for using markets to manage climate change, which requires markets to have an extremely high degree of interoperability and thus low tolerances for divergence.

Consequently, in addition to communicating particular economic values about energy use and efficiency through a price signal, markets also communicate a range of other concepts and social values. The language that permeates the markets in each region reflects common understandings of what the markets are expected to be as well as who has the authority to govern and shape them. As Donald MacKenzie (2009) suggests, material agencements (combinations of human beings, material objects, and technical systems) structure markets. It is a combination of human agency, organizational structure, and systems of meaning that constitutes markets.

The global imperative to address climate change through technical framing as generated by the UNFCCC provides both a logic of action as well as an infrastructure of organization. It suggests, and relies upon, institutional isomorphism of the markets, in particular giving a common basic framework: a cap, the allocation of permits, a commodity of trade, and comparable organizations of monitoring, verifying, and reporting. The agencies with expertise in navigating these roles take on market authority. And yet, the global imperative and isomorphic form (markets copied from the structure of the EU ETS for example) competes with economic *metis* grounded in sociocultural and political norms. For example, in China private accountancies, consultancies, and legal firms do not have the same authority as the National Development and Reform Commission to perform the functions of monitoring, verifying, and reporting. Conversely, in the United States the construction and functioning of the market lies largely in the hands of private sector accountancies, consultancies, and legal forms. The basic shape and function of the market is therefore translated through different infrastructures and the isomorphism disappears as the markets are built and enacted through different cultural logics.

Network Governance in the Context of Market Cultures

Linking network governance to market culture allows for additional insight into the dynamics that underpin climate policy. Because climate policy requires coordination across multiple organizations within the complexity of contemporary political economies, the concept of network governance can provide important analytical leverage. Provan and Kenis (2008, p. 231) define a network as three or more legally autonomous organizations working collaboratively toward a common goal. By emphasizing multiple nodes of agency, this approach allows for a more realistic engagement with climate policy, in which multiple organizational actors must be harnessed into collaborational networks to produce the profound shifts in energy production and use required to reduce carbon emissions.

All networks are not the same, however. Provan and Kenis (2008) identify two cleavages. First, networks vary by the degree of centralization. At one end—called shared governance—are completely decentralized networks in which all the participants interact and governance direction is equally shared. The underlying market cultures shape their function and organization. At the other end of the spectrum—termed lead organization—are networks in which one or a small number of organizations take the lead directing governance and contact between the other organizations in the network is limited. A second cleavage concerns the source of leadership. At one end are networks that are governed by their own members. At the other are networks that are governed by organizations external to the network. These external governors are called network administrative organizations (NAO) (Provan & Kenis, 2008, pp. 233–234). In their comprehensive review of the network governance literature, Dal Molin and Masella (2016, pp. 497–498) identify an additional set of six factors: trust in others to take account of interests, size of network, goal consensus, leadership that support governance, embeddedness of goals in network relations, and diversity of network members. Table 9.5 below combines the typology identified by Provan and Kenis with the factors outlined by Dal Molin and Masella and applies them to the United States, Japan, European Union, China, South Korea, and Australia. Clearly, these polities differ quite dramatically. Much of this difference is due to differences in the market cultures.

Table 9.5 Network architype defining characteristics

	Archetype defining characteristics					
Country	Size	Diversity	Embeddedness	Goal consensus	Trust	Leadership
United States of America	Large—50 states with 535 elected representatives, and open campaign finance laws	Geographic interests, democratic vs republican, special interest groups and industries	Strong state-state dyadic relations based on region, very loosely connected through the federal government	High—High among the parties, low among the federal government	Low to moderate—Federal government monitored by states, competition in private sector	Federal government
	Lead or NAO	**Shared**	**NAO or Lead**	**Shared governance or lead**	**NAO**	**NAO or Shared governance**
Japan	Small	Low—Consensus that technology is the solution	High level – Joint Crediting Mechanism implies that Japanese ministries/agencies are working together to form individual plans with other countries	High – Recognition of need for "good" or "technological" market	High density	Lots of cooperation between industries
	Shared governance	**NAO**	**NAO or Shared**	**NAO**	**Shared governance**	**Shared**

(continued)

Table 9.5 (continued)

	Archetype defining characteristics					
Country	Size	Diversity	Embeddedness	Goal consensus	Trust	Leadership
European Union	Moderately large – 28 member states	Very diverse based on different member state size, cultures/history, and objectives	Strong relational embeddedness – Certain countries have a history of working well together, or share common climate change goals. Structural embeddedness – High level. Network actions are determined by dyadic relations.	Moderately high	Moderate density—Countries can place trust in the EU, which represents the whole. Don't need to necessarily trust every individual country	Single overarching government
	Lead or NAO	**NAO**	**Shared governance**	**NAO**	**NAO**	**Lead**
China	Large	High—Large country with different geographic and economic needs, but united under centralized gov.	Strong relational and structural embeddedness due to highly centralized government	Very high	Moderately high	Centralized government
	Lead or NAO	**Lead**	**Lead**	**Shared governance or NAO**	**NAO**	**Lead**

Note. Source: Design by authors

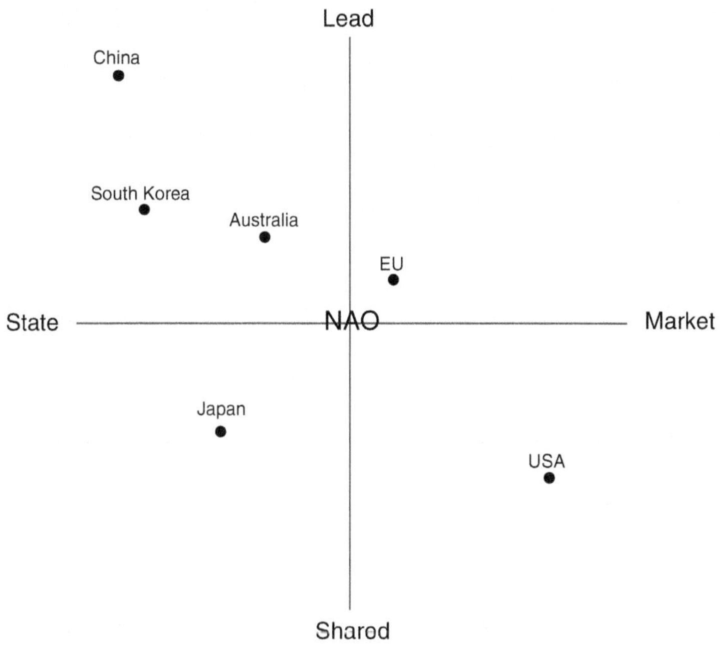

Fig. 9.1 Form of network governance plotted against market culture. Source: Design by authors

In Fig. 9.1, we present the differences in network governance mapped onto market culture.

With this figure, we clearly demonstrate the relational position of the various polities with respect to climate policy. The United States, more than any of the other countries surveyed, indicated a strong positive correlation with the idea that the capacity of market governance to reduce emissions is an advantage. The United States' positioning on the y-axis is due to the dynamics created by the two-party system and the profusion of political lobbying. Although the United States bears potential to demonstrate characteristics more akin to that of a lead organization, the checks and balances system of the US system along with strong party polarization are indicative of a form of shared governance. As such, goal consensus within the parties is high, but goal consensus within the federal government is low. Similarly, trust between the parties is low, but within each party is moderately high. Additionally, lobbyists from a variety of business industries have a large presence in American governance. Campaign contributions from wealthy individuals and special interest group have reached historical highs since the Citizens United decision, adding more voices to the American political process.

As evidenced by the figure, Australia's culture is moderately state oriented and network is moderately lead oriented. Australia respondents were of the opinion that although a private-sector-driven market solution to climate change is likely the optimal solution due to how markets incentivize innovation, it is unlikely that consensus on a global market structure will ever be reached. The Australian view is that pricing within a carbon market is fraught with challenges due to the uncertainty of pricing environmental impacts and the challenges of maintaining stable, long-term projections amidst that uncertainty. The uncertainty surrounding pricing and the unlikelihood of global consensus places Australia in favor of more state-controlled responses to climate change. In relation to governance, Australia is dominated internally by a strong private sector and is host to a similar lobbying presence like the US. However, the party in power in Australia has control over the ministries, indicative of a lead-organization style of governance.

By contrast, the EU demonstrates characteristics that align well with the NAO form of network governance. The number of member states is moderately high, with moderate trust density between member states and the EU itself and between member state dyads. The leadership style is an external network organization, the EU, overlooking participant member states. Additionally, because of the separation of the member state governments from the EU, lobbying does not affect EU governance in the same way that lobbying impacts the United States: Member-state politicians are not the same politicians as those who represent the member state at the EU level. The EU respondents indicated a strong correlation with the reliance on governance cohesion as a disadvantage for global climate markets—EU respondents acknowledged that markets must be international and interconnected to be effective, but the international governance cohesion necessary does not yet exist. The aforementioned response is reflected in their positioning towards the center of the x-axis, with a slight preference for markets.

Japan differs from the rest of the study polities in that the Japanese respondents indicated a strong preference for domestic solutions to climate change through technological development. The Japanese BOCM/JCM is structured to allow each host country to partner independently with Japan, creating highly specialized partnerships as opposed to facilitating participation in a general, global system. Goal consensus among Japanese ministries and industries is high with a low diversity of opinions, indicative of a shared form of governance. Finally, Japan's position on the x-axis is reflective of the respondents' consensus across different occupational sectors that the faults of market governance are high, coupled with their emphasis on a "good" or "technological" market as the solution to emissions reductions.

Korea sits with Australia and China in the State and Lead quadrant. Although the Korean respondents appreciate the flexibility and viability that carbon markets offer, they do not see carbon markets as a mechanism for combatting climate change that can be easily implemented at a global scale because of the differences between industries in terms of the instruments and processes that are used. The leftward positioning on the x-axis is complemented by a more lead-organization style of network governance, as the Korean ministries are highly influential within the government and contribute to a single national agenda.

Finally, China most strongly demonstrates the characteristics of a lead organization as its form of network governance, with a substantial emphasis on state governance for climate change solutions. Chinese respondents indicated a negative correlation with the idea that markets are a more efficient form of governance than the state, especially considering the amount of time it would take to develop the necessary institutions and regulatory agencies within China to monitor a carbon market. The preference towards a state solution is indicated by China's left leaning position on the x-axis. Additionally, applying the Provan and Kenis typologies of network governance while using the expanded Dal Molin and Masella characteristics, we have determined that the size, embeddedness, and leadership qualities of China are most like those of a lead organization, with the remaining characteristics of trust, diversity, and goal consensus indicating state cohesion and cooperation due to the highly centralized nature of the Chinese government.

Market Perceptions: From General Claims to Specific Cases

Market-based governance is driven by assumptions that climate change is a techno-economic rather than a sociopolitical problem and is thus amenable to universalistic neoliberal-economic policy prescriptions. However, these prescriptions interact with political and social systems in different scalar contexts, producing variations in how the market is understood and operates. With our analysis presented here, we show that climate policies cannot neglect the ways in which differences in scale and place—local, national, and regional—impact on how international mandates translate into action. Shortcomings in the global effort to address climate change can thus in part be traced to failures to appreciate how economic *metis* shapes economic practice at different scales and in different places. As suggested in this chapter, culture at the national and regional levels plays a crucial role in determining the enactment of international imperatives. Markets do not exist in institutional vacuums and failure to account for local economic knowledge means that efforts to establish governance on the basis of universalistic economic behavior are unmoored from the ways states and societies practice economics. These variations also play out across different network governance forms, making for a complex climate policy landscape.

The mix of social and governance contexts and motivations behind policy suggests that greater flexibility is needed at the international level to enable states to act on climate change. The arrival of different polities at different policy solutions reflects the distinctive sociopolitical systems in which they are formulated. In the EU, the market has been implemented quickly but has struggled to make even slight adjustments. In the US, global and popular discourses have generated periods of heightened concern over climate change, only for these to be destabilized by "economic realism." Nonetheless, well-established bureaucratic systems to manage air-quality-control problems in the 1970s and 1980s in California enabled the establishment of a technocratically managed system, albeit years after the EU ETS. In Australia, a strong bureaucratic culture propagated carbon-pricing policies only for political turmoil to threaten to unravel the system a few years later.

In China, the emissions markets have been established as political compromise to mitigate climate change but in a less stringent manner than the usual command

and control approach. Although the strategy of trialing different forms across seven regions has lent greater flexibility to the system, it to a degree has also marginalized the impact of emissions trading. In Japan, attempts by the Ministry of Environment to institute regional and domestic emissions trading systems have been met with strong resistance from industry. Cultural norms dictate that markets must have a material impact. The compromise has been to adapt the basic idea and structure of emissions market or offset trading and apply it to the transfer of technology to developing regions. Although the approach is beneficial for both energy and emissions gains, it has pulled Japan from international negotiations. In South Korea, the markets adapt well to the strong technical capacity and existing authority of various ministries. And yet, the South Koreans have struggled to scale the markets to a size and form that best fits the conditions of their relatively consolidated industrial groups. As we have demonstrated in this chapter, the idea that climate change can be addressed through a single, unified techno-economic prescription does not hold against empirical reality.

References

Andonova, L. B., Betsill, M. M., & Bulkeley, H. (2009). Transnational climate governance. *Global Environmental Politics, 9*(2), 52–73. https://doi.org/10.1162/glep.2009.9.2.52

Bailey, I., & Wilson, G. A. (2009). Theorising transitional pathways in response to climate change: Technocentrism, ecocentrism, and the carbon economy. *Environment and Planning A, 41,* 2324–2341. https://doi.org/10.1068/a40342

Betsill, M., & Bulkeley, H. (2007). Looking back and thinking ahead: A decade of cities and climate change research. *Local Environment, 12,* 447–456. https://doi.org/10.1080/13549830701659683

Clark, G. L. (1998). Stylized facts and close dialogue: Methodology in economic geography. *Annals of the Association of American Geographers, 88,* 73–87. https://doi.org/10.1111/1467-8306.00085

Dal Molin, M., & Masella, C. (2016). From fragmentation to comprehensiveness in network governance. *Public Organization Review, 16,* 493–508. https://doi.org/10.1007/s11115-015-0320-4

Eisenhardt, K. M., & Graebner, M. E. (2007). Theory building from cases: Opportunities and challenges. *Academy of Management Journal, 50,* 25–32. https://doi.org/10.5465/amj.2007.24160888

Garsten, C., & Jacobsson, K. (2007). Corporate globalisation, civil society and post-political regulation: Whither democracy? *Development Dialogue, 49,* 143–158.

Goldman, M. (2003). Partitioned nature, privileged knowledge: Community-based conservation in Tanzania. *Development and Change, 34,* 833–862. https://doi.org/10.1111/j.1467-7660.2003.00331.x

Gioia, D. A. (1998). From individual to organizational identity. In D. A. Whetten & P. C. Godfrey (Eds.), *Identity in organizations: Building theory through conversations* (pp. 17–32). Thousand Oaks: SAGE. https://doi.org/10.4135/9781452231495.n2

Gioia, D. A., Corley, K. G., & Hamilton, A. L. (2013). Seeking qualitative rigor in inductive research: Notes on the Gioia Methodology. *Organizational Research Methods, 16,* 15–31. https://doi.org/10.1177/1094428112452151

Hayek, F. A. (1945). The use of knowledge in society. *The American Economic Review, 35,* 519–530.

Hoffmann, M. J. (2011). *Climate governance at the crossroads: Experimenting with a global response after Kyoto*. Oxford, UK: Oxford University Press. https://doi.org/10.1093/acprof:oso/9780195390087.001.0001

Katzenstein, P. J. (1996). Introduction: Alternative perspectives on national security. In P. J. Katzenstein (Ed.), *The culture of national security: Norms and identity in world politics* (pp. 1–32). New York, NY: Columbia University Press.

Knox-Hayes, J. (2012). Negotiating climate legislation: Policy path dependence and coalition stabilization. *Regulation and Governance, 6*(4), 545–567. https://doi.org/10.1111/j.1748-5991.2012.01138.x

Knox-Hayes, J. (2016). *The cultures of markets: The political economy of climate governance*. Oxford, UK: Oxford University Press.

Krueger, R., & Gibbs, D. (2010). Competitive global city regions and 'sustainable development': An interpretive institutionalist account in the South East of England. *Environment and Planning A, 42*, 821–837. https://doi.org/10.1068/a42111

MacKenzie, D. A. (2009). Constructing emissions markets. In D. A. MacKenzie (Ed.), *Material markets: How economic agents are constructed* (pp. 137–176). Oxford, UK: Oxford University Press.

Newell, P., & Paterson, M. (2010). *Climate capitalism: Global warming and the transformation of the global economy*. Cambridge, UK: Cambridge University Press.

Ocampo, J. A. (2011). The transition to a green economy: Benefits, challenges and risks from a sustainable development perspective: Summary of background papers. In J. A. Ocampo, A. Cosbey, & M. Khor (Eds.), *The transition to a green economy: Benefits, challenges and risks from a sustainable development perspective: Report by a panel of experts to second preparatory committee meeting for United Nations Conference on Sustainable Development* (pp. 3–15). New York: UNDESA.

Owens, S. E., & Cowell, R. (2011). *Land and limits: Interpreting sustainability in the planning process* (2nd ed.). London: Routledge. https://doi.org/10.4324/9780203832226

Pattberg, P. H. (2007). *Private institutions and global governance: The new politics of environmental sustainability*. Cheltenham: Edward Elgar.

Provan, K. G., & Kenis, P. (2008). Modes of network governance: Structure, management, and effectiveness. *Journal of public administration research and theory, 18*, 229–252. https://doi.org/10.1093/jopart/mum015

Redclift, M. (2012). Living with a new crisis: Climate change and transitions out of carbon dependency. In M. Pelling, D. Manuel-Navarrete, & M. Redclift (Eds.), *Climate change and the crisis of capitalism: A chance to reclaim self, society and nature* (pp. 21–36). London: Routledge.

Scott, J. C. (1998). *Seeing like a state: How certain schemes to improve the human condition have failed*. New Haven, CT: Yale University Press.

Stehr, N. (2020). The atmosphere of democracy: Knowledge and political action. In J. Glückler, G. Herrigel, & M. Handke (Eds.), *Knowledge for governance* (pp. 69–91). Knowledge and Space: Vol. 15. Cham: Springer.

While, A., Jonas, A. E. G., & Gibbs, D. (2010). From sustainable development to carbon control: Eco-state restructuring and the politics of urban and regional development. *Transactions of the Institute of British Geographers, 35*, 76–93. https://doi.org/10.1111/j.1475-5661.2009.00362.x

Chapter 10
The Fight Against Corruption in Brazil: A Case of Good Governance?

Markus Pohlmann and Elizangela Valarini

Since the start of Operation Carwash (Operação Lava Jato)[1] in March 2014, politics, senior executives, and the judiciary have been in great turmoil in Brazil. Having ensnared the construction industry and Petrobras, the state-run oil company, the specter of corruption is now threatening to envelop other state-owned enterprises, such as the nuclear energy industry, with the construction of nuclear power plants as well as further Brazilian federal entities, such as the Brazilian Development Bank (BNDES), a Brazilian federal bank, the Caixa Econômica Federal, etc. Operation Carwash became the largest ever ongoing criminal investigation in Brazilian history. In special, the criminal process against the former President Luiz Inácio Lula da Silva and his political party, the Workers' party (Partido dos Trabalhadores (PT)), detonated a political tinderbox. Not only have the accusations against ex-President Lula, the impeachment against his successor Dilma Roussef, the arrest of the President of the Parliament, Eduardo Cunha of the Brazilian Democratic Movement Party (Partido do Movimento Democrático Brasileiro (PMDB)), and current proceedings against roughly 279 politicians robbed the Brazilians of their sleep, the sheer multitude of corruption scandals in the economy also keeps them in suspense.

What is going on in Brazil? And what is behind these scandals? It seems that tolerance of corruption in Brazil reached a turning point. Whereas criminal processes and sentences were once slight and rare, and only pronounced with the exclusion of the general public, they are now coming fast, hitting hard, and reaching the

[1] The name Operation Carwash have been giving from Brazilian Federal Police to the uncovered scheme of money laundering that was linked to a chain of gas stations and car wash systems. Later, the data showed that this case was linked to the Petrobras corruptions scheme.

M. Pohlmann (✉) · E. Valarini
Max-Weber-Institute of Sociology, Heidelberg University, Heidelberg, Germany
e-mail: markus.pohlmann@soziologie.uni-heidelberg.de;
elizangela.valarini@mwi.uni-heidelberg.de

J. Glückler et al. (eds.), *Knowledge for Governance*, Knowledge and Space 15,
https://doi.org/10.1007/978-3-030-47150-7_10

news. With the enacting of the Freedom of Information Act no. 12.527/2011,[2] anyone who is curious enough can easily follow the trials on official sites of the justice system, but also in the media. Also, punishment became harsher. For example, the federal court of Paraná sentenced Marcello Odebrecht, the president of one of the biggest business groups in Brazil, and a descendant of a German immigrant family, to 19.4 years on the charges of bribery, participating in organized crime, and money laundering.

In our article, we will investigate the unlawful practices of the Brazilian construction companies involved in the Petrobras plot and their relationship with the public sector. In our analysis of criminal trials, inquiries, testimonials, and examination of the Brazilian context, we attempt to demonstrate that good governance is related to the interaction of various types of institutions (cognitive, normative, and regulative), as well as embedded belief patterns shared in a society. For this purpose, we analyze the institutionalized patterns in a deviant environment and the competitive pressure in which enterprises are embedded. Furthermore, we examine internal organizational incentives and the profits from illegal practices, the hierarchical pressure, as well as the legitimation of unlawful activities, rationalization, and socialization in Brazilian organizations.

Relying on a multilevel analytical approach, we begin our analysis by describing the current Petrobras scandal and the main actors involved. Furthermore, we run our observations on the institutional level, and we analyze the change of regulations, laws, and the efforts of law enforcement agencies to investigate and prosecute corruption in Brazil. We aim to find out what kind of institutional and regulative changes were realized in Brazil and how the switch from defective to more effective regulations took place. In the next chapter, we argue that the defective regulation in the fight against corruption was one significant background and reason for the "normalization of corruption" on a corporate level. Dealing with the actors in charge of running the companies, we analyze how subcultures or social "cocoons" emerged inside the companies, providing frames of action for organizational wrongdoing. We ask whether and how so many high-ranking officials and senior managers took part in the corrupt setup. One of the stunning facts revealed by Operation Carwash's investigators is how closely cartel building and bribe paying by important members of Brazil's big business was intertwined with illegal party financing and political corruption. We thus dedicate the sixth chapter to the analysis of political corruption in Brazil as the other side of the coin. Moving back to the actor level, we then ask how many and what kind of political actors have been involved and which

[2]According to the Freedom of Information Act, enacted in November 2011, all public agencies, members of the direct administration of the Executive, Legislative, including the Audit Courts and Judiciary and Public Prosecution as well as all municipalities, public foundations, state-owned enterprises, mixed-capital-companies and other entities directly or indirectly controlled by the Federal Government, States, Federal District and Municipalities must to divulge all information related to their administration, such as income and expenses as well as produced and guarded information that are of collective interest (http://www.planalto.gov.br/ccivil_03/_ato2011-2014/2011/lei/l12527.htm).

framework of political wrongdoing has been established. Could Operation Carwash destroy the pattern of political corruption? In our subsequent conclusions, we will address whether the Brazilian case is just one of effective changes in law enforcement or a case of "good governance" as well.

The Brazilian Petrobras Plot

The Petrobras plot unveiled a complex and profound corruption scheme in Brazil. The number of political parties, enterprises, high-ranking politicians, executives, and managers as well as the sophisticated modus operandi discovered in the context of Operation Carwash was unprecedented in Brazilian society. The case's magnitude, but also the change in the methods and procedures applied on Operation Carwash, makes it a crucial case for social scientists.

Since Operation Carwash was launched in March 2014, its investigators have discovered US$2 billion of bribery payments from private companies to Petrobras senior executives, politicians, and political parties. More than 760 cases are now in court, involving politicians, senior executives, owners of Brazil's largest enterprises, and other actors. 182 people have been convicted, penalties add up to 1809 years.[3]

To understand the Petrobras case, we shall pay attention to four core investigated groups. The first core group consists of senior executives occupying key positions at three of Petrobras's business units: Supply, International, and Corporate and Services. Petrobras's executive directors, supported by political parties, ensured access to the expensive Petrobras contracts for a distinct group of Brazilian private construction companies. In return, the construction companies paid them somewhere between 1% and 3% on top of the contract's value. In this context, it is also important to account for the composition of the executive boards at Brazilian state-owned enterprises. Generally, the directors are appointed by government officials, the President, and/or members of the political parties. In the case of Petrobras, the directors of the executive board were supported by political parties of the current ruling coalition: PT, PMDB, and PP (Progressive Party/Partido Progressista). The second main group in the Petrobras case was the so-called *black market dealers*. These intermediaries served as a link between Petrobras executives and managers of the construction companies, as well as between the government and party members. Furthermore, they were responsible for finding means to launder the money from bribery payments. The third group involved in this bribery scheme consists of the largest Brazilian construction companies. They were organized in a cartel to obtain Petrobras's major public bids related to their business areas and allocate the contracts internally. The cartel dealt with Petrobras directors and makes it possible to restrict and exclude new bidders interested in Petrobras projects. Thanks to their "agreement" with Petrobras' senior managers, they ensured that the bid winner

[3] See http://www.mpf.mp.br/grandes-casos/lava-jato (accessed 23.02.2018).

always came from within the cartel. The fourth main group of suspects uncovered by Operation Carwash investigators is made up of politicians, government officials, and political parties, in other words: the bribe takers.[4] Leaders of political parties, politicians, such as state and federal deputies, senators, and ministers financed their political parties, campaigns, and political alliances through payment of bribes. Even the former republic President Michel Temer is suspected of receiving bribes from large enterprises,[5] as was the former President Luiz Inacio Lula da Silva, who was sentenced to 12 years in prison.[6]

According to Petrobras's internal audit (Petrobras, 2014), perpetrators violated a series of the company's internal rules and general laws about public bids.[7] The government has regulated public bids for hiring a service provider through strong legislation: Lei No. 8.666/1993. For the success of the bribery scheme, the participation of the business unit "Corporate and Services" was strategic, because its members were in charge of the bid procedures and regulations, project price calculation, as well as selection and accreditation of partner companies. The board of Corporate and Service alone could not approve new contracts, because the bid process demanded several business units, departments, and the monitoring of board members, but according to the former Petrobras president between 2012 and 2015, Maria das Graças Foster,[8] the expertise and professional competence of the director of this unit was appreciated. In this respect, the business units' directors had a large scope to influence or even "manipulate" the results of bid proceedings in favor of their interests.

[4] See trials for administrative improbity: 500662892.2015.4.04.7000/PR; 5006717-18.2015.4.04.7000/PR; 5006695-57.2015.4.04.7000/PR; 5006675-66.2015.4.04.7000/PR; 5011119-11.2016.4.04.7000/PR; 5006694-72.2015.4.04.7000/PR; and criminal trials: 5083351-89.2014.4.04.7000/PR; 504524184.2015.4.04.7000/PR; 508325829.2014.4.04.7000/PR;5083401-18.2014.4.04.7000/PR;5083376-05.2014.4.04.7000/PR;503652823.2015.4.04.7000/PR;508336051.2014.4.04.7-000/PR;501233104.2015.4.04.7000/PR;507344138.2014.4.04.7000/PR; 5027422-37.2015.4.04.7000/PR; 502698037.20-16.4.04.7000/PR).

[5] Inquiry number 4.483/DF, documents 162339/2017/GTLJ-PGR and 162973/2017/GTLJ-PGR.

[6] Criminal trial 5046512-94.2016.4.04.7000/PR. The sentence and result of the appeal court decision has not yet been issued.

[7] Public bid is the set of laws to regulate the contracts between public and private sector, but also is used to administer contracts for purchasing materials and supplies by public entities. The public bid law aims to ensure that public entities receive the best possible price for the supplies or public works.

[8] Maria das Graças Foster's testimony https://www.youtube.com/watch?v=9TkDOJZlLOg

Institutional and Regulative Changes in Brazil: From Defective to Effective Institutions and Regulations?

On the institutional level, we are analyzing the changes in legislations and law enforcement over the last two decades, concerning political and economic corruption in Brazil. We are focusing mainly on the constitutional reality and the policy outcome, that is, on the efficiency and effectiveness of law enforcement in Brazil. Not long ago, there was a lack of law enforcement and monitoring systems for private and public corruption (Chaves, 2013), creating an institutional environment prone to deviance and corruption. For example, Brazilian legislation does not yet cover private corruption[9] (Pascolati Junior, 2016), some politicians have legal immunity, and thus, under the constitution, due this legal status they can be just investigated and prosecuted in special courts, such as the supreme court. According to the Carwash' prosecutors, although the penalties for corporate crime and political corruption have become more severe, the probability of being detected and incarcerated remained low (Dallagnol, 2017). What we call "defective institutions and defective regulations" have been part of the Brazilian reality concerning the fight against corruption.

In our discussion, we differentiate between institutions and regulations. We define institutions as cognitive, normative, and regulative structures and/or activities that run on stability, but also constrain, regularize, and give meaning to social behavior, but assume an impersonal and objective form of reality (Scott, 2014). They incorporate both symbolic systems and regulative processes that orient social action and organizational behaviors in a defined social group and/or society (Bathelt & Glückler, 2014; Scott, 2014). Furthermore, institutions are determined by locality, cultural patterns, social conditions, etc., and constituted and reproduce by ongoing social interaction existing in a specific social context (Glückler, Suddaby, & Lenz, 2018). "Although institutions may represent the intentions inherent in rules" (Bathelt & Glückler, 2014, p. 356), they differ from rules, laws and regulations. These can regulate social interaction from organizations and actors independently if those behind social beliefs or a common knowledge. In that sense, rules, laws and regulations can regulate social practice through their sanction systems (Bathelt & Glückler, 2014). Thus, introduced laws, rules and regulations can lead to different outcomes, depending the existing institutions, with them the new regulation will be interact (Glückler & Lenz, 2016).

Institutions and regulations are defective when laws and directives, such as anti-corruption laws, do exist but these either fail to set sufficient standards or the administrative enforcement thereof is patchy or inadequately guaranteed. In this case, the constitutional legal bases are not undermined, but the purpose of the regulation is largely missed. The fight against corruption may be only "nice to have," as one has

[9] Bill 236/2012 has been in process since 2012 in the Brazilian Federal Senate to implement the penal legislation with legal instrument to define and punish private corruption (Pascolati Junior, 2016, p. 148).

repeatedly witnessed over the past years. We believe that such a defective regulation creates important preconditions for the emergence of forms of deviant self-regulation at the level of enterprises. And that is exactly what happened in the case of the fight against corruption in Brazil. Defective regulations and institutions are, in this sense, functional for the maintenance of a system that systematically and legally subverts the purposes of statutory regulation. We call this "systemic corruption."

In the 1990s, the global diffusion of regulatory agencies (Braithwaite, 2008; de Rugy & Warren, 2009; Jordana, Levi-Faur, & Fernández i Marín, 2011) also spread to Brazil, a country which a long-lasting tradition of a regulatory state and regulatory capitalism (Braithwaite, 2008). Although impeded by a self-sustaining pattern of cronyism, a new phase of regulation emerged in Brazil as well, leading to the detection and prosecution of big corruption scandals. In the last decades, Brazil has committed itself, through several international agreements,[10] to fighting against corruption (Barkemeyer, Preuss, & Lee, 2015; Dantas de Araujo, 2012; de Sanctis, 2015; Richard, 2014). To do so, Brazil has implemented a national strategy against corruption, organized crime, and money laundering, the National Strategy for Combating Corruption and Money Laundering (Estratégia Nacional contra a Corrupção e a Lavagem de Dinheiro (ENCCLA)) (de Sanctis, 2015). This initiative is not only a new paradigm to combat corruption and economic and financial crimes, but it is also responsible for concrete changes aimed at involving the judicial system more strongly in the execution and investigation of corruption cases. The new Anti-Corruption Act n. 12.846/2013 is a result of the support of the ENCCLA's strategies. Its significant change consists in the accountability and punishment of Brazilian and foreign companies as legal entities. They can be addressed by the penal code and have obligations to install compliance systems (Chaves, 2013; Correia, 2015; Wohlnick & Correia, 2016).

The diffusion of international anticorruption policies in Brazil has fomented a series of changes related to new regulation, such as enacting of laws and rules to regulate and control the public administration and to generate more transparency of public spending and public activities (Lei No. 8.112/1990; Lei No. 8.429/1992; Lei Complementar No. 101/2000; Lei Complementar No. 131/2009; Lei No. 12.527/2011), the business relation between the public and private sector (Lei No. 8.666/1993; Lei No. 10.520/2002), and so forth. (Filgueiras & Araújo, 2014). Some authors have also pointed to changes related to the perception of Brazilian society about corruption. With the redemocratization process, the corruption scandals have also become public. Most recent changes on the institutional level in Brazil related to laws and regulations, such as the Anti-Corruption Act, may be the product of civil society pressure (Filgueiras & Araújo, 2014). Even if the perception about corruption in Brazilian society over time seems to be changing, small and large corruption is still affecting the relation between actors, organizations, and government.

[10] United Nations Convention against Transnational Organized Crime; Convention on Combating Bribery of Foreign Public Offices in International Business Transactions (OECD); Convention on Combating Bribery (UK) (2003); United Nations Convention against Corruption (Mérida Convencion); Inter-American Convention against Corruption; and so forth.

The regulative changes have been impressive and well acknowledged all over the world as good governance in the fight against corruption. Yet it is more precise to call it "good law enforcement" and to keep open whether it leads to a mode of good governance, in which various interest groups and societal actors engage in the fight against corruption.

The "Normalization of Corruption" on a Corporate Level

In our view, defective regulations in the fight against corruption are one important precondition for the "normalization of corruption" on a corporate level. By analyzing the cartel of construction companies, we discovered that only 30% of these enterprises had a code of conduct to regulate the business relations in- and outside of the company before Operation Carwash began. Our analysis revealed that although these enterprises do have a code of conduct, the terms and clauses prescribing rules of doing business ethically and those declaring illegal and unlawful behavior to be punishable are phrased vaguely. Furthermore, in the code of conduct was lacking a precise description of no acceptable practices and behaviors inside of the organization, in its relationship with the political field and with other enterprises as well as a concretely description of sanctions and penalties against the wrongdoers was missing. Thus, in the code of conduct, this is considered a first step to establishing a compliance system. Most of the companies under investigation by the Operation Carwash had no compliance measures installed. They only started to establish a compliance system when they moved into the spotlight of law enforcement. Only those companies with major business activities abroad or with international investors ran an internal compliance system. However, lawyers and compliance professionals have strongly challenged the legitimacy and effectiveness of the internal compliance (Pizarro, 2015). If we extend our analysis beyond Brazil to the Latin America context, we verify that the introduction of compliance systems to regulate corporate behaviors are a relatively young practice in Latin American countries. According to Passas and Kleinhempel (2016), although 35% of the existing compliance systems were introduced between 2005 and 2010, more than 50% were implemented in the last few years. These authors have pointed out that there is a gap in employee awareness about the importance of such programs. An average of 19% of employees did not know about the introduction of a compliance program in their company (p. 82).

Far more important is that companies like Odebrecht ran departments dedicated to carrying out illicit activities, such as the department Odebrecht called "departamento de operações estruturadas" (department for structured operations) which had existed for decades. According to the former senior manager Hilberto Mascarenhas, the chief officer of Odebrecht's "departamento de operações estruturadas," the company had also established an internal bonus system that provided incentives for paying bribes. In his case, the annual bonus was as high as one third of his annual salary of US$1.2 million and was part of a provision system for bribe payers.

> What was agreed was this: I will pay you a bonus on the basis of your outcomes. And we will negotiate it annually. So, you may do anything that you need to achieve it [bonus]. So if you ask someone who wants to increase the revenue of his project, but this requires the approval of an 'aditivo' [additional investments] that is still on the table of someone who does not want to approve of it, this person may do everything for the approval of his 'aditivo' to increase the value of his project. (Hilberto Mascarenhas, Odebrecht former senior manager, 2017 testimony[11])

Moreover, the paid bribes were subtracted from individual bonuses gained through the project, so that the responsible manager had autonomy to fix the sum of the bribe. The existence of a permissive code of conduct and the lack of a working compliance system that may prevent the unlawful activities inside of the organization, therefore, were both elements that gave rise to internal incentives for resorting to illegal practices; both contributed to the establishment of an organizational culture in which such practices were accepted and normalized. In their testimonies, the executives displayed a self-image as professionals that achieved sound performances. In other words, "they did their job."

Of course, by now all these companies have changed their policies concerning the illicit dimensions of their business activities. The "departamentos de operações estruturadas" have been wound up, and the bonus systems terminated. The company Odebrecht repented by running apologies in the print media for its wrongdoing.[12]

The Corporate Level Actors: Entrepreneurs and Top Managers

The number of entrepreneurs, managers, chief executives, and middle managers involved in Operation Carwash increased with every new discovery in the investigation. We concentrated our analysis only on the construction companies, which were accused and convicted of crimes related to cartel formation and their illegal access to the Petrobras contracts. To date, prosecutors have opened eleven criminal trials against sixteen construction companies and 64 executives at the Federal Court of Paraná,[13] investigating crimes against the state-controlled company Petrobras. Just five of these criminal trials were concluded; the other six are still under investigation. Presently, 33 executives and owners have been convicted of corruption, cartel building, and money laundering, and only nine managers have been absolved, due to lack

[11] https://g1.globo.com/politica/operacao-lava-jato/noticia/executivos-da-odebrecht-superfaturavam-obras-para-ganhar-bonus-da-empresa.ghtml

[12] See: https://economia.estadao.com.br/noticias/geral,odebrecht-fecha-acordo-de-r-6-8-bilhoes-e-pede-desculpas-ao-pais-por-pratica-de-corrupcao,10,000,091,979.

[13] The eleven criminal trials against executives, owners, and managers of the construction companies that participated in the price-fixing cartel: 5036528-23.2015.404.7000; 5036518-76.2015.4.04.7000; 5027422-37.2015.4.04.7000; 5025847-91.2015.4.04.7000; 5083351-89.2014.404.7000; 5083376-05.2014.404.7000; 5083360-51.2014.404.7000; 5083401-18.2014.404.7000; 5083258-29.2014.404.7000; 5011926-65.2015.4.04.7000.

of evidence. On average, the convicted executives were sentenced to 15 years in prison on top of substantial fines.[14]

The results of the fight against corruption show that the case involves not just a few people, but a large number of entrepreneurs and chief executives in top positions of the construction companies. Both the rank and number of those who perpetuated price-fixing, money laundering, and bribery indicate the institutionalization of a strong subculture or a social cocoon. In the cartel case, the establishment of a deviant subculture can be self-sustaining. For decades the companies have developed a common modus operandi with the public sector: They decide which projects to pursue and at what price to run them, but they also decide, on the basis of the sum of their campaign donation, if their business activities can be perpetuated.

The analysis of court documents, especially the testimonies, not only provided strong evidences for the development of a social cocoon, as well as a subculture inside the organizations, but also gave us an insider's look at the cartel and the relationship among these companies. The social cocoon can be considered as part of the organizational culture, because it thrives and shapes the managers' and employees' values, norms, and attitudes in support of the corrupt practices. The social cocoon differentiates itself from a negative social perspective about corruption, which group members consider as common practices that are necessary for the organization to achieve its objectives (Campbell & Göritz, 2014). This subculture within the organization is developed over time through three main mechanisms: institutionalization of systematically used corrupt practices; rationalization by plausible social justification for the use of these practices; and socialization of the unwritten rules and informal interpretations within the organization.

Operation Carwash investigators have revealed that the cartel's activities were supported by the top management at the enterprises involved. The senior executives and managers who sustained these practices were on average 69 years old and had worked at the company for 23 years. Thus, socialization and insider recruitment became important factors in establishing and perpetuating the subculture. Moreover, the ways in which illegal practices were normalized, so that the actors involved did not question them, supply strong evidence for the establishment of a social cocoon. The majority of actors questioned during the investigation avoided the official definition of their practices—such as bribery, fraud, and price-fixing—and preferred to use words such as "commission" or "toll" to explain how, when, and why they decided to act corruptly.

Another important element in understanding the corrupt action from the organizational culture's perspective is the benefits for the organization. According to available court documents, the managers and entrepreneurs participating in the Petrobras plot were aiming primarily at the success of the organization. In one case, a senior executive at a construction company explained that the paid bribes may cause disadvantages in the final project costs and consequently this would decrease his

[14] Not included here are the 77 executives of Odebrecht who entered into plea bargains.

bonus.[15] This case is of course not common, but at the same time it is not common either among the cartel companies to identify personal gains as the rule of the game. More than 80% of the construction companies' implicated senior executives and managers were paying bribes to the Petrobras senior executives, because that was the rule of the game. The unwritten rules were widely known in the construction industry, rather than limited to the entities that participated in the cartel.

Achieving organizational goals may also satisfy personal interest. The high-ranking managers may get promoted to higher positions, thus ensuring better status and better remuneration. Bonuses and careers play an important role in this case, not just on a personal but also on an organizational level, which awards loyal managers who are pulling the company to a better market position.

Illegal practices related to personal gains are not mentioned in the court documents, because these crimes are not charged in the indictment. But in the course of the investigation investigators also revealed that some managers and senior managers of the investigated enterprises indulged in private deviations. In this case, the managers kept a part of the bribes that were to be delivered to a "black market dealer" or to a politician. Sometimes, they received offers from politicians and other public agents to "help" [pay bribes] in the political campaigns and in retour; they may receive a percentage of the sum. Other executives also misappropriated part of bribes that should be payed to politicians. The number of executives who have owned up to these practices is very low.

These managers' individual deviations were in part supported by the organization, because the bonus system could be understood here as a strong incentive for paying bribes. Even if they pursued personal gains, these actors were still conforming to the informal rules of the organization. Thus, the individual deviance, which is driven by the organization, is an individual expression of collective practices.

But Operation Carwash has changed this landscape. At present, investigators have discovered bribery payments of US$2 billion. One-hundred-and-eighteen suspects have been arrested, and fifteen top managers and entrepreneurs were given an average sentence of 15 years. A total of 77 criminal prosecutions against 289 individuals, as well as eight corruption charges against sixteen major construction companies and 50 individuals, are being processed. These accusations are directed against the top management of the largest private construction companies and the state-owned oil company Petrobras. Their presidents, managers, and employees face charges of corruption (article 33, penal code, § 4), money laundering with document forgery (article 1, caput, inciso V, Lei No. 9.613/1998), and the participation in a criminal organization (article 2, Lei No. 12.850/2013).

---

[15] Executive sentenced in the criminal procedure no. 5036518-76.2015.4.04.7000.

The Normalization of Political Corruption

Teori Zavascki, one of the eleven Ministers of the Supreme Federal Court of Brazil, died on January 19th, 2017, in an aircraft crash. Teori Zavascki was the justice presiding over the trials resulting from Operation Carwash. He was in charge of the legal proceedings of the Petrobras scandal and the intertwining relations between economic and political corruptions in Brazil it helped to unearth. His successor, Luiz Edson Fachin, has assumed the mantle.

One of the first testimonies under the framework of the "Leniency Agreement" with Odebrecht came from Claudio Melo Filho,[16] a senior manager at the construction's groups department for "institutional relations." His testimonies—whatever judicial value they may turn out to have—suggest that in Brazil, the relations between big companies and politicians who hold the lever of state power go far beyond bestowing simple favors and minor legal backings. According to Melo Filho, Odebrecht and other companies were able to ensure that certain laws were bended in their favor. Melo Filho argues that, as part of this Brazilian favoritism in politics and legislation, companies such Odebrecht returned favors from influential politicians, candidates, and parties in their clique by serving their political interests and fulfilling personal wishes. Claudio Melo Filho described how, over a 12-year period, his company managed to build a network of inroads to cover many high-ranking politicians.

Melo Filho listed payments by the Odebrecht construction group to various politicians who are now the subject of many investigations and legal proceedings. According to his testimonies, these payments, to the tune of millions of US dollars, were supposed to change certain laws in favor of Odebrecht and other companies. This included, for example, a modification to the legislation on tax preferential treatment of corporate profits of Brazilian companies operating abroad (MP 627/13[17]). Claudio Melo Filho also mentioned nine other legislative reforms and legislative proposals that have been influenced by political donations and payment of bribes over the last decade.

These testimonies against political corruption in Brazil, among others, make it clear that all major political parties have systematically used this method of funding politicians and parties. According to the current investigation, 51% of the accused and convicted come from the Partido Progressista (PP) party, 22% from the PMDB party, 18% from the PT party, around 3% from the Partido da Social Democracia Brasileira (PSDB), and 6% from other parties. Admittedly, these results are now changing on a daily basis[18] (Netto, 2016). It is now known that Odebrecht

[16] Inquiry document of Odebrecht's Leniency Agreement: Anexo pessoal from Claudio Melo Filho.

[17] The provisional measure (MP 672/13) was enacted in 2014 and converted into the law 12.973/2014. It regulated the tax payment of Brazilian enterprises that operate abroad (see: http://www.planalto.gov.br/ccivil_03/_ato2011-2014/2014/Lei/L12973.htm)

[18] The number of involved politicians per political party changes daily according to the investigation.

construction group handed out approximately US$14.4 million to 15 different political parties during the 2014 presidential and gubernatorial elections alone.[19] João Vaccari, a former accountant of the PT party, claimed that he had received roughly US$200 million for the PT party alone between 2003 and 2013.[20] As evidenced by the previous investigations, the Petrobras bribery system paid out somewhere between US$3 and 6 billion as political donations and bribes over the last 10–15 years.

So far, Operation Carwash investigators have shown that political corruption served various purposes in Brazil. It facilitated personal enrichment, funded election campaigns and political parties, and influenced legislations. The case of José Dirceu of the PT party, who was the former presidential minister and has been sentenced to more than 23 years in prison on charges of bribery, money laundering, and membership in a criminal organization, highlights the connections between personal enrichment and party interest. In one of these criminal trials, it became apparent that Petrobras's then-chairman of the board had bribed him to the tune of US$ 3.9 million, part of which he then handed over to the PT party. However, he retained around US$1 million for himself, which he then invested, for example, in real estate properties.[21]

Since the beginning of Operation Carwash, dozens of Brazilian politicians, members of government, and high-ranking officials, as well as top executives and entrepreneurs of many big companies, have been investigated. 193 inquiries, seven criminal processes against 100 people, and 212 agreements are currently going up in front of the Brazilian Supreme Court. Eight of President Michel Temer's 28 ministers have been investigated. Furthermore, twelve governors, 24 senators, and 39 deputies have been the target of investigations and legal proceedings in the Supreme Court and in the Electoral Tribunal. More than 200 politicians are involved in total.[22] The Leniency Agreements and testimonies of implicated corporate managers and executives are shedding more and more light onto the dubious entanglements between politics and businesses in Brazil.

The Political Actors: Political Leaders

Brazil's corrupt political environment is neither associated with a single party, nor is grounded in particular family or social backgrounds. Politicians and members of the government who were investigated and arrested belong to different political parties and come from different interest groups, which in turn have different political

[19] http://meucongressonacional.com/eleicoes2014/empresa/15102288000182

[20] See: criminal trials: 5045241-84.2015.4.04.7000, 5013405-59.2016.404.7000.

[21] Criminal Process 5,045241-84.2015.4.04.7000/PR.

[22] http://congressoemfoco.uol.com.br/noticias/sai-a-lista-de-fachin-ministro-do-stf-autoriza-investigacao-de-quase-100-politicos-na-lava-jato/

goals and social conceptions. However, all seem to use similar practices to maintain their political positions and the primacy of their parties. As soon as a good relationship and a possible "money source" appears on the horizon, they snatch to bankroll their campaign and party, as well as to line their own pockets.

But politicians and their parties alone cannot maintain this politically deviant environment in Brazil. Big business also plays an important role here. However, it is difficult to say who is driving the system or who began it. Is it the political system that fails to show the company any way to be economically successful without corruption? Or are companies exploiting the political system's weakness in order to benefit themselves? In any case, it is the big companies that keep this system of corruption running. They not only have the necessary capital to fund the parties and politicians; they are also the ones who have the capacity to commit bribery and corruption schemes through government contracts. They are not only the means, but also the ends. As a quid pro quo, they received preferential treatment both in winning fat contracts and benefitting from legislative amendments for their enterprises.

The investigations of Operation Carwash make it clear that the arrests of, and harsh penalties given to, companies and politicians have not prevented them from continuing their corrupt practices. Once new detection measures are adopted, actors devise new strategies for fraud, bribery, and money laundering. Sooner or later, however, most of them land in an investigation. This is exactly the case of former Republic President Michel Temer, who was vehemently committed to impeaching the former President Dilma Rousseff. He is charged with corruption, money laundering, and the formation of a criminal organization. His "money source," the big Brazilian company J&F Investimentos, which is currently being investigated in several corruption scandals, has reached an agreement with the public prosecution authority and has testified against almost all of his accomplices.[23]

Joesley Batista, the now resigned president of the J&F Investimentos, presented the public prosecution authority with four pieces of what he claimed to be taped conservations with high-ranking politicians, including a conversation with former president Michel Temer.[24] The recorded conversation clearly indicates that J&F Investimentos was trying to prevent, with Temer's support, further investigations by the judiciary. To prevent Eduardo Cunha, the former President of the Chamber of Deputies, who has been arrested since October 2016, from testifying before the court, Joesley Batista takes care of the well-being and financial affairs of the Cunhas family.[25] Batista also ensured that the testimonies from those currently in custody remained in tune with each other.

The Brazilian judiciary has, undoubtedly, been much more efficient in the investigation and prosecution of corruption cases in recent years. Not only have several regulative and institutional changes that have taken place over the last decades played a part, so have the Brazilians' changed perception and tolerance of

[23] Testimony n. 7003.

[24] Inquiry n. 4483/STF.

[25] Inquiry n. 4489/STF.

corruption. However, many politicians in Brazil still seem to believe that in the end it will remain business-as-usual. The conviction of former President Luiz Inacio Lula da Silva (in July 2017, sentenced to nine-and-a-half years in jail by the Federal Court of Paraná[26]), the indictment against President Michel Temer, the investigations of eight state ministers, 24 senators, 39 members of the Chamber of Deputies, and eight governors, as well as the convictions and arrests of dozens of other politicians and top managers, do not seem to be enough to stop the hidden political corruption.

A Case of Good Governance?

Those who have hitherto been ignorant or have had no capacity to estimate the extent of political corruption are now being provided with more and more evidence by the Brazilian judiciary: A system of political corruption existed in Brazil, the extent of which is becoming increasingly evident with each additional legal proceeding. It is striking that none of Brazil's political leaders is willing to take the political responsibility. All are looking on and protesting their own innocence, until they are either taken to court or happily escape condemnation. This is also part of a political system which, as is now becoming more and more apparent, is characterized by organized irresponsibility and graft.

Joesley Batista and former president Michel Temer did not stop their conversation at the obstruction of justice, but showed clearly that the bribery system remains valid despite all the excitement and attention of the media, authorities, and the general populace. In this sense, the fear of being uncovered seemed to have been much lower than the fear of leaving the deeply institutionalized old corrupt patterns. Those transforming a political system characterized by corruption, such as Brazil's, must take time to consolidate the effect of institutional changes in its structure. Although policymakers have completed major regulative changes and institutional changes over the last decades, which have made an enormous difference in terms of improving transparency in public administration and strengthening press freedom and justice, they have not (yet) been able to eradicate systemic corruption. However, all these changes have contributed to the fact that the previous corruption scandals are now accessible online to all and well-publicized in the press. It is not enough to stop systemic corruption, but it is a first and important step in a long-term process. Further institutional changes on the political level are just as important. The introduction of new regulations and the law enforcement create a new possibility for changing attitudes in Brazilian business and politics, and they can also support the creation of new informal institutions, such as new expectations, beliefs, and perception, of ethics and good governance in Brazilian society. In that sense, our analysis has made clear that regulations alone cannot generate good governance attitudes,

[26] Criminal trial n. 5,046512-94.2016.4.04.7000/PR, sentence 948, July 12th, 2017.

but require the institutionalization and legitimation of a new belief system, that should be shared in different groups such as business people, politicians, citizens, and organizations in a social and economic context. To conclude, the new era of law enforcement may be characterized as effective, with significant regulative and institutional changes. Law enforcement has been "good," shoving powerful interest groups aside. However, it remains unclear to what extent various interest groups and societal actors really engage in the fight against corruption. Parts of civil society have been influenced by the political parties and their interests. Most political parties are still in a state of shock and are not supporting the new era of law enforcement in Brazil If good governance is based on significant participation by various interest groups and societal actors in the fight against corruption, it is still missing in Brazilian society.

References

Article 33, Penal Code, § 4, Law Decree No. 2.848 [Art. 33 do Código Penal- Decreto Lei 2848/40]. (1940, December 7). Brazil. Retrieved from https://www.jusbrasil.com.br/topicos/10636569/artigo-33-do-decreto-lei-n-2848-de-07-de-dezembro-de-1940

Barkemeyer, R., Preuss, L., & Lee, L. (2015). Corporate reporting on corruption: An international comparison. *Accounting Forum, 39,* 349–365. https://doi.org/10.1016/j.accfor.2015.10.001

Bathelt, H., & Glückler, J. (2014). Institutional change in economic geography. *Progress in Human Geography, 38,* 340–363. https://doi.org/10.1177/0309132513507823

Braithwaite, J. (2008). *Regulatory capitalism: How it works, ideas for making it work better.* Cheltenham: Edward Elgar.

Campbell, J.-L., & Göritz, A. S. (2014). Culture corrupts! A qualitative study of organizational culture in corrupt organizations. *Journal of Business Ethics, 120,* 291–311. https://doi.org/10.1007/s10551-013-1665-7

Chaves, A. C. S. (2013). A corrupção privada no Brasil [Private corruption in Brazil]. *Revista Jurídica ESMP-SP, 4,* 231–260. https://doi.org/10.13140/2.1.2858.2084

Correia, I. D. (2015). *Os principais aspectos da Lei Anticorrupção brasileira 12.846/2013 e a importância da cultura de compliance no ramo empresarial* [The main aspects of the Brazilian Anti-Corruption Law 12.846/2013 and the importance of the compliance culture in the business sector]. Rio de Janeiro: Pontifícia Universidade Católica do Rio de Janeiro. https://doi.org/10.17771/PUCRio.acad.26080

Dallagnol, D. (2017). *A luta contra a corrupção: A Lava Jato e o futuro de um país marcado pela impunidade* [The fight against corruption: Lava Jato and the future of a country earmarked by impunity]. Rio de Janeiro: Primeira Pessoa.

Dantas de Araujo, F. (2012). Uma análise da Estratégia Nacional Contra a Corrupção e a Lavagem de Dinheiro (ENCCLA) por suas diretrizes [An analysis of the national strategy against corruption and money laundering (ENCCLA) by its directives]. *Revista Brasileira de Políticas Públicas,* 2(1), 53–82. https://doi.org/10.5102/rbpp.v2i1.1649

de Rugy, V., & Warren, M. (2009). Expansion of regulatory budgets and staffing continues in the new administration: An analysis of the U.S. budget for fiscal years 2009 and 2010. Retrieved from https://www.mercatus.org/system/files/Regulators-Budget-Report-Final-Version-October-29.pdf

de Sanctis, F. M. (2015). Voice and accountability: Improving the delivery of anticorruption and anti–money laundering strategies in Brazil. In J. Wouters, A. Ninio, T. Doherty, & H. Cisse' (Eds.), *Improving delivery in development: The role of voice, social contract, and accountability* (pp. 391–413). The World Bank Legal Review: Vol. 6. Washington, D.C.: World Bank Group. https://doi.org/10.1596/978-1-4648-0378-9

Filgueiras, F., & Araújo, M. M. (2014). A política anticorrupção e o marco legal no Brasil [Anti-corruption policy and the legal landmark in Brazil]. In I. F. Cunha & E. Serrano (Eds.), *Cobertura jornalística da corrupção política: Sistemas políticos, sistemas mediáticos e enquadramentos legais* (pp. 36–71). Lisboa: Alêtheia Editores.

Glückler, J., & Lenz, R. (2016). How institutions moderate the effectiveness of regional policy: A framework and research agenda. *Investigaciones Regionales, 36,* 255–277.

Glückler, J., Suddaby, R., & Lenz, R. (2018). On the spatiality of institutions and knowledge. In J. Glückler, R. Suddaby, & R. Lenz (Eds.), *Knowledge and institutions* (pp. 7–40). Knowledge and Space: Vol. 13. Cham: Springer. https://doi.org/10.1007/978-3-319-75328-7

Jordana, J., Levi-Faur, D., & Fernández i Marín, X. (2011). The global diffusion of regulatory agencies: Channels of transfer and stages of diffusion. *Comparative Political Studies, 44,* 1343–1369. https://doi.org/10.1177/0010414011407466

Lei Complementar No. 101 [Tax Liability Law No. 101]. (2000, May 4). Brazil. Retrieved from http://www.planalto.gov.br/ccivil_03/leis/lcp/lcp101.htm

Lei Complementar No. 131 [Tax Liability Law No. 131]. (2009, May 27). Brazil. Retrieved from http://www.planalto.gov.br/ccivil_03/leis/lcp/lcp131.htm

Lei No. 8.112 [Law No. 8.112]. (1990, December 11). Brazil. Retrieved from http://www.planalto.gov.br/ccivil_03/leis/l8112cons.htm

Lei No. 8.429 [Law of Administrative Improbity No. 8.429]. (1992, June 2). Brazil. Retrieved from http://www.planalto.gov.br/ccivil_03/leis/l8429.htm

Lei No. 8.666 [Bidding Law No. 8.666]. (1993, June 21). Brazil. Retrieved from http://www.planalto.gov.br/ccivil_03/leis/l8666cons.htm

Lei No. 9.613 [Law of Money Laundering No. 9.613]. (1998, March 3). Brazil. Retrieved from http://www.planalto.gov.br/ccivil_03/Leis/L9613compilado.htm

Lei No. 10.520 [Bidding Law No. 10.520]. (2002, July 17). Brazil. Retrieved from http://www.planalto.gov.br/ccivil_03/LEIS/2002/L10520.htm

Lei No. 12.527 [Freedom of Information Act, Law No. 12.527]. (2011, November 18). Brazil. Retrieved from http://www.planalto.gov.br/ccivil_03/_ato2011-2014/2011/lei/l12527.htm

Lei No. 12.846 [Anti-Corruption Act, Law No. 12.846]. (2013, August 1). Brazil. Retrieved from http://www.planalto.gov.br/ccivil_03/_Ato2011-2014/2013/Lei/L12846.htm

Lei No. 12.850 [Law of Criminal Organizations No. 12.850]. (2013, August 2). Brazil. Retrieved from http://www.planalto.gov.br/ccivil_03/_Ato2011-2014/2013/Lei/L12850.htm

Netto, V. (2016). *Lava Jato: O juiz Sergio Moro e os bastidores da operação que abalou o Brasil* [Lava Jato: Judge Sergio Moro and the backstage of the operation that shook Brazil]. Rio de Janeiro: Sextante.

Pascolati Junior, U. A. (2016). Corrupção privada do direito brasileiro [Private corruption in the Brazilian Law]. In A. J. C. da Cunha Filho, G. R. B. de Araújo, R. Livianu, & U. A. Pascolati Junior (Eds.), *48 Visões sobre a corrupção* (pp. 147–163). São Paulo: Quartier Latin.

Passas, N., & Kleinhempel, M. (2016). Transparency and compliance policies of multinational companies in Latin America. In A. Stachowicz-Stanusch (Ed.), *Corporate social performance in the age of irresponsibility: Cross national perspective* (pp. 79–88). Contemporary Perspectives in Corporate Social Performance and Policy: Vol. 21. Charlotte: Information Age.

Petrobras. (2014). Comissão Interna de Apuração: Protocolo de Registro na Segurança Empresarial n.°031/2014 [Internal Investigation Commission: Protocol for Registration in Business Security] (DIP DABAST 71/2014). Retrieved from https://politica.estadao.com.br/blogs/fausto-macedo/wp-content/uploads/sites/41/2018/01/3_COMP115.pdf

Pizarro, L. (2015, July 11). Programas anticorrupção no Brasil são 'para inglês ver' [Anti-corruption programs in Brazil are 'for the English to see']. *O TEMPO (Contagem).* Retrieved from http://www.otempo.com.br/capa/economia/programas-anticorrup%C3%A7%C3%A3o-no-brasil-s%C3%A3o-para-ingl%C3%AAs-ver-1.1068058

Richard, M. (2014). Brazil's landmark anti-corruption law. *Law and Business Review of the Americas, 20,* 141–148. Retrieved from https://scholar.smu.edu/lbra/vol20/iss1/9/

Scott, W. R. (2014). *Institutions and organizations: Ideas, interests, and identities* (4th ed.). Los Angeles: SAGE.

Wohlnick, A., & Correia, I. D. (2016). Korruptionsbekämpfung und Compliance in Brasilien: Das Antikorruptionsgesetz 12.846/2013 und seine Bedeutung für in Brasilien tätige Unternehmen [Anti-corruption and compliance in Brazil: The Anti-Corruption Law 12.846/2013 and its significance for companies operating in Brazil]. *Zeitschrift für Internationales Wirtschaftsrecht (IWRZ), 5,* 210–214.

Chapter 11
Lateral Network Governance

Johannes Glückler

The Puzzle of Governing Networks

When firms organize themselves into interorganizational networks, they need to agree on common goals and a viable way of working with each other. How should networks be designed and governed to help partners achieve their common goals efficiently? This question is particularly salient for organizations that make an effort to surmount collective action dilemmas at the local level as well as across geographical scales to establish legitimate practices of cooperation and compliance in diverse contexts, such as in environmental (Bulkeley, 2005; Liverman, 2004; Scott, 2015) and regional governance (Feiock, 2007; MacLeod & Goodwin, 1999; Montero & Chapple, 2019), the governance of local commons (Hardin, 1968; Ostrom, 1999), and in new—often local—organizational arrangements of so-called diverse economies (Gibson-Graham, 2008) and alternative economic practices (Sánchez-Hernández & Glückler, 2019; Seyfang, 2006). Here, research on human geography, organization studies, the political, social, and environmental sciences, as well as network analysis intersects, with researchers working to conceive solutions for successful governance among diverse sets of actors. Although scholars have established an understanding of networks as an alternative mode of governance vis-à-vis markets and hierarchies (Powell, 1991; Williamson, 1985), little is known about how network organizations should actually be governed. More accurately, what is needed is an understanding of which forms of network governance are suitable in which contexts of interorganizational cooperation: "We are thus left with an understanding of why networks may be a superior mode of governance but not of how they are themselves governed" (Provan, Fish, & Sydow, 2007, p. 504).

J. Glückler (✉)
Department of Geography, Heidelberg University, Heidelberg, Germany
e-mail: glueckler@uni-heidelberg.de

J. Glückler et al. (eds.), *Knowledge for Governance*, Knowledge and Space 15,
https://doi.org/10.1007/978-3-030-47150-7_11

It is premature to call the field of inquiry into network governance[1] a coherent research framework (Ansell & Torfing, 2016; Glückler, Dehning, Janneck, & Armbrüster, 2012; Jung, Krebs, & Teubner, 2015; Keast, 2016; Maggetti & Gilardi, 2014; Prota, 2016), even more so because the term governance has been used rather broadly across the social sciences, such as in political science (Ansell & Gash, 2008; Crouch, 2005; Rhodes, 2007), institutional economics (Williamson, 2005), sociology (Podolny & Page, 1998; Powell, 1991; Rowley, Behrens, & Krackhardt, 2000), and human geography (Allen & Cochrane, 2007; Charron, Dijkstra, & Lapuente, 2014; Macleod & Goodwin, 1999). The dearth of concepts of and empirical insights into network governance stems primarily from the egalitarian aspirations of partners cooperating in lateral networks. For unlike so-called *strategic networks* (Sydow, Schüßler, & Müller-Seitz, 2016) or vertically structured networks such as global value chains (Crang, Hughes, Gregson, Norris, & Ahamed, 2013; Gereffi, Humphrey, & Sturgeon, 2005; Glückler & Panitz, 2016; Grabs & Ponte, 2019) or global production networks (Coe, Dicken, & Hess, 2008; Levy, 2008; Yeung & Coe, 2015), in which governance functions are often assumed by a focal lead firm, lateral networks consist of members who see themselves as autonomous and equal to everyone else. Therefore, members of lateral networks are often wary of outside control: "[T]here is little empirically grounded work researching how organizations without permanent bosses and followers, in which all members ultimately have formally equal say in running operations or exercising control, are able to operate" (Lazega, 2001, p. 1).

In this article, I would like to advance the concept of lateral network governance in a specific empirical context of interorganizational networks that has received only limited attention in governance research: the organized network. An organized network is a voluntary and deliberate association of members that directs multilateral cooperation between a limited number of legally and economically independent organizations towards a shared economic goal (Glückler & Hammer, 2015). Expectations of autonomously exercising control in partnership with each other are either inconsistent with rigid decision-making hierarchies or impossible to meet, for weak governance cannot ensure compliance. This article centers on the paradox of having equals commit themselves to coordinating their actions. To solve this fundamental problem, I propose analyzing network governance as a structure for the legitimate delegation of decision-making. The lateral governance that arises through such transfer takes account of both the contextuality and the malleability of networks and allows one to give formal and informal governance mechanisms equal and simultaneous consideration. The structure for legitimate transfer of decision-making, as the empirical analysis will illustrate, builds the foundation for efficient and legitimate governance of organized networks.

In the first section, I propose a framework of four elements—context, object, mechanism, and agency of governance—that together account for the empirical variation in governance forms. In the second section, I build on the notion of relational distributions of legitimacy to develop the concept of lateral network

[1] In this article, I use the term *network governance* as a superordinate term to include both *network control* and *network regulation*.

governance. It helps to identify legitimate agents of governance by capturing the delegation of discretionary authority among equals. In the third section, I develop a method for empirically assessing the legitimate transfer of decision-making authority by drawing on concepts from social network analysis. In the fourth section, I explore two organized networks of small and medium-sized enterprises (*SMEs*)—one in the area of management consulting, the other in dental technology—to empirically show how sharply the actually observed relational distribution of legitimacy can diverge from the formal governance structure. A concept of lateral network governance and the analysis of relational structures of legitimacy suggest a more holistic understanding of network governance that integrates formal positions, offices, and contracts with informal social relations of authority and legitimacy.

The Governance Challenge

One reason for the diversity of research on network governance is the large variation in empirical contexts and phenomena researchers usually study. A more general theory of network governance, therefore, requires consistent terminology and needs to capture the particular empirical context that distinguishes networks from hierarchical organizations such as corporations or bureaucracies. Forms of network governance vary according to at least four key elements that together constitute the complex of network governance: These are (i) the context of governance, that is, the conditions framing collaboration among actors; (ii) the object of governance, in other words, the common interest and goals of collaboration; (iii) the mechanisms of governance, such as contracts, power, trust, and so on; and (iv) the agency of governance, in other words, actors and bodies granted the authorization or legitimacy to exercise governance.

The Context of Governance: Network Cooperation

Generally, the process of governance is intended to reduce variance in an existing system and to exert more than just piecemeal influence on autonomous dynamics, events, and interactions (Sydow, 2000). When it comes to organized networks as defined above, some of the most important contextual factors are the expectations of legally independent and equal partners in a voluntary group (Staber, 2000). This is all the more germane in the context of horizontal networks, in which business firms operate at equal or similar stages of the value chain. Horizontal networks represent arenas for potential or actual competition between the members on their respective factor and commodity markets. Legal autonomy, potential competition, and mutual independence regarding each other's resources constitute a highly distinct context. This constellation of factors poses a great challenge to network governance as it precludes hierarchical principles as control mechanisms (Lazega, 2000; Lazega & Krackhardt, 2000). The voluntary and autonomous basis of action makes it difficult to issue hierarchical directives, which the members *can* accept but by no means must. The same goes for sanctions. Even if it is contractually agreed that undesirable behavior in the network may be sanctioned, applying these formal sanctions may jeopardize the network insofar as members could leave it, an exit that could eventually destabilize it (Glückler & Hammer, 2017; Lazega, 2000). Voluntary membership is based on the benefit a person expects from it (Windeler, 2001). Network members tie the extension of their membership to their assessment of whether the benefits it bestows can outweigh its costs. The will to have a say is thus relatively high, and the network's objective must undergo constant discussion in which members build consensus. Ring and van de Ven (1994) argue that network members must develop a common, not an individual, concept of their motivation, their feasible investments, and the likely uncertainties—a process typical of the negotiations stage in maturity models of network evolution. Although empirical studies do not necessarily substantiate the validity of such a maturity model (Melot de Beauregard, Németh, & Glückler, 2012), negotiation of consensus on objectives and governance have an important bearing on the understanding of the network.

The Object of Governance

The second element of network governance is the object to be collectively governed. In their review of the literature, Park and Ungson (2001) conclude that cooperation is more likely to fail than to succeed in the long run. Accordingly, they are interested in the factors leading to a network's breakdown and failure, not in those facilitating successful group cooperation. By calling attention to the "don'ts" more than the "musts", the reasoning that these two scholars pursue in their conceptual model considerably enriches the literature. They confine their discussion to bilateral alliances, but it is both eminently possible and helpful to apply their arguments to multilateral networks. Park and Ungson (2001) see the main challenge of cooperation as lying in the rivalry between the participating partners and in the complexity of what is to be governed amid the uncertainties and vagueness of the management strategies involved. For them, the most important objects of network governance are the controlling (the monitoring of costs and benefits accruing from interfirm cooperation), the balancing of interests, and the coping with dissimilar management styles of the member firms. Governance is hampered by the necessity of minimizing its costs so as to avoid eating into either the savings or the gains of cooperation (Park & Ungson, 2001). Sydow (2000) and Windeler (2001) discriminate more precisely between the objects of governance in the context of what they call network regulation. Focusing on the modalities of the governance framework for conducting business within networks, they delineate six objects: selection, allocation, evaluation, systems integration, position configuration, and boundary-setting. This framework represents a detailed typology of the range of governance objects in network governance. Its authors focus particularly on the selection of members and the delimitation of the network vis-à-vis its environment; the allocation of resources and the distribution of benefits; the definition of common goals, process control, and the evaluation of their achievement. Network regulation is thus geared to various strategic and operational matters, and its total cost must never exceed the savings or benefit that the network partners reap from their cooperation. So-called network coordinators are in charge of network regulation, and their legitimacy entitles them to act on behalf of the network.

The Mechanisms of Governance

The third element of a network governance framework is the mechanisms through which the objects of governance are actually achieved in a context of collective action. Conventionally, scholars distinguish two types of mechanisms: Whereas formal regulations stipulate nominal bodies and offices for the persons enforcing them, informal control mechanisms are rooted in social institutions such as trust, reciprocity, or reputation (Glückler, Suddaby, & Lenz, 2018). Informal governance is conceived of as socially practiced governance, in which social mechanisms such as power, trust, sympathy, and the exercise of influence are regarded as primary instruments of control. This perspective rests on voluntary integration, which predicates at least some degree of familiarity among the partners in the network. In this case, control of the network is assumed by actors whose legitimacy for this function originates not in formal, legal offices but rather in social practice. Social mechanisms such as trust, reputation, conventions, and the formation of a macro or network culture, predominate as forms of coordination in networks and guide the economic behavior of the actors (Glückler & Armbrüster, 2003; Jones, Hesterly, & Borgatti, 1997; Keast, 2016). In addition, Bachmann (2001) notes the relevance of power and trust as controlling mechanisms capable of affecting each other and of coordinating the actions at various levels of a network.

Researchers studying formal governance usually focus on justiciable rules and contracts on which network members agree in writing. Cooperation agreements and association statutes are examples. Formal rules lay down what rights, obligations, and sanctions are applicable in the network. They vary depending on the objectives and context of the network in question and range from rather loose articles of association—for instance, among medical practices to improve local health service delivery—to binding legal contracts in a joint-venture network. Drawing on the concept of state, Schäfer (2009) refers to a formal regulatory system within networks as a *network constitution*. Organizations, too, have their own regulations that describe the basic legal structure of their corporate governance. Although these instances of corporate governance can help regulate bilateral collaboration between an organization and its partner, they are not efficacious to rule an organized network as a whole and beyond the focal interest of one corporation. This is where network governance is necessary to close the gap between individual (corporate) and collective levels of collaboration. Schäfer intends to mitigate the problem of collective action by creating a legitimate entity capable of enforcing the rules agreed upon. The formal governance design of a network thus enhances its problem-solving capacity.

This conclusion underscores the need for research on the question of how organized networks can acquire legitimacy and clout. Unfortunately, researchers have so far studied formal and informal control mechanisms separately or have even distinguished between them as mechanisms typical of different maturity levels. According to maturity models of network governance, young and relatively small networks are characterized by informal governance mechanisms, whereas mature and larger networks draw on formal mechanisms such as contracts, organs, and nominal management authorities (Provan & Kenis, 2008; Ring & van de Ven, 1994). As Provan et al. (2007) observe, "relationships between organizations in a network are either informally maintained, through the structure of the network and norms of reciprocity and trust, or formally maintained, through the existence of contracts, rules and regulations" (p. 503). Rather than juxtaposing formal and informal mechanisms as a dualism, I conceive relational structures of legitimacy to include the simultaneity and duality of formal and informal sources of lateral governance.

The Agency of Governance

The fourth and final element of network governance refers to the locus of control and the organizational entities imbued with legitimate authority to rule. Provan and his colleagues propose a typology of governance forms based on where the formal agency of governance is located (Provan et al., 2007; Provan & Kenis, 2008; Provan & Milward, 1995). They distinguish self-governance by the members from external governance by a network administration organization (*NAO*). Self-governance can also be exercised centrally by a mandated representative of the network (an arrangement called *lead-organization governance*) or decentrally through mutual coordination by many or all of the members (*shared governance*). This decentralized form of self-governance (also called a *lateral control regime*) is encountered in law offices, consulting firms, and other partnerships (Lazega, 2000, 2001; Lazega & Krackhardt, 2000). The combination of these two dichotomies—self-governance versus external governance and lead-organization versus shared governance—is the basis for three, albeit not pure and exclusive, forms of governance. Provan and Kenis (2008) point out that an NAO can exist in both shared governance and in lead-organization governance. Moreover, they suggest that network performance depends on the suitable choice of each form of governance in particular conditions, such as network size, the existence of trust and skills, and the degree of goal consensus among the members. While self-governance is suitable for small networks with bedrock trust among its members, expanding membership and growing network competence may require a transition to the other two forms of governance. Other authors, including Ring and van de Ven (1994), likewise attribute to networks a dynamic trajectory along which interactions per se and their quality change. Apart from the useful distinction of these governance forms, the model's authors are rather vague about the sources of authority to exert control. Although Provan and Kenis (2008) focus on the networks' formal design, any governance authority must be acknowledged as legitimate and representative by the other members to have the sway to coordinate a network effectivley (Windeler, 2001). Such legitimacy is pivotal to the development of the theory of lateral governance in the next section. Few studies' authors have delved into the special significance that the inception of legitimacy has for the success of a network.

Lateral Network Governance

The Logic of Negotiation and the Regime of Lateral Control

So far, the literature's authors have set out from an isolated consideration of formal and informal governance mechanisms, without looking much at their empirical interdependence. Such dualism obscures the fact that organizations actually exploit both dimensions of network governance (Lazega, 2000). Forms of governance such as centralized network control (lead-organization governance) can operate formally and informally alike. Conversely, forms of contractually shared and informal governance are conceivable. In practice, there are forms of governance that cannot be called either purely informal or formal. For instance, an honor code, which in many corporate networks is agreed upon in writing, is not informal but is not readily actionable in legal terms. The empirical parallelism and interdependence of formal and informal forms of coordination therefore call for an alternative conception of network governance—one in which the two dimensions are not isolated but integrated by an encompassing principle based on the legitimacy of influence and control. This is the objective guiding the concept of lateral governance.

Lazega (2001) has focused on the parallel existence of formal and informal control mechanisms in law firms of equal partners. This organizational context is similar to the governance context of organized networks, which is of key interest in this article. Equity partners of a law firm as well as business firms associated as partners in an organized network all have equal legal standing and responsibility for the common outcome. In his empirical study, Lazega (2001) found that in cases of conflict the partners of a law firm tended to avoid formal intervention in order to limit the risk of open confrontation or escalation. At the same time, however, they tried to minimize the costs of informal governance. A lateral control regime thereby emerges at the collective level (Lazega & Krackhardt, 2000), so formal mechanisms such as official meetings are definitely used as arenas of informal micropolitics. Although the strategic orientation in this context is shaped by the consensus of all the participants, micropolitical and, hence, informal strategies figure strongly in forging that consensus. Based on their status, prestige and trust, a particular type of actors, the so-called oligarchs, figure prominently in facilitating the emergence of the consensus. As a consequence, the network's strategic questions are discussed by everyone involved but informally shaped by a few (Lazega, 2001). This insight appreciably advances our understanding of the interdependence of formal and informal processes of network governance. They are "scarcely possible to determine or plan from the outset, not least because of the loose coupling of the system's elements, and are instead always outcomes of (partly) autonomous processes that the initiator intentionally triggers but cannot totally control" (Sydow, 2002, p. 248).

An alternative access to conceiving an inclusive mechanism of network governance is found in the dominant logic of action. Equality before the law and the principle of voluntariness ordain that all members be regarded as equally warranted decision-makers (Lazega, 2000; Mayntz, 1993). Moreover, all decisions are to be made to the satisfaction of the network actors so as to preclude the departure of any member who feels disadvantaged against his or her will and to preserve goal consensus. Writing from the perspective of actor-centered institutionalism in political science, Mayntz and Scharpf (1995) posit that negotiation is the logic of action in these situations. They see the superiority of negotiation in the fact that pure exchange leads to unintentional negative aggregate effects and that hierarchical control breaks down anyway because the members expect to have a voice in decisions. The logic of negotiation predicates the interest in a joint result and facilitates agreement that accommodates the wishes of all network members. Whereas the principle is definitely practicable and has proven in game theory to be effective in small networks, goal consensus and decision-making become ever more difficult as group size increases (Scharpf, 2006). The effectiveness and efficiency of decision-making are not all that suffers in large groups. There is also the growing danger that minorities will be outvoted and that the cohesion will diminish. In line with Provan and Kenis (2008), hence, shared governance becomes problematic beyond a certain network size (Mayntz, 1993).

The Concept of Lateral Network Governance

How, then, can legitimate governance of relatively large networks be achieved? To answer the question, it is useful to look at other realms of society that face similar problems. Perhaps the most instructive case is the discussion of democratic systems in political theory. In democracies, which by definition comprise no hierarchical directives, decisions must be taken among independent peers, just as in networks. To make collective decisions feasible, the solutions vary on a continuum between symmetric and asymmetric ways to forge agreement. In symmetrical approaches, each individual's vote has equal weight in the group decision, whereas in asymmetrical approaches, discretionary authority is transferred to one or a few representatives to make binding decisions for everyone. Between these two extremes can lie schemes that define specific majorities, each arrangement having its merits and drawbacks. Whereas asymmetrical procedures offer the advantage of quick decision-making, they suffer from the danger of minority boycott. Conversely, symmetrical procedures hold the promise of consensus-based decisions, but the necessary negotiations referred to by Mayntz (1993) can drag on and on. To solve this dilemma, democratic systems delegate decision-making power, concentrating it in a small group of individuals who represent the interests of their electorate and enjoy authorization to negotiate decisions applying to everyone. Decisions can thereby be taken legitimately and efficiently by a small number of people. For all the dissimilarities between a public bureaucratic system and privately organized networks, the principle of delegating power offers a promising point of departure for responding to the question of dealing with governance issues in networks.

I assume that successful coordination in organized networks is based on the principle of delegating decision-making power. Such delegation is a prerequisite and principal origin of the concept of lateral network governance, in which all members legitimately share. The perspective of lateral network governance can be situated within the four elements of the governance framework explained above. First, it does justice to a context of governance in which the members have an equal right to participate. Second, it can be applied to any object of governance within a network that pursues common goals or collective goods. Third, lateral governance is based on a form of legitimation that can be both formally and informally rooted, and these two mechanisms can be integrated and simultaneously observed. Concentrating on the intensity and sharing of legitimacy in a network overcomes the dualism of formal and informal network governance both theoretically and empirically. Such a perspective is supported by Human and Provan (2000), who also stress the importance of the legitimacy of actors, actions, and structures, without separating their causes into informal and formal aspects. In addition, Weber (1978) also points out that legitimacy can come from informal practice such as tradition, feelings, or values as well as from formally accepted systems of legality. Therefore, I here adopt an approach with which I refrain from discriminating between formal and informal sources of legitimacy and aim to assess the relational distribution of legitimacy across the agents of governance within an organized network. Fourth, and consequently, the agent of governance is explicable not in terms of formal or informal facets but rather in the degree of its legitimacy. Delegating legitimate decision-making power therefore theoretically satisfies both demands of network governance elaborated above: coordination based on a partnership of equals and the minimization of transaction costs.

Locally and Globally Legitimate Agents of Governance

Legitimate delegation of decision-making can be distributed across the network's various interest groups. Depending on the relational structure of clusters of interests, subgroups might emerge that jointly grant themselves legitimate representation of their interests and might, in their mutuality, clearly set themselves apart from other subgroups. These subgroups structure the network into what I call legitimate

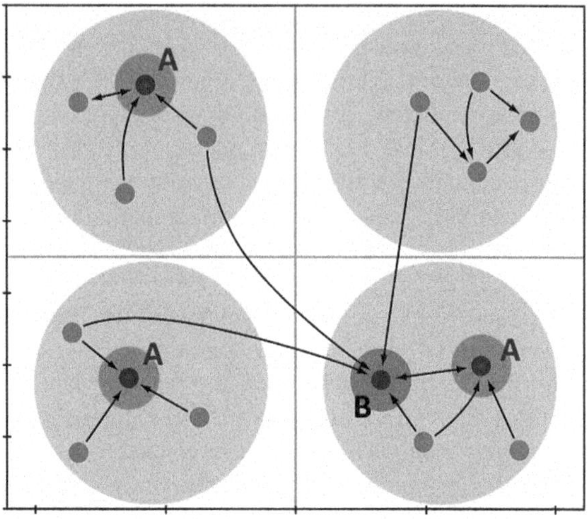

Fig. 11.1 Hypothetical network of the legitimate transfer of decision-making authority. Source: Design by author

factions. Figure 11.1 depicts a hypothetical distribution describing the transfer of legitimate decision-making authority between the members of a network. In this model, four members each have at least three votes from the network and possess the greatest legitimacy as measured by the number of times he or she has been named. However, member types A and B differ in the structure of those votes. The votes for type A all come from one local faction; those for type B, from all factions of the network. Whereas type A members are locally legitimate, type B members are globally legitimate. With this example, I illustrate the import that the specific structure of legitimacy has when comparable degrees of legitimacy are conferred on individual members of a network by their colleagues. If a member receives many votes yet exclusively from one faction (type A), then that person has local legitimacy. By contrast, global legitimacy enables a member to contribute much more to decisions acceptable throughout the network. As far as a network logic of action is concerned, the actors enjoying that kind of legitimacy (type B) can enter into negotiations and thereby make collective decisions more readily than the entire network can. Of course, whether type B actors succeed in achieving consent ultimately depends on the severity of conflict between the factions (as maintained by their type A representantives) and type B's legitimacy and ability to mediate their interests and eventually reconcile their conflict.[2]

Efficient network governance depends on a particular distribution of legitimacy in the network. Relational distributions of legitimate decision-making power vary with the strength and fragmentation of delegated decision-making power. An

[2]Type A actors exclusively represent the interests of their own factions and will thus enter into negotiation with a clear stake. In contrast, as actors of type B draw legitimacy from several factions, they may run the risk of losing part or all of that legitimacy in cases of conflict if the individual factions feel that their interests are not sufficiently supported. Therefore, a type B position per se is not a sufficient condition to solve conflicts within a network.

empirical analysis of governance structures thus centers on the following research questions: First, and methodologically, how can relational structures of legitimacy be assessed? Second, and empirically, how do observed patterns of legitimate governance relations (the empirical distribution of legitimacy) overlap with or diverge from the formal governance structures in an organized network? Third, and effectively, how does an empirical distribution of legitimate delegation of authority affect the network outcome? I will address these questions in a comparative case study of two organized networks of small and medium-sized business firms in Germany.

Research Design

Two Organized Networks: Management Consulting and Dental Technology

My first case study was on ConsultingNet,[3] a regional network of 23 management consultants that operated primarily to promote the professional exchange between its members by means of regular events and get-togethers. ConsultingNet was a registered association whose governance structure was written into its statutes (Table 11.1). The management board consisted of two members and was responsible for the management of the network activities. The association's limit on the term of office and the election of the two members during the annual general meeting arguably render the board a rather weak formal agent of governance. Despite the rules laid down in such documents, there was, legally speaking, relatively little provision made for clout and division of labor in an association, especially if, as in this instance, no recourse to sanctions had been worked out. A member who would break the code of conduct could be expelled from the association, but other, less drastic ways and means to sanction behavior had not been formalized. The governance model within ConsultingNet therefore corresponded closest to the model of shared governance (Provan & Kenis, 2008).

My second case study was on Dentis,[4] a network of 27 small and medium-sized dental laboratories located across Germany. Dentis was a limited liability company

Table 11.1 Formal governance of two corporate networks

Elements	ConsultingNet	Dentis
Legal form	Registered association	Limited liability
Control mechanism	Management board	NAO,* supervisory board
Electorate	General assembly of members	Shareholders' meeting
Form of governance	Shared governance	Shared governance, NAO

Note. *Network administration organization. Source: Design by author

[3] ConsultingNet is a pseudonym.

[4] Dentis is a pseudonym.

whose charter stipulates the election of five company representatives to the supervisory board at the shareholders' meeting (Table 11.1). Responsibility for the network's operations falls to a management office, which functions as an NAO (Human & Provan, 2000). Because of the member firms' geographic separation and the different, in some cases intense, multilateral cooperation between them, the network held quarterly shareholder meetings, at which the managing director, the members of the supervisory board, and the shareholders negotiate, decide on, and evaluate joint activities. High membership dues for individuals and the remarkable amount of work and time invested characterize Dentis as a very active and long-term network. The objectives of Dentis were to achieve networked production and joint development of new concepts and solutions in marketing, distribution, IT, quality standards, training, and other areas. Its governance structure typifies shared governance with a jointly operated NAO (Provan & Kenis, 2008).

I researched both organized networks according to the research procedure SONA—situational organizational network analysis (Glückler & Hammer, 2015; Glückler, Panitz, & Hammer, 2020)—and evaluated them for an extended period. SONA includes qualitative observation during personal and group interviews as well as quantitative data gathered with a standardized network survey and evaluated with methods of social network analysis. I have based the following analysis of governance structure on numerous interviews and, for each network, a survey covering more than 70% of the members (see Table 11.2).

The two networks fit into the foregoing classification of network governance. The context was the same for both. Of course, the market situation varies from one member firm to the next because each firm is affiliated with a different economic sector, but in both networks firms that were at least potential rivals had banded together to improve their competitiveness. The member firms in the two networks were legally autonomous and economically independent from each other. Regarding the object of governance, the divergence between the networks was greater because of what they did. Whereas ConsultingNet engaged in rather soft activities that were not capital intensive for its members (e.g., events), Dentis pursued more ambitious goals of collectively developing and investing in networked production and joint marketing and sales strategies. The ensuing complexity of governance in that network was evident from its use of an NAO. The networks also had similar governance mechanisms and agents of governance. Their regulations were formal, as were the concomitant formal authorities of governance. In both networks, however,

Table 11.2 SONA: mixed method research design and database

Instrument	ConsultingNet	Dentis
Preliminary discussion	1	1
Number of personal interviews	4	5
Network Survey response (rate)	19 (82.6%)	20 (74%)
Group discussion (validation)	1	1
Number of network members	23	27

Note. Source: Design by author

the personal interviews indicated awareness of a parallel dimension of control that was not formally regulated, one without which neither network would have been feasible.

Methodology: Measuring the Legitimate Delegation of Decision-Making Authority

Through preliminary discussions as well as the personal interviews with members of the two networks, I learned that the networks' governance was repeatedly linked with critical decision-making situations at the network level. For purposes of empirical observation, the delegation of decision-making power between the members of an organized network was operationalized with the following question item:

> Imagine that an important decision has to be taken in the network and that you cannot take part in it. What other network members would have to be present when the decision is made so that you could accept the outcome?

This question item facilitates an independent measurement of each actor's legitimacy as an agent of governance, for the multilateral assessment of a member by others cannot be affected by that member at the time of the survey. Although hypothetical questions are more disputable than questions intended to reconstruct actual events, the instructive studies by Lazega (2000, 2001), for example, prove that hypothetical questions can indeed capture valid subjective representations of social structures of influence. With a network survey, I could thus collect the complete structure of all votes on the delegation of authority to make decisions. To characterize the resulting distribution of legitimacy, two parameters are important: the strength and the structure of legitimacy within the network.

Strength. The strength of legitimacy is measured by the indegree. The indegree d_I of a member n is the number of the votes (mentions), v that a member receives from the other members j so that he or her can participate in decisions as the legitimate representative in their absence: $d_I(n_i) = \sum_j v_{ij}$. The maximal indegree for each network of size N is n - 1.

Structure. The structural dimension refers to the specific distribution of the delegation of decision-making authority. Strong legitimacy can rest on votes from a unitary faction or in votes from members from various factions. To discover the specific distribution of decision-making delegations, I use the external-internal (*E-I*) index (Krackhardt & Stern, 1988). It measures the ratio of legitimacy relations across factions (external legitimacy, or EL) to legitimacy relations within factions (internal legitimacy, or IL). The *E-I* index is calculated as follows: $E\text{-}I = (EL - IL) \div (EL + IL)$. However, because the *E-I* can be calculated only in symmetrical networks, the direction of the vote between the members cannot be dispensed with. The *E-I* thus does not reveal whether a transfer of decision-making authority goes from A to B or rather from B to A. It calculates only the extent to which each member is involved in internal or external factional relations of legitimacy. Theoretically, the

E-I varies between −1 (solely internal factional relations) and + 1 (solely external factional relations).[5]

I then use the strength and structure of the legitimacy distribution as dimensions of a model of lateral network governance. Together, they yield a matrix with four quadrants subdivided by the median value of the indegree and the null value of the E-I index. The model permits the formal definition of three types of an actor's legitimacy (see Fig. 11.2):

Low legitimacy. Members with below-average indegree are less likely than members with above-average indegree to gain sufficient acceptance of their decisions in the network.

Local legitimacy. Members in the upper-left quadrant have above-average legitimacy, but it comes mostly from their own faction. Like the head of a faction, they can gain acceptance for decisions within their factions.

Global legitimacy. Members in the upper-right quadrant have above-average legitimacy stemming from various factions. Speaking on behalf of several factions, these members have the greatest opportunity to win the acceptance of decisions in the entire network.

Consequently, the point of empirically analyzing the structure of legitimacy distribution in the ConsultingNet and Dentis networks is to ascertain how much that structure corresponds to the formal model of governance and how much it creates overall legitimacy.

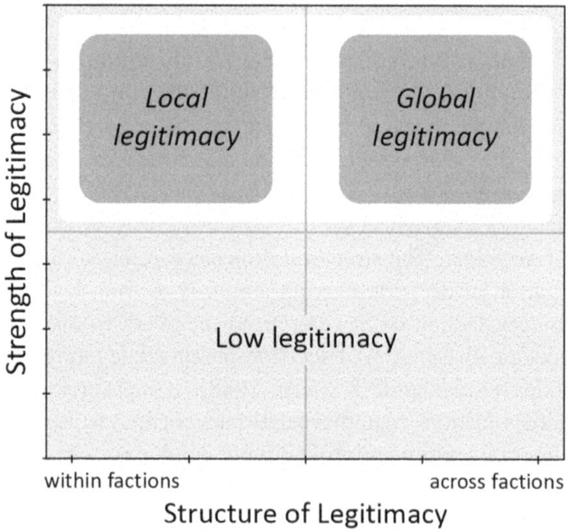

Fig. 11.2 Model of lateral governance: legitimate delegation of decision-making authority. Source: Design by author

[5] *E-I* values approaching +1 are unrealistic because the prior factional analysis of the legitimacy network classifies members into coherent network-based factions. Factional affiliation therefore makes the tendency toward factional internal relations likely.

Empirical Structures of Lateral Network Governance

Planned Versus Practiced Governance

Because of the democratic principle of delegating decision-making authority, the formally planned governance structure need not match the actual distribution of legitimacy in the network. Formal positions such as those of managing director or management board member, and formal bodies such as the advisory or supervisory board, set forth responsibilities and decision-making authority, but they do not preclude the legitimacy of other network members. Some members can be seen as important (legitimate) for certain decisions in the network even though they do not hold an office legally granting decision-making authority. Instead, other members' initiative and influence on the decision-making process can even be expected or required.

To test this conjecture, I compared the formally planned and the actually practiced forms of network governance. Figure 11.3 juxtaposes the ideal-type diagram of formal governance with the empirically documented distributions of legitimacy. The structure of legitimacy distribution in the two networks is depicted by nodes representing the members of the respective networks and by edges representing the individual relations between the surveyed members involved in the delegation of legitimate decision-making authority.

In both case studies, there is only moderate statistical relation between planned and practiced governance (with coefficients of 0.18 in the case of ConsultingNet and 0.23 in the case of Dentis). I could thus confirm the expectation and, from the

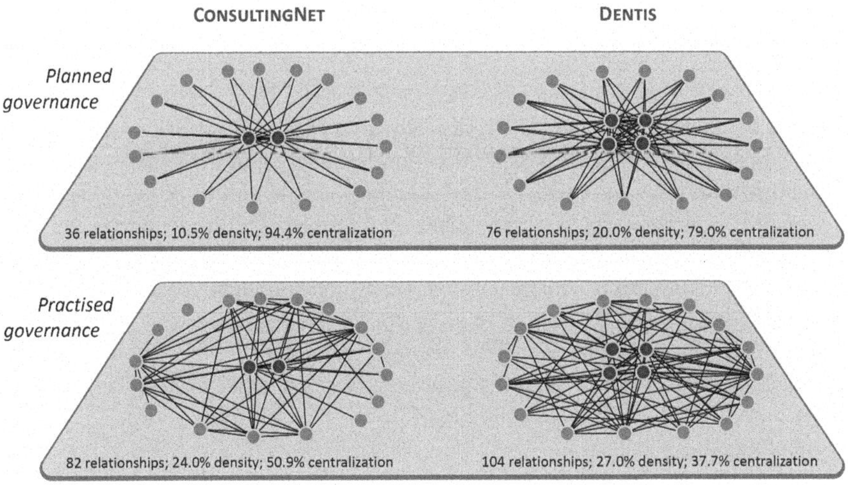

Fig. 11.3 A comparison between planned and practiced governance in two networks. Grey nodes = network members; black nodes = network members on the board of directors. Source: Design by author

networks' perspective, the hope that the planned structure of governance is contained within the actual distribution of legitimacy: Formal offices are held by legitimate actors. However, the actual distribution of legitimacy strays from the planned structure in notable ways, as revealed by the two parameters with which I measure this facet of the networks—density and centralization. With the network's density, I learn the percentage of theoretically possible votes that were received. With centralization, I gauge the network's tendency to concentrate all delegation relations on only one actor. Unanimity—100% centralization—would register as a perfect star. If delegation relations were completely distributed between the members, centralization would be 0. With my analysis, I learned that practiced governance in both networks is distributed across many more members than was formally planned. The structures that were actually used had greater density than foreseen, and the degree to which power was delegated was less centralized.

Thus, the actual number of actors legitimately positioned to affect the process of making acceptable decisions substantially exceeds the stipulated number. The low degree of centralization with governance as actually practiced reflects the tendency of lateral governance not to reduce to the formal agents of governance. By virtue of social legitimacy, informal members also figure in governance as practiced in the networks, becoming part of the de facto agent of network governance. The modeling of a formal governance structure does not capture the reality of network governance. Focusing on formal membership in executive boards alone can thus quickly prove inadequate. Legitimate members could feel disregarded or could later contest decisions that have been taken.

Local Versus Global Legitimacy: Structures for the Delegation of Decision-Making Authority

In the first part of my analysis, I revealed the divergence between the practiced and planned governance in lateral networks. However, the concept of lateral network governance makes for an even more discerning evaluation of the express structures of the distribution of legitimacy. In this section, I use the model of lateral network governance developed above, which makes it possible to set up a matrix in which the legitimacy is plotted for every network member according to strength (indegree) and structure (*E-I* index). With this matrix, I can single out three roles of legitimacy: low, local, and global. The role of low legitimacy is characterized by the marginal degree of legitimacy conferred by the network partners (see Fig. 11.4, bottom left and right quadrants). Low legitimacy is registered as a below-average indegree ($d_I(n_i) < d_i$), which reflects the number of times an actor in the network is named by other members.

All members with above-average legitimacy are represented in the upper quadrants of the matrix and are the analytical agent of governance. They are highly relevant to governance but differ—some of them considerably—in their position

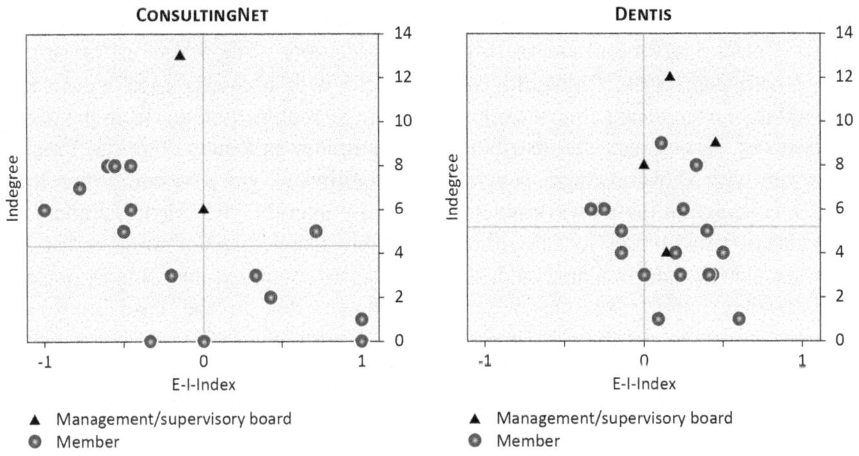

Fig. 11.4 Empirical distribution of legitimacy in the matrix of lateral network governance. Source: Design by author

within the distributive structure through which legitimate decision-making authority is bestowed. The second role of legitimacy—the local—appears in the upper left quadrants (Fig. 11.4). Its above-average legitimacy is granted by a more or less unitary subgroup describable as a faction because it has an E-I smaller than zero in each instance. By contrast, the third role of legitimacy, the global, appears in the upper right quadrant (Fig. 11.4). It is characterized by an E-I greater than zero, has its legitimacy from the network as a whole, that is, primarily from outside a unitary faction. Members with local legitimacy, such as in ConsultingNet, advocate decisions for a homogenous part of the network, with those decisions not necessarily enjoying support throughout the network. Members with global legitimacy, however, unite above-average and network-wide acceptance by the other members and have a greater likelihood of being able to work toward decisions that are more capable of building consensus and winning acceptance than would otherwise be the case. Such members serve as representatives of the network as a whole, such as in the Dentis network.

With Fig. 11.4, I illustrate the distribution of members in the matrix for both networks. The empirical models suggest three important insights. First, both networks have an agent of governance not adequately represented by their formal offices. In both networks, above-average decision-making authority is delegated to a handful of members whose legitimation is considerably more than purely formal. In ConsultingNet, in addition to the two board members formally elected, seven more members were named as legitimate representatives. In Dentis, another four were found as equally legitimate as the four formal board members. Second, the place that the formal offices occupy in both networks is evident. Although there are significant similarities between the planned and the actually practiced structures of legitimation, both networks had numerous other members with above-average legitimacy. The identified agent of lateral network governance therefore delineates the actual governance structure and complements the work of the supervisory board. Third, ConsultingNet and Dentis differed considerably in the composition of the two governance roles. ConsultingNet members with above-average legitimacy derived their acceptance almost exclusively from a local faction, not from the overall network. In contrast, the distribution of legitimacy in Dentis comprised many members with above-average, network-wide legitimacy and who could therefore gain acceptance in the network much more easily than the prominent members of ConsultingNet with only local legitimacy. Notably, ConsultingNet nosedived at the time the survey data was analyzed, whereas Dentis continued investing in professionalization. Giving reasons for the problems cited in the case study on ConsultingNet, its members described matters affecting the entire network that had brought on a torrent of objections. Apparently, the legitimacy of the responsible decision-makers had thus only been partial or insufficient. A large share of network members had meanwhile left it. Dentis, by contrast, had no serious trouble with its governability and ability to take action and has continued to successfully purse its common goals.

Conclusion

In this article I have elaborated on a theory of lateral network governance, with which I conceptually surmount the divide between formal and informal governance by focusing on the relational distribution of the legitimacy to rule and control in a governance agreement among equals. With this concept of *lateral network governance* I seek to reconcile the voluntary nature of network membership and the expectation of nonhierarchical cooperation on the one hand with an interest in effective agents of governance and compliance with its normative and procedural standards on the other. For this vision of lateral network governance I am drawing on the culture of democratic decision-making, in which the smallest possible, sufficiently representative group of delegates negotiates consensus-based decisions that meet with network-wide acceptance and promote long-term consensus on objectives and identification. Although I have focused predominantly on legitimacy in this article, future research should further explore the role of power in such governance arrangements. The nexus to concepts of power constitutes a promising link between economic geography and political geography (Allen, 2011; Allen & Cochrane, 2007; Grabs & Ponte, 2019; Griffin, 2012).

Methodologically, I have drawn on methods of social network analysis to turn this concept into empirical practice. While network methods have become increasingly recognized and used in human geography (Giuliani, 2007; Glückler, 2007; ter Wal & Boschma, 2009), network researchers have simultaneously acknowledged the role of space and place in real social networks (Doreian & Conti, 2012). Future research on local and network governance will benefit from further cross-fertilization between network studies and geography, for instance, by adopting positional, dynamic, and multi-level approaches (Glückler & Doreian, 2016; Glückler, Lazega, & Hammer, 2017; Lazega & Snijders, 2016).

Empirically, my analysis of two case studies reveals that concentrating solely on the formal structure of governance while neglecting the invisible spheres of actually practiced governance can lead to conflicts and mounting resistance in network collaboration. Instead, the analysis of the relational distribution of legitimacy helps actors examine and, if necessary, adapt network governance by identifying the most legitimate actors and discrimentating the locus—local or global—of their legitimacy for governance.

A theory of lateral network governance does not imply static architectures of organization but serves as a compass guiding daily network governance. The actual design of a governance structure depends not only on the four building blocks—context, object, mechanism, and agency of governance—but also on its geographical and institutional context in order to convey appropriate, legitimate, and effective practice of network governance.

Acknowledgements This chapter is an extended and modified version of an article published in the special issue on Gobernanza, Redes y Territorio in the Revista de Geografía Norte Grande (Glückler, 2019). I draw on research that I have partly realized in collaboration with Schila Nemeth and Ingmar Hammer. I am grateful to them for their valuable support in collecting and analyzing the data of the case studies and in discussing the concepts developed here. I also gratefully acknowledge the financial support the German Ministry of Education and Research (BMBF) has provided.

References

Allen, J. (2011). Topological twists: Power's shifting geographies. *Dialogues in Human Geography, 1,* 283–298. https://doi.org/10.1177/2043820611421546

Allen, J., & Cochrane, A. (2007). Beyond the territorial fix: Regional assemblages, politics and power. *Regional Studies, 41,* 1161–1175. https://doi.org/10.1080/00343400701543348

Ansell, C., & Gash, A. (2008). Collaborative governance in theory and practice. *Journal of Public Administration Research and Theory, 18,* 543–571. https://doi.org/10.1093/jopart/mum032

Ansell, C., & Torfing, J. (2016). *Handbook on theories of governance*. Cheltenham: Edward Elgar. https://doi.org/10.4337/9781782548508

Bachmann, R. (2001). Trust, power and control in trans-organizational relations. *Organization Studies, 22,* 337–365. https://doi.org/10.1177/0170840601222007

Bulkeley, H. (2005). Reconfiguring environmental governance: Towards a politics of scales and networks. *Political Geography, 24,* 875–902. https://doi.org/10.1016/j.polgeo.2005.07.002

Charron, N., Dijkstra, L., & Lapuente, V. (2014). Regional governance matters: Quality of government within European Union member states. *Regional Studies, 48,* 68–90. https://doi.org/10.1080/00343404.2013.770141

Coe, N. M., Dicken, P., & Hess, M. (2008). Global production networks: Realizing the potential. *Journal of Economic Geography, 8,* 271–295. https://doi.org/10.1093/jeg/lbn002

Crang, M., Hughes, A., Gregson, N., Norris, L., & Ahamed, F. (2013). Rethinking governance and value in commodity chains through global recycling networks. *Transactions of the Institute of British Geographers, 38,* 12–24. https://doi.org/10.1111/j.1475-5661.2012.00515.x

Crouch, C. (2005). *Capitalist diversity and change: Recombinant governance and institutional entrepreneurs*. New York, NY: Oxford University Press. https://doi.org/10.1093/acprof:oso/9780199286652.001.0001
Doreian, P., & Conti, N. (2012). Social context, spatial structure and social network structure. *Social Networks, 34,* 32–46. https://doi.org/10.1016/j.socnet.2010.09.002
Feiock, R. C. (2007). Rational choice and regional governance. *Journal of Urban Affairs, 29,* 47–63. https://doi.org/10.1111/j.1467-9906.2007.00322.x
Gereffi, G., Humphrey, J., & Sturgeon, T. (2005). The governance of global value chains. *Review of International Political Economy, 12,* 78–104. https://doi.org/10.1080/09692290500049805
Gibson-Graham, J. K. (2008). Diverse economies: Performative practices for 'other worlds'. *Progress in Human Geography, 32,* 613–632. https://doi.org/10.1177/0309132508090821
Giuliani, E. (2007). The selective nature of knowledge networks in clusters: Evidence from the wine industry. *Journal of Economic Geography, 7,* 139–168. https://doi.org/10.1093/jeg/lbl014
Glückler, J. (2007). Economic geography and the evolution of networks. *Journal of Economic Geography, 7,* 619–634. https://doi.org/10.1093/jeg/lbm023
Glückler, J. (2019). Gobernanza lateral de redes: Legitimidad y delegación relacional de la autoridad decisoria [Lateral network governance: Legitimacy and the relational delegation of decision-making authority]. *Revista de Geografía Norte Grande, 74,* 93–115. https://doi.org/10.4067/S0718-34022019000300093
Glückler, J., & Armbrüster, T. (2003). Bridging uncertainty in management consulting: The mechanisms of trust and networked reputation. *Organization Studies, 24,* 269–297. https://doi.org/10.1177/0170840603242004
Glückler, J., Dehning, W., Janneck, M., & Armbrüster, T. (2012). *Unternehmensnetzwerke: Architekturen, Strukturen und Strategien* [Business networks: Architectures, structures and strategies]. Heidelberg: Springer. https://doi.org/10.1007/978-3-642-29531-7
Glückler, J., & Doreian, P. (2016). Editorial: Social network analysis and economic geography: Positional, evolutionary and multi-level approaches. *Journal of Economic Geography, 16,* 1123–1134. https://doi.org/10.1093/jeg/lbw041
Glückler, J., & Hammer, I. (2015). Cooperation gains from network goods. In S. Jung, P. Krebs, & G. Teubner (Eds.), *Business networks reloaded* (pp. 22–40). Farnham: Ashgate.
Glückler, J., & Hammer, I. (2017). Connectivity in contiguity: Conventions and taboos of imitation in co-located networks. In J. Glückler, E. Lazega, & I. Hammer (Eds.), *Knowledge and networks* (pp. 269–290). Knowledge and Space: Vol. 11. Cham: Springer. https://doi.org/10.1007/978-3-319-45023-0_13
Glückler, J., Lazega, E., & Hammer, I. (2017). *Knowledge and networks*. Knowledge and Space: Vol. 11. Cham: Springer. https://doi.org/10.1007/978-3-319-45023-0
Glückler, J., & Panitz, R. (2016). Relational upgrading in global value networks. *Journal of Economic Geography, 16,* 1161–1185. https://doi.org/10.1093/jeg/lbw033
Glückler, J., Suddaby, R., & Lenz, R. (2018). *Knowledge and institutions*. Knowledge and Space: Vol. 13. Cham: Springer. https://doi.org/10.1007/978-3-319-75328-7
Glückler, J., Panitz, R., & Hammer, I. (2020). SONA: A relational methodology to identify structure in networks. *Zeitschrift für Wirtschaftsgeographie*. https://doi.org/10.1515/zfw-2020-0003
Grabs, J., & Ponte, S. (2019). The evolution of power in the global coffee value chain and production network. *Journal of Economic Geography, 19,* 803–828. https://doi.org/10.1093/jeg/lbz008
Griffin, L. (2012). Where is power in governance? Why geography matters in the theory of governance. *Political Studies Review, 10,* 208–220. https://doi.org/10.1111/j.1478-9302.2012.00260.x
Hardin, G. (1968). The tragedy of the commons. *Science, 162,* 1243–1248. https://doi.org/10.1126/science.162.3859.1243
Human, S. E., & Provan, K. G. (2000). Legitimacy building in the evolution of small-firm multilateral networks: A comparative study of success and demise. *Administrative Science Quarterly, 45,* 327–365. https://doi.org/10.2307/2667074

Jones, C., Hesterly, W. S., & Borgatti, S. P. (1997). A general theory of network governance: Exchange conditions and social mechanisms. *Academy of Management Review, 22,* 911–945. https://doi.org/10.5465/AMR.1997.9711022109

Jung, S., Krebs, P., & Teubner, G. (2015). *Business Networks Reloaded.* Farnham: Ashgate.

Keast, R. (2016). Network governance. In C. Ansell & J. Torfing (Eds.), *Handbook on theories of governance* (pp. 442–453). Cheltenham: Edward Elgar. https://doi.org/10.4337/9781782548508

Krackhardt, D., & Stern, R. N. (1988). Informal networks and organizational crises: An experimental simulation. *Social Psychology Quarterly, 51,* 123–140. https://doi.org/10.2307/2786835

Lazega, E. (2000). Rule enforcement among peers: A lateral control regime. *Organization Studies, 21,* 193–214. https://doi.org/10.1177/0170840600211003

Lazega, E. (2001). *The collegial phenomenon: The social mechanisms of cooperation among peers in a corporate law partnership.* New York, NY: Oxford University Press. https://doi.org/10.1093/acprof:oso/9780199242726.001.0001

Lazega, E., & Krackhardt, D. (2000). Spreading and shifting costs of lateral control among peers: A structure analysis at the individual level. *Quality and Quantity, 34,* 153–175. https://doi.org/10.1023/A:1004759418226

Lazega, E., & Snijders, T. (2016). *Multilevel network analysis for the social sciences: Theory, methods and applications.* Methodos Series: Vol. 12. Cham: Springer. https://doi.org/10.1007/978-3-319-24520-1

Levy, D. L. (2008). Political contestation in global production networks. *The Academy of Management Review, 33,* 943–963. https://doi.org/10.5465/AMR.2008.34422006

Liverman, D. (2004). Who governs, at what scale and at what price? Geography, environmental governance, and the commodification of nature. *Annals of the Association of American Geographers, 94,* 734–738.

MacLeod, G., & Goodwin, M. (1999). Space, scale and state strategy: Rethinking urban and regional governance. *Progress in Human Geography, 23,* 503–527. https://doi.org/10.1191/030913299669861026

Maggetti, M., & Gilardi, F. (2014). Network governance and the domestic adoption of soft rules. *Journal of European Public Policy, 21,* 1293–1310. https://doi.org/10.1080/13501763.2014.923018

Mayntz, R. (1993). Policy-Netzwerke und die Logik von Verhandlungssystemen [Policy networks and the logic of negotitation systems]. In A. Héritier (Ed.), *Policy-Analyse: Kritik und Neuorientierung* (pp. 39–56). Politische Vierteljahresschrift: Vol. 24. Opladen: Westdeutscher Verlag. https://doi.org/10.1007/978-3-663-01473-7_2

Mayntz, R., & Scharpf, F. W. (1995). *Gesellschaftliche Selbstregelung und politische Steuerung* [Societal selfregulation and political governance]. Schriften des Max-Planck-Instituts für Gesellschaftsforschung Köln: Vol. 23. Frankfurt: Campus.

Melot de Beauregard, P., Németh, S., & Glückler, J. (2012). Rechtsformen und Governance von Unternehmensnetzwerken [Legal forms and governance of interorganizational networks]. In J. Glückler, W. Dehning, M. Janneck, & T. Armbrüster (Eds.), *Unternehmensnetzwerke: Architekturen, Strukturen und Strategien* (pp. 121–137). Heidelberg: Springer Gabler. https://doi.org/10.1007/978-3-642-29531-7_7

Montero, S., & Chapple, K. (2019). Peripheral regions, fragile governance: Local economic development from Latin America. In S. Montero & K. Chapple (Eds.), *Fragile governance and local economic development: Theory and evidence from peripheral regions in Latin America* (pp. 1–18). Regions and Cities: Vol. 129. Abingdon-on-Thames: Routledge. https://doi.org/10.4324/9781315101569

Ostrom, E. (1999). Coping with tragedies of the commons. *Annual Review of Political Science, 2,* 493–535. https://doi.org/10.1146/annurev.polisci.2.1.493

Park, S. H., & Ungson, G. R. (2001). Interfirm rivalry and managerial complexity: A conceptual framework of alliance failure. *Organization Science, 12,* 37–53. https://doi.org/10.1287/orsc.12.1.37.10118

Podolny, J. M., & Page, K. L. (1998). Network forms of organization. *Annual Review of Sociology, 24,* 57–76. https://doi.org/10.1146/annurev.soc.24.1.57

Powell, W. W. (1991). Neither market nor hierarchy: Network forms of organization. In G. Thompson, J. Frances, R. Levačič, & J. Mitchell (Eds.), *Markets, hierarchies and networks: The coordination of social life* (pp. 265–276). London: SAGE.

Prota, L. (2016). Toward a polanyian network analysis: Market and non-market forms of coordination in the rice economy of Vietnam. *Journal of Economic Geography, 16,* 1135–1160. https://doi.org/10.1093/jeg/lbw039

Provan, K. G., Fish, A., & Sydow, J. (2007). Interorganizational networks at the network level: A review of the empirical literature on whole networks. *Journal of Management, 33,* 479–516. https://doi.org/10.1177/0149206307302554

Provan, K. G., & Kenis, P. (2008). Modes of network governance: Structure, management, and effectiveness. *Journal of Public Administration Research and Theory, 18,* 229–252. https://doi.org/10.1093/jopart/mum015

Provan, K. G., & Milward, H. B. (1995). A preliminary theory of interorganizational network effectiveness: A comparative study of four community mental health systems. *Administrative Science Quarterly, 40,* 1–33. https://doi.org/10.2307/2393698

Rhodes, R. A. W. (2007). Understanding governance: Ten years on. *Organization Studies, 28,* 1243–1264. https://doi.org/10.1177/0170840607076586

Ring, P. S., & van de Ven, A. H. (1994). Developmental processes of cooperative interorganizational relationships. *The Academy of Management Review, 19,* 90–118. https://doi.org/10.2307/258836

Rowley, T., Behrens, D., & Krackhardt, D. (2000). Redundant governance structures: An analysis of structural and relational embeddedness in the steel and semiconductor industries. *Strategic Management Journal, 21,* 369–386. https://doi.org/10.1002/(SICI)1097-0266(200003)21:3<369::AID-SMJ93>3.0.CO;2-M

Sánchez-Hernández, J. L., & Glückler, J. (2019). Alternative economic practices in Spanish cities: From grassroots movements to urban policies? An institutional perspective. *European Planning Studies, 27,* 2450–2469. https://doi.org/10.1080/09654313.2019.1644295

Schäfer, A. (2009). *Netzwerk-Verfassungen: Institutionelle Lösungsansätze zur Überwindung kollektiver Handlungsprobleme in Verbundnetzwerken* [Network constitutions: Institutional approaches to solve collective action Problems] (Doctoral dissertation). Universität Oldenburg, Oldenburg.

Scharpf, F. W. (2006). *Interaktionsformen: Akteurzentrierter Institutionalismus in der Politikforschung* [Games real actors play: Actor-centered institutionalism in policy research] (O. Treib, Trans.). Wiesbaden: VS Verlag für Sozialwissenschaften.

Scott, M. (2015). Re-theorizing social network analysis and environmental governance: Insights from human geography. *Progress in Human Geography, 39,* 449–463. https://doi.org/10.1177/0309132514554322

Seyfang, G. (2006). Sustainable consumption, the new economics and community currencies: Developing new institutions for environmental governance. *Regional Studies, 40,* 781–791. https://doi.org/10.1080/00343400600959173

Staber, U. (2000). Steuerung von Unternehmensnetzwerken: Organisationstheoretische Perspektiven und soziale Mechanismen [Network governance: Perspectives of organization theory and social mechanisms]. In J. Sydow & A. Windeler (Eds.), *Steuerung von Netzwerken: Konzepte und Praktiken* (pp. 58–87). Wiesbaden: Westdeutscher Verlag.

Sydow, J. (Ed.). (2000). *Steuerung von Netzwerken: Konzepte und Praktiken* [Network governance: Concepts and practices]. Opladen: Westdeutscher Verlag.

Sydow, J. (2002). *Strategische Netzwerke: Evolution und Organisation* [Strategic networks: Evolution and organization]. Wiesbaden: Gabler.

Sydow, J., Schüßler, E., & Müller-Seitz, G. (2016). *Managing inter-organizational relations: Debates and cases.* London: Palgrave Macmillan.

ter Wal, A. L. J., & Boschma, R. A. (2009). Applying social network analysis in economic geography: Framing some key analytic issues. *The Annals of Regional Science, 43*, 739–756. https://doi.org/10.1007/s00168-008-0258-3

Weber, M. (1978). *Economy and society: An outline of interpretive sociology* (Vols. 1–2). Berkeley, CA: University of California Press.

Williamson, O. E. (1985). *The economic institutions of capitalism: Firms, markets, relational contracting*. New York: Free.

Williamson, O. E. (2005). The economics of governance. *American Economic Review, 95*(2), 1–18. https://doi.org/10.1257/000282805774669880

Windeler, A. (2001). *Unternehmungsnetzwerke: Konstitution und Strukturation* [Corporate networks: Constitution and structuration]. Wiesbaden: Westdeutscher Verlag.

Yeung, H. W., & Coe, N. M. (2015). Toward a dynamic theory of global production networks. *Economic Geography, 91*, 29–58. https://doi.org/10.1111/ecge.12063

Part III
How Governance Affects Learning and Innovation

Chapter 12
Knowledge and the Deliberative Stance in Democratic Systems: Harnessing Scepticism of the Self in Governing Global Environmental Change

Simon Niemeyer

Deliberative democracy is characterized as an approach to governance that valorizes the operation of reason (Chambers, 2012; Cohen, 2007). Although there is a danger of this interpretation implying hyper-rationalism, as some researchers have suggested, considerable scope exists for understanding the relationship between knowledge, reason, and governance in a broader sense using a deliberative lens (see Bächtiger, Niemeyer, Neblo, Steenbergen, & Steiner, 2010). To begin, the emphasis on deliberation implies that the mere fact of knowledge is insufficient to derive legitimacy for any particular action. An actor may use knowledge to demonstrate the existence of climate change, for example, but the choice of what action to take involves normative questions about what the polity values, which can only be addressed with reference to citizens.

At the same time, the relationship between citizens, knowledge, and collective choices is not improving, and may possibly even be deteriorating (Capstick, Whitmarsh, Poortinga, Pidgeon, & Upham, 2015). The problem is not ignorance per se, nor a lack of baseline will (O'Brien, 2012). When surveyed, most citizens endorse environmental sustainability, but the message is often lost in political translation. However, it is important to understand the broader dynamics of knowledge, and the processes whereby these are translated into action. Take for example the "governance trap," where basic acceptance of the fact of climate change fails to translate into action, because citizens and the government each attribute responsibility for such action to the other, thus ossifying inaction (Pidgeon, 2012). In this chapter, I seek to develop an understanding of these processes through a deliberative framework.

To this end, I begin the chapter with a survey of the challenge of translating knowledge into political action, demonstrating how a governance lens, particularly

S. Niemeyer (✉)
Centre for Deliberative Democracy and Global Governance, University of Canberra, Canberra, ACT, Australia
e-mail: simon.niemeyer@canberra.edu.au

J. Glückler et al. (eds.), *Knowledge for Governance*, Knowledge and Space 15,
https://doi.org/10.1007/978-3-030-47150-7_12

a deliberative one, can illuminate these dynamics. I then illustrate them using a case study that compares the effect of knowledge gains of climate impacts to the absence or presence of deliberation, via relative changes to discourses of climate change governance. I argue that an important feature influencing the impact is not just the role of knowledge, but also the changing disposition of citizens, which is impacted by the deliberative context. I examine this dispositional effect using the concept of the deliberative stance, then develop these findings into broad possibilities for improving the relationship between knowledge and governance in deliberative systems, before revisiting the overall findings in a concluding section.

Knowledge and Governance Systems in Climate Change

Researchers have known the basic science underpinning anthropogenic climate change since the latter part of the nineteenth century. Originally heavily contested, the scientific consensus has moved on from debating its existence to yielding increasingly alarming revisions of earlier findings in respect to its pace and impacts (Oreskes, 2018).

Yet, despite the growing evidence there is a growing gap between the scale of the problem and the need for action that it implies (e.g., Dunlap, 2013). Self-interest at the national level is part of the problem, particularly where politics is dominated by the fossil fuel and associated industries (e.g., Christoff, 2013), but this is only part of the picture. In any system of governance there remains a thread of legitimization between the public and decision makers responding to (or at least managing) popular demand for action on climate change. The problem remains that this level of demand has varied considerably over time, even declining in some countries (Capstick et al., 2015).

Although a liberal view might support the right of citizens to choose as they wish, a deliberative democratic view of governance demands that such a deficit be understood. If it reflects a systematic dislocation between generalizable public interest (such as maintaining a healthy environment; see Dryzek, 1990) and expressed preferences, remedies must be sought to realign the two.

In recent history, researchers have viewed these dynamics through an *information deficit* prism, whose proponents hold that the public, once appraised of the climate change risk, will follow suit and demand action. To be sure, climate change knowledge is correlated to action (Bord, O'Connor, & Fischer, 2000), and there exists a knowledge deficit in relation to climate change among members, evidenced for example by a tendency to conflate localized air pollution, chlorofluorocarbons (CFCs), and ozone depletion as drivers of climate change (Capstick et al., 2015). However, this relationship does not contribute very much to understanding the overall picture.

Closer inspection of knowledge dynamics through the lens of governance suggests a richer dynamic. For a start, knowledge itself is limited. Definitive knowledge in respect to climate change is increasingly unobtainable, given the complexities of the issue. Acquiring knowledge in respect to climate change concerns the degree of specialization required to understand all its extant dimensions—to a point where it is impossible for any given individual to be fully conversant. This complexity leads to the problem of the *knowledge-ignorance paradox*, in which the need for specialization induces ignorance even among experts. Among citizens, it also serves to remove knowledge ever further and decrease inclination to engage with complex issues such as climate change. Its complexity, in combination with a lack of easily communicable metaphors to translate it into a "hot" issue, makes it difficult to hold public attention (Ungar, 2000).

Consequently, governance systems require knowledge specialization, which is predicated on trust, but this increased knowledge paradoxically has the potential to undermine this trust. The greater demands on collective knowledge for complex issues such as climate change confront a countervailing need for cognitive closure—closing the mind to information inconsistent with prior beliefs (Kruglanski & Boyatzi, 2012)—that is driven by greater demands on citizens' limited time and cognitive resources. Even the most knowledgeable citizen—including climate scientists—will need to resort to trust in the wider knowledge system in which they are situated. Many citizens appear to do just that: They accept the scientific consensus in respect of anthropogenic climate impact and support action even if they do not individually possess detailed knowledge (van der Linden, Leiserowitz, Feinberg, & Maibach, 2015). Where the message's implications diverge from prior beliefs, however, there is a danger of ever-increasing complexity driving a wedge between the scientist and the citizen—that is, in the absence of creating a deliberative context.

Even so, the overall relationship between knowledge and behavior is complex and mediated by a range of factors, including worldview, norms (social and personal), intentions, and habits.[1] Knowledge, in the form of belief in the human causes and consequences of climate change, appears to be a requisite (Bord et al., 2000) but not a sufficient condition for action, which also requires acceptance of responsibility for action (Capstick et al., 2015). Ever more sophisticated models to predict environmental behavior (Klöckner, 2013) still only account for a small proportion of variation in actual behavior, and, even if accurate, inform remedies that are piecemeal at best (Shove, 2010).

A key contention here—which suggests both a diagnosis and a cure—is that these relationships, between knowledge and behavior, are at least partly constructed by the governance context in which such behaviors occur, as well as being part of the processes that form that context. One example of this context shaping the role of knowledge is that of the prevailing ideological setting (e.g., McCright, Dunlap, & Marquart-Pyatt, 2016). Researchers well understand the role of worldview and ideology in determining views on climate change, and they can be powerful forces that influence the views of scientists themselves within knowledge systems (e.g., Carlton, Perry-Hill, Huber, & Prokopy, 2015).

These forces also play a role in attempts to reconstruct politics in ways that either support or work against the uptake of knowledge in political systems. McCright and

[1] The knowledge dimension in environmental psychology models predicting behavior tends to follow the form of awareness about the consequences associated with action, and awareness of individual responsibility for that action. For a review, see Klöckner (2013).

Dunlap (2010), for example, have described an ideologically driven attempt to work against the ideal of reflexive modernism (Beck, 1992) and critical self-evaluation and confrontation of the consequences of human actions on the environment via the centrality of *impact science*—as opposed to *production science* focused on economic output—and social movements. Many of these strategies involve attacking the institutions as part of the machinery of reflexivity, as well as framing decision choices to induce "non-decisions" that maintain a status quo more consistent with their interests. This includes strategies modeled on casting enough doubt on the science of climate change to forestall public mobilization (Oreskes & Conway, 2010; see also Chap. 3 by Scott).

Reflexive modernists advocate remedies for the problem of knowledge distorting practices in political systems such as those used by "merchants of doubt" (Oreskes & Conway, 2010) at the macro-institutional level via the privileging of impact science and advocacy by interest groups (e.g., McCright & Dunlap, 2010). The approach is aimed at government: balancing the impact of knowledge against unreflexive forces that often seek to maintain an advantageous status quo. However, the effectiveness is limited to the extent that it fails to engage with the effect of these processes on citizens, delegitimizing a reflexive approach just enough to maintain the status quo.

Deliberative Democracy: Reflexive Systems, Reflexive Citizens

Deliberative democracy offers a different frame from which to view the uptake of knowledge in governance systems, as well as how that process might be improved. It implies an approach to governance that engages citizens in a manner that encourages political discourse and reflection of a deliberative quality, that such a process should be broadly inclusive, and that the considered will of citizens find a mechanism inducing consequences for both specific decisions and the operation of the system of governance as a whole (Dryzek, 2009).

Deliberative democrats speak of a relationship between reflexive and deliberative modes of governance at the wider "systems" level (e.g., Dryzek & Pickering, 2017; Stevenson & Dryzek, 2012), but it is also important to think of how citizens function as part of these systems, and how individual-level dispositions impact on larger scale dynamics (Owen & Smith, 2015). Habermas (2006, p. 418), for example, draws this link when he refers to the reflexive character of the public sphere as a function of how all citizens "can revisit perceived public opinions and respond to them after reconsideration."

Less developed than theories of deliberative democracy, however, are well constructed and practical mechanisms for understanding the knowledge-action dynamics among citizens. Researchers face the challenge of empirically capturing the nature of these dynamics and translating them into mechanisms for boosting the reflexivity of the system in ways that account for both systems level and citizen level processes.

A relevant consideration involves what actual features of deliberation yield the hypothesized benefits. Although deliberative democracy researchers have tended to focus on the quality of deliberation as a process (Bächtiger et al., 2010) and the ideal features that it embodies—such as openness, sincerity, respect, reciprocity, and intersubjectivity—there is a growing interest in how deliberation is experienced by citizens and in the demands that it makes of them.

Deliberation cannot be divorced from a set of (reflexive) capabilities that are implicit in deliberative theory via characteristics such as open-mindedness. Owen and Smith (2015) have described something akin to the activation of these capabilities when they advance the idea of a "deliberative stance":

> Part of the political ideal of deliberative democracy is that its (normative) stability is generated by citizens being able intelligibly to conceive of (adopt a stance towards) themselves as equals engaged in a process of public reasoning oriented to a shared practical judgment, where such a process involves citizens reflectively taking up each other's standpoints. (p. 219)

The ideal of the deliberative stance reflects a predilection for *truth seeking* and contestation (Curato, Niemeyer, & Dryzek., 2013), an openness to arguments, and a capacity for empathy (Morrell, 2010), taking seriously alternative arguments while simultaneously treating them with judicious scepticism (Kruglanski & Boyatzi, 2012). It implies Socratic humility, an open-mindedness that continually resists the need for cognitive closure, inviting ongoing negotiation of the relationship between the self and knowledge and its integration into thought.

Knowledge, Information, and Deliberation: A Case Study

One factor that improves the uptake and integration of knowledge is that of context (Kruglanski & Webster, 1996), including the governance setting in which politically relevant knowledge is acquired (e.g., Pierce, Lovrich, & Dalton, 2000).

Although the term "context" can cover a multitude of possibilities (see e.g., Shove, 2010), here the governance context is primarily viewed through a discursive lens, in which the public sphere, which is a product of discourses reproduced among individuals as discursive sites, where discourses are broadly understood in similar terms to Dryzek and Niemeyer (2008, p. 481) as "a set of categories and concepts embodying specific assumptions, judgments, contentions, dispositions, and capabilities." Discourses enable the mind to process sensory inputs into coherent accounts, which can then be shared intersubjectively.

This definition of discourse provides a clue to their capacity to shape the dynamics of knowledge uptake in a governance context as well as shaping that context. These dynamics can be demonstrated via an Australian study on public responses to climate change and governance in the Australian Capital Region (from here on, the ACR study; see Hobson & Niemeyer, 2011). The study involved investigating discourses around the question of climate change, adaptation, and governance and the relative impact of exposure to climate change scenarios and participation in a deliberative process.

The empirical approach—described in detail in Hobson and Niemeyer (2011)—involved surveying responses to 33 statements covering themes dealing with beliefs about climate change, trust in knowledge sources, beliefs about the capacity of the system of governance to respond to the challenge, and demands about what should be done to improve that capacity (see Table 12.3 in the Appendix).

Researchers recorded two kinds of responses. The first, non-discursive, approach involved a standard Likert-scale response to the statements, from "strongly disagree" to "strongly agree" on an 11-point scale (−5,5), permitting a broad analysis of aggregate changes. The second, discursive approach, required participants to sort/rank statements into a quasi-normal distribution across the same scale for analysis based on Q methodology to identify the climate governance discourses reported below.

The survey responses were recorded at the following four stages:[2]

Stage 1: Baseline survey before exposure to the climate change scenario;
Stage 2: Following exposure to the High Emissions climate change scenario;
Stage 3: Prior to participation in a three-day deliberative process on climate change adaptation and policy, 3 months after the climate change scenario interviews; and
Stage 4: Immediately following deliberation.

Following the surveying of baseline perspectives (Stage 1), participants in Stage 2 experienced modelled climate impacts for the region as part of an extended, individual interview process. The researchers communicated the scenario via a series of PowerPoint slides illustrating modelled climate impacts associated with a "high emissions" scenario in the years 2050 and 2100.[3] Participants were asked to reflect on their real-world reaction to exposure to the scenario and to re-perform the surveys.

Stage 3 involved re-performing the surveys months following the scenario interview, immediately prior to participation in a three-day deliberative process. The process began with group activities designed to activate norms consistent with the deliberative stance, followed by 2 days of presentations from witnesses speaking to a wide range of issues associated with climate change and group discussion (see Hobson & Niemeyer, 2011). The final day involved group deliberation developing policy recommendations dealing with climate change and adaption, which was followed by the post-deliberative survey (Stage 4). The scenario phase of the study

[2] An additional stage was included in the original study involving exposure to a "medium" emissions climate change scenario, but is not reported here (Hobson & Niemeyer, 2011).

[3] The scenarios were developed using the Commonwealth Scientific and Industrial Research Organisation's (CSIRO) OzClim model, which contains patterns of regional changes in climate projected from 23 global climate models run by and other research centres and archived at the Program for Climate Model Diagnosis and Intercomparison (PCMDI). The model enables users to select from six SRES scenarios (taken from the 'Special Report on Emissions') and two commonly used CO_2 concentration stabilisation scenarios to generate projections for any of the available global climate models (Năkićenović et al., 2000). An additional "medium emissions" scenario was included in the study, but not reported here.

collected viable data from 103 participants, 34 of whom—out of the 40 selected—went on to complete the deliberative process (Stages 3 and 4).

The list of statements used for the Q-study component of the survey can be found in the appendix's Table 12.3. As per Q method, these were drawn from a database of real language statements pertaining to relevant dimensions of climate change and adaptation discourses in the public sphere (Hobson & Niemeyer, 2011). An abridged selection of these most relevant for the discussion here is shown in Table 12.1 below.

The average responses to the statements using the first (Likert) method are shown in the four columns to the right of the statements for each stage of the study, for the subset of deliberative participants. Results for the larger set of participants in the scenario study are shown in square brackets. The significance of change for each result compared to the baseline is indicated by an asterisk, with an additional column showing the significance of change for the deliberative phase (Stages 3–4).

Table 12.1 contains significant and dramatic changes in relation to beliefs following exposure to the climate change scenarios (statements 1, 3, and 25) and increased trust in scientific knowledge (6), but a decrease in information from the media (12). There was no change in belief regarding the quality of overall governance response to the issue (2), but an increased tendency to outsource this issue and to lay blame at the feet of government (33), rather than seek a broader response within the community. As will be seen below, it makes sense that there is no increase in the already high demand for education in climate change, particularly as it was understood by those most in favor of it at the baseline stage (climate sceptics, see below). This contrasts to the increasing demand that the government take responsibility (32).

The average responses in Table 12.1 for Stage 3 (3-month follow-up, prior to participation in the deliberative process) reveal that in most cases the effect of information is not sustained, as I further discuss below The standout exception to this trend is declining trust in information from the government (7) and, to a lesser extent, the media (12)—although in this case, as for the other changes at this stage, the effect is a residual from changes that occur in the scenario stage.

Deliberation, by contrast, had a qualitatively different impact. Some dramatic changes occurred, the most significant being the belief that there will be positive response to climate change (2) and demand for more educational programs (4). Beliefs about the reality of climate change grew during deliberation (1, 3, and 25; although the latter two deliberative changes fell just short of achieving significance). And, although participants did not complete a follow-up survey after participation in deliberation, a number did participate in meetings 6 months after the event, as well as a number of semi-structured interviews during which most reported the enduring nature of the views formed during deliberation. This included a belief that climate change is real and problematic, despite a recent break occurring to a long-term drought (Hobson & Niemeyer, 2011).

For almost every survey item in Table 12.1, the observed effect from the scenario stage (information effect) among the deliberative cohort is similar to the group as a whole, except in the case of Item 8 (need for strong political leadership). However, there is an overall trend of more conservative results among the wider group, likely

Table 12.1 Statement responses—knowledge, trust, governance (Likert Scale)

No	Statement	Stage 1. Base-line	Stage 2. High scenario	Stage 3. Pre-Delib	Stage 4. Post-Delib	
		←*Information Effect*→				
		←*Sustained Information Effect*→		←*Deliberative Effect*→		Significance of overall effect
		←*Overall Effect*→				
	Beliefs about climate change					
1	There is enough evidence proving climate change	2.41 [1.99]	3.91*** [3.42]***	2.57	3.65*	*
3	Climate variation is normal, so there is no problem	−1.97 [−1.84]	−3.29*** [−3.27]***	−2.74*	−3.21**	
25	Australia is particularly vulnerable	2.79 [2.65]	3.74* [3.65]*	3.43*	3.79*	
	Trust in Knowledge Sources					
6	Trust what scientists say	1.82 [1.89]	2.09* [2.23]*	1.42	3.04*	**
7	Trust what government says	0.12 [0.55]	−0.76 [0.47]	−1.11**	−0.88*	
12	Trust what media says	−2.79 [−2.18]	−1.35*** [−1.15]***	−1.89*	−2.55	
	Governance capacity					
2	The response to climate change will be positive	0.15 [−0.07]	0.41 [0.22]	−0.63	−1.87***	***
33	Failure to address climate change is the fault of political leaders	0.82 [0.76]	1.97*** [1.65]***	2.21*	1.85*	
	Governance demands					
4	Need more educational programmes	3.62 [3.27]	3.65 [2.91]	3.57	4.26*	***
8	Need strong political leadership	3.91 [3.88]	4.59** [4.10]	3.90	4.57*	**
32	Government should take responsibility	3.21 [2.82]	3.79* [3.87]*	3.61	3.90*	*

Note. Source: Design by author. Figures show the average response for that stage
Significance levels indicate the significance of change compared to the previous stage (apart from overall change indicated in the RHS column) based on T-scores where H_0: no change in average response; $p < 0.05$*, $p < 0.01$**, $p < 0.001$***

due to self-selection bias—something also observed in relation to other studies (e.g., Curato & Niemeyer, 2013; see also Jennstål, 2018)—where a particular type of usually conservative and deep climate sceptic participant withdraws after agreeing to participate in deliberation. I will revisit this issue, and the implications for the overall findings, below.

Discursive Transformation

The ability to identify relevant "types" of individual, such as the deliberative-shy deep sceptics, was facilitated by a second type of analysis, using the forced distribution scores, based on Q methodology (Brown, 1980). This *Q analysis* helped to discern different themes in the responses to statements among participants—called factors; although here I use the term "discourses" in an informal sense (see Dryzek, 1990). The approach is more consistent with the "sense-making" approach of discursive psychology (Niemeyer, 2019)—although there are limits to this interpretation of the method (Danielson, 2015)—where the associations between items emerge as part of the analysis, rather than a priori.

The analysis, which is described in detail in Hobson and Niemeyer (2011), resulted in four discourses, reflecting different levels of agreement/disagreement with the entire set of 33 statements, within which the positions of most participants could be located:

A. Government Scepticism
B. Government Imperative
C. Pragmatism
D. Alarmed Defeatism

I report the "typical" responses to the statements, which are used to interpret the discourses, in Table 12.3 in the appendix. In order to more easily communicate the features of each discourse—and later locate participants within them—I have located abridged versions of the survey statements schematically in Fig. 12.1 among the four overlapping discourse spheres. The statements paraphrased in Table 12.1 are shown in bold.

Discourse A represented a particular form of climate scepticism—where a dedicated analysis revealed a number of variations (Hobson & Niemeyer, 2013)—with much lower levels of trust in knowledge sources (scientists, government, media) compared to the other discourses. Individuals associated with this discourse tended to believe climate change is an important issue—as did those in all the discourses. However, the risk is overstated for this cohort, who tended to ascribe climate change to natural climate variation.[4]

[4]The study included a large proportion of climate sceptics, a number of who held deeply sceptical positions that could not easily incorporated into the main discourse analysis and do not quite fit

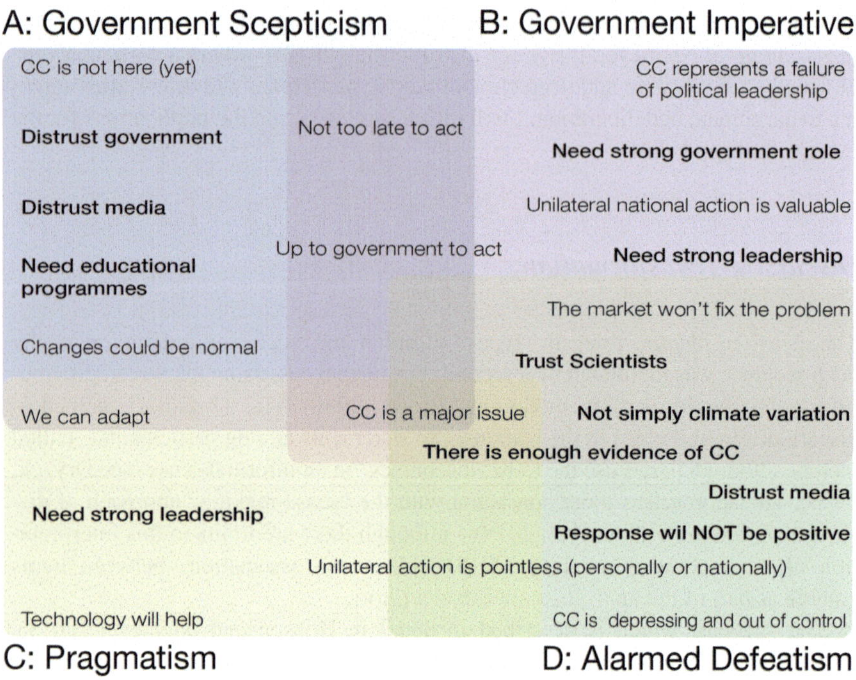

Fig. 12.1 Discourses of climate change adaptation and governance. Source: Design by author

Proponents of Discourse B are much more concerned about climate change, with greater levels of trust in scientific knowledge. But there is a strong element of displeasure in the government's response to the issue, emphasizing centralized action in ways that may not ultimately prove particularly adaptive in the absence of additional recognition of individual responsibility (Hobson & Niemeyer, 2013)—which is consistent with social psychological models used to predict pro-environmental behavior (Klöckner, 2013).

Discourse C, although a very small discourse in terms of influence among participants, represents a pragmatist view whose proponents are optimistic that solutions can be found without causing too much disruption. Discourse D, by contrast, is entirely maladaptive, representing a retreat from any form of cooperative attempt to govern in manner that improves the collective response (see Hobson & Niemeyer, 2011). This is not due to a lack of knowledge, but to acceptance that dramatic change is occurring.

The study's participants can be located schematically on the discourse map—using their factor loadings, indicating how strongly they correlate to the standard responses to the survey statements represented by the discourse (see Hobson & Niemeyer, 2011). The baseline (and high emissions) locations are indicated by the

---

into Discourse A. A separate analysis was conducted separately for the entire climate sceptic cohort, and is reported in Hobson and Niemeyer (2013).

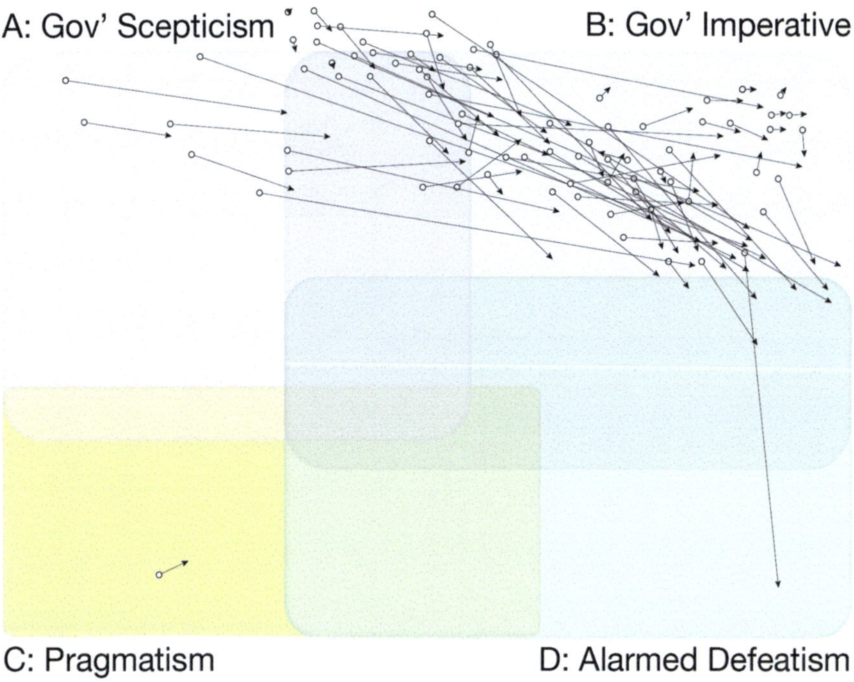

Fig. 12.2 Migration across discourses: stages 1–2 (Baseline-High Emissions Scenario). Source: Design by author

arrows in Fig. 12.2, where the points at the beginning of each arrow reflect the baseline position of participants, prior to exposure to the climate change scenarios.[5] At the study's baseline stage, most participants could be located within Discourse B.

In Fig. 12.2, I capture the resulting movement across the discourse map from exposure to the climate change scenario. The scenario intervention induced a strong overall movement in the direction of Government Imperative (Discourse B), reflecting less an increase in reflexivity than a redirection from merely blaming government on the issue to demanding it take action. But it also involved a movement toward Alarmed Defeatism (D), which is an unreflexive perspective that represents a wholesale retreat from a desire to act on the issue (see Hobson & Niemeyer, 2011). As anticipated, improved knowledge about climate change impacts thus did not appear to improve adaptive capacity and reflexivity.

As discussed above in relation to the Likert responses, the acquisition of knowledge about climate change had no lasting effect. Figure 12.3 shows the impact of the knowledge gains between Stages 1 and 2 following exposure to the climate change scenarios (lighter arrows with closed head) compared to the subsequent movement between Stages 2 and 3 in the follow-up prior to participation in the deliberative

[5] The deep climate sceptics who do not fit into this discourse schema are shown outside the map at a point reflecting those discourse they are most closely (but not significantly) associated with.

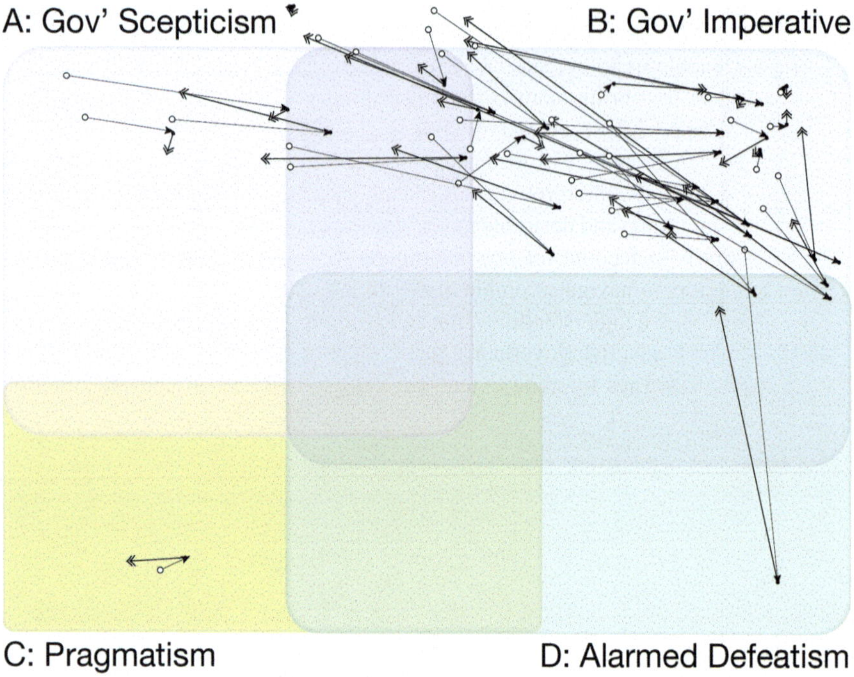

Fig. 12.3 Migration across discourses: High Emissions Scenario Follow-Up. Source: Design by author

process (heavier arrows with open, double head). The figure reveals that most participants return to a position approaching their baseline after the initial impact of information. As is often the case for other studies, the effect of information alone is short lived (see Hobson & Niemeyer, 2011; Howell, 2011).

Deliberation, on the other hand, had a very different impact—both quantitatively and qualitatively. The Likert response analysis in Table 12.1 indicates strong changes during deliberation, but looking at the changes through the lens of the discourses reveals the substantive story. The impact not only included a move across the existing discourse map; it involved a reconfiguration of the map itself and the way in which participants perceived possibilities for governance in relation to climate change. I schematically describe these post-deliberative discourses in Fig. 12.4 in the same manner as the scenario discourses in Fig. 12.2. They include:

A′. Accommodating Scepticism
B′. Governance and Engagement Imperative
E. Collective Action Imperative
F. Adaptive Reassurance

Overall, deliberative engagement increased the desire for more inclusive approaches to governance, rather than for outsourcing to government—observed in

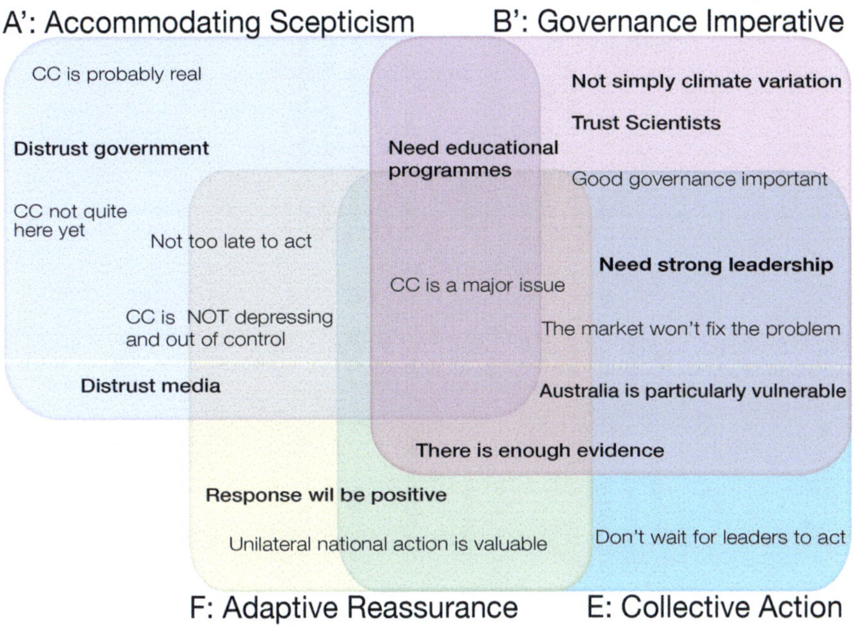

Fig. 12.4 Climate change Discourses: post-deliberation. Source: Design by author

respect to other deliberative processes (Niemeyer, Batalha, & Dryzek, 2013), whose researchers commonly report increases in internal political efficacy (Morrell, 2005). This transformation is reflected in the change in nomenclature from "government" to "governance." Sceptics became more accommodationist in their perspective (Discourse A′).

Following deliberation, most participants were ultimately located in the modified Discourse A′ or the new Discourse E (Collective Action), but the content of these new discourses is at least as important as the distribution of the participants. The overall post-deliberative discursive landscape represents greater adaptive possibilities compared to the other research stages (Hobson & Niemeyer, 2011).

Of particular interest here is the relationship between knowledge and governance. With my analysis associated with Table 12.1, I have already alluded to a changing context in terms of beliefs about climate change, trust in knowledge sources, beliefs about governance capacity, and demand for changes to governance. I have summarized these features for both sets of discourses in Table 12.2 below, including the overall assessment of compatibility with creating the governance conditions for improved climate change adaptation as analyzed in Hobson and Niemeyer (2011).

In Table 12.2, I report the major features pertaining to governance, knowledge, and adaptation for the three main discourses in both discourse maps, as well as

Table 12.2 Discourse knowledge and governance dimensions

Component	A: Self-Assured scepticism	B: Government imperative	D: Alarmed defeatism	A′: Accommodating scepticism	B′: Governance & engagement imperative	E: Collective action imperative	Overall change
Beliefs about CC	Sceptical; others have too simplistic a view	Strong belief	Strong belief	Accepting	Strong belief	Strong belief	Increased belief in evidence for CC
Trust in Knowledge Sources	Distrust all sources (less so science)	Trust scientists	Strongly trust scientists	Distrust all sources (less so science)	Strongly trust scientists	Neutral	Increased trust in scientists
Governance capacity	Neutral	Partly pessimistic	Pessimistic	Neutral	Less pessimistic	Partly optimistic	Increased belief in positive response
Demands for change	Education (strong demand)	Leadership, government role	Leadership	Education, government role	Education, leadership, government role	Leadership	Increased demand for education, leadership and role of government
Overall adaptive capacity[a]	LOW	MEDIUM/ LOW	LOW	LOW	MEDIUM	HIGH	

Note. Source: Design by author
[a]From Hobson and Niemeyer (2011)

summarize the overall changes occurring during deliberation reported in Table 12.1.[6] These must be assessed together because, although there is a significant overall change, the specific nature and source of the change varies between discourses. For example, the increased belief in evidence supporting climate change occurs mainly among the less entrenched sceptics associated with Discourse A, but they are less associated with the overall increase in trust in scientific sources of knowledge. That comes from participants associated with Discourses A′ and B′—and erstwhile sceptics who migrated to Discourse A' during deliberation. Discourse A accounts for most of the observed overall change—in terms of increased belief in climate change, trust in scientists, belief in capacity for a positive response, and demand for education about climate change, increased leadership, and the role of government—by virtue of most participants associated with it post-deliberation. Particularly interesting is the emergence of Discourse E, which Hobson and Niemeyer (2011) associate with the strongest potential for adaptive climate governance. It represents an increase in the factors that contribute to improved collective action via increased acceptance of the reality of climate change and personal and collective efficacy in undertaking action—important factors contributing to the likelihood of increased action (Klöckner, 2013).

Overall, the findings point to deliberation contributing to substantial discursive transformation of the sort that is consistent with the kind of reflexivity demanded by the climate change issue. Researchers in another climate change adaptation and deliberation study, in this case conducted in Sydney, Australia, have also observed overall transformative dynamics very similar to those reported above (Schlosberg, Collins, & Niemeyer, 2017).

Knowledge, Reflexivity and Deliberative Systems

With my findings in relation to the ACR case study, I illustrate the relationship between governance settings, knowledge uptake, and the potential behavioral response to climate change, which in turn feeds back into the governance context. Mere exposure to alternative knowledge or positions is not itself enough to induce reflection.

Worse, exposure to information in the absence of deliberation may actually be counterproductive if the response induces cognitive closure in the face of increasing dissonance via similar dynamics to those discussed earlier in respect to increasing issue complexity. The existence of high and in some cases increasing levels of climate scepticism owes some of its genesis to this phenomenon (Capstick et al., 2015).

A key factor influencing these dynamics is the citizens' disposition in their interaction with that knowledge. And that disposition is, in turn, influenced by the context. Accessing alternative arguments in a non-deliberative context—such as online

---

[6] I have omitted the fourth discourse for both maps for reasons of space. In both cases, the number of individuals associated with these discourses is very small and they do not constitute a major feature in the overall analysis.

(Twitter), where the broader politics is characterized by hyper-partisanship (Bail et al., 2018)—can actually lead to a decline in openness to information. By contrast, changing that context to a more deliberative one induces a qualitatively different interaction with knowledge inputs, as evidenced by the ACR climate change case study.

To be sure, deliberation does not fundamentally change citizens, who still have roughly the same set of capabilities as before. But it serves to activate a set of capabilities—referred to earlier under the "deliberative stance" rubric—that is otherwise not ordinarily activated in political settings. Part of this activation involves switching modes of cognition from shallow to deeper (see Niemeyer, 2011). But other kinds of activation also occur, consistent with the deliberative stance. The ideals of openness and intersubjectivity, valorized by deliberative democrats, stand in contrast to social psychological phenomena, such as cognitive closure (Kruglanski, 2013) and motivated reasoning (Taber & Lodge, 2006).

The deliberative stance as described here implies a form of scepticism of the self: an acceptance that views may be incomplete or in need of updating. It is as much an affective disposition as a cognitive one, where high levels of anxiety can inhibit the processing of knowledge, particularly where it involves confronting implications and/or high levels of complexity. I observed this in the emergence of the Alarmed Defeatism discourse (D) in the ACR case study, which would likely undermine any effort to find collective solutions to the issue—with comparable effects observed elsewhere (e.g., Stoll-Kleemann, O'Riordan, & Jaeger, 2001).

Another kind of anxiety can be observed in the form of deep climate sceptics, who usually hail from a demographic that feels threatened by structural changes necessary to meet the challenge, as well as holding certain (usually conservative) political values (Whitmarsh, 2011). As for the Twitter example (Bail et al., 2018) in the ACR case study, exposure to the climate change scenarios absent of deliberation sometimes entrenched existing views rather than transforming them (Hobson & Niemeyer, 2013).

There are limits to this effect. Although participation in deliberation changed these dynamics for many of the sceptics, those most deeply sceptical opted instead to exit the process altogether (Hobson & Niemeyer, 2013), demonstrating the limits to inducing self-scepticism and a deliberative stance for these groups. But, as I will discuss below, even these dynamics might be addressed to some extent if the issues of knowledge and environmental communication and inducing the deliberative stance were to be dealt with through a systemic lens.

Systemic Deliberative Dynamics

The findings so far reinforce the idea that the dynamics of environmental knowledge and behavior must be viewed as part of a wider governance context (Burgess, Harrison, & Filius, 1998). But this is not to suggest that improving knowledge uptake (or in the case of climate change, acceptance) requires exclusive focus at the macro level. The effects must be understood in systemic terms—neither macro- nor micro-level, but rather operating at both levels as part of an overall systemic effect. Despite the onus on individual-level capacity in the form of the deliberative stance herein, exclusive focus on micro-level, individual knowledge dynamics at best is only likely to yield gains incommensurable to the challenge (Shove, 2010) and more of the same kind of linear modeling of environmental behavior that fails to escape the orbit of a knowledge deficit paradigm (O'Brien, 2012). Such an approach also reinforces the previous assertion regarding the incomplete nature of the *impact science* strategy advocated by reflexivity scholars, which fails to account for the processes that contribute to its uptake.

The nature of systems thinking required to address these shortcomings is informed by the nature of the dynamics observed above. The deliberative context of the ACR project impacted on the manner in which individuals interacted with the information provided. It reconfigured the citizens' orientation toward the knowledge system—accepting a division of labor in its production (Christiano, 2012)—ameliorating the effects of the knowledge ignorance paradox (Ungar, 2000) as part of a changing discursive landscape. It also inured citizens to the forces of blandishment represented by anti-reflexive political forces and strategies of "merchants of doubt" (Oreskes & Conway, 2010), something that researchers have observed in relation to a number of issues with similar knowledge dimensions (Niemeyer, 2011).

The key here is the changing discursive landscape within which individual capacities in the form of the deliberative stance were activated. An important contextual factor facilitating these changes is not just the mere fact of citizens deliberating, but the evolving language that was deployed as part of developing a shared understanding of the issue, which is partly captured by the discursive changes reported in Fig. 12.4.

These changes constitute systemic effects, even though for the ACR they are induced in fairly managed, micro-deliberative settings. Once the nature of these effects is understood, the challenge remains in terms of the mechanisms whereby they might be "scaled up" to a wider audience. There are possibilities, although much needs to be done to develop them (Niemeyer, 2014; Niemeyer & Jennstål, 2018). Shove (2010), for example, advocates an approach to understanding human relationships with the environment through the lens of "practices" rather than behaviors, and a deepening of deliberative practices in a wider democratic context seems to fit well with this approach.

A democratic deepening of this sort yields potential systemic effects on the governance system, such as avoiding the climate change governance trap (Pidgeon, 2012)—as researchers observed in the Australian Capital Region climate change and governance study, it led to higher responsiveness at the individual and community level, as well as the demand for collective action. Such changes orient the system as a whole toward a high-order form of reflexivity (Dryzek & Pickering, 2017), with improved trust in knowledge as well as a higher disposition toward translating that knowledge into action. The activation of a deliberative stance is neither a cause nor an effect of this shift, but an inherent component of an improved democratic practice.

To the extent that such practices can be scaled up, it is just possible that a polity that is more deliberative not only responds to climate change and the associated knowledge more constructively and in ways that reflect the inner desires of its citizens, but will also impact the anxious dynamics that induce citizens' anti-reflexive forces. These dynamics do not necessarily lead to increased knowledge on the part of citizens—although there is good evidence that this would also be the case (Andersen & Hansen, 2007; Grönlund, Setälä, & Herne, 2010; Luskin, Fishkin, & Jowell, 2002)—but they do appear to hold a good deal of promise in addressing the pathologies of knowledge that characterize much of the climate change debate and policy response in the prevailing governance context.

Conclusion

That knowledge alone is insufficient to ensure action on complex and challenging governance issues such as climate change is demonstrable on both normative and empirical grounds. Here, I have attempted to demonstrate the manner in which the context plays a role in this process from the perspective of deliberative governance, particularly in respect to the discursive landscape that frames the modes in which knowledge may be accepted, rejected, or simply ignored. This discursive context includes the changing climate itself, which has the power to alter the discursive landscape and, paradoxically, work against the operation of reflexivity in governance systems.

Deliberation works on a small scale, by activating capacities that exist in most citizens but are not activated in prevailing governance settings. These capacities, captured by the concept of the deliberative stance, include a more normatively appealing scepticism of the self and a willingness to judiciously consider and absorb new information. This does not demand complete comprehension, because a deliberative system is based on trust that distributes knowledge generation and synthesis.

Building capacity to capture these dynamics beyond the mini-public could involve reforming public engagement informed by deliberative principles in ways that scale up these effects. And these reforms could involve harnessing the power of mini-publics, to the extent that they can be articulated with the wider governance system (Niemeyer & Jennstål, 2018). A good deal of work remains to be done on such approaches and their effectiveness.

In spite of these gaps, in this chapter I have demonstrated the utility of conceptualizing knowledge uptake through the lens of governance. The adoption of systems thinking beyond micro-level information deficit approaches or macropolitical strategies is also of relevance. A discursively understood system of governance cannot separate the citizen from the governance context, and it is here that productive solutions might be found.

Appendix

Table 12.3 shows the statements used in the Q sort for the CCPS case study, along with the *z*-scores representing the typical level of agreement/disagreement for each statement associated with each of the discourses.

Table 12.3 CCPS case study statements and factor scores

No.	Statement	Discourses (Factor z-scores)							
		A	B	C	D	A′	B′	E	F
1	There is not enough information to definitively say that climate change is real.	0.21	−1.61	−1.97	−1.87	−0.77	−1.92	−0.98	−0.52
2	The response to climate change is not going to be positive. The same mistakes will keep happening.	−0.48	−0.14	−0.79	0.83	−0.21	−0.29	−0.36	−1.52
3	Climate variation is normal, so why should this be a problem?	0.20	−1.14	0.13	−1.09	−0.50	−1.09	0.02	−0.35
4	More educational programmes are needed to increase public awareness about climate change.	1.29	0.99	0.96	−0.24	1.17	1.80	0.88	0.74
5	Climate change will not be a problem because there will be technological solutions available.	0.06	−0.80	1.76	−0.70	0.53	−0.23	−0.60	−1.30
6	I don't trust what scientists say about climate change.	0.21	−1.28	−1.97	−1.09	0.64	−1.46	0.02	−0.26
7	I don't trust what I hear about climate change from government.	0.94	−0.05	−0.75	−1.09	1.22	−0.14	−0.12	0.22
8	We need strong political leadership to do something about climate change.	0.53	1.68	1.25	0.37	0.96	1.78	1.19	0.69
9	I think it is safe to say climate change is here.	0.88	1.11	1.71	1.99	0.98	1.23	0.53	1.30
10	I'm not going to do anything to address climate change because it is not a major issue.	−2.00	−1.41	−1.08	−1.40	−1.79	−1.40	−1.12	−1.34
11	There's not much point in me doing anything to fix this. No-one else is going to.	−1.82	−0.79	−0.57	−0.25	−1.90	−0.75	−1.50	−1.39
12	It's difficult to trust what comes out in the media on the issue of climate change.	1.60	0.05	0.17	1.14	1.46	0.62	0.40	1.34
13	It is already too late to do anything, as any action to stop climate change will take a long time to take effect.	−1.37	−0.75	0.98	0.68	−0.96	−0.35	−1.64	−1.56
14	I'm not concerned enough to do anything drastic about this, such as participate in political action.	−0.95	−1.01	−0.60	0.37	−0.76	−1.21	−0.22	−0.95
15	It is unfair that we are going to leave the climate in a mess for future generations.	1.29	1.17	0.85	1.29	1.35	1.11	0.88	1.60
16	We should pay for greenhouse emissions.	−0.41	0.97	−0.04	−0.55	0.20	1.02	0.33	1.00

(continued)

Table 12.3 (continued)

No.	Statement	Discourses (Factor z-scores)							
		A	B	C	D	A′	B′	E	F
17	We can adapt to the coming changes.	1.10	−0.27	1.87	−1.87	0.84	0.00	0.53	0.87
18	It is clear that we are already entering the zone of dangerous climate change.	−0.54	1.25	0.41	1.45	−0.52	0.86	1.25	0.78
19	I care about the planet.	2.24	0.83	−0.11	1.91	1.79	1.03	1.43	1.95
20	I don't know what to do. I'm very concerned and would like to do something, but I don't have a realistic shortlist of things that would really make a difference.	−0.90	−0.16	0.21	0.52	−0.51	0.01	−0.22	−1.43
21	Australia does not owe it to the rest of the world to reduce emissions and suffer economically.	−0.06	−1.05	−0.67	−0.17	0.18	−0.97	−0.16	−0.39
22	If Australia reduces greenhouse gases it won't make a difference. That will just shift Australian jobs to other countries.	−0.04	−0.94	−0.96	−1.25	0.27	−0.69	−0.12	−0.56
23	This is so depressing and is so out of our control.	−1.05	−0.02	−1.33	1.29	−1.47	−0.25	−1.74	−0.22
24	I believe that the difference we can have as an individual, in Australia, is so minimal that our actions are worthless.	−1.39	−0.66	1.20	0.52	−1.45	−0.73	−1.88	−1.26
25	Australia is particularly vulnerable to climate change, and it is in our interest to help find an effective global solution.	0.91	1.36	0.97	0.06	0.86	1.08	1.39	0.82
26	We need laws addressing climate change because people are not going to volunteer to change.	−0.10	1.31	−0.55	0.07	0.21	0.79	1.77	0.39
27	I want to do something, but it is too big and too hard.	−1.30	−0.26	0.38	0.29	−1.29	−0.32	−0.40	−0.65
28	When I read in the paper that climate change is not true, I start to have doubts about whether it is changing.	−0.25	−0.90	−0.46	−0.01	−0.86	−0.86	−0.88	−0.17
29	Doing something to reduce emissions feels a bit hopeless but I just want to feel that I'm doing the most I can.	0.63	0.37	0.03	0.29	−0.78	−0.07	0.50	1.39
30	The fate of the planet is too important to be left to market forces.	0.67	1.42	0.62	−0.94	0.33	0.71	1.63	0.78
31	Australia's emissions are tiny, so it's not up to us to act.	−0.68	−1.28	−1.08	−0.48	−0.61	−1.36	−0.98	−0.17
32	Governments should take a far greater role in preparing towns and cities to adapt to the impacts of climate change.	0.54	1.34	0.12	−0.47	1.11	1.38	0.91	0.39
33	Failure to address climate change is the fault of political leaders.	0.06	0.67	−0.67	0.36	0.29	0.68	−0.74	−0.22

Note. Source: Design by author

References

Andersen, V. N., & Hansen, K. M. (2007). How deliberation makes better citizens: The Danish deliberative poll on the euro. *European Journal of Political Research, 46,* 531–556. https://doi.org/10.1111/j.1475-6765.2007.00699.x

Bächtiger, A., Niemeyer, S., Neblo, M., Steenbergen, M. R., & Steiner, J. (2010). Disentangling diversity in deliberative democracy: Competing theories, their blind spots and complementarities. *Journal of Political Philosophy, 18,* 32–63. https://doi.org/10.1111/j.1467-9760.2009.00342.x

Bail, C., Argyle, L., Brown, T., Bumpus, J., Chen, H., Fallin Hunzaker, M. B., Lee, J., Mann, M., Merhout, F., & Volfovsky, A. (2018). Exposure to opposing views can increase political polarization: Evidence from a large-scale field experiment on social media. https://doi.org/10.17605/OSF.IO/4YGUX

Beck, U. (1992). *Risk society: Towards a new modernity*. London: SAGE.

Bord, R. J., O'Connor, R. E., & Fischer, A. (2000). In what sense does the public need to understand global climate change? *Public Understanding of Science, 9,* 205–218. https://doi.org/10.1088/0963-6625/9/3/301

Brown, S. R. (1980). *Political subjectivity: Applications of Q methodology in political science*. New Haven, CT: Yale University Press.

Burgess, J., Harrison, C. M., & Filius, P. (1998). Environmental communication and the cultural politics of environmental citizenship. *Environment and Planning A: Economy and Space, 30,* 1445–1460. https://doi.org/10.1068/a301445

Capstick, S., Whitmarsh, L., Poortinga, W., Pidgeon, N., & Upham, P. (2015). International trends in public perceptions of climate change over the past quarter century. *Wiley Interdisciplinary Reviews: Climate Change, 6,* 35–61. https://doi.org/10.1002/wcc.321

Carlton, J. S., Perry-Hill, R., Huber, M., & Prokopy, L. S. (2015). The climate change consensus extends beyond climate scientists. *Environmental Research Letters, 10,* 094025. https://doi.org/10.1088/1748-9326/10/9/094025

Chambers, S. (2012). Deliberation and mass democracy. In J. Parkinson & J. J. Mansbridge (Eds.), *Deliberative systems: Deliberative democracy at the large scale* (pp. 52–71). New York, NY: Cambridge University Press.

Christiano, T. (2012). Deliberation among experts and citizens. In J. Parkinson & J. J. Mansbridge (Eds.), *Deliberative systems: Deliberative democracy at the large scale* (pp. 27–51). NewYork, NY: Cambridge University Press.

Christoff, P. (2013). Climate discourse complexes, national climate regimes and Australian climate policy. *Australian Journal of Politics & History, 59,* 349–367. https://doi.org/10.1111/ajph.12020

Cohen, J. (2007). Deliberative democracy. In S. W. Rosenberg (Ed.), *Deliberation, participation and democracy: Can the people govern?* (pp. 219–236). Basingstoke: Palgrave Macmillan.

Curato, N., & Niemeyer, S. J. (2013). Reaching out to overcome political apathy: Building participatory capacity through deliberative engagement. *Politics & Policy, 41,* 355–383. https://doi.org/10.1111/polp.12015

Curato, N., Niemeyer, S. J., & Dryzek, J. S. (2013). Appreciative and contestatory inquiry in deliberative forums: Can group hugs be dangerous? *Critical Policy Studies, 7,* 1–17. https://doi.org/10.1080/19460171.2012.758595

Danielson, S. B. (2015). Q method as discourse analysis: Three cautionary tales. Retrieved from http://debitage.net/academic/Qdiscourse-webdraft.pdf

Dryzek, J. S. (1990). *Discursive democracy: Politics, policy, and political science*. Cambridge, UK: Cambridge University Press.

Dryzek, J. S. (2009). Democratization as deliberative capacity building. *Comparative Political Studies, 42,* 1379–1402. https://doi.org/10.1177/0010414009332129

Dryzek, J. S., & Niemeyer, S. J. (2008). Discursive representation. *American Political Science Review, 102,* 481–494. https://doi.org/10.1017/S0003055408080325

Dryzek, J. S., & Pickering, J. (2017). Deliberation as a catalyst for reflexive environmental governance. *Ecological Economics, 131,* 353–360. https://doi.org/10.1016/j.ecolecon.2016.09.011

Dunlap, R. E. (2013). Climate change skepticism and denial: An introduction. *American Behavioral Scientist, 57,* 691–698. https://doi.org/10.1177/0002764213477097

Grönlund, K., Setälä, M., & Herne, K. (2010). Deliberation and civic virtue: Lessons from a citizen deliberation experiment. *European Political Science Review, 2,* 95–117. https://doi.org/10.1017/S1755773909990245

Habermas, J. (2006). Political communication in media society: Does democracy still enjoy an epistemic dimension? The impact of normative theory on empirical research. *Communication Theory, 16,* 411–426. https://doi.org/10.1111/j.1468-2885.2006.00280.x

Hobson, K. P., & Niemeyer, S. J. (2011). Public responses to climate change: The role of deliberation in building capacity for adaptive action. *Global Environmental Change, 21,* 957–971. https://doi.org/10.1016/j.gloenvcha.2011.05.001

Hobson, K. P., & Niemeyer, S. J. (2013). "What sceptics believe": The effects of information and deliberation on climate change scepticism. *Public Understanding of Science, 22,* 396–412. https://doi.org/10.1177/0963662511430459

Howell, R. A. (2011). Lights, camera … action? Altered attitudes and behaviour in response to the climate change film "The Age of Stupid." *Global Environmental Change, 21,* 177–187. https://doi.org/10.1016/j.gloenvcha.2010.09.004

Jennstål, J. (2018). Deliberative participation and personality: The effect of traits, situations and motivation. *European Political Science Review, 10,* 417–440. https://doi.org/10.1017/S1755773918000024

Klöckner, C. A. (2013). A comprehensive model of the psychology of environmental behaviour: A meta-analysis. *Global Environmental Change, 23,* 1028–1038. https://doi.org/10.1016/j.gloenvcha.2013.05.014

Kruglanski, A. W. (2013). *The psychology of closed mindedness.* Essays in Social Psychology: Vol. 1. London: Taylor and Francis. https://doi.org/10.4324/9780203506967

Kruglanski, A. W., & Boyatzi, L. M. (2012). The psychology of closed and open mindedness, rationality, and democracy. *Critical Review, 24,* 217–232. https://doi.org/10.1080/08913811.2012.711023

Kruglanski, A. W., & Webster, D. M. (1996). Motivated closing of the mind: "Seizing" and "Freezing." *Psychological Review, 103,* 263–283. https://doi.org/10.1037/0033-295X.103.2.263

Luskin, R. C., Fishkin, J. S., & Jowell, R. (2002). Considered opinions: Deliberative polling in Britain. *British Journal of Political Science, 32,* 455–487. https://doi.org/10.1017/S0007123402000194

McCright, A. M., & Dunlap, R. E. (2010). Anti-reflexivity. *Theory, Culture & Society, 27,* 100–133. https://doi.org/10.1177/0263276409356001

McCright, A. M., Dunlap, R. E., & Marquart-Pyatt, S. T. (2016). Political ideology and views about climate change in the European Union. *Environmental Politics, 25,* 338–358. https://doi.org/10.1080/09644016.2015.1090371

Morrell, M. E. (2005). Deliberation, democratic decision-making and internal political efficacy. *Political Behavior, 27,* 49–69. https://doi.org/10.1007/s11109-005-3076-7

Morrell, M. E. (2010). *Empathy and democracy: Feeling, thinking, and deliberation.* University Park, PA: Pennsylvania State University Press. https://doi.org/10.5325/j.ctt7v16x

Năkićenović, N., Alcamo, J., Davis, G., de Vries, B., Fenhann, J., Gaffin, S., Gregory, K., Grübler, A., Jung, T. Y., Kram, T., La Rovere, E. L., Michaelis, L., Mori, S., Morita, T., Pepper, W., Pitcher, H., Price, L., Riahi, K., Roehrl, A., Rogner, H.-H., Sankovski, A., Schlesinger, M., Shukla, P., Smith, S., Swart, R., van Rooijen, S., Victor, N., & Dadi, Z. (2000). *Emissions scenarios: Special report on emission scenarios: A special report of working group iii of the Intergovernmental Panel on Climate Change.* Cambridge, UK: Cambridge University Press.

Niemeyer, S. J. (2011). The emancipatory effect of deliberation: Empirical lessons from mini-publics. *Politics & Society, 39,* 103–140. https://doi.org/10.1177/0032329210395000

Niemeyer, S. J. (2014). Scaling up deliberation to mass publics: Harnessing mini-publics in a deliberative system. In K. Grönlund, A. Bächtiger, & M. Setälä (Eds.), *Deliberative mini-publics: Involving citizens in the democratic process* (pp. 177–202). Essex: ECPR.

Niemeyer, S. J. (2019). Analysis of deliberative transformation: A multi-level approach incorporating Q methodology. In S. Elstub & O. Escobar (Eds.), *Handbook of democratic innovation and governance* (pp. 540–557). Cheltenham: Edward Elgar. https://doi.org/10.4337/9781786433862.00049

Niemeyer, S. J., Batalha, L., & Dryzek, J. S. (2013). Changing orientations toward Australian democracy. In L. Carson, J. Gastil, J. Hartz-Karp, & R. Lubensky (Eds.), *The Australian citizens' parliament and the future of deliberative democracy* (pp. 133–145). University Park, PA: Pennsylvania State University Press.

Niemeyer, S. J., & Jennstål, J. (2018). Scaling up deliberative effects: Applying lessons of mini-publics. In A. Bächtiger, J. S. Dryzek, M. E. Warren, & J. J. Mansbridge (Eds.), *Handbook of deliberative democracy* (pp. 329–347). Oxford, UK: Oxford University Press.

O'Brien, K. (2012). Global environmental change iii: Closing the gap between knowledge and action. *Progress in Human Geography, 37,* 587–596. https://doi.org/10.1177/0309132512469589

Oreskes, N. (2018). The scientific consensus on climate change: How do we know we're not wrong? In E. A. Lloyd & E. Winsberg (Eds.), *Climate modelling: Philosophical and conceptual issues* (pp. 31–64). Cham: Springer International.

Oreskes, N., & Conway, E. M. (2010). *Merchants of doubt: How a handful of scientists obscured the truth on issues from tobacco smoke to global warming* (1st U.S. edition). New York: Bloomsbury.

Owen, D., & Smith, G. (2015). Survey article: Deliberation, democracy, and the systemic turn. *Journal of Political Philosophy, 23,* 213–234. https://doi.org/10.1111/jopp.12054

Pidgeon, N. (2012). Public understanding of, and attitudes to, climate change: UK and international perspectives and policy. *Climate Policy, 12,* 85–106. https://doi.org/10.1080/14693062.2012.702982

Pierce, J. C., Lovrich, N. P., & Dalton, R. J. (2000). Contextual influences on environmental knowledge: Public familiarity with technical terms in nuclear weapons production in Russia and the United States. *Environment and Behavior, 32,* 188–208. https://doi.org/10.1177/00139160021972496

Scott, M. (2020). Knowledge of governance as knowledge for governance: Spatialized techniques of neutralization. In J. Glückler, G. Herrigel, & M. Handke (Eds.), *Knowledge for governance* (pp. 51–67). Knowledge and Space: Vol. 15. Cham: Springer.

Schlosberg, D., Collins, L. B., & Niemeyer, S. J. (2017). Adaptation policy and community discourse: Risk, vulnerability, and just transformation. *Environmental Politics, 26,* 413–437. https://doi.org/10.1080/09644016.2017.1287628

Shove, E. (2010). Beyond the ABC: Climate change policy and theories of social change. *Environment and Planning A: Economy and Space, 42,* 1273–1285. https://doi.org/10.1068/a42282

Stevenson, H., & Dryzek, J. S. (2012). The discursive democratisation of global climate governance. *Environmental Politics, 21,* 189–210. https://doi.org/10.1080/09644016.2012.651898

Stoll-Kleemann, S., O'Riordan, T., & Jaeger, C. C. (2001). The psychology of denial concerning climate mitigation measures: Evidence from Swiss focus groups. *Global Environmental Change, 11,* 107–117. https://doi.org/10.1016/S0959-3780(00)00061-3

Taber, C. S., & Lodge, M. (2006). Motivated skepticism in the evaluation of political beliefs. *American Journal of Political Science, 50,* 755–769. https://doi.org/10.1111/j.1540-5907.2006.00214.x

Ungar, S. (2000). Knowledge, ignorance and the popular culture: Climate change versus the ozone hole. *Public Understanding of Science, 9,* 297–312. https://doi.org/10.1088/0963-6625/9/3/306

van der Linden, S. L., Leiserowitz, A. A., Feinberg, G. D., & Maibach, E. W. (2015). The scientific consensus on climate change as a gateway belief: Experimental evidence. *PLoS One, 10,* 1–8. https://doi.org/10.1371/journal.pone.0118489

Whitmarsh, L. (2011). Scepticism and uncertainty about climate change: Dimensions, determinants and change over time. *Global Environmental Change, 21,* 690–700. https://doi.org/10.1016/j.gloenvcha.2011.01.016

Chapter 13
Nurturing Adaptive Governance Through Environmental Monitoring: People, Practices, Politics in the Kruger to Canyons Biosphere Region, South Africa

Lisen Schultz, Simon West, and Cláudia Florêncio

How can governance become more adaptive to rapidly evolving knowledge? How can it address problems spanning different sectors, scales, and interests in a holistic manner? And how can it navigate novel challenges where no obvious response exists? These are core questions of the literature on adaptive governance, which brings together studies on natural resource management, environmental governance, and collaborative governance, in pursuit of pathways to sustainability in the context of rapid social-ecological change. In this chapter, we briefly introduce the concept of adaptive governance to a broader audience, identify two key research frontiers in the literature, and address them through an empirical case study of the Environmental Monitors (EM) program in the Kruger to Canyons Biosphere Region (K2C) in South Africa—exploring the *people, practices, and politics* of adaptive governance.

What Is Adaptive Governance?

Adaptive governance refers to governance in the context of complexity and uncertainty (Dietz, Ostrom, & Stern, 2003). In adaptive governance, actors are linked across scales (local, regional, global) and sectors (state, private, civil society), in learning-based approaches that emphasize monitoring and experimentation, with the aim of responding to evolving challenges, issues, and threats (Schultz, Folke, Österblom, & Olsson, 2015). Although the concept of adaptive governance has been used in several fields, including political science and health research (see Chaffin, Gosnell, & Cosens, 2014), in this paper we refer to adaptive governance as applied to sustainability issues. In the field of sustainability science,

L. Schultz (✉) · S. West · C. Florêncio
Stockholm Resilience Centre, Stockholm University, Stockholm, Sweden
e-mail: lisen.schultz@su.se; simon.west@su.se

J. Glückler et al. (eds.), *Knowledge for Governance*, Knowledge and Space 15,
https://doi.org/10.1007/978-3-030-47150-7_13

adaptive governance emerged from social-ecological systems perspectives, where people are seen as *inextricably interwoven* with the natural environment (Folke, Hahn, Olsson, & Norberg, 2005). On the one hand, researchers have used adaptive governance as a descriptive concept to analyze how governance systems have responded (or not) to complex sustainability challenges, including wetland degradation (Olsson, Folke, Galaz, Hahn, & Schultz, 2007), climate change (Boyd & Juhola, 2014), deforestation (Boyd, 2008), and overfishing (Österblom & Sumaila, 2011; Valman, Österblom, & Olsson, 2015). On the other hand, they have utilized adaptive governance as a normative concept, to prescribe how governance *should* be structured in an *anthropocene era* where surprises, cascading effects, and tipping points will be the norm (Berkes, 2017; Duit & Galaz, 2008; Galaz, 2014). As we will show, the distinctions between descriptive and normative uses of adaptive governance are often blurred in practice.

Collaboration, learning, and bridging organizations are key features of adaptive governance (Karpouzoglou, Dewulf, & Clark, 2016). Collaboration draws attention to the formal and informal partnerships and networks that connect actors operating in different domains, and potentially work to enhance the *institutional fit* between governance systems and the problems they seek to address (Folke, Pritchard, Berkes, Colding, & Svedin, 2007). Learning captures the importance of monitoring, experimentation, and multiple sources of knowledge and *ways of knowing*, for developing more holistic understandings of sustainability challenges and for facilitating ongoing adaptation to changing contexts (Cundill, Leitch, Schultz, Armitage, & Peterson, 2015). Bridging organizations have been identified as central to initiating and sustaining adaptive governance over time (Folke et al., 2005; Hahn, Olsson, Folke, & Johansson, 2006; Schultz, 2009). Bridging organizations work to connect different actors, facilitate dialogue, share information, resolve conflict, and build trust, lowering the transaction costs of collaboration for participants. However, the characteristics that enable bridging organizations to play this role—such as their flexibility and lack of allegiance to one particular *type* of authority—mean that they are often ephemeral organizations that suffer from a lack of consistent funds and high turnover of staff (Moss, Medd, Guy, & Marvin, 2009), and experience tensions and challenges in their attempts to be "all things to all people" (Parker & Crona, 2012, p. 263).

Research Frontiers in Adaptive Governance Literature

Over the past 15 years, adaptive governance literature has evolved into a vibrant body of scholarship exploring governance in the context of rapid social-ecological change. More recently, researchers articulating new theoretical perspectives (Boyd, Ensor, Broto, & Juhola, 2014; Leach, Scoones, & Stirling, 2010; van Kerkhoff & Lebel, 2015) and attempting to review and synthesize the literature (Chaffin et al., 2014; Karpouzoglou et al., 2016) have identified a number of gaps, grey areas, and pressing questions for the next wave of adaptive governance scholarship to address. In this chapter, we address one empirical and one theoretical research frontier.

An empirical challenge concerns the emergence of adaptive governance—how does it 'come about' (Chaffin, Folke, & Hahn, 2014)? So far, researchers have pointed towards a range of factors, including the role of leadership (Olsson, Folke, & Hahn, 2004), the creation of "vertical" and "horizontal" networks (Österblom & Folke, 2013), and the building of trust between different actors including, for instance, scientists, policymakers, and citizens (Schultz et al., 2015). Moreover, analysts have focused on how *windows of opportunity* (produced by, e.g., a policy change, the creation of a new organization, or a biophysical perturbation) and *perceptions of crisis* may stimulate shifts to adaptive governance (Olsson et al., 2006). These factors are indicative, but raise a number of questions: What is perceived as a crisis, by whom? Whose interpretations "matter" and provide a sufficiently compelling vision to provoke change? What kinds of knowledge are important for the emergence of adaptive governance, and how is this knowledge produced, collected, and used? And how do practices associated with the emergence of adaptive governance, such as network-building and information-sharing, fit within, and connect to existing practices of governance? Moreover, the existing empirical literature is weighted to the Global North and implies that particular kinds of financial, human, and technological resources are necessary for the emergence of adaptive governance (e.g., Galaz, 2014). It is therefore important to empirically explore if and how adaptive governance might emerge with the different kinds of resources available in developing countries and the Global South (Karpouzoglou et al., 2016).

A theoretical challenge is to develop accounts that are grounded in the experiences and practices of the people "doing" adaptive governance in messy, real-world contexts (e.g., van Kerkhoff & Lebel, 2015; Wyborn, 2015). The *first wave* of adaptive governance scholars highlighted the value of individual leadership and interpersonal networking, and emphasized the importance of establishing "the right links, at the right time, around the right issues" (Westley, 2002, p. 357; Folke et al., 2005; Olsson et al., 2006). These studies move towards more *agential* accounts of adaptive governance, as compared to purely *structural* accounts whose authors focus on network patterns and information flows (for a combination of these approaches, see Berdej & Armitage, 2016). But they provide rather functional explanations of adaptive governance, where individual action is motivated by the pursuit of self-evidently "better-functioning" governance systems, and potentially idealistic accounts of individual "heroes" (Leach et al., 2010; Stirling, 2016). These functional explanations do not enable us to explain *how* workable solutions are arrived at by actors who—like all of us—bring many different meanings, preferences, and interests to their activities, and whose everyday work is guided by multiple responsibilities, rationales, and imperatives. For instance, what constitutes the "right" issue, link, or time is a fundamentally interpretive question that depends on the meanings ascribed to particular situations (and desirable means and ends) by those involved. Therefore, if the literature is to account for how adaptive governance is enacted—and how it helps practitioners to navigate complexity (if indeed it does)—it will be necessary to develop theoretical understandings that can account for the everyday experiential contexts in which adaptive governance takes shape. In these situations, the 'best' way forwards is always to some extent unclear and undecided, and all kinds of decisions need to be made about, for example, the kinds of knowledge to produce, the types of interpersonal and inter-organizational connections to establish, and the sorts of information to share (e.g., West, Schultz, & Bekessy, 2016). In developing such approaches, researchers will explore how adaptive governance emerges through the daily decisions and practices of imperfect people, struggling to get work done in confusing and demanding situations.

Towards a People, Practices, and Politics *Perspective on Adaptive Governance*

In this paper, we address these research frontiers by sketching out an analytical lens centered around *people, practices, and politics* (3P) and applying it to an empirical case of the potential emergence of adaptive governance. We intend *people, practices and politics* to be useful primarily as broad "sensitizing concepts," providing a general sense of guidance and "directions along which to look" rather than strictly applicable definitive concepts (Blumer, 1954, p. 7). Indeed, the broad nature of the 3P scheme means that it will probably be of most use in the interdisciplinary context of sustainability science rather than, perhaps, governance studies or political science, where these concepts may be taken for granted. In developing our 3P lens, we draw on theoretical approaches in deliberative and decentered governance. Deliberative governance emerged as a way of capturing the shift from ideas of *government* to *governance* in the 1990s and the accompanying focus on deliberation through distributed decision-making, citizen participation, informal social networks, and cross-scale connections between agencies and organizations (Hajer & Wagenaar, 2003). Researchers have recently articulated decentered governance as a more critical perspective on these shifts, noting that ideas of 'governance' have become a new orthodoxy and emphasizing more radical democratic possibilities (Griggs, Norval, & Wagenaar, 2014). What both share, and what distinguishes theories of deliberative and decentered governance from adaptive governance (apart from the focus in AG on complex social-ecological change), is their rootedness in interpretive theories that situate accounts of governance in terms of the production of meaning through, for example, experience, discourse, and practice.

By prioritizing "people" in our 3P scheme, we wish to highlight that the rather abstract-seeming qualities of adaptive governance—including scientific monitoring, information-sharing, and network building—are enacted by people, with their particular (and unique) capabilities, experiences, emotions, hopes, and desires. In more theoretical terms, here we aim to draw out the *agential* nature of governance work, which entails the creative, contingent construction of meaning by those involved (Griggs et al., 2014; Westley et al., 2013). With the term *practice* we seek to emphasize the active nature of "doing" adaptive governance and draw attention to the way that adaptive governance emerges through acting on the situation at hand, which involves physical engagement with, for instance, tools and artefacts, material environments, and colleagues. More theoretically, we seek to make clear that people involved in adaptive governance are not free to simply construct their own meanings, but that these meanings are enabled and constrained within collective fields of activity including, for example, organizational routines, policies and imperatives, social habits, and technologies (Hajer & Wagenaar, 2003). Finally, through the emphasis on *politics* we aim to nurture greater sensitivity to the plurality of beliefs, allegiances and values that, as Griggs et al., (2014, p. 9) put it, actors bring to "the spaces in which collective problem solving takes place." This means that enactments of adaptive governance are always political, in the sense that they involve the articulation of, and deliberation between, different interests, identities, intentions, and visions. This first attempt at articulating a 3P lens inevitably requires more work and we invite others to build on and critique this approach. In the rest of this chapter, we develop this broad lens though a case study of the K2C Biosphere Region in South Africa.

Methods

Previous researchers have suggested that UNESCO Biosphere Reserves (BRs) are particularly well placed to play a bridging role within adaptive governance (Hahn et al., 2006; Schultz et al., 2018). We identified the K2C as a particularly interesting site in this regard because participants are actively working with concepts of adaptive management and governance as means to address pressing sustainability issues in a highly contested social-ecological context. Within the K2C we focused particularly on the implementation of the Environmental Monitors (EM) program. The EM program is a national governmental initiative that seeks to enhance governance responses to rhino poaching and also address high unemployment rates around Protected Areas (PAs). To enact the EM program the K2C has employed local people within *Host Institutions* (HIs) that are part of the K2C partnership network—including local government, Non-Governmental Organizations (NGOs), research organizations, and private game reserves—to conduct environmental monitoring, protected area patrols, environmental education, and generally help HIs to fulfill their organizational mandates. To develop our 3P lens, we adopted a qualitative, broadly interpretive research approach. Interpretive approaches are considered particularly appropriate for exploring the experiences of people in enacting policies, projects, and programs, and how these experiences shape their practices.

In total, we conducted 40 semi-structured qualitative interviews. Twenty-four of these were conducted in 2013 (13 with EM participants and 11 with participants in the broader K2C stakeholder network), and 16 were conducted as follow-up interviews in 2015/2016. This longitudinal approach enabled us to capture experiences from the first 3-year phase of the project. We selected respondents to ensure coverage of the range of HIs participating, as well as the range of roles within the project, including HI managers, data collators, and the EMs themselves, and transcribed the interviews verbatim. Although we did not develop a formal thematic analysis for this chapter, we adopted a broadly thematic approach to derive insights from our interview data, employing primarily deductive strategies (Braun & Clarke, 2006). Firstly, we focused on the aspects of the EM project most relevant to features of adaptive governance—knowledge generation, information-sharing, networking and collaborating, and responding to change. Secondly, we employed our 3P lens to enhance our sensitivity to particular aspects of our interviewees' accounts of their work relating to these features, including experiences, emotions, and meanings (people); routines, technologies and patterns of activity (practices); and interests, visions and values (politics). To reflect the holistic nature of interviewee experiences, we have chosen to interweave people, practices, and politics throughout our presentation of the case material, rather than address them separately for each adaptive governance feature. The 3P lens therefore flows through our analysis, rather than being rigidly applied in each section.

The Case: Towards Adaptive Governance in the K2C Region

The Central Lowveld and Escarpment region lies in South Africa's northeastern corner, stretching from the savannah ecosystems of the iconic Kruger National Park (KNP) in the East to the afro-montane forest of the Blyde River Canyon and Drakensberg Escarpment in the West. The landscape is strikingly demarcated along biophysical, jurisdictional, socio-economic, and ethnic lines. The KNP and the many private nature reserves that border it form a large network of PAs—initially created by white Afrikaans and British settlers in the nineteenth and twentieth centuries by forcibly evicting black African communities—that cater to a wealthy national and international tourist market (Carruthers, 1995). During the apartheid era (1948–1994), black communities of various tribal and geographic origins were forced into "homelands" bordering the PA network, which effectively functioned as ghettoes and suffered from chronic state neglect, lack of economic opportunity, and high levels of poverty (Pollard, Shackleton, & Carruthers, 2003). The homelands were abolished with the advent of democracy in 1994, but their legacy remains imprinted on the landscape. The population in the former homeland areas in the region is circa 1.5 million and rising, with a density sometimes exceeding 300 people per km^2 (Pollard et al., 2003; Pool-Stanvliet, 2013) and a landscape consisting of settlements (including the towns of Bushbuckridge and Acornhoek) set amongst communal rangelands (Coetzer, Erasmus, Witkowski, & Reyers, 2013). The broader region also contains wealthier settlements, such as Hoedspruit, that cater to the PA-related tourist industry, as well as commercial agriculture, mining, and—towards the Drakensberg escarpment—areas of plantation forestry (Coetzer et al., 2013).

The democratic transition in the early 1990s prompted a profound shift in the governance of the region. Single-party rule and the homelands system made way for a new system of provinces and municipalities, and dominant political rhetoric shifted towards reconciliation, equity, and broad-based economic empowerment

(Ramutsindela & Simon, 1999). Reflecting international shifts in environmental policy, the natural resources sector adopted a suite of policies and legislation emphasizing integrated catchment and ecosystem-level management, cooperative governance, and equitable distribution of resources (Colvin et al., 2008). At the same time, the KNP began to recognize the importance of the wider landscape to the integrity and sustainability of park ecosystems, and shifted from an "inward-looking, isolationist" management approach to a complexity-oriented, social-ecological perspective emphasizing learning and collaboration between the broad range of stakeholders in the region (Pollard, du Toit, & Biggs, 2011; Venter, Naiman, Biggs, & Pienaar, 2008). In practical terms, this shift in philosophy has led to the dropping of fences between the KNP and the neighboring private nature reserves, and to the creation of a number of social projects that seek to engage the former homelands communities.

In this context, various conservation and development actors began to support the idea of a Biosphere Reserve (BR) in the region as a means for, as one respondent put it, "different communities to reach out over borders," and ensure a future for biodiversity conservation in the region while also making sure that conservation contributed to equitable and sustainable socio-economic development. BRs are intended to function as *learning sites for sustainable development*, with three types of zoning that correspond with three thematic functions: *core zones* that emphasize nature conservation, *buffer zones* of limited human use that support scientific research, monitoring and education, and *transition areas* with larger human populations that foster sustainable development (Fig. 13.1; UNESCO, 1996). Designated in 2001, the K2C Biosphere Region—named *region* to avoid the *guns and guards* connotation of *reserves*—bridges the provinces of Limpopo and Mpumulanga, and spans two district municipalities and four local municipalities (covering approx. 2.6 million hectares in total). In the K2C, the BR zones were not applied in the stylized sense of model BRs but according to the existing landscape mosaic (Coetzer et al., 2013). The KNP and the Blyde River Nature Reserves form the core areas, the private nature reserves the buffer zones, and the areas of rangeland, settlements, agriculture, and mining the transition zones (Fig. 13.2).

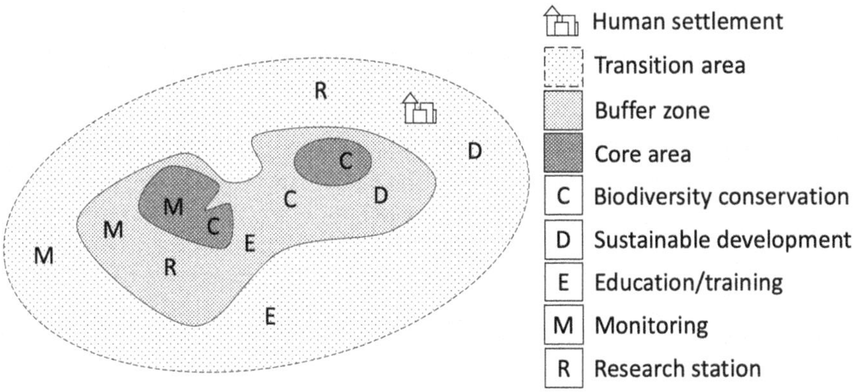

Fig. 13.1 A stylized representation of the Biosphere Reserve zonation. Source: Design by authors

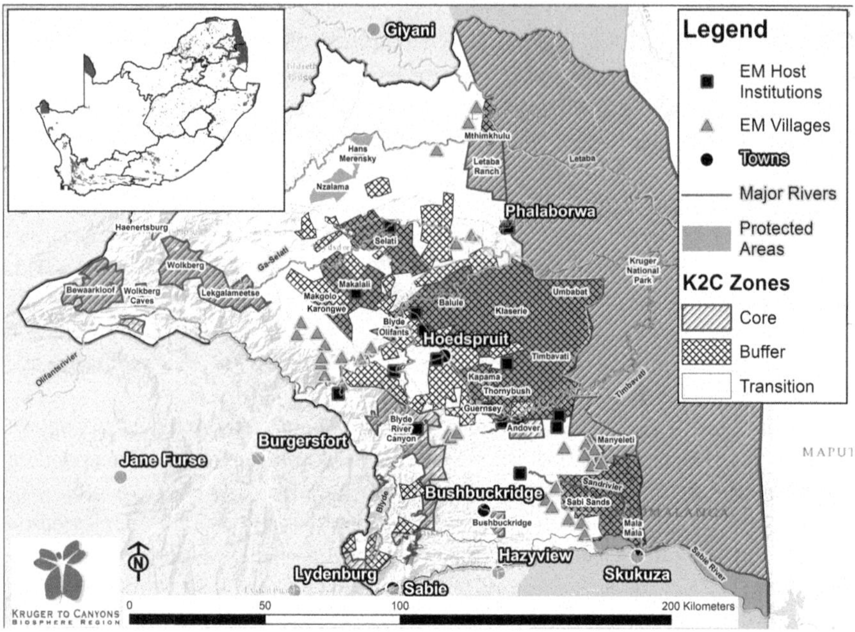

Fig. 13.2 The BR zonation as applied to the K2C. Copyright 2018 by Wehncke van der Merwe. Reprinted with permission

Although the K2C represents geographic space, it is also a non-profit company tasked with pursuing actions to fulfill the BR mandate. As an organization, the K2C's attempts to nurture partnerships and collaboration, and implement and support projects, with a view to reconciling biodiversity conservation and sustainable development. In particular, the K2C seeks to address a range of interlinked sustainability challenges in the region, including lack of economic opportunity, severe inequality, poaching (especially of rhino), high levels of poverty and HIV/AIDS, droughts, habitat degradation (Coetzee, Biggs, & Malan, 2012), and an influx of refugees (Coetzer, Erasmus, Witkowski, & Bachoo, 2010). However, as with the BR network more generally, the K2C has—until recently—received relatively little legislative or financial backing. Interviewees recalled that in the first decade of operation, the K2C largely consisted of several committed individuals attempting to attract funding and generate activity to fulfill the designation. In the late 2000s, however, the so-called 'Anyway Group', a "long-standing informal stakeholder network" of conservation, government, and community actors (many of whom were influential in the original biosphere designation process), re-mobilized under the auspices of the K2C (Coetzee et al., 2012, p. 4). Their intention was to simultaneously ensure the alignment of a number of incoming sustainability initiatives to the region—including USAID's RESILIM program, and the national government's EM program and *Wildlife Economy* concept—while also attracting further international funding through the Global Environment Facility (GEC). In the words of one interviewee, the K2C label was used to "[form] networks in order to obtain collective outcomes." This led in turn to the reinvigoration and formalization of the K2C stakeholder network involving many government, NGO, business and community actors in the region, the creation of forums convening actors under the themes of *Environmental Education*, *Natural Resource Management Projects*, *Lowveld Protected Areas*, and *GEF Small Grants* (GEF, Global Environment Facility), and the inauguration of the *K2C Network Coordination Unit* intended to align all of these initiatives. This institutional work has been heavily informed by complex social-ecological systems thinking, including strategic adaptive management and adaptive governance (Coetzee et al., 2012).

This brief overview, although inevitably simplified and incomplete, suggests that the K2C is becoming an increasingly important bridging actor and governance hub in the region. Although we do not aim to evaluate in this chapter whether adaptive governance *is* or *is not* happening in the K2C region, the interplay between formal and informal institutions—see, for instance, Volume 13 on institutions in this series (Glückler, Suddaby, & Lenz, 2018), emphasis on learning and collaboration, and response to complex social-ecological issues—certainly reflect key adaptive governance principles and create fertile ground for more adaptive forms of governance to emerge. In the remainder of this chapter, we "zoom in" to explore how one aspect of the K2C nexus of activity, the EM program, may be supporting key aspects of adaptive governance in the region, and highlight the challenges faced by participants in their everyday work enacting the program.

Results: Nurturing Adaptive Governance Through the Environmental Monitors Program

The South African Department of Environmental Affairs (DEA) designed the EM program in response to two crises: an explosion in rhino poaching from 2010 onwards, and a widespread lack of employment opportunities in areas adjacent to PAs (DEA, 2014). The program, initiated in 2012, is part of the Expanded Public Works Programme's (EPWP) suite of projects, including Working for Water, Working for Wetlands, and Working for the Coast, where the aim is to alleviate poverty by providing temporary work and skills development for the unemployed. The specific aim of the EM program is to enhance the integrity of PAs by combatting poaching through monitoring programs, patrols, and environmental education (DEA, 2014). The Biodiversity Social Projects (BSP) wing of South African National Parks (SANParks) was designated as the implementing authority, and the initial intention was to roll out the program within national parks and private nature reserves. The Kruger region inevitably became a primary focus, with the KNP being the most concentrated site of rhino poaching in the world (Lunstrum, 2014), and the K2C—despite, as a sustainability organization outside the PA network, constituting something of an anomaly in the program as a whole—was consequently selected to participate. The implementation manager at the time recalled that because of

> … the strategic placement of the K2C, and the upcoming GEF funding, we zoomed in on the K2C itself and then decided we wanted to—because it is such a huge knowledge hub, and there is so much pressure on this whole Bushbuckridge area as well, we decided we [would] establish a program there.

In turn, the coordinator of the K2C saw the EM program as a way to strengthen the recently re-mobilized K2C stakeholder network, and to fulfill the K2C goal of reconciling biodiversity conservation with socio-economic development:

> [T]he catalyst for the Anyway Group*, or the groupings of the K2C Biosphere network, was about preparing for GEF [funding], but it is also a method of linking to one another and [creating] the synergy where it is more effective than the sum of its parts…. The EM program was the practical way for us to do both these things: the network and to empower people.

Indeed, in its enactment through partnerships and networks of diverse actors, the EM program at the K2C has taken on much more significance—for adaptive governance in particular—than EPWP-type projects in general and has become a nationally recognized "flagship" site in the EM program as a whole.

The K2C EM program is managed by the K2C coordinator from the K2C offices in Hoedspruit, supported by four "data collators" responsible for training and supporting EMs, collating the data produced and conducting administrational tasks. The EMs themselves are placed within Host Institutes (HIs) participating in the K2C stakeholder network, including public and private nature reserves, NGOs, community-based organizations, and research organizations. The Maeba Group, a leadership group of 12 EMs and the data collectors, functions as a link between the data collators and the wider EM group, and devises a monthly learning framework for all EMs (consisting of Health and Safety, Vocational, and Life Skill themes). At the end of the program's first stage in March 2016, the K2C was employing 265 people, which constitutes 19.5% of those employed in the EM program across South Africa (SANParks BSP & K2C, 2016). The program has subsequently been extended for another 3 years. The K2C EM program has exceeded participant expectations of what can be achieved through public works style programs, despite—or perhaps because of—the fact that its parameters were not strictly delineated beforehand. As a SANParks-BSP manager notes,

> I think that probably the biggest challenge was when we started [in 2012] there wasn't really any clear direction from [the DEA] as to what they wanted to do with the program. For example, the concept document around the program was only developed in 2014. And that was a year later, so that was a struggle to clearly outline how the program should be implemented.

Indeed, the precise form that the EM program has taken at the K2C—and its contribution to adaptive governance in the region—has emerged largely through day-to-day negotiation between the actors involved. In the following sections, we employ our 3P lens to explore how the challenges of enacting the EM program at the K2C can shed light on key aspects of adaptive governance: generating knowledge, sharing information, networking and collaborating, and responding to change (see Table 13.1).

Table 13.1 Using the 3P lens to explore aspects of adaptive governance supported by the Environmental Monitors program at the K2C

	Generating knowledge	Sharing information	Networking and collaborating	Responding to change
People Experiences, meanings, capabilities, hopes, desires	Enthusiasm and willingness to learn new techniques Boredom in the field	Mutual respect, trust and friendship important for information sharing Creation of communication channels Fear and reluctance to share data	Weariness of meetings, meetings 'such a ball' Celebrations to maintain morale	Self-confidence to facilitate discussions, suggest ideas, pursue self-defined goals Belief in ideals of programme
Practices Routines, patterns of activity and communication, technologies	Collecting data on e.g. sand mining, invasive alien species, water quality, animal species (e.g. cattle and game counts), family planning and firewood collection Standardizing data, maintaining flexibility Patrolling Training to use technical equipment, learn scientific techniques Synchronizing data collection in smaller collaborations	Preparation and 'cleaning' of data Creation of common database Spontaneous sharing of data Maintaining the integrity of data to avoid inappropriate use	K2C facilitating discussions between different actors Administration routines (time sheets, contracts, application forms)	Shifting work patterns of HIs (capacity freed up among permanent staff to address emerging issues) Revised management strategies of HIs (invasive species) Shifting patterns of employment and hiring (e.g. treating all employees professionally), promoting capacity building
Politics Plural interests, values, imperatives, identities and visions	Different expectations among HIs about how to use EMs (common data collection vs other tasks specific to the HIs) Different research interests among scientific HIs	Contrasting visions of a 'successful' project (e.g. shaping ideas about database) Various perspectives on appropriate land management (limiting willingness to share information)	Worries that government might use strong networks to roll out projects without consultation Fears of government relying on partners to do their work for them	Different imperatives and interests (person-days worked, data collection, skills development) K2C as more prominent political 'actor' in the region

Note. Abbreviations in Table 13.1: *K2C* Kruger to Canyons Biosphere Region, *HI* Host institution, *EM* Environmental Monitor
Source: Design by authors

Generating Knowledge

In the early days of setting up the EM program, managers from the HIs met several times to sketch out its parameters. A key concern was what the EMs would be required to do—particularly in terms of data collection. The more scientifically-oriented HIs suggested that there should be a core set of indicators that all EMs should monitor, whereas others—particularly community-based organizations—argued for more flexibility. A scientific HI representative remembers, "there were some that felt that they must contribute to a broader program, and others that said, 'no we are employing them and we can use them for what we like.'" In the end, flexibility won out. The DEA assesses SANParks' implementation of the program primarily on person-days worked, so SANParks did not have a strong incentive to ensure common data collection requirements in the program. This left it up to the scientific HIs to advocate for and nurture practices of data standardization in the program, and they have not done so systematically given their limited human and financial resources. Currently, some EMs do not collect data at all, but carry out various tasks related to the mandate of the HI such as environmental and health education and patrols (this flexibility is considered a strength of the EM program by the community-based organizations). Those that collect data do so under five themes: PAs; rangelands; fresh water (Fig. 13.3); health; and socio-economic trends in rural communities. There are no standardized indicators or variables within or across the entirety of these five themes. Rather, the EMs collect data according to the interests and needs of their HIs—for instance, the locations of snares and holes

Fig. 13.3 Monitoring rainfall. Source: Photography by Cláudia Florêncio

in the fences of game reserves—and on emerging issues or perceived potential threats in the region (e.g. sand mining, invasive alien species, family planning and firewood collection). Nevertheless, pockets of collaboration in data collection have emerged informally. For instance, one research-based HI is collaborating with several game reserves to synchronize variables monitored. One researcher noted that such collaboration was—perhaps counter-intuitively—easier with non-research-based organizations:

> ... the most collaborative ones have been the least scientific ones! The guys running nature reserves, they are quite open to us coming with a predefined list of [variables and] data collection methods. But the [other research organizations] have their own research ideas and so they are not so open to some other researcher coming in.

Ensuring the standards of the data collected has also been a key concern for the research-based HIs. As the EM program is primarily a poverty relief program, ensuring data standards is directly connected with the skills development component. The hired EMs have had relatively little formal education (they are not trained scientists) and may not necessarily have been particularly interested in environmental data collection prior to employment in the program. Training is therefore essential. This has been relatively extensive, with training conducted by the HIs and by partners in the network—for instance in the identification of species (Fig. 13.4) and the use of new monitoring technology such as CyberTracker software. Nevertheless, the EM program does not expressly provide funding for training and equipment. Although in theory money for training could be included in the budget, this would mean reducing funding to actually employ EMs (and as a poverty relief program,

Fig. 13.4 Learning to identify species at a scientific HI. Source: Photography by Cláudia Florêncio

employment is inevitably the priority). This means that training is unevenly distributed among the HIs. As one scientific HI participant noted, “I think that’s where the whole idea has slacked a little bit—there’s some strong host institutions and then there’s some slack ones, and it’s not necessarily the institution’s fault, but it’s more the resources.” The use of CyberTracker technology is accordingly limited, with clipboards also used (partly due to skepticism of the technology by some HIs), and there is no funding available to employ a data manager to train EMs and “clean” the data collected.

Almost all HIs expressed regret that they could not spend more time training and supervising their EMs during their long hours in the field. The lack of supervision and low pay of some of the EMs has posed challenges of motivation and discipline, with HIs reporting that the boredom some EMs experience sometimes prompts them to “spend a lot of time on Facebook.” For several of the data collators, this is a problem because it “demotivates other EMs” who are working hard. Nevertheless, despite these challenges, a significant amount of data has been collected, and many HIs expressed astonishment at the EMs’ commitment and willingness to learn, and mastery of scientific data collection in a relatively short time span. Overall then, data collection is not the only objective of the program, which also includes patrols, environmental and health education, and generally helping the HIs to fulfill their mandates—under the overarching goals of job creation and skills development. Therefore, the type and quality of the data produced is affected by the other imperatives operating in the program, and the practical trade-offs imposed by time and resources. Although these are familiar issues for practitioners working in such programs, these issues have yet to be worked into the adaptive governance literature, and used to develop theoretical and practical understanding of what it takes to nurture adaptive governance and what it might achieve.

Information-Sharing

The intricacies of sharing information have been a recurring theme in the EM program. As one participant notes,

> The biggest challenge is, we have got all this data, we need to [collate] all this data, we need to utilize this data, and it has to be distributed. So we have had it on the agenda for quite some time, [and] we actually need to find a proper way forward.

Throughout the lifespan of the project so far, scientific participants in particular have raised the idea of constructing a common database to share data and information between participating HIs (and amongst the K2C stakeholder network more broadly) with the aim of coordinating activities and making data available for broader use and analysis. However, this has taken several years to materialize, with work starting on the database in 2017. In the meantime, data sharing has been ad hoc: “No formal sharing is happening. It is more spontaneous, like ‘by the way, do you have this?’ And then the person shares it or not.” Participants offered different and sometimes contradictory explanations for the challenges in constructing a database, which speak to broader tensions about the utility of knowledge for governance and the practicalities and politics of "real-world" adaptive governance.

Participating HIs have been reluctant to, in the words of one participant, "just start handing the data out," for various reasons. Scientific institutions have been wary of relinquishing intellectual property rights: "many of them are like, 'you are going to collect this data and then you will go and publish it and we will get nothing.'" Meanwhile, representatives of private game reserves are wary of data on, for example, rhino movements falling into the hands of poachers, or worried that data will be used to critically scrutinize their land management practices. Moreover, the sharing of information by participating health institutions is regulated by legal frameworks safeguarding patient confidentiality. There are also a number of practical issues in constructing a database. There are very different kinds of data being collected by the different HIs, on different topics and variables, with different protocols, and of varying quality. There would therefore be a significant amount of work needed to prepare and "clean" data, as well as providing detailed information on the methods and assumptions used to collect each type of data:

> ...because just columns of numbers are pretty useless ... if people don't understand what those variables mean, how they were collected, what the limitations of the data are ... If people don't have a deep understanding of the actual data, the quality of the data, [then] the data can be used in ways that are inappropriate.

There is no money provided within the EM program to employ a trained consultant, or train the data collators or EMs to conduct these technical and time-intensive tasks.

These complex considerations have led to a range of perspectives on the issue. For some, a database is unnecessary, and peripheral to the overall value of the EM program. As one participant argues, "you just can't really collect it in one interface. Maybe for me it's about respect you know. So if somebody wants these data that you are collecting now, then they need to go to you."

Some participants add that the degree of collaboration renders the database pointless: "... we know where [the data] is, and we can ask for it [if we need it]." Indeed, there has been informal sharing throughout the first program cycle, and as another participant describes, "99.99% of the time they are willing to share, no problem. Because you are friends with them, and you know each other professionally from the area." Nevertheless, for others, the idea of a shared database of some kind is an important criterion for the program's worth and success—to ensure that partners coordinate actions, avoid duplication of work, and maximize the use and value of the data collected. Indeed, one actor's description of current practice serves as an illustration of why some participants continue to advocate for the database:

> So all of that data is basically sitting on my computer at the moment, it's not on a common server. I think I emailed some of the data, just the spreadsheets to K2C, for the data collators, but I don't know what they have done with it.

Through the initiative of one HI that has sourced additional external funding, work started in 2017 on a "meta-database" with a provisional understanding that it will identify what kind of data is being collected and stored by whom, rather than providing immediate access to the data itself. Whatever the shape and form of the database that eventually emerges, it will be conditioned by all of these considerations, and likely coexist with all sorts of informal and issue-driven data sharing. All this is to show that the precise forms of AG that emerge in particular places arise through the intricate, often messy negotiation of people, practices, and politics.

Networking and Collaborating

Although formal mechanisms of information sharing have been slow to materialize, the EM program has significantly strengthened networking and collaboration among the K2C stakeholder network in a more general sense. The multi-sectoral partnership approach to program implementation adopted by the K2C was novel for national SANParks BSP staff with long experience of working with EPWP programs: "[W]hen we started the Working for Water program, networking was important, but it was different government departments that needed to be involved. It wasn't like a partnership program which the K2C is very effective in." This has resulted in new connections established between actors across scales, with one national actor noting that, "I would never have spoken to Balule or Timbavati [private nature reserves] before, but now I do and because of the EM program we created a platform." The program has also stimulated new connections across domains. In particular, participants from private game reserves reported that they had established new lines of communication and trust building with communities "beyond the fences" by employing local people as EMs, and by collaborating with community-based HIs in the program. Many of these connections have been established through everyday work on the program, and through informal and chance meetings—although the data collators and the Maeba group hold monthly meetings to share experiences, there are no formal meetings between HI representatives. This lack of formal meetings among HI representatives was experienced by participants both as a positive and a negative. Although many HIs expressed a desire to initiate such meetings—particularly in order to, for example, share experiences and coordinate and standardize the data collection (see above)—they were also aware that they were already overstretched with attending meetings across a huge geographic region. As one interviewee observed, meetings are "such a ball." Indeed, K2C staff have explicitly tried to avoid creating meeting fatigue, and participants often take the chance to discuss EM-related issues at other K2C events. There is also an annual "EM day" that functions as a celebration and an update on work within the EM program, and a "K2C Partners' Day" that situates the EM program within the K2C network more broadly (Fig. 13.5).

Many participants perceived the distinct role and broad mandate of the K2C as essential to fostering engagement and collaboration in the EM program. Indeed, the K2C's "intermediary" status, linking the KNP with the broader landscape, enabled

Fig. 13.5 Participants at the K2C Partners' Day. Source: Photography by Cláudia Florêncio

K2C representatives to capture the funding for the EM program in the first place, and through its broad mandate the K2C has the ability to connect very different organizations. As one participant notes,

> … the K2C is like a net thrown on the entire area … they go from helping local communities with the tribal authorities, to the traditional healers, to the schools, to waste management, to the game reserves, to education, health. I mean it is a very broad spectrum and that is what makes it so good.

Although the K2C has had this mandate for some time, the EM program has provided a tangible, concrete initiative to incentivize collaboration. In this sense, the K2C coordinator notes that the program has "galvanized networking because there was a central pivot point … with this common program." In particular, by taking responsibility for the administration of the program, the K2C substantially lowered transaction costs for organizations to participate. As one HI participant explains,

> … the time sheets and all of that, you know? The fact is that if I had to do that, I wouldn't have taken on EMs because I don't have an administrator. It [would be] taking time away from writing papers, doing funding grants, writing reports. If I had to do all of that admin, there is no way I could have taken on the EMs.

Nevertheless, participants also situated their involvement in the program within a broader political context. Some interviewees raised the concern that the strong networks in the K2C region meant that it was increasingly seen as an attractive place for the national government to roll out projects, without necessarily undergoing the degree of public scrutiny that would happen elsewhere. Others noted that the emergence of collaboration in the region was occurring in the context of weak government presence and capacity (that they were "stepping in" to fulfill roles that could potentially be done by government).

Responding to Change

Revision of management actions and strategies

The use of monitoring data to revise and inform management actions and governance strategies is a key element of adaptive governance. In some cases of adaptive governance, members of a central node collect and analyze data, and then use the data to inform formal decision-making at a broad regional or (inter-)national scale (e.g., Österblom & Folke, 2013). This is not, at the moment, the case within the South African EM program as a whole. The national manager explains that

> … from an implementation perspective [SANParks is] more interested just in, 'are people being employed?' 'what are they doing?' and 'what is the ratio?' because these are our objectives and deliverables linked to the program. The K2C would be more interested in the data.

Although the K2C functions as a central node in the K2C EM program, it does not yet have the capacity to process and analyze the data collected across all participating HIs (see discussion above), or the mandate to use the data to inform or revise formal decision-making in local, regional, or national government. Rather, the data use in the K2C network has so far been uneven and distributed, with the HIs using the data in different ways. In the private nature reserves, the EMs have collected data on snares and holes in the game fences, which have been used to inform anti-poaching strategies, as well as participating in regular monitoring activities such as game counts and vegetation surveys, which inform ongoing management activities. EMs have also collected data on invasive species in the K2C region, which has been used to inform the Working for Water "bush clearing" teams:

> So if it is for the bush clearing team, as soon as we collect the data, we give it to the people in charge of the bush clearing and then they will have to see what action needs to be planned, if the place is big we will need a month to go there, this is the type of herbicides that we will need, we need how many workers, how many days…

In addition, data collected by EMs on soil erosion spurred the creation of a Rehabilitation Project to address the problem. The data collected on socio-economic trends in local communities have not yet been used to inform management activities but have provided a better understanding of potential emerging issues: "[F]rom what we collected we learned a lot from it and now we have an understanding of what is going on in our villages, what are the social problems or the economic problems in our villages." Nevertheless, the lack of capacity, time, and resources to analyze much of the data collected has proved a persistent challenge. As one research-based HI explains, some data collected by EMs has been used in Masters projects, but "to be honest, some of the other data hasn't necessarily been kind of used or analyzed."

Fulfilling organizational mandates

The EM program has provided HIs with the resources to better achieve their individual mandates. For instance, participating environmental education organizations noted that the manpower provided by the EMs enabled them to reach a much bigger audience:

> … nine schools is huge learning that would not have taken place without [the EM program]. I would still be going to one school once every two weeks… [Our] impact would be a couple of hundred kids rather than 1200 kids.

This increased organizational capacity has enabled HIs to better respond to emerging threats—not only through the EMs themselves, but also by allocating EMs to tasks that will "free up" other staff. For instance, the recent explosion in rhino poaching prompted many national parks and game reserves to redirect resources from traditional conservation management activities to anti-poaching. Although the EMs do conduct some anti-poaching activities (and indeed there is an armed EM unit), they have also provided the KNP and some private nature reserves with the manpower to continue to perform their more traditional land management activities. Likewise, research organizations have used EMs to collect data and support research into emerging environmental threats: "We are using the EMs to collect data for certain research questions that otherwise we wouldn't have the resources to collect the data ourselves." Perhaps most strikingly, the EM program has enabled the K2C itself to fulfill its own mandate of pursuing activities that reconcile conservation and development through partnerships. Many interviewees reported learning about the K2C—and appreciating its value—for the first time through the EM program: "[F]or me, the environmental monitors program has actually opened my eyes [to the fact that there] is a biosphere reserve, basically." Indeed, interviewees felt that the EM program had enhanced the K2C's reputation as a trustworthy "broker" linking different actors in the region.

Human development and capacity building

Participants reported that the EM program has been particularly successful in developing the skills and enhancing the capacity of those participating (Fig. 13.6). Many EMs reported having developed self-confidence through their work on the program:

> … the confidence to stand in front of people and talk, I think that is the biggest impact. Because before I didn't have that confidence to stand, I was shaking, nervous. But now yeah, I think I have more self-confidence to stand in front of people and facilitate talks.

Fig. 13.6 Collecting data on cattle health in the rangelands. Source: Photography by Cláudia Florêncio

Skills development is a stated aim of all EPWP projects, but they have also been frequently criticized for their limited success in this regard (e.g., Mccord, 2005). Indeed, it is arguably the enactment of the EM program through the networks (and with the coordination of) the K2C that has strengthened the skills development and training component in this case. This has occurred through, for example, mobilizing the network to provide training to EMs, exchange visits to other EM sites, and initiatives like the Maeba Group, where particular committed EMs gather to pursue self-defined learning goals:

> … the people that were actually selected to be Maebas because of their potential, their willingness, and their commitment to the program. So they were chosen to be Maebas to supervise and support the EMs and train them and build capacity among the EMs.

Nevertheless, many respondents expressed concern that the EM program did not offer permanent positions, or sufficient opportunities for job progression within the program—both wider problems characteristic of EPWP programs in general (McConnachie, Cowling, Shackleton, & Knight, 2013). However, the training, skills, and experience acquired had enabled some EMs to find better job opportunities outside the program. The K2C EM program has also built capacity among HIs, especially the community-based organizations with little previous experience of participating in government-run programs:

> … out here in the rural areas previously if you needed someone to work, you drove past, you picked them up and off you went. This project has made the Host Institutions realize that you have to, regardless of the status of the job and the salary, you need to treat that person as an employee that you would treat like any other employee.

Although human development and capacity-building are not necessarily classic aspects of adaptive governance as referred to in the literature so far, they appear absolutely essential for scientific forms of adaptive governance to emerge at a grassroots level in the context of a developing country. The data gathered may not necessarily always be of the highest scientific quality, but this may not be the main point of the exercise. Rather, the process of gathering the data is immensely valuable in itself, as participants learn to navigate the people, practices, and politics that the pursuit of sustainability entails.

Concluding Discussion

Early articulations of adaptive governance emerged through researchers' attempts to describe what they were observing in the evolving management and governance of complex sustainability issues. Nowadays, adaptive governance, and related ideas of complex social-ecological systems, have entered into the conceptual vocabulary of governance practitioners and are being actively used in attempts to mold and shape governance practices. In this paper, we have presented an empirical case study of the K2C Biosphere Region, where ideas of strategic adaptive management and adaptive and multi-level governance are being actively used to shape governance and program implementation. In an attempt to ground our analysis in the everyday experiential environment of practitioners, we have developed an analytical lens based around 3Ps—people, practices, and politics—and used it to explore the ongoing implementation of the EM program. This has enabled us to portray adaptive governance as it is "in the making" (rather than a retrospective account of a particular outcome), thus highlighting the everyday decisions and choices that bring life to abstract ideas such as "networking" and "information-sharing." This shift in perspective enables us to provide a number of useful contributions to the adaptive governance literature.

First, our study supports previous findings of the importance of bridging organizations and their networks to initiate and sustain adaptive governance over time, while adding details from real-world accounts. Importantly, the experience of the K2C indicates the need for bridging organizations to provide a specific reason or incentive for other actors to engage with them (thus enabling them to "bridge"). In the K2C's case, the implementation manager at the BSP wing of SANParks and the K2C coordinator identified a window of opportunity to situate EMs within the K2C stakeholder network. The flexible, partnership-based approach taken by the K2C in enacting an initially rather traditional "public works" style program constitutes an innovative approach to nurturing more adaptive and collaborative governance in the region. Through the EM program, the K2C has initiated connections and facilitated discussions between many actors who had not previously worked together, and has nurtured these connections by, for example, attempting to avoid meeting fatigue and organizing celebrations to motivate engagement. Our study also reveals the centrality and importance of everyday administrative tasks to successful "bridging," including, in the EM program's case, taking responsibility for time-sheets, employment contracts, and so on across many different kinds of Host Institution.

Second, our findings problematize and add detail to the idea that adaptive governance is formed by coordinating actors around a shared vision, prompted by a perception of crisis. The creation of the EM program as a whole was prompted by a perception within the conservation sector and the national government of rhino poaching as a "crisis." However, although rhino poaching is certainly a major issue within the K2C region and the K2C BR stakeholder network, it is only one of a range of pressing, interlinked sustainability issues. Indeed, the EM program at the K2C was conceived as a means of addressing a much broader suite of issues, captured under the broad K2C vision of reconciling biodiversity conservation and sustainable development through partnerships. And although the vast majority of participating HIs in the EM program would agree with this broad vision, our results highlight the innumerable perspectives among participants on what this vision means for their work, how they might help to fulfill that vision, and what it is possible and/or desirable to do given practical possibilities and constraints. Within attempts at coordination around a broadly shared vision, then, our study highlights the many different kinds of negotiation around particulars. For instance, activity within the program was shaped by negotiation between different versions of success, including person-days worked (SANParks), scientific data and a shared database (some scientific HIs), and training, capacity-building, and skills development (K2C). More generally, our results highlight the interplay of meaning and action—within innumerable everyday decisions—that shape the way adaptive governance emerges in practice.

Third, by focusing on adaptive governance in the making we emphasize that adaptive governance does not simply emerge in a particular place and replace all other kinds of governance, but rather that adaptive practices may evolve within and infuse a governance landscape that carries many influences. Moreover, it is not necessarily immediately apparent what practices are useful and will enhance adaptiveness in a particular context—indeed, "adaptive" is not a universal criterion, certain practices are more or less adaptive *in relation* to existing arrangements. A good example of this lies in the discussions around the proposed common database within the EM program. Although a shared database to synthesize information and use it to inform relevant decision-making would perhaps be considered more quintessentially adaptive in the literature, not all participants in the EM program necessarily saw it as desirable. Even if the resources were there to create and maintain a database, the K2C does not have formal decision-making authority in the sense of a government authority to act upon the data gathered. In this context, the networking and collaboration between the diverse set of HIs in the region—initiated by the K2C—might be considered even more valuable, by nurturing a web of formal and informal information sharing and enhancing "adaptiveness" more generally across the landscape.

Finally, our study provides a valuable complement to an adaptive governance literature weighted to "high-level" governance contexts in the Global North, including multinational corporations, international institutions, and so on (Galaz, 2014; Österblom, Jouffray, Folke, & Rockström, 2017). In focusing on how adaptive governance might be nurtured and strengthened through a poverty alleviation program in the Global South, we suggest that the developing the capacity and skills to perform the "knowledge work" required for adaptive governance (gathering, synthesizing, and using information) may potentially be an appropriate development strategy in certain contexts. Indeed, interviewees noted that human development and capacity-building have been key outcomes of the program.

To conclude, adaptive governance is no longer just a descriptive academic concept, but also a normative goal and a lived experience of an increasing number of people across the globe. Continuously bringing this lived experience into the evolution of the concept is crucial for making ideas of adaptive governance meaningful and useful, for practitioners and academics alike.

Acknowledgments We would like to thank all participants of the study for sharing their insights with us, and the K2C Biosphere Region for making this research possible. We would also like to thank the Knowledge for Governance symposium for the opportunity to share the insights in this chapter with a broader audience. This chapter is an output of the GLEAN project (A Global Survey of Learning, Participation and Ecosystem Management). The GLEAN project was funded by Vetenskapsrådet (Grant No. 2011-1837), Ebba och Sven Schwartz Stiftelse, and Mistra through a core grant to the Stockholm Resilience Centre. Simon West's work on the chapter was also partly funded by a Formas Mobility Starting Grant (Grant No. 2017-01631).

References

Berdej, S. M., & Armitage, D. R. (2016). Bridging organizations drive effective governance outcomes for conservation of Indonesia's marine systems. *PLoS ONE, 11*(1), e.0147142. https://doi.org/10.1371/journal.pone.0147142

Berkes, F. (2017). Environmental governance for the anthropocene? Social-ecological systems, resilience, and collaborative learning. *Sustainability, 9,* 1232. https://doi.org/10.3390/su9071232

Blumer, H. (1954). What is wrong with social theory? *American Sociological Review, 19,* 3–10. https://doi.org/10.2307/2088165

Boyd, E. (2008). Navigating Amazonia under uncertainty: Past, present and future environmental governance. *Philosophical Transactions of the Royal Society B: Biological Sciences, 363,* 1911–1916. https://doi.org/10.1098/rstb.2007.0023

Boyd, E., Ensor, J., Broto, V. C., & Juhola, S. (2014). Environmentalities of urban climate governance in Maputo, Mozambique. *Global Environmental Change, 26,* 140–151. https://doi.org/10.1016/j.gloenvcha.2014.03.012

Boyd, E., & Juhola, S. (2014). Adaptive climate change governance for urban resilience. *Urban Studies, 52,* 1234–1264. https://doi.org/10.1177/0042098014527483

Braun, V., & Clarke, V. (2006). Using thematic analysis in psychology. *Qualitative Research in Psychology, 3*(2), 77–101. https://doi.org/10.1191/1478088706qp063oa

Carruthers, J. (1995). *The Kruger National Park: A social and political history.* Pietermaritzburg, South Africa: University of Natal Press.

Chaffin, B. C., Gosnell, H., & Cosens, B. A. (2014). A decade of adaptive governance scholarship: Synthesis and future directions. *Ecology and Society, 19*(3), 56. https://doi.org/10.5751/ES-06824-190356

Coetzee, M., Biggs, H. C., & Malan, S. (2012). *Sharing the benefits of biodiversity: A regional action plan to nurture and sustain the contribution of biodiversity and ecosystem services to livelihoods and resilient economic development within the Kruger to Canyons Biosphere.*

[Report]. Nelspruit: Mpumalanga Tourism and Parks Agency. Retrieved from https://www.kruger2canyons.org/01-21%20-%20Benefits%20of%20Biodiversity.pdf

Coetzer, K. L., Erasmus, B. F. N., Witkowski, E. T. F., & Bachoo, A. K. (2010). Land-cover change in the Kruger to Canyons Biosphere Reserve (1993–2006): A first step towards creating a conservation plan for the subregion. *South African Journal of Science, 106*(7/8), 1–10. https://doi.org/10.4102/sajs.v106i7/8.221

Coetzer, K. L., Erasmus, B. F. N., Witkowski, E. T. F., & Reyers, B. (2013). The race for space: Tracking land-cover transformation in a socio-ecological landscape, South Africa. *Environmental Management, 52,* 595–611. https://doi.org/10.1007/s00267-013-0094-9

Colvin, J, Ballim, F, Chimbuya, S, Everard, M, Goss, J, Klarenberg, G, Ndlovu, S, Ncala, D., & Weston, D. (2008). Building capacity for co-operative governance as a basis for integrated water resource managing in the Inkomati and Mvoti catchments, South Africa. *Water SA, 34*(6), 681–689.

Cundill, G., Leitch, A. M., Schultz, L., Armitage, D., & Peterson, G. (2015). Principle 5—Encourage learning. In R. Biggs, M. Schlüter, & M. L. Schoon, (Eds.), *Principles for building resilience. Sustaining ecosystem services in social-ecological systems* (pp. 174–200). Cambridge: Cambridge University Press. https://doi.org/10.1017/CBO9781316014240.008

Department of Environmental Affairs (DEA). (2014). Draft National Environmental Monitors Concept.

Dietz, T., Ostrom, E., & Stern, P. C. (2003). The struggle to govern the commons. *Science, 302*(5652), 1907–1912. https://doi.org/10.1126/science.1091015

Duit, A., & Galaz, V. (2008). Governance and complexity—emerging issues for governance theory. *Governance, 21,* 311–335. https://doi.org/10.1111/j.1468-0491.2008.00402.x

Folke, C., Hahn, T., Olsson, P., & Norberg, J. (2005). Adaptive governance of social-ecological systems. *Annual Review of Environment and Resources, 30,* 441–473. https://doi.org/10.1146/annurev.energy.30.050504.144511

Folke, C., Pritchard, L., Berkes, F., Colding, J., & Svedin, U. (2007). The problem of fit between ecosystems and institutions: Ten years later. *Ecology and Society, 12*(1), 30. https://doi.org/10.5751/ES-02064-120130

Galaz, V. (2014). *Global environmental governance, technology and politics: The Anthropocene gap.* Cheltenham: Edward Elgar.

Glückler, J., Suddaby, R., & Lenz, R. (Eds.). (2018). *Knowledge and institutions*. Knowledge and Space: Vol. 13. Cham: Springer. https://doi.org/10.1007/978-3-319-75328-7

Griggs, S., Norval, A. J., & Wagenaar, H. (Eds.). (2014). *Practices of freedom: Decentred governance, conflict and democratic participation.* New York, NY: Cambridge University Press.

Hahn, T., Olsson, P., Folke, C., & Johansson, K. (2006). Trust-building, knowledge generation and organizational innovations: The role of a bridging organization for adaptive comanagement of a wetland landscape around Kristianstad, Sweden. *Human Ecology, 34,* 573–592. https://doi.org/10.1007/s10745-006-9035-z

Hajer, M. A. & Wagenaar, H. (2003). *Deliberative policy analysis: Understanding governance in the network society.* Cambridge, UK: Cambridge University Press.

Karpouzoglou, T., Dewulf, A., & Clark, J. (2016). Advancing adaptive governance of social-ecological systems through theoretical multiplicity. *Environmental Science & Policy, 57,* 1–9. https://doi.org/10.1016/j.envsci.2015.11.011

Leach, M., Scoones, I., & Stirling, A. (2010). *Dynamic sustainabilities: Technology, environment, social justice.* Pathways to Sustainability: Vol. 1. London: Earthscan.

Lunstrum, E. (2014). Green militarization: Anti-poaching efforts and the spatial contours of Kruger National Park. *Annals of the Association of American Geographers, 104,* 816–832. https://doi.org/10.1080/00045608.2014.912545

McConnachie, M. M., Cowling, R. M., Shackleton, C. M., & Knight, A. T. (2013). The challenges of alleviating poverty through ecological restoration: Insights from South Africa's "Working for Water" program. *Restoration Ecology, 21,* 544–550. https://doi.org/10.1111/rec.12038

Mccord, A. (2005). A critical evaluation of training within the South African national public works programme. *Journal of Vocational Education and Training, 57,* 563–586. https://doi.org/10.1080/13636820500200302

Moss, T., Medd, W., Guy, S., & Marvin, S. (2009). Organising water: The hidden role of intermediary work. *Water Alternatives, 2*(1), 16–33. Retrieved from https://www.semanticscholar.org/paper/Organising-water%3A-The-hidden-role-of-intermediary-Moss-Medd/5ee01546753ea997b66d817603884832c907d724

Olsson, P., Folke, C., Galaz, V., Hahn, T., & Schultz, L. (2007). Enhancing the fit through adaptive comanagement: Creating and maintaining bridging functions for matching scales in the Kristianstads Vattenrike Biosphere Reserve, Sweden. *Ecology and Society, 12*(1), 28. https://doi.org/10.5751/ES-01976-120128

Olsson, P., Folke, C., & Hahn, T. (2004). Social-ecological transformation for ecosystem management: The development of adaptive co-management of a Wetland Landscape in Southern Sweden. *Ecology and Society, 9*(4), 2. https://doi.org/10.5751/ES 00683 090402

Olsson, P., Gunderson, L. H., Carpenter, S. R., Ryan, P, Lebel, L., Folke, C., & Holling, C. S. (2006). Shooting the rapids: Navigating transitions to adaptive governance of social-ecological systems. *Ecology and Society, 11*(1), 18. https://doi.org/10.5751/ES-01595-110118

Österblom, H., & Folke, C. (2013). Emergence of global adaptive governance for stewardship of regional marine resources. *Ecology and Society, 18*(2), 4. https://doi.org/10.5751/ES-05373-180204

Österblom, H., Jouffray, J-B., Folke, C. & Rockström, J. (2017). Emergence of a global science-business initiative for ocean stewardship. *Proceedings of the Natural Academy of Sciences, 114,* 9038–9043. https://doi.org/10.1073/pnas.1704453114

Österblom, H., & Sumaila, U. R. (2011). Toothfish crises, actor diversity and the emergence of compliance mechanisms in the Southern Ocean. *Global Environmental Change, 21*(3), 972–982. https://doi.org/10.1016/j.gloenvcha.2011.04.013

Parker, J., & Crona, B. (2012). On being all things to all people: Boundary organizations and the contemporary research university. *Social Studies of Science, 42*(2), 262–289. https://doi.org/10.1177/0306312711435833

Pollard, S., du Toit, D., & Biggs, H. C. (2011). River management under transformation: The emergence of strategic adaptive management of river systems in the Kruger National Park. *Koedoe. African Protected Area Conservation and Science, 53*(2), 1–14. https://doi.org/10.4102/koedoe.v53i2.1011

Pollard, S., Shackleton, C., & Carruthers, J. (2003). Beyond the fence: People and the lowveld landscape. In J. T. du Toit, K. H. Rogers, & H. C. Biggs (Eds.), *The Kruger experience. ecology and management of Savanna heterogeneity* (pp. 422–446). Washington, D.C.: Island.

Pool-Stanvliet, R. (2013). A history of the UNESCO Man and the Biosphere Programme in South Africa. *South African Journal of Science, 109*(9/10), 1–6. https://doi.org/10.1590/sajs.2013/a0035

Ramutsindela, M. F., & Simon, D. (1999). The politics of territory and place in post-apartheid South Africa: The disputed area of Bushbuckridge. *Journal of Southern African Studies, 25,* 479–496. https://doi.org/10.1080/03057070.1999.11742770

SANParks BSP & K2C. (2016). Environmental Monitor Programme Close Out Report 2013–2016.

Schultz, L. (2009). *Nurturing resilience in social-ecological systems: Lessons learned from bridging organizations* (Doctoral dissertation). Stockholm: Stockholm University Press. Retrieved from www.diva-portal.org/smash/get/diva2:214516/FULLTEXT01.pdf

Schultz, L., Folke, C., Österblom, H., & Olsson, P. (2015). Adaptive governance, ecosystem management and natural capital. *Proceedings of the Natural Academy of Sciences, 112,* 7369–7374. https://doi.org/10.1073/pnas.1406493112

Schultz, L., West, S., Bourke, A. J., d'Armengol, L., Torrents, P., Hardardottir, H., Jansson, A., & Roldán, A. M. (2018). Learning to live with social-ecological complexity: An interpretive analysis of learning in 11 UNESCO Biosphere Reserves. *Global Environmental Change, 50,* 75–87. https://doi.org/10.1016/j.gloenvcha.2018.03.001

Stirling, A. (2016). Knowing doing governing: Realizing heterodyne democracies. In J. P. Voß & R. Freeman (Eds.), *Knowing governance: The epistemic construction of political order* (pp. 259–289). London: Palgrave Macmillan.
UNESCO. (1996). Biosphere reserves.:The Seville strategy & the statutory framework of the World Network. Retrieved from http://unesdoc.unesco.org/images/0010/001038/103849Eb.pdf
Valman, M., Österblom, H., & Olsson, P. (2015). Adaptive governance of the Baltic Sea—Lessons from elsewhere. *International Journal of the Commons, 9*(1), 440–465. https://doi.org/10.18352/ijc.532
Van Kerkhoff, L. E., & Lebel, L. (2015). Coproductive capacities: Rethinking science—Governance relations in a diverse world. *Ecology and Society, 20*(1), 14. https://doi.org/10.5751/ES-07188-200114
Venter, F. J., Naiman, R. J., Biggs, H. C., & Pienaar, D. J. (2008). The evolution of conservation management philosophy: Science, environmental change and social adjustments in Kruger National Park. *Ecosystems, 11*(2), 173–192. https://doi.org/10.1007/s10021-007-9116-x
West, S., Schultz, L., & Bekessy, S. (2016). Rethinking social barriers to effective adaptive management. *Environmental Management, 58*(3), 399–416. https://doi.org/10.1007/s00267-016-0721-3
Westley, F. R., Tjornbo, O., Schultz, L., Olsson, P., Folke, C., Crona, B., & Bodin, Ö. (2013). A theory of transformative agency in linked social-ecological systems. *Ecology and Society, 18*(3), 27. https://doi.org/10.5751/ES-05072-180327
Westley, F. (2002). The devil in the dynamics: Adaptive management on the front lines. In L. H. Gunderson & C. S. Holling (Eds.), *Panarchy: Understanding transformations in human and natural systems* (pp. 333–360). Washington, D.C.: Island.
Wyborn, C. A. (2015). Connecting knowledge with action through coproductive capacities: Adaptive governance and connectivity conservation. *Ecology and Society, 20*(1), 11. https://doi.org/10.5751/ES-06510-200111

Chapter 14
Ex Ante Knowledge for Infectious Disease Outbreaks: Introducing the Organizational Network Governance Approach

Jörg Raab, Patrick Kenis, Marleen Kraaij-Dirkzwager, and Aura Timen

The Importance of Knowledge in Infectious Disease Outbreaks

The world is regularly and increasingly confronted with the outbreaks of infectious diseases (Smith et al., 2014). In the Netherlands, for example, infectious diseases are a clear health risk (Volksgezondheidenzorg.info, n.d.). In 2010, more than 30,000 hospital admissions were related to infectious diseases and almost 18,000 patients were treated in outpatient care facilities. Despite structural control measures (e.g., immunization programs, clean water, hygiene, and sanitation efforts), new infectious diseases emerge due to factors such as increased trade through globalization, migration, and adaptation of microorganisms and can become serious public health issues. In addition to the morbidity and mortality of infectious diseases, outbreaks cause societal distress and large societal costs.

The large quantity of unknown factors makes it impossible to assess such an occurrence's timing, duration, or path of development in advance. Given the

J. Raab (✉)
Department of Organization Studies, Tilburg University, Tilburg, The Netherlands
e-mail: j.raab@uvt.nl

P. Kenis
Tilburg School of Economics and Management, Tilburg University, Tilburg, The Netherlands
e-mail: P.N.Kenis@tilburguniversity.edu

M. Kraaij-Dirkzwager
Department for Environmental Health, Aftercare and Security (VLH), Dutch National Institute for Public Health and the Environment (RIVM), Bilthoven, The Netherlands
e-mail: marleen@artsmg-advies.nl

A. Timen
National Coordination Centre for Outbreak Management (LCI), Dutch National Institute for Public Health and the Environment (RIVM), Bilthoven, The Netherlands
e-mail: aura.timen@rivm.nl

J. Glückler et al. (eds.), *Knowledge for Governance*, Knowledge and Space 15,
https://doi.org/10.1007/978-3-030-47150-7_14

uncertainty and complexity, the question is to what extent and how policymakers can prepare to control a large—possibly cross-border—infectious disease outbreak. In the present paper, we take up this task by advancing an approach to gain relevant knowledge to control the outbreak before it has taken place. This is an exceptional challenge for at least two reasons. First, outbreaks of infectious diseases very often show high complexity in the sense that they are characterized by incomplete, contradictory, and changing requirements that are often difficult to recognize and change during their development. Second, a large number of actors are likely to become active in case of an outbreak of infectious diseases, because they believe they have a stake in the outbreak. Other relevant actors, however, might not become active even though others think they should have a stake in the outbreak management. Consequently, an orderly response is unlikely and an unclear and even confusing set of actors will probably form. Improving capacity to respond to outbreaks of infectious diseases therefore requires researchers to gain knowledge on the evolving actor set and its interdependencies.

Asserting that knowledge is simply unavailable—given the wicked conditions mentioned above—is not an option considering what is at stake in controlling infectious disease threats (just think of the 11,000 persons who died because of an Ebola infection; Medaglini, Santoro, & Siegrist, 2018; Munjita, Chileshe, & Mutemwa, 2015). Consequently, in the present paper we develop an approach to the production of useful knowledge for preparing the control of infectious disease outbreaks. We label it the *organizational network governance approach,* which we build on three main arguments: First, the response to infectious disease outbreaks can best be conceptualized as an *organizational network response*; second, we can describe and analyze the structure of an organizational network using *network analytical methods;* and third, we can assess these networks in terms of their effectiveness in controlling disease outbreaks using *governance concepts*. We will demonstrate our approach by introducing two infectious disease threat scenarios and show the value of conceptualizing them as crisis-response networks, analyzing them as social networks and assessing them from a governance perspective. In this way, we create available knowledge for an effective response to infectious disease outbreaks.

The Research Context

Two fictitious but realistic scenarios of infectious disease outbreaks in the Netherlands

This study is based on two fictitious but realistic outbreak scenarios in the Netherlands: an outbreak of a *New Asian Corona Virus (NAC)* and an outbreak of the *West Nile Virus (WNV)*. The *New Asian Corona Virus* is a fictive coronavirus but falls in the category of viruses emanating from Asia causing serious respiratory illnesses such as SARS (severe acute respiratory syndrome) or MERS (Middle East respiratory syndrome). There are no vaccinations or other preventative medicines to date. The virus is transmissible between humans via airborne infection or direct contact. There is no evidence for transmission via feces. Coronaviruses can cause various diseases in animals as well as people, varying from the common cold to severe respiratory syndromes. In humans, coronaviruses cause about 15–20% of upper respiratory infections. Disease symptoms vary by virus type, but coronaviruses often produce symptoms such as fever, coughing, fatigue, shortness of breath, and gastrointestinal symptoms like diarrhea. Some coronaviruses cause very serious symptoms, such as pneumonia, acute respiratory distress syndrome (ARDS), and multiorgan failure.

Much remains unknown about coronaviruses because they are hard to isolate. The first detection of an animal coronavirus in the laboratory was in 1937. Human coronaviruses were discovered in 1960. An outbreak of SARS-CoV (severe acute respiratory syndrome-related coronavirus) occurred in 2003–2004, causing over 8000 reported patients worldwide, of which about 10% died. In June 2012, scientists identified the first case of MERS-CoV infection in Saudi Arabia. The number of infected persons has exceeded 1800 cases globally, of which 640 have died. National and European systems are in place to notify and monitor important cases of MERS-CoV. In the Netherlands, MERS-CoV occurred in 2014 in two persons who travelled back from Saudi-Arabia.

The West Nile Virus is an arbovirus, which is transmitted from animals to humans or horses via mosquitoes (Bellini, Zeller, & van Borte, 2014; Weaver & Reisen, 2010). Wild birds are the primary enzootic cycle of WNV, with mosquitoes transmitting the virus amongst these wild birds (Bellini et al., 2014). When (climate) conditions permit, virus circulation may increase and spillover transmission via mosquitoes to humans or horses can occur. When transmission occurs, humans and horses usually serve as a dead-end host, meaning that not enough viral load is built up to infect mosquitoes. However, human-to-human transmission is possible following blood or organ donation from an infected donor. Although no symptoms occur in most human infections, in 20–30% of infections symptoms such as sudden onset of fever, headache, fatigue, and myalgias arise (Lim, Koraka, Osterhaus, & Martina, 2011), as well as gastrointestinal complaints with the risk of dehydration.

WNV can affect all ages, with high incidences among younger individuals, and among the elderly and immunocompromised. In addition, both susceptibility and the severity of the infection increase with age (Lim et al., 2011). Elderly people are therefore at higher risk of developing neuroinvasive disease, which may result in encephalitis, meningitis, or a poliomyelitis-like syndrome (Sejvar, 2014). Such outcomes are seen in less than 1% of infections, but are significantly more debilitating and lead to long-term outcomes in over 50% of cases (Lim et al., 2011; Sejvar, 2014).

There is no available treatment for WNV in humans, other than supportive care (Sejvar, 2014), which highlights the impact the disease's introduction may have on a country (Rizzoli et al., 2015). For horses, on the other hand, vaccines are available to protect them from developing West Nile Fever and other WNV-related outcomes (Bowen et al., 2014; Iyer & Kousoulas, 2013).

National and European systems are in place to notify and monitor cases/the epidemiology of WNV (ECDC, 2013). A vast majority of European countries have reported either human or animal cases of WNV in the past, for example Greece, 2010 (ECDC, 2010); Turkey, 2010–2011 (Kalaycioglu et al., 2012); Croatia, 2012 (Pem-Novosel et al., 2014); Italy, 2012 (Barzon et al., 2012). To date, the Netherlands has had no autochthonous infections of WNV (i.e., infections acquired within the country) (Chancey, Grinev, Volkova, & Rios, 2014).

The two fictitious outbreaks would call for immediate responses or what is called outbreak management. Outbreak management is partially context specific, as control measures are related to the pathogen involved (characteristics of the virus or bacterium), the route of transmission (through inhalation, direct contact, sexual contact, oral intake), and the risk groups (related to many factors: age, immune response, and—very importantly—specific behavior). Risk groups can be those who become more easily infected due to exposure, or groups at larger risk of developing complications after an infection. It takes expertise to recognize an outbreak (understand the epidemiology and determine the source of infection, the mode of transmission, and the risk groups) and to develop effective and timely control measures.

In the Netherlands, it is the *Public Health Act* which regulates the response to events threatening public health in the Netherlands, including outbreak management (Wet publieke gezondheid, 2017). The National Coordination Center Communicable Disease Control (in Dutch: *Landelijke Coordinatie Infectieziektebestrijding*, LCI), a department of the National Institute for Public Health and the Environment (in Dutch: *Rijksinstituut voor Volksgezondheid en Milieu,* RIVM) (later labeled the National Coordination Authority), is charged with coordinating actors within the response system if an outbreak involving different Dutch regions occurs (RIVM, 2019). The coordination of relevant actors is necessary in order to control the risks associated with an outbreak as effectively (less morbidity, mortality, and societal unrest) and efficiently (efficient use of human and financial resources) as possible.

To facilitate the formulation and implementation of control measures at the population level, an infrastructure for analysis and decision making is established (RIVM, 2012). The director of the Centre for Infectious Disease Control can invite the members of the Outbreak Management Team (OMT) to convene. The OMT is formed by a group of "fixed" experts, invited based on their personal expertise (e.g., communicable disease specialists, infectiologist, microbiologist, epidemiology, general practice; in case of a zoonotic disease, veterinary partners attend). The OMT is expanded based on pathogen- or context-specific needs (e.g., specific knowledge about risk groups, including specific veterinary expertise). The OMT advises the Board of Administrative Executives (in Dutch: *Bestuurlijk Afstemmings Overleg*), directed by the Director-General of the Ministry of Health. The BAO advises the Minister of Health on legal, financial, and political aspects of the proposed control measures. The minister of Health will interact with other ministers if collective control measures have an effect on, for example, trade, schools, or airports. Once the decision on collective control measures has been taken, the Minister of Health requests that the National Coordinating Authority to support actors in implementing the control measures with information and coordination as needed for an effective response.

Although there is a clear response system in place, it is evident that given the potentially broad societal and economic impact of the described scenarios, a myriad of actors within and outside the public health field will become involved.

The infectious disease outbreaks described are characterized by complexity in the sense that they entail incomplete, contradictory, and shifting requirements that are often difficult to recognize and change during their development. Moreover, numerous organizations, agencies, and other actors are likely to be involved in significant ways. Consequently, response patterns are emergent rather than routine or planned (Majchrzak, Jarvenpaa, & Hollingshead, 2007).

The Research Challenge and Theoretical Approach

Given the empirics of the diseases described above, actors search for relevant knowledge to control an outbreak before it takes place. More specifically, we concentrate on the question whether investigating the multiplicity of actors related to a crisis as well as the relational pattern of knowledge seeking and sharing between these actors provides a useful knowledge base for controlling outbreaks of infectious diseases (Borgatti, Everett, & Johnson, 2018). The questions become: Who are the actors related to an outbreak of a specific disease, what are the patterns of relationships between these actors, and what information do these possess that are relevant for controlling an infectious disease outbreak? Before presenting our findings based on an analysis of two fictitious disease outbreaks, we discuss our research's theoretical foundation.

We propose conceptualizing a situation of an infectious disease outbreak as an *organizational network response*, in which a myriad of actors will become active while others stay inactive (nevertheless others expecting them to become active). Different types of relationships will or will not develop between these actors, resulting in a system of information sharing, command, collaboration, and so forth (Glückler & Panitz, 2016). The assumption is that this so-called organizational network decisively influences the response's development and quality. We base such analysis of crisis response from an *organizational network-response* perspective on the following assumptions. First, we begin our organizational network-response from a realist perspective and do not a priori include or exclude certain organizations that should (or should not) be part of the response. We thus widen our lens to include possible peculiarities of the crisis leading to improper measurement or missing important knowledge (see, e.g., Weick's (2006) study on the WNV incidence in new NYC in 1999). Indeed, stakeholder analysis has become important in crisis response analysis, but few studies exist whose researchers examine crisis response from an overall network perspective (compared to ego-centric perspectives, with stakeholders as alters). Second, such an approach opens a perspective for studying the network positions and interactions between the different actors in the networks, thus producing knowledge about the network's dispersion, information flow, leading organizations, the presence of peripheral groups, and so on (Glückler & Doreian,

2016). Third, such an approach serves as a basis for better understanding how collaboration and communication between actors can be improved (Moon et al., 2015; Swaan et al., 2017; Vinck et al., 2011), linking its findings to a governance perspective. We here define governance as the structures and interactive processes that steer actors' activities towards a common goal (Ansell & Torfing, 2016; Kenis, 2016). Whereas networks describe the actors and the relational patterns between these actors, a governance perspective adds the question of whether and how these networks lead to network outcomes. Given the absence of market logic or classical hierarchical logic, the question becomes which mechanisms steer the network's functioning. We consider these mechanisms essential to understanding how a set of organizations and its relational patterns function in a network form of organization. We here define actors, relational patterns, and mechanisms as network-level governance (Glückler, Dehning, Janneck, & Armbrüster, 2012).

The organizational network governance approach proposed here is first and foremost an analytical one, whose utilizers attempt to create knowledge to improve response preparedness by conceptualizing the response system as a network based on actors and their ties. Network is thus used as an *empirical tool* (Raab & Kenis, 2009). As stated above, however, we believe that in a situation where neither market nor hierarchical mechanisms seem likely to work, as is the case for international infectious disease threats, network as a form of governance or *governance tool* is the most likely and appropriate option. Whether the most appropriate response is based on a shared governance mode, lead organization or network administrative organization (Provan & Kenis, 2008), or a mixed form (Berthod, Grothe-Hammer, Müller-Seitz, Raab, & Sydow, 2016) is an empirical question and depends on several factors, of which the formal legal framework in which the response takes place is likely an important one.

The organizational-network response approach resonates with recent observations in the field of crisis management. Recent research on organizational networks in general and their use as a tool to respond to disasters and emergencies has significantly improved our understanding of the structure, governance, functioning, and effectiveness of such systems. In addition, the field of public sector networks in general has made important progress regarding the governance of goal-directed networks (Raab, van den Oord, & Kenis, 2015). Provan and Kenis (2008), for example, have provided the field with a conceptual vocabulary and specific lens that has helped researchers to better analyze the different forms of network coordination in general (Ansell & Gash, 2008; Emerson & Nabatchi, 2015; Glückler, Lazega, & Hammer, 2017; Provan, Fish, & Sydow, 2007; Raab, Mannak, & Cambré, 2015) and emergency response systems in particular (Berthod et al., 2016; Moynihan, 2009). In addition, the 9/11 terror attacks in New York and Washington triggered a whole stream of research on interorganizational response networks (Hu, Knox, & Kapucu, 2014; Kapucu, 2009; Moynihan, 2009; Nowell & Steelmann, 2015). The perspective of an organizational network-response was recently also proposed in a report on "New Directions in Governing the Global Health Domain" (Kickbusch, Cassels, & Liu, 2016). Its authors concluded that those dealing with health challenges will in the future need to widen their lens to include actors who lie outside what has traditionally been defined as the infectious disease architecture.

This is due to considerable failures in responses (e.g., Ebola; Moon et al., 2015; Stoto & Higdon, 2015) and the increasing interdependencies in today's world. Public health specialists in the Netherlands have also recognized this situation (Huizer, Kraaij-Dirkzwager, Timen, Schuitmaker, van Steenbergen, 2014; Kraaij-Dirkzwager, Schol, Schuitmaker-Warnaar, Timen, & van Steenbergen, 2019).

How, then, can one usefully describe and analyze an organizational network-response that results in reaction to a disease outbreak? Given the fact that we conceptualize the situation as one with an a-priori unknown set of actors and among which the interaction has an important effect on controlling an outbreak as early as possible and with minimal consequences, we propose network analytical tools as an appropriate approach. The use of such tools goes beyond the more common mapping of the relevant actors and providing a generic list of all possible actors involved and will be introduced in the following.

We thus explore the potential governance system for these two fictitious but realistic infectious disease outbreaks through a network lens. Knowledge that we will acquire through this exploration will likely help in fighting future outbreaks in the following way:

1. Which actors are mobilized? Are these all the appropriate actors? Some actors are frequently not mobilized because they are not part of the core actor set within public health, but are nonetheless crucial for outbreak management.
2. To what extent are those that are deemed important willing to engage? Ideally, we would like to see actors on the diagonal, in other words, the more important, the more engaged.
3. In terms of network structure, how are the relevant actors connected? If actors form clusters and these are connected by brokers, these brokers should have the competence and capacity to function in such a connector role.
4. In terms of governance, to what extent are core health care actors well and densely connected in the center, collectively coordinating the response under the LCI's leadership and well connected (but more sparsely) to the more peripheral actors?

Public health authorities can utilize these insights to develop a relational lens with which to analyze, structure, and manage the emerging organizational network response.

Introducing Network Analytical Tools for Studying Infectious Diseases Responses

Network analytical tools are useful for describing the (in)formal relationships in an evolving crisis-response network. Insight into the outbreak networks described in the scenarios above provides parallel insights into the unfolding communication and coordination at the level of the crisis-response network.

Network analysis is a method of collecting, analyzing, and visualizing relational structures and processes. The network perspective is now more than a metaphor: It is a systemic way for researchers to study how societies with their individuals, communities, and organizations interrelate, based on and with its own theoretical statements, methods, and research findings (see Borgatti et al., 2018; Rainie & Wellman, 2012, for an overview). Proponents study and describe systems of interaction in terms of both their actors (called "nodes" or "vertices" in the language of network theory) and relationships (called "links", "ties" or "edges"). Links can be of many types, such as information exchange, trust, exchange of resources, and the like. Networks are usually represented by data matrices and related diagrams, in which the units are represented by points and lines between them, either with or without arrowheads dependent on whether they have a direction or not.

Network analysis researchers have developed a great and powerful number of ways to describe networks characteristics, which we have summarized in Table 14.1.

Measures, Data Collection, and Data Analysis

We have operationalized the data collected for the present study in order to contribute to the knowledge required to help fight future outbreaks in the following ways:

The List of Actors in the Crisis-Response Networks

In order to mirror reality as best as possible, we used two fictive scenarios as the basis for our network analyses. One scenario described the early onset of a West-Nile Virus (WNV) outbreak with several autochthonous cases among humans and horses in the Netherlands. The second scenario described a rapidly evolving outbreak of a new coronavirus (NAC) after introduction through a traveller returning from Asia. (Full descriptions of the cases are available on request.)

Because the networks in both cases included actors from inside and outside the public health field, we used exploratory interviews and two focus groups with infectious disease control experts (n = 6 and n = 7, respectively) to determine the network boundaries and to develop the questionnaire. In addition, we had to find a way to combine concrete organizational actors like the National Coordination Authority with actor groups like general practitioners, boards of academic hospitals, emergency physicians, microbiologists, infectiologists, or veterinarians. With the help of the two focus groups we defined a relevant actor as: "any organization and/or representative that has a positive or negative influence on the prevention, control, treatment, and/or decision making with regard to (the outbreak of) the infectious disease at hand." We also attributed three main characteristics to actors that influence their role during an outbreak: their level of influence on outbreak management (related to their interest in acting and potentially contributing), the amount and type of

Table 14.1 Overview of basic network descriptors

Dimension	Indicator	Rationale for use in empirical analysis
Derived attributes of actors	Centrality	Indicators of centrality identify the most important node (actors) within a network. Given the fact that "important" can mean different things (depending on the research questions) there are also different centrality indicators. For example we can look at the "importance" of an actor (e.g., who is closest to all other actors) or the importance of the actor in the cohesiveness of the network. The most commonly used centrality indicators are: degree, closeness and betweeness. Eigenvector centrality assigns relative scores to all nodes in the network based on the concept that connections to high-scoring nodes contribute more to the score of the node in question than equal connections to low-scoring nodes.
	Status, prestige, prominence	In a directed network, a tie is not a symmetric connection between two actors, but an asymmetric link, going from one actor to another leading to difference in prominence or prestige among actors. The simplest measure of prominence for directed networks simply breaks down the degree count for incoming ties (in-degree) and outgoing ties (out-degree). Many other centrality measures can be similarly adapted to directed networks and used as prestige measures as well.
Structural roles of actors	Bridge, broker, …	Apart from the absolute number of ties an actor in a network can have specific positions in a network. An actor who is connected to actors who are themselves not directly connected has opportunities to mediate between them and the actor itself or the overall network can profit from this mediation.
Structural partitions	Communities	A network is said to have a community structure if the nodes of the network can be easily grouped into (potentially overlapping) sets of nodes such that each set of nodes is densely connected internally. The inhomogeneity of connections suggests that the network has certain divisions within it. The most commonly used indicators for analyzing community structures in networks are: Clique analysis, cluster analysis, and block analysis.
Derived network attributes	Density	Density indicates the general level of cohesion in a network. It calculates the proportion of direct ties in a network relative to the total number possible. Denser networks are not by definition better than sparser networks. This is contingent on the research question and perspective. Would we like to see 'old boys networks' to be more dense or less dense?
	Centralization	Centralization refers to the overall cohesion or integration of a network. Networks may, for example, be more or less centralized around particular actors or groups of actors. Most centralized for instance is a star network.
	Cohesiveness	It refers to the minimal number of actors in a social network that need to be removed to disconnect the group. This might be important for understanding how social networks shape communities, facilitate norm maintenance, or form the basis of categorical group identity, among many other things.

(continued)

Table 14.1 (continued)

Dimension	Indicator	Rationale for use in empirical analysis
	Connectedness	The inventory of the total connections among actors is useful for getting a sense of how closely coupled the entire network is. This information can be used to understand how information moves in the network.
	Subgraphs and components	This indicates that the graph can be partitioned in certain ways. A *subgraph* is a subset of the nodes of a network, and all of the edges linking these nodes. Any group of nodes can form a subgraph. Components, on the other hand, are portions of the network that are disconnected from each other.

Note. Reprinted from "Analyzing policy-making II: Policy network analysis", by P. Kenis and V. Schneider, 2018, in M. Puppis, K. Donders, L. van Audenhove, & H. van den Bulck (Eds.), *The Palgrave handbook of methods for media policy research* (pp. 471–491), Cham: Palgrave Macmillan. Copyright 2018 by Springer Nature. Reprinted with permission

knowledge and information they have available (related to their position in the network), and the level of collaboration they engage in. We asked respondents to focus on the potential collaborations and coordination with other relevant actors in the Netherlands as the scenarios allowed. Based on the generic list of actors available at the National Coordination for Communicable Disease control, the focus groups identified 98 possibly relevant actors for the WNV scenario and 61 potential actors in the NAC scenario. We thus applied a realist strategy in determining the network boundaries.

We did not include nonorganized stakeholders such as infected individuals, travellers, recreational water/land users, hunters, farmers, and gardeners (i.e., specific risk-groups) or professionals who were already represented through other organizations (e.g., equine specialists through the Animal Health Services and Royal Netherlands Veterinary Association). To include a representative sample of all these stakeholders would mean a disproportional effort at this stage of our exploratory research. We excluded media for the same reason. We approached some stakeholders via an umbrella organization—those we judged to be extremely relevant but difficult to access directly—for example: boards of the academic hospitals (via the Netherlands Federation of University Medical Centres (NFU)) or microbiologists (via the Dutch Association of Medical Microbiology (NVMM)). By asking the associations to select five representatives from different regions to fill in the questionnaire, we argue that a proper reflection of their information flows would be captured without burdening too many people. In addition, in the case of two organizations—with formal tasks related to infectious disease control and disaster/crisis management in the health care sector—that are located throughout the Netherlands, all divisions were approached to obtain insights into potential regional differences within institutes (i.e., Municipal Health Services (GGD) and Regional Consultation on Acute Care (ROAZ)).

Data Collection and Types of Ties

We piloted the questionnaire and adapted it accordingly (questionnaire available on request in Dutch). We first asked respondents to indicate their role(s) and potential contributions (unique skills and capabilities) to outbreak control, and then introduced the scenarios. Second, we incorporated a list of control activities (approved by the focus groups of control experts) and asked respondents to indicate in which control activities they were likely to be involved. Third, the questionnaire contained a table in which we asked respondents to indicate per organization identified if they would obtain and/or provide information/advice from and to this organization. Fourth, we asked the respondents to indicate with whom of the stakeholders they would expect to have the most intense collaboration, plus their expectations of the activities the particular stakeholder would undertake. Finally, the respondents were asked to indicate their perceived level of influence over the outbreak control in the scenario provided and their level of interest in being involved in outbreak control.

We therefore based the network analysis on three types of ties: joint involvement in control activities, providing and receiving information/advice, and collaboration. For actor groups (health care professionals), we coded a tie as existent (=1) if at least 50% of respondents from an actor group—for example, 50% of the general practitioners—indicated a relationship, and did additional robustness checks, for example, 60% of respondents or at least one respondent.

We used Visone (Brandes & Wagner, 2004) to analyze and visualize the network data. We calculated general descriptive network measures for the actor level, such as degree, (flow) betweenness, closeness, and status. We further used the spring embedder algorithm implemented in Visone to identify possible clusters and visualize the two mode networks of actors and measures for outbreak control in order to analyze how the response measures and actors are connected and between which actors coordination would be required. For joint involvement we used a spring embedder analysis to identify clusters of actors around certain control measures. For information/advice provision/reception we applied flow betweenness centrality and status, and in case of collaboration we again used the spring embedder to identify cliques and brokers.

Data Collection

The questionnaires were sent to the representatives of the actors or selected respondents from actor groups. A reminder was sent 10–14 days later. If a response still had not been received after 5 days, the organization was mailed or called as an extra reminder to minimize nonresponse. In total, in the NAC case we included 43 actors or actor groups in the data analysis, which represents a response rate of 80%. For the WNV case, we included 82 actors or actor groups in the data analysis, which represents a response rate of 82%.

Data Analysis

We discussed the analyses' preliminary outcomes during a 1h focus group with communicable disease control specialists (medical doctors, nurses, and policy officers); public health specialists (medical doctors and policy officers); microbiologists; entomologists; researchers with an interest in crisis management, preparedness, and response related to infectious disease threats; and guideline developers.

Results

In the following we present the results with actors in an anonymized form (confidentiality was promised to the respondents). We present the actors who have been named relevant in the two settings, their involvement, and the distribution of information.

Actors in the Two Networks

An important question in infectious disease outbreak response is, of course, which actors should be involved to manage the outbreak. This is a tricky issue. Compared to simple or even complicated tasks in which one could specify in advance what the perfect task division would be to get the task done, this is quite different when we are dealing with "wicked problems" (Head, 2008) or those as complex as explained above. Consequently, the first question researchers using an organizational network response approach ask is: Who is or should be part of the response system? Often, actors who are not part of the core of the public health system might need to be involved as early as possible to limit the impact of the outbreak.

The larger number of actors in in the WNV scenario is due to the fact that transmission happens via mosquitos and animals and therefore a large group of actors becomes relevant that does not belong to the traditional (human) public health field such as hunters, veterinarians, water management associations, or the Ministry that deals with agricultural and nature issues in the Netherlands. Different actors constitute the core of the response network in the NAC and the WNV scenario. For instance, an organization with the mandate to control mosquitoes and coordinate specific veterinary measures are prominent in the core of the network controlling the WNV outbreak.

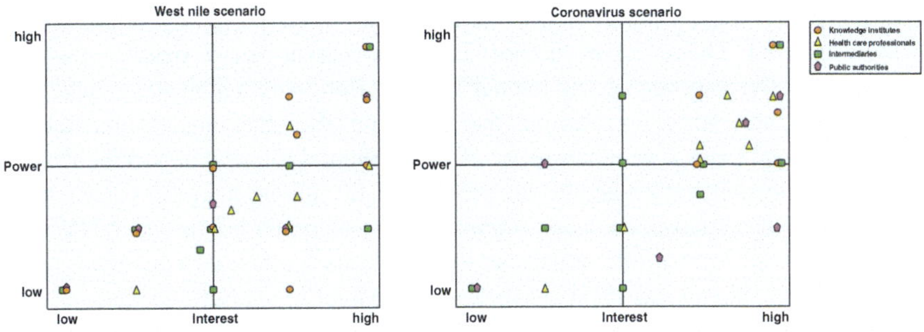

Fig. 14.1 Power interest matrix for the two scenarios. Source: Design by authors

Actor Involvement in the Two Networks

Here we focus on to what extent the actors have an interest in actively participating in the impact of the outbreak. This is an indicator for whether they see a role for themselves in the organizational network response. In order to answer this question, we created a power-interest matrix (see Fig. 14.1 above). A power-interest matrix combines two types of information: whether the actors see a role for themselves in the response and whether they are interested in playing that role. The first tells us something about the task allocation and the second tells us something about how rewarding they consider their participation or, in other words, how interested they are in becoming active in the response. The crosstabulation of these two dimensions proves particularly helpful, because it can point to the discrepancy or tension between the necessary task allocation and the degree of interest in performing one's task.

In these figures, we show different types of actors (knowledge institutes, health care professionals, intermediaries, and public authorities) and their specific position in the power interest matrix. Some points in the matrix represent more than one actor of the same type. Actors are willing to indicate their self-perceived interest and power and differentiate their role in the scenarios (e.g., health care providers estimate a larger role for themselves in the NAC scenario than in the WNV scenario). For coordinating parties in the response system, this provides an interesting mirror for their own expectations of actor involvement, a starting point for dialogues with actors/groups about possible fulfillment of their role in a specific context.

The power-interest matrix (above) also provides an interesting mirror for the actors' own expectations of their involvement and a starting point for dialogues with actor(group)s, including about possible rewards and incentives to fulfill their role in a specific case and context.

In both scenarios, we see a general linear relationship between self-perceived power and interest, with the north-west quadrant remaining empty. This signals that many actors who think they have the most power to intervene in the transmission also indicate a strong interest in participating. Interestingly, there are also actors with self-perceived low power who (might) have crucial skills, assets, or capacities at their disposal. The actors in the northeastern quadrant can be described as the core actors in the response system in their self-description with high power and high interest. It is interesting, for example, that an organization that engages in extermination of mosquitos is very much aware of their potential role and importance in the WNV scenario, and that the health care professionals (yellow triangles) voice strong interest in both scenarios but assess their power to be much less in the WNV scenario (most health care professionals are located in the southeastern quadrant in the WNV scenario, but appear in the northeastern quadrant in the NAC scenario).

Actors' involvement regarding specific measures to fight and control the outbreak can also be seen as actors indicating the tasks they plan to perform in the outbreak scenarios. In the next two figures we depict the respondents' answers to the question "in which of the following outbreak control measures is your organization involved in this scenario." We had identified 28 measures in the NAC scenario and 34 in the WNV scenario as necessary for reacting to and controlling the outbreak, from which respondents could choose. Blue squares represent actors; circles the different measures, which were grouped in identification of the infection source (pink), developing guidelines and informing health care providers and risk groups (yellow), developing and implementing of control measures (green), coordination, evaluation, and research (orange). The measures clearly represent the operational level of reacting to and controlling an outbreak (Fig. 14.2 and Fig. 14.3).

For the NAC scenario, we can observe nine core actors in the middle who are involved in taking different measures and are therefore also likely or at least could potentially coordinate the application of these diverse measures. However, there is no clear clustering of measures and actors, which means that actors and measures of different types are closely connected and that there are no separate actors who deal exclusively with specific measures. Therefore, the integration of the different types of measures and the measures themselves seems quite high but not very structured. A disadvantage of this situation, therefore, could be that it requires significant conscious additional coordination, which is relatively inefficient if there is no formal structure with mandated or natural coordinating actors (other than the National Coordination Authority/NCA). Ideally, some actors would simultaneously be involved in certain measures, and others would be specialized and involved only in a limited number of other measures.

In the NAC scenario, these nine core actors represent some of the most important public health actors in fighting an outbreak, with other actors scattered more at the periphery. This analysis implies that when it comes to the coordination of measures, it is mainly these nine important public health actors that need to reconcile their strategies and actions on the operational level.

In the WNV scenario, the network is characterized more by a bifurcated structure with some actors on one side and the others on the other, connected by different measures regarding information delivery and (public) communication.

As far as the task allocation is concerned, therefore, many operational tasks are performed in the two organization network scenarios and similar tasks are often performed by several organizations. These initially look like rather uncoordinated

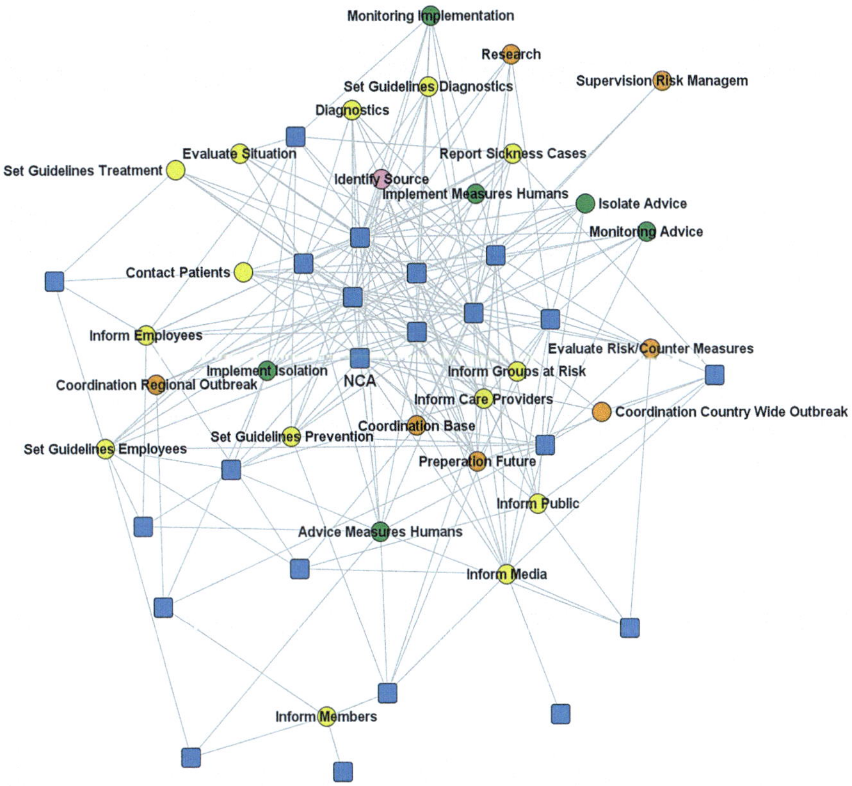

Fig. 14.2 Two-mode network based on involvement in reactive measures (NAC). Source: Design by authors

systems, but to better understand the coordination of tasks in the systems we asked respondents to nominate up to five actors with whom they would work together most intensively in such a scenario regarding limiting transmission of the virus and its impact. This is a good indication of the division of labor which would evolve in the network.

In Fig. 14.4 and Fig. 14.5 below, we depict the structure based on the type of tie "collaboration". The ties are directed to indicate the nomination. Red undirected relations (the thicker lines) indicate that both actors have nominated each other reciprocally. These ties, in other words, connect those groups/actors one would expect to have the most intense collaboration with respect to outbreak control in the specific scenario, and also indicate that organizations hold expectations towards each other about their respective roles and contributions in outbreak management.

We have visualized the network using a spring embedder algorithm, placing nodes in such a way that connected actors are attracted to each other and placed close to each other, and actors who are not connected repel each other and are placed further apart. Through this algorithm, network clusters become visible.

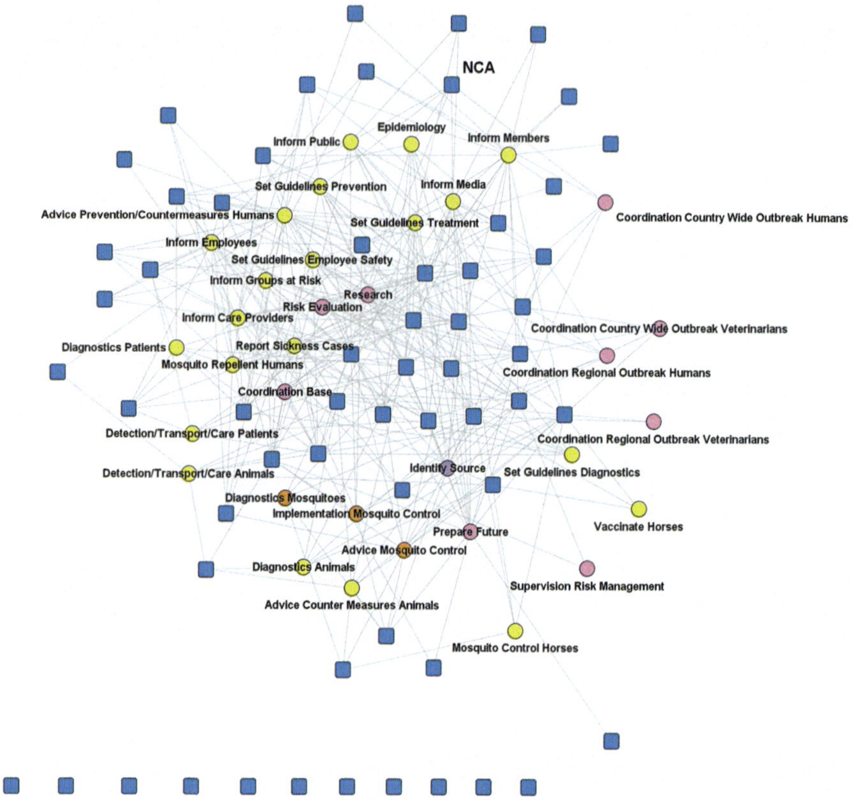

Fig. 14.3 Two-mode network based on involvement in reactive measures (WNV). Source: Design by authors

For both scenarios, one can observe that a relatively dense core exists where actors also frequently nominate each other in a reciprocal way. Due to the higher number of actors, there are also more collaboration relations in the WNV scenario. Interestingly, a lot of directed and thus unreciprocated nominations also exist at the network's core. An interesting question to follow-up is, of course, what the ratio of these hubs is and the ratio of the connections between the hubs. Are these hubs and spokes in effect confirming the law of homophily, which would imply that similar actors are more likely to collaborate with each other? This situation is not necessarily the most appropriate one in a situation of complexity, where one would expect a more integrated approach to collaboration.

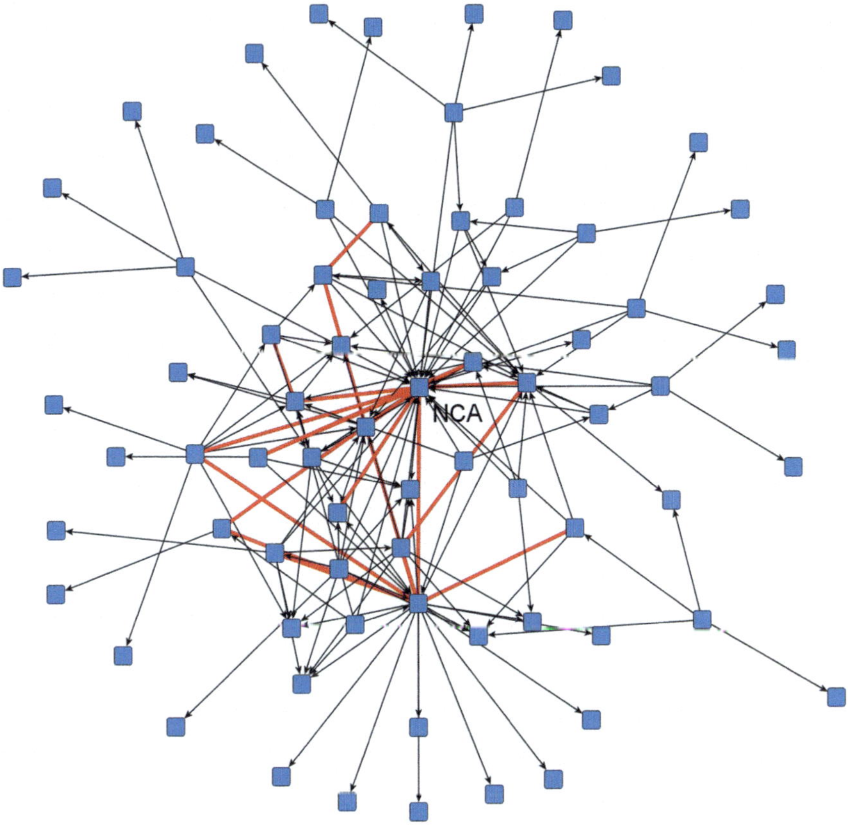

Fig. 14.4 Collaboration between actors (NAC scenario, min 1x mentioned, red lines reciprocal). Source: Design by authors

Information Distribution in the Crisis-Response Networks

Information provision is about the information actors need to perform their task and coordinate their work with others. In what follows, we do not analyze the need, but the actual nomination of which actors provide information to other actors. The question becomes to what extent the information provision reported contributes to the goal of the organizational network, in turn limiting the transmission of the infectious disease and its impact.

In Fig. 14.6, we depict the network structure in a centrality layout with regard to the flow betweenness of the actors based on the confirmed ties "giving information/ advice" in case of a virus outbreak. Actors are regarded as central if they lie on a path between any two other actors. Ties are seen as pipes through which information can flow. The more often an actor is on such a path, the more important it becomes for the transmission of information. We assume that information in principle travels

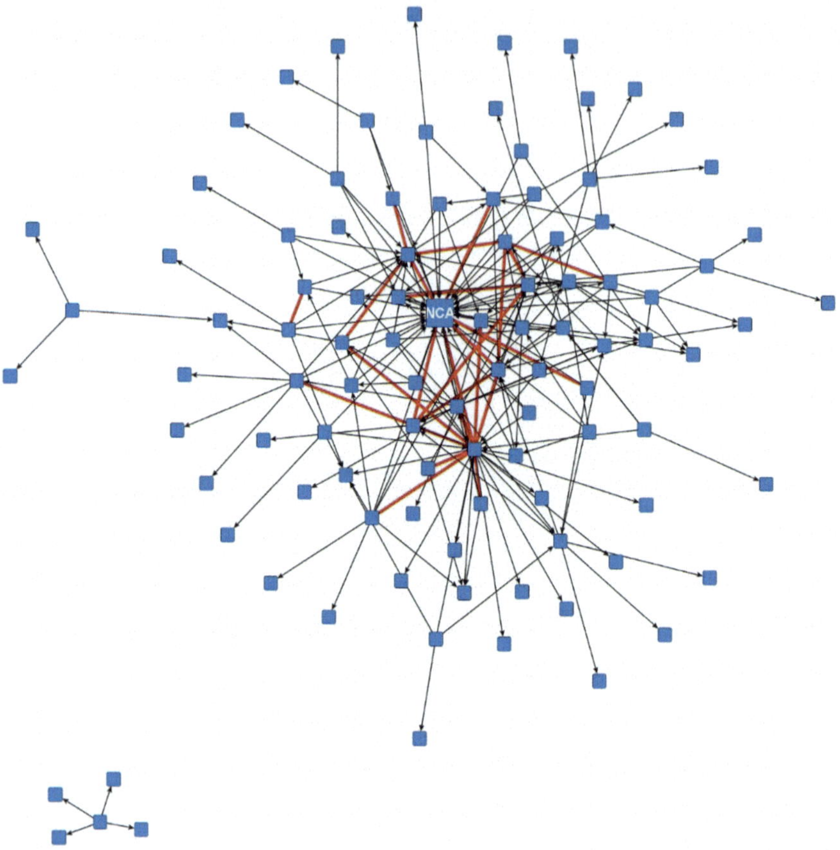

Fig. 14.5 Collaboration between actors (WNV scenario, min 1x mentioned, red lines reciprocal). Source: Design by authors

on the shortest path between any two actors but that it might also travel on longer paths (though is less likely to do so). Therefore, the shorter the paths are, the more weight they receive in calculating the flow betweenness centrality.

In Fig. 14.6, we show that the communication network is relatively centralized, with the NCA in the most central position. Although a lot more actors are involved in the WNV scenario, the structures in the two scenarios very much resemble each other, with the Ministry of Health and the Municipal Health Agencies (actors with formally described roles in the Public Health Act) in the following ranks with some distance to the NCA. The centralization of the communication structure around the NCA implies a lead organization type network in terms of governance, because the NCA is operationally involved and clearly by far the most central actor. However, there are also many linkages between the other actors.

Although links between the alters are present and encouraged in the ideal typical model suggested by Provan and Kenis (2008) to avoid information overload in the

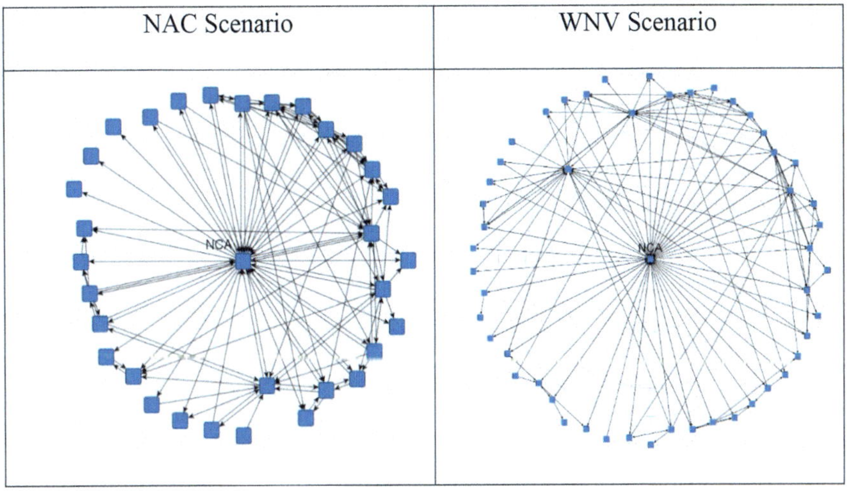

Fig. 14.6 Flow Betweenness Centrality based on giving information/advice (confirmed ties). Source: Design by authors

center and foster innovation, one might wonder how such a considerable number of links impact network governance. We believe this may prove a particular challenge when it comes to formulating a consistent message to the public and to health care professionals.

One possible solution for information provision and coordination is a network that is characterized by a central entity with more decentral hubs or brokers. With such a network structure, information provision and coordination can be achieved relatively efficiently through local hubs that each are responsible for a certain cluster or section of the network. Researchers have reported on this type of solution in earlier studies (Lemaire & Provan, 2009; Moynihan, 2009).

In Fig. 14.7 below, we depict the structure based on confirmed ties with regard to giving and receiving information or advice in terms of reacting to and controlling an infectious disease outbreak in the NAC and WNV scenarios. Compared to Fig. 14.6, the actors' prominence is not determined through flow betweenness centrality but by their status. In the two visualizations in the upper row, we have calculated status on the basis of outgoing ties; in other words, the more information ties actors have to other actors who again spread information to many other actors, the more prominently they are positioned; in other words, these can be labeled as *super spreaders*, to use a term from diffusion theory. This is reversed for the visualizations in the second row. Here, actors are the more prominent the more information they receive from organizations that already receive a lot of information. In this way, top receivers of information can be detected.

In both cases, the NCA evidently holds the top position. A set of about 10–15 organizations also exists that forms the top of the information distribution pyramid.

It becomes clear that some actors are both top information spreaders and information receivers, such as the NCA, the Ministry of Health, and the municipal health services. On the other hand, some actors are primarily receivers of information. The questions regarding the preparedness are then: Are the appropriate actors in the important positions, are the knowledge institutions well connected with the actors that make decisions and have coordination tasks, and are any actor positions in danger of information overload?

In Table 14.2, we summarize the results for the two scenarios.

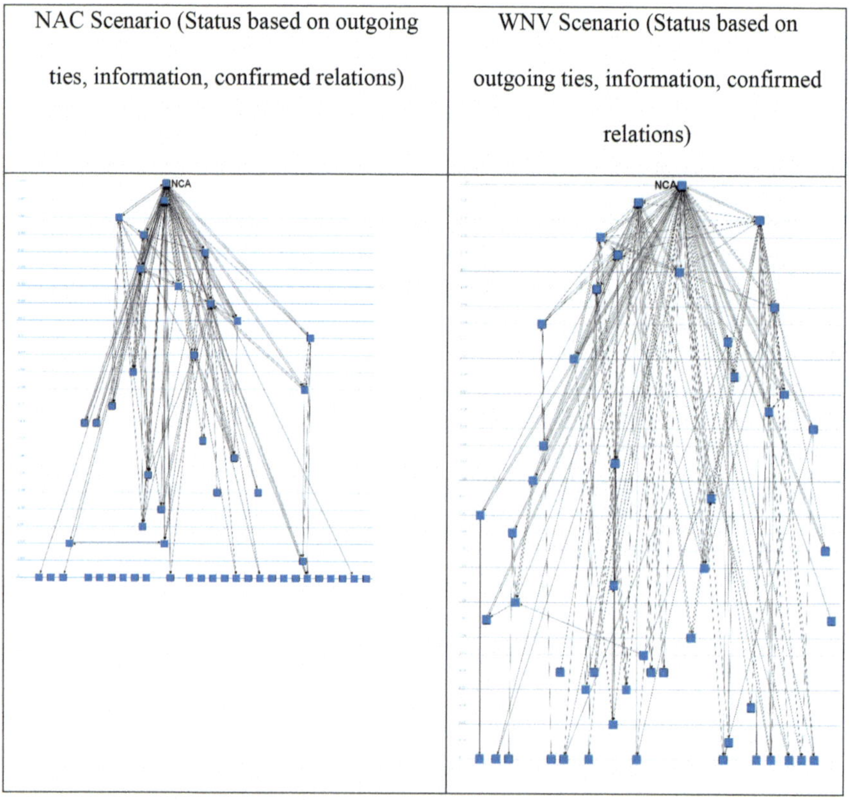

Fig. 14.7 Actor status on the basis of confirmed information ties. Source: Design by authors

NAC Scenario (Status based on incoming ties, information, confirmed relations)	WNV Scenario (Status based on incoming ties, information, confirmed relations)
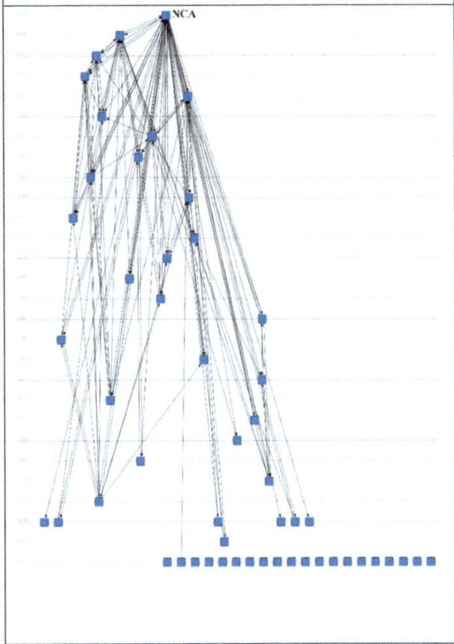	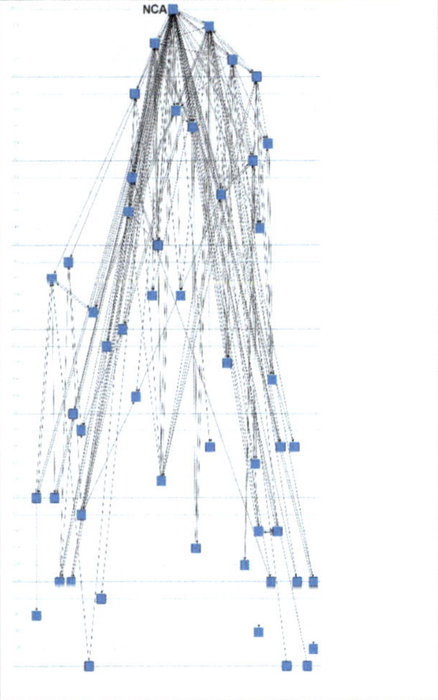

Fig. 14.7 (continued)

Discussion and Conclusion

In this paper, we present the organizational network governance approach as a way to analyze the response to a potential infectious disease outbreak. Although similar approaches have been gaining ground in the recent past to analyze organizational responses to disasters (Kapucu, 2009; Nowell, Steelman, Velez, & Yang, 2018) and to infectious disease outbreaks (Bdeir, Hossain, & Crawford, 2013) in particular, most researchers have conducted a retrospective analysis; in other words, they look at the response and the structures and governance that evolved after the fact. However, if the field wishes to improve preparedness for often uncertain threats, it must find a way to assess the capabilities and capacities of a response system to deal with a disaster or an outbreak before an incident happens. Researchers should therefore ideally gain information ex ante about the potential organizational network response to increase preparedness.

In this paper, we have developed a specific and feasible approach to produce such knowledge to limit transmission of the infectious disease and its impact. We have based this approach, on the one hand, on the observation that infectious disease threats by definition provoke an organizational network-response and, on the other

Table 14.2 Summary of results

	NAC scenario	**WNV scenario**
# Actors	61	98
Type of actors	Mainly core public health actors	Public health actors but also hunters, veterinarians, water management associations or the Ministry of Economic Affairs that deals with farming issues
Actor engagement (power-interest matrix): Self assessment, how much interest an organization has in participating in a response and its perceived influence over the outbreak control	General linear relationship between self-perceived power and interest with the north-west quadrant being empty, in other words, no actors indicating high power, but low interest in outbreak management	General linear relationship between self-perceived power and interest with the north-west quadrant being empty, that is, no actors indicating high power, but low interest. Several actors, especially groups of health care professionals indicating high interest but only medium or even low power in outbreak management
Joint involvement (spring embedder, two mode): In which control activity will your organization be involved?	28 control measures identified Similar tasks are often performed by several organizations. 9 core actors of important public health actors that are involved in taking different measures and could potentially coordinate the application of these diverse measures. However, there is no clear clustering of measures and actors, which likely requires significant conscious additional coordination	34 control measures identified Similar tasks are often performed by several organizations. Bifurcated structure with some actors on one side and the others on the other connected by different measures regarding information delivery and (public) communication
Collaboration (spring embedder): With which of the organization do you expect your organization will have the most intense collaboration?	Relatively dense core exists where actors also frequently nominate each other in a reciprocal way. However, also a lot of asymmetric nominations	Relatively dense core exists where actors also frequently nominate each other in a reciprocal way. Due to the higher number of actors, there are also more collaboration relations in the WNV scenario. However, also a lot of asymmetric nominations

(continued)

Table 14.2 (continued)

	NAC scenario	**WNV scenario**
# Actors	61	98
Information distribution: Provision/reception of information/advice (*flow betweenness*): To which of the organization will your organization provide information/advice, from which will it receive information/advice?	Structure is relatively centralized with the NCA in the most central position. Around the NCA implies a lead organization type network in terms of governance, since the NCA is operationally involved and clearly by far the most central actor. However, we also see a lot of linkages between the other actors	Structure is relatively centralized with the NCA in the most central position. Even though there are a lot more actors involved in the WNV scenario, the structures in the two scenarios very much resemble each other with the Ministry of Health and the Municipal Health Agencies (actors with formally described roles in the Public Health Act) in the following ranks with some distance to the NCA
Information distribution: Provision/reception of information/advice (*status*)	NCA holds the top position both in receiving and providing information. There is also a set of about 10–15 organizations which form the top of the information distribution pyramid. Analysis shows actors that are both top information spreaders and information receivers like NCA, the Ministry of Health, and the municipal health services. On the other hand, there are actors that are primarily receivers of information	Very similar structure compared to NAC scenario despite larger number of actors

Note. Source: Design by authors

hand, the fact that this response can fruitfully be analyzed using social network analysis. The fact that an infectious disease threat provokes an organizational network-response was confirmed by our data, which we collected for two fictitious but realistic scenarios of infectious disease outbreaks in the Netherlands, an outbreak of a New Asian Corona Virus (NAC) and an outbreak of the West Nile Virus (WNV). In both cases, respondents named a very high number of different organizations, with a substantial number of organizations outside the classical health architecture, as having a task in responding to the infectious disease threat.

Of course, such an ex ante approach with fictitious but realistic scenarios does not limit the need for, and usefulness of after-action assessments. Although collection of relational data might suffer from recall bias, especially if the incident is farther in the past, relational data based on scenarios might suffer from a social desirability bias. What one might do could differ from what would happen in reality. However, vignette studies, which are similar to our approach, have been proven to be a useful tool for research for some time and we have tried to minimize the potential bias in the relational data by basing the analysis on confirmed ties.

We continued our analysis by applying social network analysis to the scenarios of infectious disease outbreaks and arrived at several observations. We saw a large and highly differentiated network emerge in both cases. The network in the WNV scenario is much larger due to the different transmission paths (not human to human but via animals and mosquitoes). This results in the involvement of many more actors, especially those from outside the traditional (human) public health system, such as mosquito exterminators, water associations, and the ministry responsible for agriculture and nature. Attention needs to be paid to those actors that presumably have a strong influence but little interest. On the other hand, we can also identify some actors that assess themselves as having low power, although they appear quite important in the status analysis. The information structure is relatively centralized around the National Coordination Authority (NCA), which would imply a lead organization structure. However, in the analysis of the actor-measures structure, we see other actors besides the NCA as potential coordinating actors. If the NCA had an exclusive coordinating role, one would expect the NCA to serve as an exclusive connector between central measures to control the outbreak. Both networks are well integrated but not in a very structured way, which might lead to inefficiencies, difficulties creating a common information policy, and a risk of overburdening certain actors while not meeting the specific (information) needs of other actors. With our analysis, we show that given the structures based on various types of ties, the networks need additional conscious coordination efforts. But how can coordination in such an extensive fuzzy network of heterogeneous actors be ensured?

Looking particularly at the National Coordination Authority (NCA) in outbreak management reveals that respondents considered the NCA the "top receiver" and "top spreader" of information. Its position in the collaboration networks is, however, less clear. It is the central actor in the information sharing networks, but what this means for its position as the central coordinator remains a guess. This is mainly due to the fact that we identified several actors (and thus potential coordinators) in the core of those networks around the control measures. We were not able to identify the same actors as brokers in the actor networks. This might be caused by the large amount of interdependencies among the actors. In this study, we did not specifically ask the respondents about their thoughts on (needed) coordination. However, the nominations for collaborations clearly showed us that organizations have expectations towards each other, which could be further explored (e.g., Organization A expecting Organization B to perform a specific task and Organization B indicating that it has no role in this scenario). One possibility is that the NCA spreads the information actors need to perform their tasks and coordinate their work with others. From the information we have here, it is difficult to assess whether we have a "command and control approach," a "coordination and communication approach," or a "network governance approach," in terms of Moynihan's work (2009).

The difficulty with positioning any NCA in a response network coping with an infectious disease outbreak might lie in the fact that the same organization needs to fulfill both specific operational expert tasks in the organizational network and network administrative tasks (as defined by Provan & Kenis, 2008). Doing both by the same organization (and partly) or even by the same persons might lead to difficulties. These could be in only acquiring, processing, and evaluating one-sided information, perceived conflicts of interest or shortage of time to exhaustively fulfill both the operational expert as well as the monitoring and coordinating functions when it comes to task division and task allocation. This might lead to even more problems when it comes to contributing to the "integration of effort" (see Puranam, Alexy, & Reitzig, 2014). On the other hand, there are clear reasons for combining expert medical knowledge and coordination authority in an NCA. After all, it is crucial that the central player has immediate access to the best scientific knowledge and information about the outbreak. Weick (2009, Chap. 4) also recognized this dilemma in his analysis of the WNV outbreak in New York City in 1998. Ideally, the NCA is positioned in such a way in the network that it has (1) perceived authority/a mandate to coordinate the actors, (2) sufficient time/resources to pool the available information as impartially as possible, and (3) can recognize, monitor, and assess the evolving organizational network.

Looking at our results, one can argue that given the large size of the Dutch network and the lack of "natural" integration of many actors' efforts, the NCA can easily become overburdened in its attempts to facilitate all actors. There is a second risk: If the NCA is expected to inform and coordinate all individual actors, this could at least partially be in conflict with its role as an actor coordinating the organizational network.

Given the peculiarities of the evolving organizational network response in a situation of an infectious disease outbreak as well as the findings from our analysis above, another coordination structure to consider could be the core-periphery network structure as recently introduced by Nowell et al. (2018) in their discussion on the structure of effective governance of disaster response networks. Scholars have theorized a core-periphery structure as potentially providing the benefit of both the cohesion and stability of a closed network while also having the flexibility for the network to grow and recognize the importance of new actors who become involved because of the disease outbreak. It might thus be especially relevant given the fact that an organizational network response to an infectious disease outbreak can be expected to consist of many different sub processes, for example integration of scientific knowledge about the pathogen and risk groups with laboratory and epidemiological processes to understand the actual outbreak, developing and supporting the implementation of context specific control measures, (targeted) risk communication to health care professionals and general public, and so forth. The actors involved are organized along other individual and network goals: for example, patient care or water management in the case of the WNV. Consequently, the approach presented can initiate a dialogue among core actors to share information and reflect on the actual situation.

Limitations and Future Research

Looking at the recent evaluations of national and international outbreaks, we believe that the organizational network governance approach introduced here can strengthen outbreak management. At the same time, we are fully aware that this approach is far from complete and and researchers must address a number of important issues in the future.

In this study, we have in the first place focused on structure as one governance characteristic. This focus has produced important insights, as we have demonstrated in presenting our network analysis for the two scenarios. With this approach, we capture and visualize basic information about the actors as well as about the structure of the collaboration, knowledge, and information exchange. This explorative description creates a basis for practitioners and policy makers to engage in an ex ante assessment and identify and address possible bottlenecks and challenges as laid out above for the two scenarios used here. It also makes it possible to theorize about improving communication and governance of crisis management in general. In this study, we have thus understood "network" primarily as an empirical tool. But insights produced based on an organizational network governance approach used for ex ante and ex post analysis can address questions such as: Which mode(s) of governance and which network structures appear(s) to be most suitable for a rapid response to an outbreak at the (inter)national level? And which methods and data are most appropriate for assessing a response system ex ante or during the event in order to understand the needs and preferences for network-coordination?

However, the data we have collected for this study does not allow us to make any causal statements about the effectiveness of certain structures for the prevention of—or even a response to—an actual event. One way to get a step closer would be to run serious games/simulations with practitioners, in which certain structural characteristics could be manipulated and then assessed by the participants and observers in terms of the effectiveness of the information transmission and coordination. In the long run, studying organizational network responses and their outcomes together with ex ante assessments should improve the field's understanding of the relationship between structures, governance, and effectiveness of responses under different conditions and for different infectious diseases.

Future research should also unpack the relationship between governance and disasters and complexity in disasters in terms of authority and behavior. Given the absence of features related to legal autonomy and formal authority for the largest part of the organizational network, the question is what role authority plays in these situations and how this can take form. The same is the case for actors' behavior. What factors actually motivate actors to take on responsibilities in the context of an organizational network to limit the transmission of the virus and its impact, even if public health is not part of their core mission and activity?

Given new global health threats and their potentially enormous impact, it is essential that insights are gained in which governance characteristics of the organizational network produce the type of knowledge necessary for limiting the transmission of the virus and its impact. This is not only vital for coordination in the implementation of measures coping with the outbreak, but already when a situation is perceived and analyzed at the start. As Weick (2009, p. 53) states, "if we spot flaws in collective induction, then we may find an explanation for their genesis in the way people are organizing. Stated more compactly, the degree of intelligence manifest by a network of nodes may be determined by the quality, not just the quantity of its interconnectivity (Taylor and Van Every, 2000:213)."

Appendix

The results presented in this chapter are based on research that had been conducted since 2015 before the Covid-19 pandemic. Unfortunately, one of the scenario's that we used for the study, that is, the outbreak of a New Asian Corona Virus, actually came true faster than we could have imagined. The actual outbreak made us aware that our study is based on the implicit assumption that the outbreak of a Corona virus would remain a local outbreak that we would be able to contain. In March, the Dutch public health system was quickly overwhelmed by the speed and the scale of the outbreak that had been happening under the radar since late February in the Netherlands. Therefore, we think that the results are applicable only for the first two weeks in March for the actual outbreak when the system was still in a containment phase. Once the outbreak came into a situation of community spread and was recognized as such, the whole governance system changed significantly. Given the scale and severity of the situation, the response to the outbreak was located at the highest government level on a daily basis and the Dutch prime minister instituted an informal decision and coordination body that met once a week that had not been part of any planning beforehand.

However, even in a changed governance system, we could confirm some of the major findings of the study as well. First, we could see that it is essential that the National Coordination Authority

(1) has a perceived authority/a mandate to coordinate the actors, (2) has sufficient time/resources to pool the available information as impartially as possible, and (3) can recognize, monitor, and assess the evolving organizational network. In addition, we could observe the difficulties we predicted for the coordination of a highly differentiated system of tasks and actors with overlapping activities and roles. The Dutch government reacted after some time mainly by centralizing the coordination of essential tasks such as getting protective material, distribution of ICU beds or testing and tracing. What the actual outbreak also confirmed was the difficult position, the National Coordination Authority finds itself in due to the fact that it fulfills several different roles such as knowledge hub, coordination center, public information organization and the public authority that issues rules and guidelines.

Most importantly, though, we have seen in the actual outbreak that in differentiated democratic societies such as the Netherlands, coping with such outbreaks is not only an epidemiological but also a complex organizational and governance challenge. The perspective we put forward in this paper therefore is very valid and essential and should receive the necessary attention as we move forward in this pandemic and prepare for possible new outbreaks of infectious diseases in the future.

Acknowledgements The empirical part of this paper is based on the work by Lisanne Bosch and Yvonne Euverman. We thank them for their help and support in developing the questionnaire and collecting the data. We also thank Jacklien Maessen for assistance in the data analysis.

References

Ansell, C., & Gash, A. (2008). Collaborative governance in theory and practice. *Journal of Public Administration Research and Theory, 18,* 543–571. https://doi.org/10.1093/jopart/mum032

Ansell, C., & Torfing, J. (2016). *Handbook on theories of governance*. Cheltenham: Edward Elgar.

Barzon, L., Pacenti, M., Franchin, E., Lavezzo, E., Martello, T., Squarzon, L., Toppo, S., Fiorin, F., Marchiori, G., Russo, F., Cattai, M., Cusinato, R., & Palu, G. (2012). New endemic West Nile virus lineage 1a in northern Italy, July 2012. *Eurosurveillance, 17,* Article20231. https://doi.org/10.2807/ese.17.31.20231-en

Bdeir, F., Hossain, L., & Crawford, J. (2013). Emerging coordination and knowledge transfer process during disease outbreak. *Knowledge Management Research & Practice, 11,* 241–254. https://doi.org/10.1057/kmrp.2012.1

Bellini, R., Zeller, H., & van Bortel, W. (2014). A review of the vector management methods to prevent and control outbreaks of West Nile virus infection and the challenge for Europe. *Parasites & Vectors, 7,* Article323. https://doi.org/10.1186/1756-3305-7-323

Berthod, O., Grothe-Hammer, M., Müller-Seitz, G., Raab, J., & Sydow, J. (2016). From high-reliability organizations to high-reliability networks: The dynamics of network governance in the face of emergency. *Journal of Public Administration Research and Theory, 27,* 479–511. https://doi.org/10.1093/jopart/muw050

Borgatti, S. P., Everett, M. G., & Johnson, J. C. (2018). *Analyzing social networks* (2nd ed.). London: SAGE.

Bowen, R. A., Bosco-Lauth, A., Syvrud, K., Thomas, A., Meinert, T. R., Ludlow, D. R., Cook, C., Salt, J., & Ons, E. (2014). Protection of horses from West Nile virus Lineage 2 challenge following immunization with a whole, inactivated WNV lineage 1 vaccine. *Vaccine, 32,* 5455–5459. https://doi.org/10.1016/j.vaccine.2014.07.093

Brandes, U., & Wagner, D. (2004). Visone: Analysis and visualization of social networks. In M. Jünger & P. Mutzel (Eds.), *Graph drawing software* (pp. 321–340). Berlin: Springer.

Chancey, C., Grinev, A., Volkova, E., & Rios, M. (2014). The global ecology and epidemiology of West Nile virus. *BioMed Research International, 2015,* Article376230. https://doi.org/10.1155/2015/376230

ECDC (European Centre for Disease Prevention and Control). (2010). West Nile virus infection outbreak in humans in Central Macedonia, Greece: July–August 2010. Stockholm: ECDC.

ECDC (European Centre for Disease Prevention and Control). (2013). West Nile virus risk assessment tool. Stockholm: ECDC.

Emerson, K., & Nabatchi, T. (2015). Evaluating the productivity of collaborative governance regimes: A performance matrix. *Public Performance & Management Review, 38,* 717–747. https://doi.org/10.1080/15309576.2015.1031016

Glückler, J., Dehning, W., Janneck, M., & Armbrüster, T. (2012). *Unternehmensnetzwerke: Architekturen, Strukturen und Strategien* [Corporate networks: Architectures, structures and strategies]. Berlin: Springer.

Glückler, J., & Doreian, P. (2016). Editorial: Social network analysis and economic geography: Positional, evolutionary and multi-level approaches. *Journal of Economic Geography, 16,* 1123–1134. https://doi.org/10.1093/jeg/lbw041

Glückler, J., Lazega, E., & Hammer, I. (Eds.). (2017). *Knowledge and networks.* Knowledge and Space: Vol. 11. Cham: Springer.

Glückler, J., & Panitz, R. (2016). Relational upgrading in global value networks. *Journal of Economic Geography, 16,* 1161–1185. https://doi.org/10.1093/jeg/lbw033

Head, B. W. (2008). Wicked problems in public policy. *Public Policy, 3,* 101–118.

Hu, Q., Knox, C. C., & Kapucu, N. (2014). What have we learned since September 11, 2001? A network study of the Boston Marathon bombings response. *Public Administration Review, 74,* 698–712. https://doi.org/10.1111/puar.12284

Huizer, Y. L., Kraaij-Dirkzwager, M. M., Timen, A., Schuitmaker, T. J., van Steenbergen, J. E. (2014). Context analysis for epidemic control in the Netherlands. *Health Policy, 119,* 66 73. https://doi.org/10.1016/j.healthpol.2014.10.004

Iyer, A. V., & Kousoulas, K. G. (2013). A review of vaccine approaches for West Nile virus. *International Journal of Environmental Research and Public Health, 10,* 4200–4223. https://doi.org/10.3390/ijerph10094200

Kalaycioglu, H., Korukluoglu, G., Ozkul, A., Oncul, O., Tosun, S., Karabay, O., Gozalan, A., Uyar, Y., Caglayık, D. Y., Atasoylu, G., Altas, A. B., Yolbakan, S., Ozden, T. N., Bayrakdar, F., Sezak, N., Pelıtlı, T. S., Kurtcebe, Z. O., Aydın, E., & Ertek, M. (2012). Emergence of West Nile virus infections in humans in Turkey, 2010 to 2011. *Eurosurveillance, 17,* Article20182. https://doi.org/10.2807/ese.17.21.20182-en

Kapucu, N. (2009). Interorganizational coordination in complex environments of disasters: The evolution of intergovernmental disaster response systems. *Journal of Homeland Security and Emergency Management, 6*(1), Article 47. https://doi.org/10.2202/1547-7355.1498

Kenis, P. (2016). Network. In C. Ansell & J. Torfing (Eds.), *Handbook on theories of governance* (pp. 149–157). Cheltenham: Edward Elgar.

Kickbusch, I., Cassels, A., & Liu, A. (2016). New directions in governing the global health domain: Leadership challenges for WHO (Global Health Centre Working Paper No. 13). Retrieved from https://repository.graduateinstitute.ch/record/294882/files/GHPWP_13_Ilona_2016.pdf?version=1

Kraaij-Dirkzwager, M. M., Schol, L. G., Schuitmaker-Warnaar, T. J., Timen, A., & van Steenbergen, J. E. (2019). Stakeholder involvement in outbreak management: To fear or not to fear? *International Journal of Information Systems for Crisis Response and Management (IJISCRAM), 11*(2), 57–78. https://doi.org/10.4018/IJISCRAM.2019070104

Lemaire, R. H., & Provan, K. G. (2009, October). *Network governance in a publicly funded child and youth health network: Sub-network embeddedness, cohesiveness, and the role of brokers.* Paper presented at the Public Management Research Conference, Columbus, OH. Retrieved from http://citeseerx.ist.psu.edu/viewdoc/download?doi=10.1.1.522.4320&rep=rep1&type=pdf

Lim, S. M., Koraka, P., Osterhaus, A. D. M. E., & Martina, B. E. E. (2011). West Nile virus: Immunity and pathogenesis. *Viruses, 3,* 811–828. https://doi.org/10.3390/v3060811

Majchrzak, A., Jarvenpaa, S. L., & Hollingshead, A. B. (2007). Coordinating expertise among emergent groups responding to disasters. *Organization Science, 18,* 147–161. https://doi.org/10.1287/orsc.1060.0228

Medaglini, D., Santoro, F., & Siegrist, C. A. (2018). Correlates of vaccine-induced protective immunity against Ebola virus disease. *Seminars in Immunology, 39,* 65–72. https://doi.org/10.1016/j.smim.2018.07.003

Moon, S., Sridhar, D., Pate, M. A., Jha, A. K., Clinton, C., Delaunay, S., Edwin, V., Fallah, M., Fidler, D. P., Garrett, L., Goosby, E., Gostin, L. O., Heymann, D. L., Lee, K., Leung, G. M., Morrison, J. S., Saavedra, J., Tanner, M., Leigh, J. A., Hawkins, B., Woskie, L. R., & Piot, P. (2015). Will Ebola change the game? Ten essential reforms before the next pandemic:

The report of the Harvard-LSHTM Independent Panel on the Global Response to Ebola. *The Lancet, 386,* 2204–2221. https://doi.org/10.1016/S0140-6736(15)00946-0
Moynihan, D. P. (2009). The network governance of crisis response: Case studies of incident command systems. *Journal of Public Administration Research and Theory, 19,* 895–915. https://doi.org/10.1093/jopart/mun033
Munjita, S. M., Chileshe, M., & Mutemwa, S. (2015). Ebola virus disease in West Africa: A call to overhaul health systems in Sub-Saharan Africa. *International Journal of Medical Science and Public Health, 4,* 873–877. https://doi.org/10.5455/ijmsph.2015.03022015216
Nowell, B., & Steelmann, T. (2015). Communication under fire: The role of embeddedness in the emergence and efficacy of disaster response communication networks. *Journal of Public Administration Research and Theory, 25,* 929–952. https://doi.org/10.1093/jopart/muu021
Nowell, B., Steelman, T., Velez, A.-L. K., & Yang, Z. (2018). The structure of effective governance of disaster response networks: Insights from the field. *The American Review of Public Administration, 48,* 699–715. https://doi.org/10.1177/0275074017724225
Pem-Novosel, I., Vilibic-Cavlek, T., Gjenero-Margan, I., Pandak, N., Peric, L., Barbic, L., Listes, E., Cvitkovic, A., Stevanovic, V., & Savini, G. (2014). First outbreak of West Nile virus neuroinvasive disease in humans, Croatia, 2012. *Vector-Borne and Zoonotic Diseases, 14,* 82–84. https://doi.org/10.1089/vbz.2012.1295
Provan, K. G., Fish, A., & Sydow, J. (2007). Interorganizational networks at the network level: A review of the empirical literature on whole networks. *Journal of Management, 33,* 479–516. https://doi.org/10.1177/0149206307302554
Provan, K. G., & Kenis, P. (2008). Modes of network governance: Structure, management, and effectiveness. *Journal of Public Administration Research and Theory, 18,* 229–252. https://doi.org/10.1093/jopart/mum015
Puranam, P., Alexy, O., & Reitzig, M. (2014). What's "new" about new forms of organizing? *Academy of Management Review, 39,* 162–180. https://doi.org/10.5465/amr.2011.0436
Raab, J., & Kenis, P. (2009). Heading toward a society of networks: Empirical developments and theoretical challenges. *Journal of Management Inquiry, 18,* 198–210. https://doi.org/10.1177/1056492609337493
Raab, J., Mannak, R. S., & Cambré, B. (2015). Combining structure, governance and context: A configurational approach to network effectiveness. *Journal of Public Administration Research and Theory, 25,* 479–511. https://doi.org/10.1093/jopart/mut039
Raab, J., van den Oord, S., & Kenis, P. (2015, June 11–13). *Modes of network governance revisited: Where, why and how have the governance modes been used?* Paper presented at the Public Management Research Conference, Minneapolis, MN.
Rainie, L., & Wellman, B. (2012). *Networked: The new social operating system.* Cambridge, MA: The MIT Press.
RIVM. (2012). National guidance for infectious disease threats and crises. Bilthoven.
RIVM. (2019, August 8). About RIVM. Retrieved from https://www.rivm.nl/en/about-rivm/organisation/national-coordination-centre-for-communicable-disease-control
Rizzoli, A., Jimenez-Clavero, M. A., Barzon, L., Cordioli, P., Figuerola, J., Koraka, P., Martina, B., Moreno, A., Nowotny, N., Pardigon, N., Sanders, N., Ulbert, S., & Tenorio, A. (2015). The challenge of West Nile virus in Europe: Knowledge gaps and research priorities. *Eurosurveillance, 20,* Article21135. https://doi.org/10.2807/1560-7917.ES2015.20.20.21135
Sejvar, J. J. (2014). Clinical manifestations and outcomes of West Nile virus infection. *Viruses, 6,* 606–623. https://doi.org/10.3390/v6020606
Smith, K. F., Goldberg, M., Rosenthal, S., Carlson, L., Chen, J., Chen, C. C., & Ramachandran, S. (2014). Global rise in human infectious disease outbreaks. *Journal of The Royal Society Interface, 11,* Article20140950. https://doi.org/10.1098/rsif.2014.0950
Stoto, M. A., & Higdon, M. A. (2015). *The public health response to 2009 H1N1: A systems perspective.* New York, NY: Oxford University Press.
Swaan, C., Öry, A., Schol, L. G., Jacobi, A., Richardus, J., & Timen, A. (2017). Ebola preparedness in the Netherlands: The need for coordination between the public health and the curative

sector. *Journal of Public Health Management and Practice, 24,* 18–25. https://doi.org/10.1097/PHH.0000000000000573

Vinck, L., Isken, L., Hooiveld, M., Trompenaars, M., Ijzermans, J., & Timen, A. (2011). Impact of the 2009 influenza A(H1N1) pandemic on public health workers in the Netherlands. *Eurosurveillance, 16*(7). Retrieved from https://www.eurosurveillance.org/content/10.2807/ese.16.07.19793-en

Volksgezondheidenzorg.info. (n.d.). Public health and health care. In Onderwerpen (Section 2). Retrieved from https://www.volksgezondheidenzorg.info/onderwerp/english/introduction#node-public-health-and-health-care

Weaver, S. C., & Reisen, W. K. (2010). Present and future arboviral threats. *Antiviral Research, 85,* 328–345. https://doi.org/10.1016/j.antiviral.2009.10.008

Weick, K. E. (2006). Managing the unexpected: Complexity as distributed sensemaking. In R. R. McDaniel & D. J. Driebe (Eds.), *Uncertainty and surprise in complex systems* (pp. 51–65). Berlin: Springer.

Weick, K. E. (2009). *Making sense of the organization, volume 2: The impermanent organization.* Chichester: John Wiley & Sons.

Wet publieke gezondheid. [Public Health Act]. (2017, May 30). The Netherlands. Retrieved from http://www.rivm.nl/Onderwerpen/W/Wet_publieke_gezondheid.

Chapter 15
Collective Learning and Institutional Collective Action in Fragmented Governance

Serena Y. Kim, William L. Swann, and Richard C. Feiock

Fragmentation of authority presents opportunities and challenges for governance. Multiple and/or overlapping political boundaries and jurisdictions can benefit governments by expanding citizen choice, increasing policy experimentation and knowledge diffusion, and reducing public expenditures (Andersson & Ostrom, 2008; McGinnis, 1999; Ostrom, Tiebout, & Warren, 1961; Schneider, 1986). However, fragmentation and lack of integration also produces externalities, or spillovers affecting third parties, that exacerbate institutional collective action (ICA) dilemmas, creating situations where authorities' particularistic incentives are misaligned with collective interests (Feiock, 2009, 2013; Swann & Kim, 2018). For instance, in the context of air quality management, all governments in a fragmented region benefit from reduced air pollution, but each government has an incentive to free-ride and not contribute to the costs of achieving cleaner air.

ICA dilemmas hinder service delivery and common pool resource protection, and lead to fiscal problems and unsustainable outcomes (Hendrick, Jimenez, & Lal, 2011; Jimenez, 2014). Thus, finding ways for governments to overcome ICA dilemmas is important for policymakers and researchers. The ICA framework (Feiock, 2007, 2013) is a valuable analytical lens for understanding how governments use integrative mechanisms to overcome barriers to collective action and promote collaborative governance. The ICA empirical literature has explored the structural, social, political, and economic factors shaping collaborations (Andrew & Carr, 2013; Berardo & Scholz, 2010; Feiock & Scholz, 2010; Feiock, Steinacker, & Park, 2009; Gerber, Henry, & Lubell, 2013; LeRoux, Brandenburger, & Pandey, 2010),

S. Y. Kim · W. L. Swann
School of Public Affairs, University of Colorado Denver, Denver, CO, USA
e-mail: serena.kim@ucdenver.edu; william.swann@ucdenver.edu

R. C. Feiock (✉)
Askew School of Public Administration and Policy, Florida State University, Tallahassee, FL, USA
e-mail: rfeiock@fsu.edu

J. Glückler et al. (eds.), *Knowledge for Governance*, Knowledge and Space 15,
https://doi.org/10.1007/978-3-030-47150-7_15

but little attention has been given to the role of learning in ICA dilemmas. Although learning is implicitly assumed to be a critical mechanism through which governments resolve ICA dilemmas, prior research has not developed a theory to articulate how this process may work.

The importance of learning is underscored in multiple literatures, including policy diffusion (Berry & Berry, 1999; Shipan & Volden, 2008), institutional change (Ostrom, 1990, 2005), advocacy coalitions (Sabatier, 1988), and policy learning (Dunlop & Radaelli, 2018; Tamtik, 2016). Organizational theorists have also long considered how learning facilitates improved decision-making, practices, and performance (Argyris, 2003; Argyris & Schön, 1978; Cohen & Levinthal, 1990; March & Olsen, 1995; Senge, 1990). More recently, researchers have paid attention to learning in collaborative governance, as collaborators face increased uncertainty and complexity (Crona & Parker, 2012), but the relationship between learning and collaboration is complex. Interorganizational collaboration is thought to enhance *collective learning* processes and their cognitive and behavioral products (Gerlak & Heikkila, 2011; Hartley & Benington, 2006; Powell, 1998; Powell, Koput, & Smith-Doerr, 1996). However, learning is also believed to be a key facilitator of collaboration, social networking, and institutional change (Ansell, Lundin, & Öberg, 2017; Koontz, Gupta, Mudliar, & Ranjan, 2015).

To unpack this relationship, we posit a dynamic, iterative process through which collaboration leads to collective learning, which can subsequently enable governments to share and expand knowledge to better overcome ICA dilemmas in the future, or exacerbate such dilemmas if actors collectively learn to be more opportunistic. This iterative relationship lends itself to theories of collaborative governance evolution in that governments with little or no history of collaboration begin by solving smaller, “first-order” coordination problems of information exchange before taking on larger, “second-order” cooperation problems (Feiock & Scholz, 2010) such as formalizing collaborative institutions. Although the ICA literature implies a logical connection between collaboration and learning (Hawkins, Krause, Feiock, & Curley, 2017; Swann, 2017; Zeemering, 2019), extant scholarship has yet to tease out the explicit roles and implications for learning within ICA scenarios.

In this chapter, we examine the role of collective learning in employing alternative integrative mechanisms for overcoming ICA dilemmas. We ask: How does experience in collective action affect collective learning, and subsequently, how does it influence the ability of governments to overcome future barriers to collaboration? Surveying the extant literature and utilizing illustrative examples, we derive propositions for how collaborative governance may affect collective learning, and how collective learning may in turn enable governments to resolve future collaboration problems such as resource sharing, large-scale organizational reforms, and *wicked* policy problems.

In the next section, we review the ICA framework, discussing the implications of its empirical research for knowledge governance. We then define collective learning and describe how it relates to collaborative governance. Next, we advance propositions for the relationship between collective learning and integrative mechanisms for overcoming ICA dilemmas. We conclude by discussing the implications of knowledge governance for practice and future research.

ICA Foundations, Components, and Empirical Applications

The ICA framework offers an institutional explanation for the emergence and durability of collective action among public authorities such as municipalities, public agencies, public service organizations, or national governments. Over the decade, many scholars have applied this framework to explain efforts to mitigate ICA dilemmas and improve collaborative governance. This section describes ICA's intellectual foundations, components, and applications to knowledge governance.

Theoretical Foundations

ICA proponents draw on elements of five theoretical traditions to explain the emergence and durability of collaborative governance. ICA is first informed by the collective action literature, whose authors hold that individuals' incentives may or may not align with collective interests, depending on the nature of the good and contextual governing rules (Olson, 1965; Ostrom, 2005). For example, Ostrom (1990) demonstrates how principals voluntarily govern themselves to obtain joint benefits and ensure the sustainable use of natural resources in the face of individual incentives to behave opportunistically. ICA assumes that policy actors can design and cultivate institutions based on trust, reciprocity, and mutuality to overcome collective action problems through means other than regulation or privatization (cf. Glückler, Suddaby, & Lenz, 2018).

Second, ICA proponents draw from local public economies (LPEs) researchers (McGinnis, 1999; Ostrom et al., 1961), who posit that polycentricity, or "having multiple governing authorities at differing scales rather than a monocentric unit" (Ostrom, 2010, p. 552), can expand citizen choice, facilitate self-governance, and produce some public services more efficiently. However, polycentricity yields spillovers, in the form of positive and negative externalities that create collective action dilemmas, such as in air quality management and environmental policy. Feiock and Scholz (2010) classify these dilemmas as problems of "coordination" and "cooperation." The former relates to information asymmetries and the inability of actors to identify joint opportunities; the latter involves how to incentivize actors to work collectively and avoid opportunistic behavior. For example, resolving a coordination problem like information asymmetry might entail joint fact finding about air pollution sources and effects, whereas overcoming a cooperation problem, such as resource sharing or governmental reforms for climate protection, would require building trust and shared norms and values.

Third, ICA proponents consider how transaction costs are associated with obtaining information, negotiating and enforcing agreements, and aligning actors' values, preferences, and incentives (Feiock, 2007; Williamson, 1991). In the context of ICA situations, Feiock (2013) highlights two dimensions of transaction costs: *autonomy* and *decision-making* costs. ICA assumes that such costs increase as decisions become more hierarchically and/or externally imposed and deviate from local preferences. For instance, as air quality policies emanate from increasingly higher levels of government, local authorities have increasingly less discretion in their implementation and thus incur higher autonomy costs.

Fourth, ICA incorporates social network theories to explain how social capital—or the shared social resources (e.g., trust, reciprocity, and social structures) that can lead to collective action—reduces the likelihood of opportunistic behavior and lower barriers to collaboration (Burt, 1992; Coleman, 1988; Putnam, Leonardi, & Nonetti, 1993). ICA assumes that increasing *strong* and *weak* network ties among actors will enhance the emergence and durability of collaborations. For example, city governments with weak ties *bridging* them to other actors and networks may have more information available to innovate and implement air quality policies, but cities with strong, *bonding* ties built on reciprocity, trust, and shared norms would be more likely to commit to improving air quality because they are confident their partners will live up to agreements.

Finally, ICA researchers draw on policy tools (Salamon, 2002) and political markets (Feiock, Lubell, & Lee, 2014) to understand the interplay between government policy suppliers and constituent group policy demanders. ICA assumes policy choice and design are a function of political bargaining and negotiation between governmental officials and affected constituencies. In air quality management, this would imply that constituencies favoring environmental protection versus development would vie for the attention and favorable treatment of policymakers by providing their political support.

Integrative Mechanisms, Transaction Costs, and Collaboration Risk

Collaborative arrangements to overcome ICA dilemmas take a number of distinct forms. Feiock (2013) proposes a taxonomy of 12 generic integrative mechanisms, ranging from informal networks, where local actors preserve their autonomy and rely on trust and reciprocity to develop cooperative agreements, to centrally imposed authority, where consolidated governments are large enough to internalize policy spillovers and achieve economies of scale but at the expense of localized autonomy and policy variation. Figure 15.1 displays these mechanisms.

With the ICA framework, researchers posit that authorities can better link the integrative mechanism to the ICA dilemma by matching the scale and coerciveness of the integrative mechanism to the scale and nature of the policy problem. For instance, transportation planning in the United States is often delegated to regional authorities, such as metropolitan planning organizations (MPOs), to avoid particularistic decisions and better coordinate across jurisdictions, whereas informal networks have been often used for environmental management in areas like watersheds

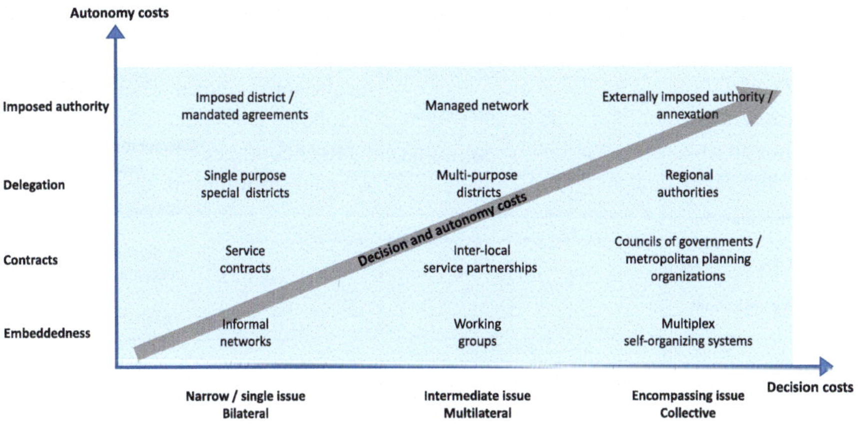

Fig. 15.1 Integrative mechanisms for overcoming ICA dilemmas. Adapted from R. C. Feiock (2013, p. 404). Copyright 2013 by John Wiley & Sons, Inc. Adapted with permission

that span boundaries and involve numerous, diffuse, and heterogenous resource users (Scholz, Berardo, & Kile, 2008).

Employing an integrative mechanism can resolve a collective action problem, but it imposes varying levels of transaction costs. Figure 15.1 illustrates how decision-making and autonomy costs relate to integrative mechanisms for institutional collective action (Feiock, 2013). The horizontal dimension captures the level of scope and complexity in collaboration, ranging from single issues on a bilateral scale to encompassing issues on a multilateral scale. The vertical dimension indicates the four levels of authority for collaborative arrangements: social embeddedness, contracts, delegation, and imposed authority. The diagonal dimension shows how autonomy and decision-making costs increase as issue complexity rises and authority becomes centralized. ICA proponents assume the preferred integrative mechanism overcomes the ICA dilemma by imposing the least costs and affording the most autonomy.

"Collaboration risk," or actors' perception of the likelihood that collaborative efforts will fail, is a key determinant of the integrative mechanism choice (Feiock, 2013). In theory, more authoritative and encompassing mechanisms will be necessary as collaboration risk increases. Although transaction costs are a function of the integrative mechanism choice, collaboration risks reflect actors' ex-ante subjective assessment of collaboration success or failure. Following Maser (1998), three types of collaboration risks are defined: *Coordination risk* refers to the inability to reconcile mutually beneficial opportunities; *division risk* concerns the difficulty of agreeing on the distribution of collaborative gains and costs; and *defection risk* is the likelihood that participants will not comply with a cooperative agreement (Feiock, 2013).

Collaboration risk emerges from the "collaboration situation," which consists of the (1) "nature of the problem" (e.g., coordination or cooperation problems, minimizing common pool resource problems, or internalizing externalities); (2) "actor

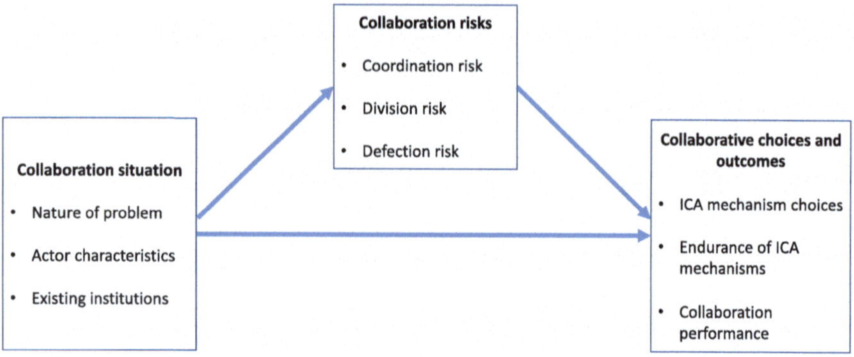

Fig. 15.2 Linkages between collaboration situation, risks, and choices and outcomes. Source: Design by authors

characteristics" (e.g., stakeholder preferences, interests, political dispositions, and resource asymmetries); and (3) "existing institutions" (e.g., rules, norms, shared values, and structural characteristics, such as actors' individual or organizational networks, influencing how they have interacted in the past). In other words, collaboration risk is conditioned by the nature of the problem, actor characteristics, and existing institutions, which in turn influences the integrative mechanism choice.

Figure 15.2 illustrates this relationship, that is, how the collaboration situation relates to collective outcomes and integrative mechanism choices. Many of the ICA empirical literature's authors have examined these relationships, mostly the direct effect of the collaboration situation on collaboration choices and outcomes. In this chapter, we add a fourth component—collective learning—to this model and argue that it has the potential to reduce collaboration risks and lead to more self-organizing integrative mechanisms, such as informal networks, that impose fewer decision-making and autonomy costs. To begin developing this theory, we first need to consider how the ICA literature informs knowledge governance.

Empirical Applications for Knowledge Governance

Although both are integral to learning, information and knowledge are different. Generally, *information* consists of facts about something or someone, but *knowledge* is gained through experience, education, and the analysis of information. In this sense, learning relates more to knowledge than information, but having (good) information should facilitate learning.

Obtaining information about collaboration can be costly but is essential (Feiock et al., 2009). So far, researchers have focused on how the informal networking of individual policymakers influences decisions to formally collaborate interorganizationally (Andrew & Carr, 2013; Hawkins, Hu, & Feiock, 2016; LeRoux et al., 2010). They claim that informal networking enables actors to build trust and shared understanding through dialogue and information exchange, which can lead to formal interorganizational collaborations. Although not explicitly mentioned, learning could be a key mechanism through which informal networking influences these partnerships, as information sharing enables actors to learn about similar projects in other jurisdictions and serve as brokers in collaboratives (Spekkink & Boons, 2016).

Information is not only a facilitator of collaboration but can also be obtained, assimilated, and disseminated through collaboration. ICA researchers investigating estuary policy networks, for example, suggest that organizational actors seek out partnerships with popular, well connected organizations to obtain informational resources (Berardo & Scholz, 2010; Scholz et al., 2008). Actors use network ties to obtain more information about other actors' credibility and to increase their technical knowledge of the policy area.

Some collaboratives, however, may inhibit information acquisition, especially over time. Governments tend to collaborate with other governments that are politically, demographically, and geographically similar (Gerber et al., 2013; Lee, 2016). Although homophily-based collaboration may be beneficial for lowering initial barriers like lack of trust, such collaboration may not be conducive to generating new information, because similar actors in the same networks likely have similar information. Thus, actors may be better off collaborating with heterogeneous partners to acquire new information (Burt, 1992).

Acquiring knowledge goes a step further than gathering information and requires combining information with analysis and experience. One way actors build knowledge in collaboration is through focusing on intermediate outcomes or "small wins" (Ansell & Gash, 2008; Huxham & Vangen, 2005). This approach is similar to what Feiock and Scholz (2010) discuss in terms of addressing first-order dilemmas such as coordinating joint efforts for planning and fact finding, before tackling second-order dilemmas such as cooperating in resource exchanges. This evolution of collaboration lends itself to collective learning: Actors build a foundation for learning as they gather information, resolve coordination issues, and achieve intermediate outcomes, and build knowledge incrementally through the cooperative application of information towards collective aims. In the following section, we draw on collective learning research to begin building a theory to explain the relation between collective learning and institutional collective action.

Learning and Collective Action

In recent years, scholars have emphasized the importance of learning in collective action (Ansell & Gash, 2008; Gerlak & Heikkila, 2011; Heikkila & Gerlak, 2013; Koontz et al., 2015; Newig, Günther, & Pahl-Wostl, 2010). Although they have defined learning in numerous ways, scholars generally agree that it involves both *cognitive* and *behavioral* changes. From this perspective, it is insufficient for one to learn by experiencing changes in the brain unless such changes are acted upon (Argyris, 2003). However, attributing policy change to learning has proved difficult

(Heikkila & Gerlak, 2013), because policy change can occur for many reasons such as mimicry and imitation, bargaining and negotiation, and exogenous influences such as economic, political, and cultural shifts, which may or may not involve learning.

Learning occurs at both the individual and collective level. Although individuals can experience cognitive and behavioral changes, these changes do not necessarily transfer to the group. Collective learning involves alteration to social structures, which are comprised of institutional (i.e., formal and informal rules, norms, and expectations) and relational (i.e., social networks) components (Newig et al., 2010). However, the process through which individual learning transfers to the group, or vice versa, remains unclear.

One possible means is through "single- and double-loop learning," in which the former involves changes based on discrepancies between actions and goals, and the latter involves changes in goals based on shifts in values (Argyris, 2003; Newig et al., 2010). Single- and double-loop learning at either level may affect learning at the other. For example, after an executive learns from performance metrics that the achievement of organizational goals is unlikely (single-loop learning), the executive engages top management and staff in a discussion about new strategies, or whether these goals are the appropriate goals to achieve at the organizational level (double-loop learning), and vice versa.

Learning may also transfer across levels through *experiential learning*, or learning from past experiences (March & Olsen, 1995). Experiential learning leads to accumulation of habits, tradition, customs, and organizational routines at the group level. According to March and Olsen, experiential learning involves three processes: (1) variation in experience, (2) selection and inference from the experience, and (3) retained experience in the form of institutions or rules. Through these processes, new ideas brought into the system compete with one another in an existing institution, which fosters adaptation of behavior through the competition of ideas. In other words, ideas proven to be more *successful* are diffused through forms of instruction and exemplification. Researchers believe that this mutual interaction between new ideas and existing institutions leads to learning.

Collective Learning Processes and Products in Governance

Following Heikkila and Gerlak (2013), we define *collective learning* as involving (1) *collective processes* such as acquiring (i.e., collecting or receiving information), translating (i.e., interpreting or applying new information), and disseminating (i.e., transferring information across groups) knowledge, and (2) *collective products* such as cognitive and behavioral changes. Heikkila and Gerlak argue that the collective processes of knowledge acquisition, translation, and dissemination influence the emergence of collective products that we typically associate with policy learning: new ideas, strategies, rules, and policies.

The relationship between collective learning processes and products may not be unidirectional, as collective learning products could influence the processes or steps through which actors learn, and vice versa. Thus, we view collective learning as a dynamic, iterative, self-reinforcing process that strengthens actors' ability to adapt and learn further through improving processes. Through collective learning processes and products, actors develop their capacity to learn in more intelligent ways and generate and share greater knowledge over time. However, it is plausible that actors can also collectively learn to be more strategic and pursue their self-interests in collaboratives, which would lead them to guard knowledge and information from others.

Although learning occurs in all sectors, we focus on the role of collective learning in producing public value (Moore, 1995), viewing the processes of learning—knowledge acquisition, sharing, translation, and adaptation—as courses of action that can cultivate mutual reciprocity and democratic values such as legitimacy and inclusivity. When it comes to learning products in governance, political and contextual knowledge about citizen preferences or building consensus are equally important as scientific and technological knowledge. Although researchers believe that these processes and products of collective learning are important for collaborative governance, extant scholarship lacks a theory for how they facilitate institutional collective action. We begin developing such a theory in the following section.

Three Pathways Linking Collective Learning to Institutional Collective Action

Figure 15.3 displays our conceptual model for how collective learning can lead to the resolution or perpetuation of ICA dilemmas and the improvement of collaborative governance. We posit three pathways linking collective learning to institutional collective action. In the first path, collaborative choices and outcomes affect collective learning: Actors either learn to be better collaborators, or more opportunistic and self-interested. "Path A" in Fig. 15.3 is informed by extant research suggesting collaboration enhances learning (Hartley & Benington, 2006; Heikkila & Gerlak, 2013; Leach, Weible, Vince, Siddiki, & Calanni, 2013).

In the second path, collective learning directly mitigates collaboration risks and in turn alters the integrative mechanism choice (Path B in Fig. 15.3). Although we acknowledge actors can collectively learn to be more strategic and pursue their own particularistic interests, we assume greater knowledge about collaboration and more information about partners enable actors to better deal with collaboration problems, heterogeneous preferences, and weak institutions for collaboration, thus reducing collaboration risk.

In the third path, collective learning moderates the relation between the collaboration situation and risks (Path C in Fig. 15.3). That is, collective learning has greater impact in high-complexity ICA situations characterized by highly fragmented, specialized, and multifaceted contexts (e.g., global climate change or international epidemic prevention) than in low-complexity ICA situations (e.g., local waste management contracts). Our logic is that because we are treating the

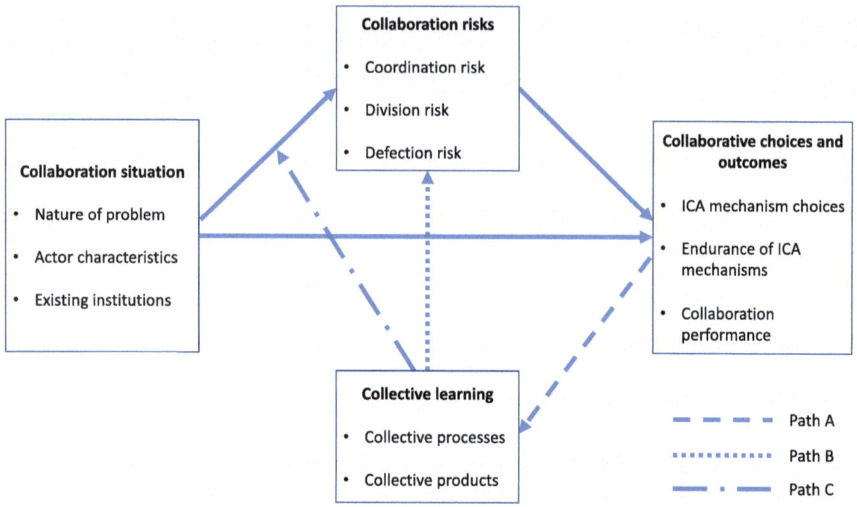

Fig. 15.3 Three pathways of collective learning in ICA situations. Source: Design by authors

collaboration situation as exogenous, collective learning has very little direct influence over the current situation or the complexity that authorities face, but rather affects the extent to which such complexity influences collaboration risks.

Path A: Collaboration Choices and Outcomes Influencing Collective Learning

In this section, we explore how collaboration choices and outcomes enhance collective learning (Path A in Fig. 15.3). Collaboration is critical for knowledge acquisition, diffusion, and development because learning often occurs through external opportunities and insights (Crona & Parker, 2012; Powell, 1998). Collaboration has been found to help policy actors discover and assimilate new knowledge that is integral for learning in governance (Gerlak & Heikkila, 2011; Hartley & Benington, 2006; Leach et al., 2013). Our first claim is that collaboration delivers new knowledge to policy actors, which in turn facilitates learning and the ability to govern fragmented governments in the future. Below, we discuss how different integrative mechanisms may have differentiated effects on collective learning, and how social networking and other information technology advancements may influence such learning.

Integrative mechanism choices and collective learning

Although collective learning is enhanced by collaboration, there may be trade-offs between different learning components (i.e., knowledge acquisition, translation, sharing, adaptation, and retainment). Research suggests that decentralized or polycentric collaboration is more conducive to knowledge generation, dissemination, and experimentation, but authoritative integrative mechanisms may be better for retaining knowledge through formalized rules (Gerlak & Heikkila, 2011; Heikkila & Gerlak, 2013; March & Olsen, 1995). Thus, we posit that different integrative mechanisms (Fig. 15.1) have differentiated advantages for collective learning. Although voluntary, informal integrative mechanisms may offer advantages in information acquisition and knowledge creation (Willem & Buelens, 2007), imposed or delegated integration, which entail greater hierarchy, may be more efficient in translating, adopting, and retaining knowledge (March, 1991; March & Olsen, 1995). Likewise, more "managed networks" (i.e., led internally or externally by a single organization) could offer advantages in processing information in a more systematic and organized manner to maintain stability, but may inhibit the ability to adapt and discover new knowledge over time (Provan & Kenis, 2008).

There are also benefits of assimilating diverse networks of actors to enhance collective learning. Although vertical integration of a learning system is conducive to coping with larger problems and reduces coordination and information-sharing costs, autonomous horizontal learning across multiple and overlapping jurisdictions fosters policy experiments, evaluation, and adaptation (Koontz et al., 2015). Thus, although researchers have acknowledged, for example, potential trade-offs in learning between exploration and exploitation (March, 1991), overlapping collaborative structures may provide some advantages for collective learning. In the ICA context, self-organizing and more authoritative mechanisms could complement each other, especially in complex situations. For example, empirical research on climate change adaptation finds the need for both authoritative, top-down and self-organizing, bottom-up collaborative networks in the production and sharing of knowledge (Homsy & Warner, 2013). Interactive, two-way collaborative learning and localized networks involving diverse stakeholders are essential for addressing complex problems (Weber & Khademian, 2008).

Proposition 1: Self-organizing integrative mechanisms are better for acquiring and disseminating new information, but more authoritative mechanisms are better for processing and retaining knowledge.

Proposition 2: Combining integrative mechanisms (e.g., self-organizing with more authoritative mechanisms) will improve collective learning and enable authorities to take on more complex, second-order ICA cooperation problems in the future.

Social networking, technology, and collective learning

Social networking and other information technology advancements reduce the cognitive distance between partners and facilitate communication and collaboration opportunities, thereby potentially enhancing shared understanding and collective learning. For example, social networking services, webinars, video conferencing, and cloud services can enhance governments' ability to absorb information and foster the shared norms and understanding traditionally cultivated by in-person dialogue. Mutual understanding and shared values are believed to emerge from frequent communication and social interactions. For example, policy dialogue in informal networks helps actors build trust and norms of reciprocity, enabling them to overcome ICA dilemmas and formalize interorganizational collaborations (Andrew & Carr, 2013; Hawkins et al., 2016; LeRoux et al., 2010). Social networking services can complement or, in some instances, replace the in-person, face-to-face dialogue that has enabled authorities to develop collaboratives.

Other information technology such as databases, cloud computing, and shared servers could also mitigate challenges in collective action for knowledge governance (Heikkila & Gerlak, 2013). Knowledge acquisition and dissemination can be facilitated through a unified information-sharing platform, allowing collaborative partners to assemble, standardize, simplify, and update information. These information-technology management tools enable organizations to better combine knowledge discovery and retention practices, and thereby manage information in a more synergistic way. For example, constructing shared servers where actors collectively encode, store, and retrieve information could enhance knowledge discovery, sharing, and retention in governance.

Whether through in-person conversation or electronic-based dialogue, collaboratives require actors to build trust, mutuality, and shared understanding through dialogue (Ansell & Gash, 2008). Current technology has expanded opportunities for dialogue, with the help of web-based information sharing and video conferencing platforms. With these advancements, it should be easier to find potential collaborators with similar interests and arrange meetings, even though actors are not geographically close. We argue that these increased opportunities for dialogue will lead to more collaboration opportunities and better outcomes, which in turn enhance collective learning and the ability to tackle higher-order cooperation problems as collaboratives mature.

> *Proposition 3: Social networking services and advances in information technology management enhance collaboration by providing capacity and facilitating dialogue, which in turn promotes collective learning and the ability to address more complex, second-order cooperation problems in the future.*

Path B: Collective Learning Influencing Integrative Mechanisms Through Collaboration Risk Mitigation

Integrative mechanism choices (Fig. 15.1) are shaped by the collaboration risks of coordination, division, and defection (Carr, Hawkins, & Westberg, 2017; Feiock, 2013). But collective learning could mitigate these risks by providing actors with greater knowledge about other actors, joint benefits (and costs), and collaborative opportunities. Thus, our second claim is that integrative mechanism choices (and presumably their endurance and outcomes) can be directly altered by mitigating collaboration risks through collective learning (Path B in Fig. 15.3). In other words, the relationship between collective learning and integrative mechanism choices is mediated by collaboration risk mitigation. Below, we discuss how collective learning relates to the three types of collaboration risks.

Coordination risk

Coordination risk emerges as participants conflict over cooperative goals or the participants' responsibilities, and is inherent in situations where actors identify the benefits of joint activity but fail to collaborate due to information incompleteness (Carr et al., 2017; Feiock, 2013). For example, interlocal collaboration for economic development may be hindered as actors fail to reach consensus because of lack of information about future economic conditions and the high uncertainty of economic ventures (Hawkins, 2009).

Coordination risk may be strongly related to resource discrepancies. For example, when affluent governments are able to implement proactive collaborative policies to protect wetlands and wildlife habitats (see Porter & Salvesen, 1995), neighboring impoverished governments may not have the same expertise and resources to take similar or complementary actions, which may exacerbate information sharing in ICA situations.

When coordination risk is high, lead network actors could also play a key role in mitigating such risk through collective learning. Researchers have widely identified knowledge sharing and information exchange as key facilitators of collaborative governance (Ansell & Torfing, 2015; Lubell, Mewhirter, Berardo, & Scholz, 2017; Shrestha, Berardo, & Feiock, 2014), and lead agencies are more likely to facilitate learning by disseminating new information and knowledge across organizations (Gerlak & Heikkila, 2011). Researchers have found that lead agencies help establish a common vision among participants in regional water management and allow for the development of norms and informal rules that are pivotal for learning (Conrad, 2015).

In addition, integrating different types of knowledge may also mitigate coordination risk (Lubell et al., 2017; Yang, 2017). For example, utilization of local knowledge (such as indigenous, traditional, or community knowledge) to make sense of scientific and political knowledge (i.e., understanding the interests of other actors) has been found to enhance the performance of interjurisdictional desertification prevention (Yang, 2017). From this perspective, we argue that utilizing different types of knowledge will help solve the conflicts over collaborative goals and reduce uncertainties associated with collaborative arrangements.

Division risk

Division risk arises when actors share general collaborative goals but are unable to agree on how to distribute the benefits and costs associated with joint activities. Division risk increases as a party to an agreement becomes relatively worse off. Here, we argue that collective learning—in the form of shared knowledge about partners' payoffs, interests, and needs—mitigates division risk.

Consider the North American Free Trade Agreement (NAFTA). Division risk could be mitigated by mutual knowledge about the social, economic, and political situations of partner countries. Let us assume, for example, that the enactment of NAFTA worsened conditions for Mexico's energy companies. This collaborative would suffer from high division risk, because Mexico could roll back the agreement under a new administration. However, NAFTA could mitigate such uncertainty through more accurate projections about expected benefits (or costs) (Zumbrun, Pérez, & Harrup, 2018). This knowledge can empower Mexican officials to persuade domestic interest groups wary of NAFTA consequences and reduce their uncertainty about economic payoffs.

Research suggests that frequent interaction mitigates the problem of dividing collaborative benefits (Lubell et al., 2017), and informal networking in frequent regional meetings leads to greater formal interlocal collaboration (Tavares & Feiock, 2017), minimizing the division risk associated with formal agreements (Hawkins et al., 2016). We posit that frequent informal interactions will help participants reach a consensus about how to divide the benefits and costs of collaborative arrangements, especially in the early stages of collaboratives.

Over time, the development of a strong history of collaboration and collective learning should have the most impact on reducing division risk. For example, Boer, and Bressers (2011) point out that interregional water governance in the Netherlands dates back to the thirteenth century, and this long history allow actors to better deal with uncertainty. Rather than controlling interregional collaboration for water governance from the top-down, which may result in decreased flexibility and autonomy, adaptive approaches based on mutual trust and a history of reciprocity have enabled participants in the Netherlands to more effectively adjust collective benefits and costs (Boer & Bressers, 2011).

Defection risk

Defection risk emerges when actors renege or free-ride on an agreement or collective effort. Defection risk differs from coordination and division risks in that participants have conflicting interests (Feiock, 2013). That is, noncompliers achieve better outcomes by free-riding or reneging, but compilers end up worse off because they bear the full cost of providing the public good. In reality, it is extremely difficult to obtain complete information about partners' motivations and commitment to achieving collective goals. Thus, proponents of transaction cost economics (TCE) argue that it is necessary to align incentives and/or design effective monitoring mechanisms. But collective learning may mitigate the need for monitoring or, in the ICA context, more authoritative integrative mechanisms by building a better understanding of partners' motives and tendencies to engage in self-interested or cooperative behavior (Nooteboom, 2000).

Generally, more authoritative integrative mechanisms are required when defection risk is high (Feiock, 2013). In transportation planning, for example, when municipalities lack trustworthiness and the capacity to make credible commitments to a collective effort, regional authorities such as MPOs are more likely to utilize vertical, top-down integration to prevent opportunistic behavior among municipalities. But collective learning may shift the integrative mechanism choice from more authoritative to more self-organizing (Fig. 15.1), as actors begin to understand more about their partners' credibility and trustworthiness. For instance, governments that have contracted with a nonprofit for decades may not need stringent monitoring mechanisms because defection risk is low, and they can rely more on relational or informal contracting (Terman & Feiock, 2016).

We summarize our arguments on how collective learning influences collaboration choices and outcomes through the mitigation of the three collaboration risks as follows:

> *Proposition 4: Collective learning directly reduces collaboration risks (i.e., coordination, division, and defection risks), thus increasing the likelihood of adopting more self-organizing integrative mechanisms for resolving ICA dilemmas in the future.*

Path C: Collective Learning Moderating the Collaboration Situation-Risk Linkage

Our third claim is that collective learning has a greater impact in high-collaboration risk situations than in low-collaboration risk situations. In other words, collective learning moderates the relation between the collaboration situation and collaboration risks (Path C in Fig. 15.3). For example, learning about actors' needs and capacities, institutional differences, and political and economic environments is relatively more critical in complex, high-risk situations (e.g., global climate change mitigation, international drug trafficking, multinational disease prevention, etc.), than in relatively less complex, lower-risk situations. Table 15.1 summarizes this contrast between high- and low-risk ICA situations.

Feiock (2013) suggests that, to address real-world problems with integrative mechanisms, the sources of collaboration risk should be diagnosed. He divides the sources of collaboration risk into three categories: (i) the nature of the problem, (ii)

Table 15.1 Low and High Collaboration Risk Situations

Collaboration situation (sources of collaboration risk)	Low collaboration risk	High collaboration risk
Nature of the problem	Jurisdictions and authority are less fragmented, more centralized	Jurisdictions and authority are highly fragmented, more decentralized
	Issue is simple (first-order coordination problem)	Issue is complex (second-order cooperation problem)
	Coordination is sufficient (e.g., information sharing, fact finding, etc.)	Cooperation is required (e.g., shared authority, institutional reform, etc.)
Actor preferences	Homogenous preferences	Heterogeneous preferences
	Reputations are well known	Little is known about actors
Existing institutions	Abundant history and institutions of collaboration	Little history and institutions of collaboration

Note. Source: Design by authors

actors' preferences, and (iii) existing institutions. Each source adds to collaboration risks and relates to collective learning in a differentiated manner.

The *nature of the problem* lies in the specific type of ICA dilemma (e.g., minimizing air pollution in a region, managing the ecosystem of a watershed, etc.) and collaboration risk varies depending on the nature of the problem, that is, the more complex the problem, the higher the collaboration risk. Interlocal shared services for solid waste management, for example, generally entail lower collaboration risk than, say, international efforts to combat global drug trafficking (see Munsing & Lamb, 2011). Although authority is fragmented in both cases, the latter issue is exponentially more complicated, as it involves more actors and has far more confounding variables at play (e.g., illegal labor markets, money laundering, macroeconomic conditions, etc.). In the public policy literature, these problems are typically described as more intractable or *wicked*. Logically, collaboratives to resolve these problems should thus involve greater risks. We argue that collective learning is more valuable in these high-risk situations because participants more strongly require knowledge and information sharing about the nature and characteristics of the problems. In high-risk situations, where other complex issues are related, and/or critical events are frequently happening, it is more vital for participants to learn the circumstances and history and to use this knowledge to adapt to fast-changing problems.

Actor preferences refer to the extent of social, economic, and political diversity or homogeneity among collaborating units (Feiock, 2013). Collaboration risk is higher when actor preferences are heterogenous and information about other actors is limited. As in the previous example, collaboration between two adjacent municipalities in solid waste management entails relatively lower collaboration risk than in an international drug enforcement network. Collective learning about other actors' behavior, capacity, and incentives is therefore more important when actor preferences are heterogenous, because actors have a greater need to develop a shared understanding of where their preferences diverge and align.

Finally, *existing institutions* entail the rules, laws, and political, economic, and social circumstances embedded within the integrative (or collaborative) mechanism (cf. Glückler et al., 2018). Collaboration risk is lower when actors have effective collaborative institutions for resolving collective action problems. For instance, two local governments with a long history of working together will have lower collaboration risk than other local governments with little previous interaction. Collective learning and shared knowledge about other governments' institutions, such as rules, norms, and practices should be be more valuable for those municipalities with few existing institutions for collaboration. In a high-risk situation with little history of collaboration, there is a greater need to learn individual actors' roles, incentives, and future contributions in and for collaboratives. Actors will also have to learn what collaborating institutions will work and what must be done to sustain collaborative benefits. We summarize our arguments on how and why collective learning may have greater impact in high-collaboration risk situations than in low-collaboration risk situations as follows:

Proposition 5: The indirect or moderating effect of collective learning on collaboration risk is greater in high collaboration risk situations than in low collaboration risk situations.

Concluding Thoughts

Learning is important in all facets of society and is especially critical in governance, where the acquisition and dissemination of knowledge is often a core function (Willem & Buelens, 2007). Fragmentation of authority can present significant barriers to knowledge sharing and collective learning. Thus, identifying how governments overcome barriers to sharing knowledge and collectively learning about collaboration is important. Combining theories of collective learning (Heikkila & Gerlak, 2013) and institutional collective action (Feiock, 2013) is one avenue for enhancing our understanding of knowledge sharing in fragmented governance. This chapter has discussed how collaboration choices and outcomes can improve collective learning, which in turn lowers barriers to fostering better collaborative arrangements by mitigating collaboration risks. The model proposed in this chapter suggests multiple pathways and propositions to be tested in future research.

It is important to note why *institutional* collective action has significant implications for collective learning. Collective learning may not have an enduring influence unless such learning is *institutionalized* in organizations. That is, in order for the learning-collaboration feedback loop to work, collective learning should change how actors collaborate; they become either more cooperative or more self-interested through collaboration. In this way, the mutual reinforcement between learning and collaboration can enhance long-term institutional resiliency, or "the degree to which political institutions lessen the significance of crisis, turbulence, and challenges using technologies, networks, or policy coalitions for knowledge governance" (March & Olsen, 1995, p. 184). More relevant knowledge acquired through "learning-by-doing" and "trial-and-error" helps build long-lasting, adaptive, and resilient institutions and enhances actors' ability to anticipate the future (Koontz et al., 2015). We expect that the more collaborative institutions are exposed to challenges, the better they can gauge what knowledge and information is valuable and worth absorbing and utilizing. Even if participants opt out of or fail in collaboration, the competition, conflicts, and difficulties in collective action can provide valuable insights, assuming actors are not encouraged to behave more opportunistically based on what they learn.

Our model also implies that governments may be able to more effectively utilize collective learning for collective action by acknowledging partners' heterogeneous experiences and ideas as sources of acquiring new information and knowledge; utilizing social networking services and information technology advancements for collaboration and collective learning; and understanding that collaboration must be developed over time, through multiple iterations. Actors in collaboratives will inevitably make mistakes, but better collaborators will learn from these mistakes and, in doing so, enhance their ability to collaborate in the future.

Although we have noted the many benefits of collective learning for collaborative governance, political institutions bear some inherent limitations. As March and Olsen (1995, p. 238) point out, political systems easily forget: Lessons that are not coded into rules, traditions, and standard operating procedures are lost through turnover and the passage of time. Conversely, political institutions remember some things too well or too poorly: Rules can take on their own justification and build beliefs that sustain them far beyond their original intent; institutional memories can be adulterated by self-interested desires, tactics, and reinterpretations of experiences. Moving forward, it is thus important to consider the context in which collective learning takes place and to be careful about generalizing across sectors or cultures.

Secondly, our model of collective learning in ICA situations (Fig. 15.3) assumes actors are learning to be more cooperative than self-interested and opportunistic. This is a big assumption, and actors could very likely learn to be more opportunistic and guarded in collaboratives, especially in the case of failures or dysfunctions. Empirically teasing out when actors learn to be cooperative and when they learn to be noncooperative is a key challenge for future research.

Finally, although many researchers over the decades have offered numerous and fruitful ways of measuring learning in social contexts, there is much work to be done. The literature appears to be moving away from convenient yet less-than-desirable proxy measures of learning, such as policy diffusion and change, performance improvements, strategic redirections, and so forth, and paving the way through process-based and behavioral approaches to understanding collective learning (Heikkila & Gerlak, 2013). This is a step in the right direction and greater experimentation in approaches will likely lead to better measures of collective learning.

References

Andersson, K. P., & Ostrom, E. (2008). Analyzing decentralized resource regimes from a polycentric perspective. *Policy Sciences, 41*(1), 71–93. https://doi.org/10.1007/s11077-007-9055-6

Andrew, S. A., & Carr, J. B. (2013). Mitigating uncertainty and risk in planning for regional preparedness: The role of bonding and bridging relationships. *Urban Studies, 50*, 709–724. https://doi.org/10.1177/0042098012455718

Ansell, C., & Gash, A. (2008). Collaborative governance in theory and practice. *Journal of Public Administration Research and Theory, 18*, 543–571. https://doi.org/10.1093/jopart/mum032

Ansell, C., Lundin, M., & Öberg, P. O. (2017). Learning networks among Swedish municipalities: Is Sweden a small world? In J. Glückler, E. Lazega, & I. Hammer (Eds.), *Knowledge and networks* (pp. 315–336). Knowledge and Space: Vol. 11. Cham: Springer. https://doi.org/10.1007/978-3-319-45023-0

Ansell, C., & Torfing, J. (2015). How does collaborative governance scale? *Policy & Politics, 43*, 315–329. https://doi.org/10.1332/030557315X14353344872935

Argyris, C. (2003). A life full of learning. *Organization Studies, 24*, 1178–1192. https://doi.org/10.1177/01708406030247009

Argyris, C., & Schön, D. A. (1978). *Organizational learning: A theory of action perspective*. Reading: Addison-Wesley.

Berardo, R., & Scholz, J. T. (2010). Self-organizing policy networks: Risk, partner selection, and cooperation in estuaries. *American Journal of Political Science, 54*, 632–649. https://doi.org/10.1111/j.1540-5907.2010.00451.x

Berry, F. S., & Berry, W. D. (1999). Innovation and diffusion models in policy research. In P. A. Sabatier (Ed.), *Theories of the policy process* (pp. 169–200). Boulder: Westview.

Boer, C. L., & Bressers, J. T. A. (2011). *Complex and dynamic implementation processes: The renaturalization of the Dutch Regge River*. Enschede, The Netherlands: University of Twente.

Burt, R. S. (1992). *Structural holes: The social structure of competition*. Cambridge, MA: Harvard University Press.

Carr, J. B., Hawkins, C. V., & Westberg, D. E. (2017). An exploration of collaboration risk in joint ventures: Perceptions of risk by local economic development officials. *Economic Development Quarterly, 31*, 210–227. https://doi.org/10.1177/0891242417710325

Cohen, W. M., & Levinthal, D. A. (1990). Absorptive capacity: A new perspective on learning and innovation. *Administrative Science Quarterly, 35*(1), 128–153. https://doi.org/10.2307/2393553

Coleman, J. S. (1988). Social capital in the creation of human capital. *American Journal of Sociology, 94*, S95–S120. Retrieved from https://www.jstor.org/stable/2780243

Conrad, E. (2015). Bridging the hierarchical and collaborative divide: The role of network managers in scaling up a network approach to water governance in California. *Policy & Politics, 43*, 349–366. https://doi.org/10.1332/030557315X14350819637950

Crona, B. I., & Parker, J. N. (2012). Learning in support of governance: Theories, methods, and a framework to assess how bridging organizations contribute to adaptive resource governance. *Ecology and Society, 17*, Article 32. https://doi.org/10.5751/ES-04534-170132

Dunlop, C. A., & Radaelli, C. M. (2018). The lessons of policy learning: Types, triggers, hindrances and pathologies. *Policy & Politics, 46*, 255–272. https://doi.org/10.1332/030557318X15230059735521

Feiock, R. C. (2007). Rational choice and regional governance. *Journal of Urban Affairs, 29*(1), 47–63. https://doi.org/10.1111/j.1467-9906.2007.00322.x

Feiock, R. C. (2009). Metropolitan governance and institutional collective action. *Urban Affairs Review, 44*, 356–377. https://doi.org/10.1177/1078087408324000

Feiock, R. C. (2013). The institutional collective action framework. *Policy Studies Journal, 41*, 397–425. https://doi.org/10.1111/psj.12023

Feiock, R. C., Lubell, M., & Lee, I. W. (2014). The political market framework and policy change (National Science Foundation Grant No. 0350799). Retrieved from https://localgov.fsu.edu/sites/g/files/upcbnu1196/files/Politcal_Market__Policy_Change_2010_5_16_ML.pdf

Feiock, R. C., & Scholz, J. T. (Eds.). (2010). *Self-organizing federalism: Collaborative mechanisms to mitigate institutional collective action*. New York, NY: Cambridge University Press.

Feiock, R. C., Steinacker, A., & Park, H. J. (2009). Institutional collective action and economic development joint ventures. *Public Administration Review, 69,* 256–270. https://doi.org/10.1111/j.1540-6210.2008.01972.x

Gerber, E. R., Henry, A. D., & Lubell, M. (2013). Political homophily and collaboration in regional planning networks. *American Journal of Political Science, 57,* 598–610. https://doi.org/10.1111/ajps.12011

Gerlak, A. K., & Heikkila, T. (2011). Building a theory of learning in collaboratives: Evidence from the Everglades Restoration Program. *Journal of Public Administration Research and Theory, 21,* 619–644. https://doi.org/10.1093/jopart/muq089

Glückler, J., Suddaby, R., & Lenz, R. (Eds.). (2018). *Knowledge and institutions*. Knowledge and Space: Vol. 13. Cham: Springer. https://doi.org/10.1007/978-3-319-75328-7

Hartley, J., & Benington, J. (2006). Copy and paste, or graft and transplant? Knowledge sharing through inter-organizational networks. *Public Money & Management, 26*(2), 101–108. https://doi.org/10.1111/j.1467-9302.2006.00508.x

Hawkins, C. V. (2009). Prospects for and barriers to local government joint ventures. *State and Local Government Review, 41*(2), 108–119. https://doi.org/10.1177/0160323X0904100204

Hawkins, C. V., Hu, Q., & Feiock, R. C. (2016). Self-organizing governance of local economic development: Informal policy networks and regional institutions. *Journal of Urban Affairs, 38,* 643–660. https://doi.org/10.1111/juaf.12280

Hawkins, C. V., Krause, R., Feiock, R. C., & Curley, C. (2017). The administration and management of environmental sustainability initiatives: A collaborative perspective. *Journal of Environmental Planning and Management, 61,* 2015–2031. https://doi.org/10.1080/09640568.2017.1379959

Heikkila, T., & Gerlak, A. K. (2013). Building a conceptual approach to collective learning: Lessons for public policy scholars. *Policy Studies Journal, 41,* 484–512. https://doi.org/10.1111/psj.12026

Hendrick, R. M., Jimenez, B. S., & Lal, K. (2011). Does local government fragmentation reduce local spending? *Urban Affairs Review, 47,* 467–510. https://doi.org/10.1177/1078087411400379

Homsy, G. C., & Warner, M. E. (2013). Climate change and the co-production of knowledge and policy in rural USA communities. *Sociologia Ruralis, 53,* 291–310. https://doi.org/10.1111/soru.12013

Huxham, C., & Vangen, S. (2005). *Managing to collaborate: The theory and practice of collaborative advantage*. Abingdon: Routledge.

Jimenez, B. S. (2014). Externalities in the fragmented metropolis: Local institutional choices and the efficiency-equity trade-off. *The American Review of Public Administration, 46,* 314–336. https://doi.org/10.1177/0275074014550703

Koontz, T. M., Gupta, D., Mudliar, P., & Ranjan, P. (2015). Adaptive institutions in social-ecological systems governance: A synthesis framework. *Environmental Science & Policy, 53,* 139–151. https://doi.org/10.1016/j.envsci.2015.01.003

Leach, W. D., Weible, C. M., Vince, S. R., Siddiki, S. N., & Calanni, J. C. (2013). Fostering learning through collaboration: Knowledge acquisition and belief change in marine aquaculture partnerships. *Journal of Public Administration Research and Theory, 24,* 591–622. https://doi.org/10.1093/jopart/mut011

Lee, Y. (2016). From competition to collaboration: Intergovernmental economic development policy networks. *Local Government Studies, 42,* 171–188. https://doi.org/10.1080/03003930.2015.1096267

LeRoux, K., Brandenburger, P. W., & Pandey, S. K. (2010). Interlocal service cooperation in US cities: A social network explanation. *Public Administration Review, 70,* 268–278. https://doi.org/10.1111/j.1540-6210.2010.02133.x

Lubell, M., Mewhirter, J. M., Berardo, R., & Scholz, J. T. (2017). Transaction costs and the perceived effectiveness of complex institutional systems. *Public Administration Review, 77,* 668–680. https://doi.org/10.1111/puar.12622

March, J. G. (1991). Exploration and exploitation in organizational learning. *Organization Science, 2*(1), 71–87. https://doi.org/10.1287/orsc.2.1.71

March, J. G., & Olsen, J. P. (1995). *Democratic governance*. New York: Free.

Maser, S. M. (1998). Constitutions as relational contracts: Explaining procedural safeguards in municipal charters. *Journal of Public Administration Research and Theory, 8,* 527–564. https://doi.org/10.1093/oxfordjournals.jpart.a024395

McGinnis, M. D. (Ed.). (1999). *Polycentricity and local public economies: Readings from the workshop in political theory and policy analysis*. Ann Arbor, MI: University of Michigan Press. https://doi.org/10.3998/mpub.16061

Moore, M. H. (1995). *Creating public value: Strategic management in government*. Cambridge, MA: Harvard University Press.

Munsing, E., & Lamb, C. J. (2011). *Joint interagency task force-south: The best known, least understood interagency success*. Institute for National Strategic Studies, Strategic Perspectives: Vol. 5. Washington, D.C.: National Defense University Press. Retrieved from https://ndupress.ndu.edu/Portals/68/Documents/stratperspective/inss/Strategic-Perspectives-5.pdf

Newig, J., Günther, D., & Pahl-Wostl, C. (2010). Synapses in the network: Learning in governance networks in the context of environmental management. *Ecology and Society, 15,* Article 24. https://doi.org/10.5751/ES-03713-150424

Nooteboom, B. (2000). Learning by interaction: Absorptive capacity, cognitive distance and governance. *Journal of Management and Governance, 4*(1–2), 69–92. https://doi.org/10.1023/A:1009941416749

Olson, M. (1965). *Logic of collective action: Public goods and the theory of groups*. Cambridge, MA: Harvard University Press.

Ostrom, E. (1990). *Governing the commons: The evolution of institutions for collective action*. Cambridge, UK: Cambridge University Press.

Ostrom, E. (2005). *Understanding institutional diversity*. Princeton, NJ: Princeton University Press.

Ostrom, E. (2010). Polycentric systems for coping with collective action and global environmental change. *Global Environmental Change, 20,* 550–557. https://doi.org/10.1016/j.gloenvcha.2010.07.004

Ostrom, V., Tiebout, C. M., & Warren, R. (1961). The organization of government in metropolitan areas: A theoretical inquiry. *American Political Science Review, 55,* 831–842. https://doi.org/10.2307/1952530

Porter, D. R., & Salvesen, D. A. (Eds.). (1995). *Collaborative planning for wetlands and wildlife: Issues and examples*. Washington, D.C.: Island.

Powell, W. W. (1998). Learning from collaboration: Knowledge and networks in the biotechnology and pharmaceutical industries. *California Management Review, 40,* 228–240. https://doi.org/10.2307/41165952

Powell, W. W., Koput, K. W., & Smith-Doerr, L. (1996). Interorganizational collaboration and the locus of innovation: Networks of learning in biotechnology. *Administrative Science Quarterly, 41*(1), 116–145. https://doi.org/10.2307/2393988

Provan, K. G., & Kenis, P. (2008). Modes of network governance: Structure, management, and effectiveness. *Journal of Public Administration Research and Theory, 18,* 229–252. https://doi.org/10.1093/jopart/mum015

Putnam, R. D., Leonardi, R., & Nonetti, R. Y. (1993). *Making democracy work: Civic traditions in modern Italy*. Princeton, NJ: Princeton University Press.

Sabatier, P. A. (1988). An advocacy coalition framework of policy change and the role of policy-oriented learning therein. *Policy Sciences, 21*(2), 129–168. https://doi.org/10.1007/BF00136406

Salamon, L. M. (Ed.). (2002). *The tools of government: A guide to the new governance*. New York, NY: Oxford University Press.

Schneider, M. (1986). Fragmentation and the growth of local government. *Public Choice, 48,* 255–263. https://doi.org/10.1007/BF00051622

Scholz, J. T., Berardo, R., & Kile, B. (2008). Do networks solve collective action problems? Credibility, search, and collaboration. *The Journal of Politics, 70,* 393–406. https://doi.org/10.1017/s0022381608080389

Senge, P. M. (1990). *The fifth discipline: The art and practice of the learning organization.* New York: Doubleday.

Shipan, C. R., & Volden, C. (2008). The mechanisms of policy diffusion. *American Journal of Political Science, 52,* 840–857. https://doi.org/10.1111/j.1540-5907.2008.00346.x

Shrestha, M., Berardo, R., & Feiock, R. (2014). Solving institutional collective action problems in multiplex networks. *Complexity, Governance & Networks, 1*(1), 49–60. https://doi.org/10.7564/14-CGN7

Spekkink, W. A. H., & Boons, F. A. A. (2016). The emergence of collaborations. *Journal of Public Administration Research and Theory, 26,* 613–630. https://doi.org/10.1093/jopart/muv030

Swann, W. L. (2017). Examining the impact of local collaborative tools on urban sustainability efforts: Does the managerial environment matter? The *American Review of Public Administration, 47,* 455–468. https://doi.org/10.1177/0275074015598576

Swann, W. L., & Kim, S. Y. (2018). Practical prescriptions for governing fragmented governments. *Policy & Politics, 46,* 273–292. https://doi.org/10.1332/030557318X15230058720979

Tamtik, M. (2016). Institutional change through policy learning: The case of the European Commission and research policy. *Review of Policy Research, 33*(1), 5–21. https://doi.org/10.1111/ropr.12156

Tavares, A. F., & Feiock, R. C. (2017). Applying an institutional collective action framework to investigate intermunicipal cooperation in Europe. *Perspectives on Public Management and Governance, 1*(4), 299–316. https://doi.org/10.1093/ppmgov/gvx014

Terman, J. N., & Feiock, R. C. (2016). The effect of formal and informal contracting mechanisms on implementation performance in the US federalist system. *Local Government Studies, 42,* 309–331. https://doi.org/10.1080/03003930.2015.1110522

Weber, E. P., & Khademian, A. M. (2008). Wicked problems, knowledge challenges, and collaborative capacity builders in network settings. *Public Administration Review, 68,* 334–349. https://doi.org/10.1111/j.1540-6210.2007.00866.x

Willem, A., & Buelens, M. (2007). Knowledge sharing in public sector organizations: The effect of organizational characteristics on interdepartmental knowledge sharing. *Journal of Public Administration Research and Theory, 17,* 581–606. https://doi.org/10.1093/jopart/mul021

Williamson, O. E. (1991). Comparative economic organization: The analysis of discrete structural alternatives. *Administrative Science Quarterly, 36,* 269–296. https://doi.org/10.2307/2393356

Yang, L. (2017). Collaborative knowledge-driven governance: Types and mechanisms of collaboration between science, social science, and local knowledge. *Science and Public Policy, 45*(1), 53–73. https://doi.org/10.1093/scipol/scx047

Zeemering, E. S. (2019). An agenda for comparing local governance and institutional collective action in Canada and the United States. *Urban Affairs Review, 55,* 858–886. https://doi.org/10.1177/1078087417738094

Zumbrun, J., Pérez, S., & Harrup, A. (2018, April 24). Energy, a bright spot in NAFTA talks, bogged down by dispute over rule change [Article]. The Wall Street Journal (New York). Retrieved from https://www.wsj.com/articles/energy-a-bright-spot-in-nafta-talks-bogged-down-by-dispute-over-rule-change-1524562201

Chapter 16
The Remapping of Forest Governance: From Shareholder to Stakeholder

Roger Hayter and Alex Clapp

Resource conflicts are a widespread feature of contemporary globalization, and resource peripheries have become contested battlegrounds that are challenging demands for sustainable development defined in both social and environmental terms (Hayter, Barnes, & Bradshaw, 2003). As an expression of these contests, in recent decades forest conflicts have proliferated on all continents, in the peripheries of rich and poor countries alike (Gritten, Mola-Yudego, Delgado-Matas, & Kortelainen, 2012). Not surprisingly, given the highly varied nature of forest ecologies and governance around the world, forest conflicts and approaches to their resolution involve diverse actors and motivations, and they vary considerably in nature (Moran & Ostrom, 2005). Yet drawing on Westoby's (1989, p. 196) insight, the realization of the *non-wood benefits* of forests, meaning their environmental and cultural values, has been a significant stimulus underlying forestry conflicts. In this regard, researchers have proposed remapping as an umbrella concept that refers to both a revaluation of forest resources that reforms the dominance of large-scale industrial uses to privilege environmental and cultural priorities, *and* to implement new land-use plans and forms of governance based on new inventories, resource maps, science, and zoning (Clapp, 2004; Hayter, 2003). In turn, forest remappings are interpreted as part of broader paradigmatic transformations of society and economy that are driven by interacting scientific, technological, cultural, political, and historical forces. In evolutionary terms, contemporary remapping is an attempt to transform the commodity-driven and shareholder-oriented forest management associated with Fordism into more locally diverse forms of governance as part of a post-Fordist or information and communication techno-economic paradigm

R. Hayter (✉) · A. Clapp
Department of Geography, Simon Fraser University, Burnaby, BC, Canada
e-mail: hayter@sfu.ca; aclapp@sfu.ca

J. Glückler et al. (eds.), *Knowledge for Governance*, Knowledge and Space 15,
https://doi.org/10.1007/978-3-030-47150-7_16

(Clapp, Hayter, Affolderbach, & Guzman, 2016). This transformation is not straightforward, but conflict-ridden and contingent.

Researchers have variously defined forest conflicts as disputes over the access and management of resources, as incompatible activities aimed at restricting one another, or as clashes among diverse institutional interests over the control and use of forests as economic, environmental, or cultural resources (Gritten et al., 2012; Hayter et al., 2003). Although these definitions are related, the focus on forest conflicts as institutional clashes captures the sense of remapping as a search for paradigmatic change. In practice, conflicts vary considerably in scope and severity and can, for example, involve minor disputes over (perceived) violation of minor regulations or of inappropriate behavior in the context of existing rules and customs. Agents of conflict-driven remapping, however, specifically seek to fundamentally change existing policies, regulations, "habits of thought" (Veblen, 1899), and the "rules of the game" (North, 1990). In practice, the forces of contemporary remapping are especially evident in peripheries where forest resources have been primarily exploited by corporations for industrial benefits and the forest sector deemed a principal engine of export-led economic growth, such as in British Columbia (BC), Tasmania, and New Zealand (Affolderbach, Clapp, & Hayter, 2012; Clapp, 2004; Hayter, 2003; Hayter & Barnes, 2012).

The drivers and direction of contemporary remapping are contentious, caught up in the contesting impulses of neoliberalism and new forms of locally contingent stakeholder governance. On the one hand, Roche (1990) and McCarthy (2006) see neoliberalism as a dominating influence over forest policies in New Zealand and BC respectively, variously expressed in terms of privatization, deregulation, enchantment with MNCs, forest commodification, and more intangibly as a mind-set committed to market solutions. Among poorer countries, Tsing (2005, p. 7) interprets neoliberalism as a "universal" force in constant "friction" with local resource practices. From this perspective, neoliberalist adherents' faith in the priority of market forces to achieve economic efficiencies at the global scale is a powerful impulse in forest peripheries, forming the basis of what we summarily label as a *shareholder model* or approach to remapping. Yet Roche and McCarthy are highly critical of neoliberal policies and, as implied by Tsing's (2005) friction metaphor in resource peripheries, these policies have been resisted. Indeed, limits to the implementation of neoliberalism have been recognized in large part because of opposition from newly empowered stakeholders and the development of an alternative *stakeholder model* of governance towards the remapping of forest peripheries (Clapp et al., 2016; Hayter & Barnes, 2012). In this approach, formerly marginalized actors, such as environmental nongovernmental organizations (ENGOs), indigenous peoples, and other local community actors, become formal stakeholders who gain access to and influence in decision-making processes hitherto dominated by the vested interests of big business and governments. Creators of stakeholder models of governance imply reductions or modifications in the autonomy of both government and business, constraining though not necessarily rejecting market forces to emphasize environmental values and the goals of local communities, including indigenous peoples. Conflicts arise as new stakeholders seek new, often radical definitions, land zones, and legal rights to replace entrenched legally defined resource tenures and the commodity values they represent.

In this chapter, we assess the concept of remapping primarily as a transition from shareholder to stakeholder thinking, for positive and normative reasons. Utilizers of the stakeholder model of remapping capture important trends in forest use in rich export-driven peripheries and are closely aligned to the development of what Ostrom (2010, p. 641) labels "polyvalent governance" within common property models derived from experiences in many poor countries (Moran & Ostrom, 2005). As an ideal type, stakeholders engaged in *successful* remapping can identify the need for new institutional arrangements and innovations, both to facilitate dialogue among parties in conflict, and to establish durable rules and organizations that enact cooperative forms of sustainable development (Affolderbach et al., 2012; Clapp et al., 2016). In this virtuous cycle, sustainable development generates localized external economies and incorporates public goods and ecosystem services as well as the material values of forests. Such an evolution in resource peripheries may be seen as an extension of Amin and Thrift's (Amin & Thrift, 1995) institutional thickness, originally conceived to help researchers understand urbanization processes. In contrast, shareholders seeking private sector deregulation produce institutional *thinning,* while underlining the contentious nature of remapping.

Stakeholder-driven remapping is neither inevitable nor inevitably favorable. Cooperative stakeholder arrangements are necessarily experimental, especially so in peripheral regions whose inhabitants are trying to re-invent themselves. These experiments face uncertainties both political and epistemological, rooted in the nature of bargaining among new and old stakeholders and in new processes of science and learning. The engagement of more stakeholders in turn implies more diverse views and hopes for the future, with the future itself of ambiguous length. Further, institutional thickening may mean more bureaucracy without local development, and institutional thinning may imply the reverse. As Bestor (1998) articulates, the idea of markets is ambiguous, comprising a diverse range of actors and relationships, and if market actors are privileged in shareholder approaches, they also feature as local stakeholders.

Nevertheless, stakeholder and shareholder models are useful contrasting starting points for understanding remappings of resource peripheries with which broader debates over the meaning of globalization can be expressed (Hirst & Thompson, 1996; Kelly, 1999). Thus, stakeholder-remappings are part of globalizing processes that shift policy making from hierarchical control by governments and markets to more diffuse, democratic forms of governance in which new social forces gain leverage (Bevir, 2012; Jessop, 1998; Mayntz, 2003). In this thinking, if neoliberalism is a global force shaping resource use (Tsing, 2005), so too are the environmental movement (O'Riordan, 2001; Zimmerer, 2006) and the surge in demands for indigenous rights (Mander & Tauli-Corpuz, 2006), and all feature local adaptations and resistances that interact with one another. Policy formulation is complex and cannot be easily reduced to neoliberal formulations, a view consistent with the theme of the limits of neoliberalism (Hayter & Barnes, 2012; Johnston & Glasmeier, 2007; Weller & O'Neill, 2014).

We have organized the rest of this chapter into three sections in which we progressively elaborate stakeholder-remapping in its broader societal and policy context, in the natures it produces, and in how it might be evaluated (Fig. 16.1). First, forest remappings are discussed as expressions of socio-economic paradigmatic changes, stimulated by crisis and conflict and in which actors view contemporary remapping as an uncertain policy challenge to the markets and hierarchy governance model of Fordist forestry with its emphasis on commodification. In the second section, we distinguish stakeholder and shareholder impulses in contemporary remapping and elaborate the former by highlighting institutional innovation and thickening in emerging forms of governance. In the last, most speculative section, we explore the meaning of *good governance* in remapping and how stakeholder-remapping may be assessed. We recognize that forest transitions in one form or another are globally widespread (Mather, 1992) and that particular forms of forest conflict, remapping, and conflict resolution are contingent and varied. Although we draw primarily on BC's attempts to remap and resolve forest conflicts, we hope that

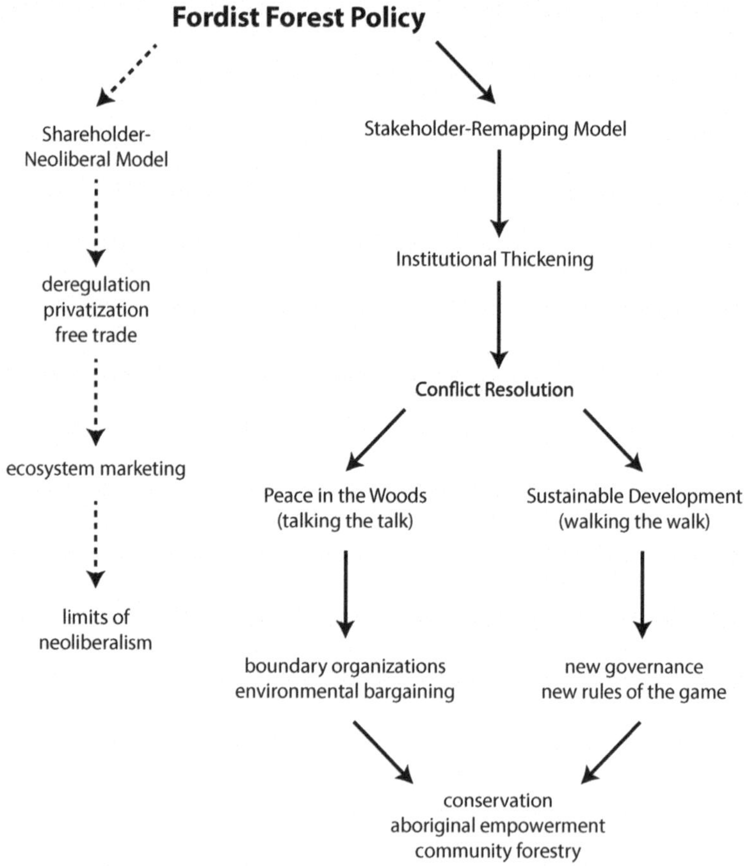

Fig. 16.1 Fordist forest policy. Source: Design by authors

the discussion will resonate more broadly in other forest regions and to spatially distributed resources in general. Although the latter concern is beyond the scope of this chapter, many key resources coexist in the same peripheries and share in broad outline the challenges posed by changing social values.

Remapping Fordist Forestry as Paradigm Change

Theorists of economic transformation and paradigmatic change, such as Freeman and Louça (2001), emphasize the role of innovation and root their work in the experience of core countries and *leading edge* sectors. Similarly, researchers with more spatially sensitive complementary approaches also focus on core regions and agglomerations (Storper, 1997), as do the authors of the original idea of institutional thickness (Amin & Thrift, 1995). Yet the implications of paradigmatic change for resource peripheries and industries are profound (Hayter, 2000). Indeed, forest policy everywhere is shaped by and evolves in accordance with changing social attitudes and forms of industrialization or innovation-driven paradigms. Reference to paradigmatic change helps illuminate the role of crisis and conflict as catalysts, and the broader evolutionary forces shaping and challenging contemporary remapping.

Drawing on North American experience, three broad forestry remappings can be identified in relation to paradigm changes since the nineteenth century (Clapp et al., 2016; Franklin, Berg, Thornburgh, & Tappeiner, 1997; Hayes & Glendenning, 2005). First, nineteenth century industrialization and the rise of largely unfettered market forces heralded an era of deforestation (1850–1910), driven by colonization, dispossession, and speculation, in which old-growth forests, once the common property of indigenous peoples, were remapped as state- or privately-owned resources to facilitate business investment. In tandem with the expansion of the factory system, the scale of forestry and wood-processing activities increased rapidly, stimulated by innovations in steam-powered technology, machinery, and wood pulping, implementers of the latter targeting the coniferous forests of northern regions, including the Pacific Northwest and BC (Rajala, 1998).

Subsequently, policy-makers' searches for more sustainable forest industries and forest-based communities ushered in a second Fordist era of remapping, in accordance with a new "scientific" forestry (1915–1970, peaking after 1945). They based this remapping on sustained yield principles and related silvicultural practices, largely developed in Europe, that supported policies that remained primarily committed to facilitating the industrial use of forests. As part of Fordism's mass production culture, forest sector activities increased in scale, and were linked within newly emerging horizontally and vertically integrated MNCs that owned or controlled vast tracts of forest. Land ownership and large-scale, secure timber tenures were expected to promote sustained yield forest rotations that would allow replacement after harvesting (Hayter, 1976), but the structural incentives for resource depletion were revealed in ambitious and front-loaded timber harvest targets (Clapp, 1995; Marchak, Aycock, & Herbert, 1999). During Fordism, technological developments also enabled the use of hardwoods in pulp and paper, leading to large-scale pine and eucalyptus forest monoculture plantations in Australia, Brazil, Chile, New Zealand, South Africa, the southern United States, and other semi-tropical and warm temperate regions (Clapp, 2001).

Although a central principle, sustained yield proved problematical for industry, while the nonindustrial benefits of forests were also undermined. In response, a third remapping has taken place since 1970, part of what is variously labeled as post-Fordism: the information and communication techno-economic paradigm (ICT), or globalization. Forestry experts suggest that this remapping has been dominated by demands for adaptive, flexible *ecosystem-based management* aimed at sustainability—ecologically, culturally, and economically. Meanwhile, technological changes featuring micro-electronic technologies have deepened economies of scale and scope in forest product manufacturing.

Environmental opposition to forest commodification has been an important driver of conflict. In Tasmania, for example, as early as 1885, enactors of a pioneering Forest Act explicitly recognized the problem of environmental degradation and created a conservator of forests who subsequently reported forestry practices as "chaotic," even if not much was to change for some time (Carron, 1985; Gee, 2001). In North America, Widick (2009) has documented the rise of environmentally-driven forest conflicts in California since the latter part of the nineteenth century, and Lee and Field (2005) and Langston (2005) in relation to Washington and Oregon, and Wilson (1998) for BC, have documented the role of professionally organized environmental nongovernmental organizations (ENGOs) in challenging prevailing industrial forest practices, especially large-scale clearcutting. The participants of the resulting debates have challenged the scientific wisdom of the Fordist forestry practices that became institutionalized in North America, and whose one-size-fits-all advice "imposed a rational, uniform and simplistic order on the complexities of localized ecological systems" (Lee & Field, 2005, p. 3). However, as Langston (2005) argues, Fordist sustained yield and silviculture experienced significant problems even for industry. Industrialized forestry practices aimed at maximizing processing efficiencies go hand in hand with a variety of economic risks related to: changes in the species mix to less economically desirable trees, along with the widespread practice of even-aged clear-cutting; increased damage due to disease and insects, fire-proneness, and erosion; and intractable conflicts with other users. Moreover, as mill capacities increased as a result of improvements in efficiency, declining forest productivity associated with the maturation of the resource cycle has led inevitably to the closure of mills, a well-documented trend in Oregon, Washington, and British Columbia (Hayter & Edenhoffer, 2016; Robbins, 2006).

Remappings resonate beyond western settler societies. Fordist forestry, in one form or another, became widespread, expressed in many developing countries by investments in big, export-oriented, foreign-owned pulp and paper mills, often using plantation wood. In Brazil and Chile, for example, direct foreign investment in the forest sector was a key, often controversial aspect of national development strategies in the 1960s and 1970s (Clapp, 1995; Dauvergne, 1997; Marchak, 1995). Indeed, forest conflicts are globally widespread, and collectively inspire urgent pleas for changes in forest governance, typically to better address local development needs simultaneously with promoting environmental sustainability.

Fordist Forestry in British Columbia

Utilizing an outline of the evolution, nature, and challenge to Fordist forestry in British Columbia (BC), we can better illustrate the general as well as the unique processes of remapping (Hayter, 2000). In BC, consistent with practices across much of Canada, the provincial government has controlled forest resources since joining Confederation in 1871. Although sporadic ventures into industrial forestry had previously occurred, the arrival of the transcontinental railroad in Vancouver in 1885 stimulated a rapid, entrepreneurial-led growth of logging and the forest industries. In support of this growth, the provincial government privatized some forests as railroad land grants and introduced various licenses and leases to permit logging. But forestry became frenzied and speculative, with no concern for renewal, and in response to fears of *forest liquidation*, the licensing boom was halted in 1905. Following the advice of a Royal Commission established in 1909, the provincial government passed the Forestry Act of 1912 to further limit alienation of forest lands, while retaining the principle of public ownership. Following another Royal Commission, full commitment to Fordism awaited the 1947 Forest Act Amendment. This Amendment and subsequent amendments created large-scale tenures that were offered to corporations, often multinational, over long and renewable time horizons, in exchange for major investments in integrated forest product activities. The tenures were intended to be large enough to allow sustained-yield logging, and a new appurtenancy clause required that timber be processed in local mills to ensure the retention of local employment and a corresponding promotion of development within the province. Essentially, the provincial government sought resource bargains with MNCs to stimulate development and set low rates of stumpage—the fee paid for timber cut—to reflect the costs and uncertainties of exporting commodities from remote areas to distant markets.

For Fordist BC, forest policy was inseparable from industrial policy. With timber valued for its industrial uses alone, old growth forests were routinely classified as decadent or over-mature, calling for rapid harvest before their economic value dropped further from the inexorable advance of heart rot and decay. From an industrial perspective, the particular form of Fordism that developed in BC could be labelled "permeable" (Jenssen, 1989), featuring a high level of foreign ownership, with corresponding profit leakage, reliance on imported technology, and limited development of value-added options. Such truncated development stands in contrast to Scandinavian experience where forest industries remained under domestic control, with multiplier effects captured within the region (Raumolin, 1985). Nevertheless, BC's forest industries boomed after 1945, generating many spin-offs and high-income union employment, while spreading growth throughout the province. Moreover, this growth was relatively stable, and during Fordism average incomes in resource communities in BC's periphery were as high if not higher than in the Vancouver metro core (Davis & Hutton, 1989). Since the 1980s, however, BC's resource communities have experienced considerable instability, labor replacement, and downsizing (Hayter & Edenhoffer, 2016; Markey, Halseth, & Manson, 2012).

Fordist forestry in BC started to unravel in the 1970s as growth trajectories leveled off and became more volatile amid a series of increasingly deep recessions, culminating in the severe crisis of the early 1980s. This crisis proved a turning point in the fortunes of the forest industry and in public conceptions of forest governance. The recession exposed the industry's declining competitiveness: Factories had become technologically obsolescent, especially on the coast, and timber supplies had declined in quality and accessibility. In 1981, the government first recognized the onset of the *fall-down* effect, in which old-growth forests were replaced by lower-yielding second-growth forests, and timber yields per hectare plummeted. Even as industry members began to downsize and restructure, this recession sparked deep-seated and long-lasting "wars in the woods" exacerbated by intersecting environmental, aboriginal, and trade conflicts (Hayter, 2003). ENGOs argued that government and industry were sustaining neither the economy nor the environment, and indigenous groups were alarmed that the forest resource would be downgraded prior to the settlement of their treaty claims. At the same time, a trade war over BC's lumber exports to the US erupted when US sawmillers, also threatened by the recession, believed their problems to be caused by cheap lumber imports from BC, and accordingly sought protection through the creation of the Coalition for Fair Lumber Imports (CFLI). This initiated 35 years of duties, litigation, export restrictions, and managed trade, which continued in newly aggravated form in 2017.

In BC, as elsewhere, economic crisis should not be considered "the" cause of remapping, but rather an important catalytic event occurring as part of longer-run technological, economic, political, and social forces of change. By the 1970s, attitudes across western economies were becoming more pro-environment, whereas established industries were experiencing productivity problems, both trends being evident in BC's forest economy. Furthermore, in contrast to the rest of Canada and other western settler societies, BC's failure to sign treaties with indigenous peoples had long been a source of concern waiting to be sparked. These conflicts empowered environmental and aboriginal interests, not because of a shared view of solutions, but because of their shared opposition to existing forest policy and shareholder entitlements.

Policy Uncertainty

Anticipating the direction and outcomes of remapping is hard, especially when assessments are made during periods of crisis and conflict. Crisis is important for provoking remapping, as it is for society-wide paradigmatic transformations, because it reveals the need for change, and weakens vested industrial interests through their downsizing, failure, and declining power. At the same time, crisis energizes new stakeholders by strengthening their arguments and social legitimacy. Yet successive waves of investment in equipment and infrastructure, managerial know-how and labor skills, and attitudes towards economic growth leave established resource policies and industries deeply ingrained. These forces of inertia are powerful, metaphorically captured by the ideas of the staple trap and the resource cycle; equally challenging is the difficulty of distinguishing short-term or cyclical recessions from secular or turning point recessions in resource sectors in which a history of boom and bust is "normal" and the end of natural abundance is inconceivable. Moreover, cost-reduction strategies in response to immediate survival needs can obscure the need to adapt by investing in R&D, innovation, and a more skilled workforce from industry members.

Forest conflicts similarly shape the contours of remapping. Widely publicized protests—logging blockades, civil disobedience, market boycott campaigns, disruptions of corporate meetings, and the shaming of environmental culprits—have become symbols of these conflicts. In tandem, litigation has become a vital tool of legitimizing and empowering protest, and of shaping remapping. As Langston (2005, p. 72) argues of forest conflicts in Oregon: "What mattered most about litigation was that it forces a variety of stakeholders, with multiple voices, multiple stories, and multiple perspectives to communicate with each other." Litigation has also been an important tool in promoting aboriginal rights in peripheries, not least in BC, where the Supreme Court of Canada has made several decisions since 1997 that have significantly increased aboriginal control over traditional territories. In Australia and New Zealand, aboriginal rights in resource peripheries have also gained judicial recognition. In the case of Tasmania, where the indigenous population was eliminated, contemporary logging is forbidden wherever artifacts representing the region's aboriginal heritage, such as evidence of stone tools, are present (Hayter & Barnes, 2012).

Forest conflicts have been generated by newly empowered stakeholders to initiate remapping processes, and their resolution has required new institutions. Indeed, institutional innovation is essential to implement compromises among established and new stakeholders with different motivations and mandates. Such innovation in turn produces new uncertainties and policy dilemmas, rooted in its experimental nature and the range of interests that have to be accommodated. For example, environmental bargaining over forests, especially old growth forests, is inherently difficult because adversaries have fundamentally different views of resource values and no common goal or easy metric to resolve conflicts (Affolderbach, 2011). What might be described as *cultural bargaining* over aboriginal rights, self-determination, and resource tenure is no less fraught with conflicting value systems. Furthermore, forest policy involves expectations over long-term horizons that are hard to define, can vary among participants, and over the past decade have become more complicated by the implications of climate change, which themselves vary from place to place. In this regard, the ecological implications of sustainable forestry will require increased scientific knowledge of local conditions and ecosystems that can be anticipated to shape forest policy in ways yet unforeseen as they evolve. Faced with these dilemmas, forest remapping is inescapably uncertain.

Globalization and Forest Conflict Resolution: Shareholder and Stakeholder Models

The destabilization of Fordism since the late 1970s and the onset of a contemporary, more volatile period of globalization have posed significant policy challenges, stimulating debates about appropriate forms of governance from global to local levels (Jessop, 1998; Mayntz, 2003). These debates have resonated strongly throughout forest peripheries where the models of stakeholder or shareholder remapping provide alternative starting points to remapping (Fig. 16.1). The shareholder model's proponents are allied to neoliberal thinking, emphasizing markets, hierarchies, and the economic motivations of shareholders. In this approach, environmental values can be achieved through privatization and deregulation—that is, by the establishment of appropriate property rights and market exchange rather than by regulation (Anderson & Leal, 2015). In contrast, the proponents of the stakeholder model of remapping promote more broadly-based decision making driven by new stakeholders, including those committed to principles of sustainability, and whose empowerment comes at the expense of vested market interests and entrenched government and corporate hierarchies. Put another way, the shareholder's share is predominantly, sometimes exclusively, economic, whereas the stakeholder's stake may also be cultural, ethical, environmental, or intrinsic.

In idealized form, shareholder and stakeholder remapping represent different, contentious ideologies and policy prescriptions. Yet these processes are problematic to disentangle and both impulses may exist in the same periphery, both subject to context and contingency. In an Indonesian context, specifically the forest and mining activities of Kalimantan, Tsing (2005) argues that the impact of neoliberalism as a universal force is modified by local frictions in the form of local cultural practices, contributing to hybrid neoliberal forms. At the same time, locally rooted stakeholders typically seek economic development that involves serving markets, and formulators of local plans cannot ignore global connections, whether perceived as threats or opportunities. Further, the environmental and cultural interests underlying new stakeholder models can be seen as either countervailing responses to neoliberal-inspired stakeholder remapping or as quasi-autonomous alternatives with distinctive roots and mandates. Indeed, the evolution of environmentalism is intimately linked with industrialization, becoming a powerful expression of contemporary globalization (or universal, in Tsing's terms) in response to the implications of Fordism's rapidly escalating demands for resources in support of mass production and consumption. In turn, environmentalists have sought to impose their views, sometimes against resistance from local peoples who wish to preserve local practices (Stevens, 2014).

As mentioned in our introductory comments, it is tempting to link stakeholder and shareholder approaches respectively with institutional thickening and thinning. In BC's Great Bear Rainforest, for example, stakeholder remapping required institutional innovation and thickening both to bring together opposing parties to exchange views and develop respectful relationships (*talking the talk*), and to implement new forms of governance (*walking the walk*) that allow actors to use forests in mutually acceptable ways (Clapp et al., 2016). More generally, common property forest tenures around the globe involve cooperation among diverse participants (Moran & Ostrom, 2005). Ostrom's notion (Ostrom, 2010) of "polyvalent governance" emerges from stakeholder remapping, and with it she underlines the importance of institutional innovation in conflict resolution. In contrast, the privatization of forests in Chile and New Zealand and the reduction in ownership regulations meant the removal of constraints on market-driven interests (Clapp, 1995; Roche, 1990). The blurred distinctiveness of shareholder and stakeholder remapping is captured in the terms themselves: *Stakeholder* once referred to a miner or logger who had established a stakes for private profit; now it implies participation in more diversified, democratic models of governance.

In practice, remapping is context-driven, shaped by government policies that prioritize development. However, it is often pragmatic, opportunistic, and difficult to categorize, even within the same region. In BC, for example, studies have emphasized neoliberal (shareholder) influences on forest remappings (McCarthy, 2006), the limits and "hybridization" of these influences, and alternative more democratic, shareholder models of governance (Hayter & Barnes, 2012). This opacity is reflected in broad policy shifts. Thus, during the 1980s, forest policy under a right-wing provincial government became confused, responding effectively to neither the recession nor the wars in the woods. During the 1990s, until 2001, initially under the slogan of bringing "Peace in the Woods," a left-wing government introduced a battery of regulations concerning environmental issues, aboriginal rights, tenure reform, regional planning, and timber taxation that are hard to relate to a neoliberal agenda. Since 2001, right-wing governments have deregulated forestry to some degree, including the removal of appurtenancy, the requirement that logs be processed in local mills. Most recently, in 2017 a new minority left-wing government's attempts to return to a remapping agenda have faced considerable challenges created by continued industry downsizing and concerns for profitability and by the diversity of stakeholder interests. In some respects, stakeholder perspectives have been retained or enhanced, such as in the adoption of the community forest model (McIlveen & Bradshaw, 2009). Even the controversial removal of appurtenance could be seen as a practical (not ideological) response to declining timber supplies insufficient to support existing mill capacity. In the case of the iconic Great Bear Agreement of 2016, a paradigmatic case of the stakeholder model, although the roots of this accomplishment were established in the left-wing forestry regulation of the 1990s, its realization has occurred under right-wing government since 2001. And if forestry conservation has been a powerful trend in BC, its continuance is not automatic: For example, Tasmania's Nature Conservation Act 2002 is in jeopardy to federal electoral politics (Affolderbach, 2011).

However blurred, the shareholder-stakeholder distinction usefully identifies distinct impulses that can be found, in varying mixes, in forest peripheries around the globe. However, if impulses are defined in terms of strong, sometimes ideological driving or motivating forces, they are not autonomous. Although shareholder thinking is supported by extremely powerful economic and political institutions, stakeholder models are increasingly widespread, a trend desirable for enhancing local interests and empowerment in the wise use of resources.

Shareholder Impulses

The implications of neoliberal imperatives of privatization and deregulation for forest peripheries were most dramatically visible in the burgeoning plantation forest sectors of Chile and New Zealand. Indeed, Chile may be regarded as a neoliberal pioneer, privatizing its plantation forests in the 1970s, predating the emergence of Reaganism and Thatcherism as synonyms for neoliberalism in the 1980s. In practice, however, the expansion of Chile's plantation sector has been heavily subsidized and promoted by government, notably in the 75% reforestation subsidies established by Decree Law 701 in 1974 (Clapp, 1995). Indeed, the plantations were privatized twice, as the government re-acquired much of the resources following the crisis of the early 1980s, only to then re-auction the plantations and forest lands, in the latter case stimulating a considerable degree of foreign investment. Neoliberalism continues to exert lasting influence in Chile, where ownership of plantations provides collateral and thereby secures access to credit.

In New Zealand's case, in response to economic crisis, the implementers of a new national government's embrace of neoliberalism to provide strategic direction for the economy targeted the forest sector (Roche, 1990). In 1987, the New Zealand Forest Service was replaced by the Forestry Corporation, a state-owned enterprise, that in 1988 began to privatize the country's plantation forests, first established in the 1920s. Privatization, mandated to increase efficiency and reduce the national debt, redefined New Zealand's forest sector. By 1995, over 90% of plantation forests were under private control, and MNCs became major players controlling more than half of forest production. Nevertheless, New Zealand's policymakers combined their commitment to the neo-liberal shareholder model with a new approach to conservation. The remaining native forests, most located in mountainous regions, were designated for conservation purposes in 1987, and their management turned over to a newly created Department of Conservation. New Zealand's productive plantation forest *lands* have been leased rather than sold outright, and it remains possible that they could be subject to Maori land claims, even if the latter have been rendered difficult by privatization (Coombes, 2003). Meanwhile in BC, forest lands have remained under public control, and environmental and aboriginal interests have opposed any suggestion of privatization, despite the strong pro-market, neoliberal tendencies routinely associated with right-wing provincial governments.

Stakeholder Impulses

In contrast to shareholder thinking, stakeholder impulses to remapping Fordist forestry are driven by diverse motivations, initiatives, and understandings, with priority given to conservation, sustainable development, and stronger commitments to local uniqueness, values, and control. Importantly, the crafters of stakeholder models do not exclude markets, and their evolutionary dynamics are highly varied, predating, coevolving with, and responding to Fordist forestry. In Japan, for example, during the 1950s and 1960s—the heyday of Fordist forestry—small-scale log auctions were expanded throughout the country, reducing transaction costs for local private wood-lot owners and small-scale sawmills (Reiffenstein & Hayter, 2006). These arrangements, rooted in a long history of cooperative forestry (Totman, 1989), are now in slow decline, because of significant opportunity costs outside of forestry for landowners and workers, coupled with low cost imports. Nevertheless, their continued operation reflects interlinked local concerns for jobs, community, and the environment. In Europe, various forms of small-scale forestry cooperatives exist to similarly maintain the multi-purpose attributes of forests (Weiss, 2011). In many developing countries where communities are highly dependent on forest resources, the actors of various forms of cooperative polyvalent governance manage forests as common property (Ostrom, 2010).

The most visceral conflicts occur where proponents of stakeholder approaches seek to provide an alternative to shareholder models, and to replace implanted Fordist forestry. In these situations, stakeholder models are both more democratic and messier because they diversify and decentralize decision making influence over forest use. Furthermore, the point of a stakeholder model is to demand paradigmatic change in forest use, and to challenge conventional scientific wisdom in forest use that has emphasized the one-size-fits-all approach embodied in the so-called *normal forest*. In a North American context, for example, Rajala (1998) argues that forestry science evolved as a handmaiden to industry, its scientists privileging the market values of timber and emphasizing a standardized approach to harvesting as well as the rotational clear cutting of large blocks that were expected either to regenerate naturally or by reforestation with a preferred species monoculture. In this view, "over-mature" forests lost value with each year they stood to rot, and needed to be cut quickly, before they became "worthless" (Percy, 1986). However, as previously noted, proponents of this view ignored or under-estimated both environmental and technological risks (Langston, 2005). In contrast, environmental and local proponents of stakeholder models emphasize ecological science aimed at preserving the multiple attributes of forests and their management in locally contingent ways.

Stakeholder remapping of Fordist forestry practices requires the coevolution of learning and bargaining processes among existing vested interests and new stakeholders. In general terms, learners must embrace knowledge of local ecologies and cultural practices and bargaining to embrace actors representing, at least in part, noneconomic interests. In BC's Great Bear Rainforest, for example, institutional innovations enabled the warring parties to talk with one another and exchange views in a constructive manner. The Joint Solutions Project (JSP) brought industry members and environmentalists face to face to find common ground, and to develop proposals for stakeholder negotiations. The Coast Information Team (CIT), a scientific boundary organization, and the Turning Point Initiative, a protocol between First Nations and the provincial government, were other crucial institutional innovations that facilitated discussion and bargaining toward the final agreement reached in 2016 (Affolderbach et al., 2012; Clapp et al., 2016; Price, Roburn, & MacKinnon, 2009). The JSP agreement between industry and ENGOs stopped the latter's market campaigns against the companies in return for a moratorium on logging in 30 watersheds and led to the creation of the CIT. As a boundary organization, the CIT compiled data on the ecology of the GBR in an Ecological Spatial Analysis, a GIS that was constructed by and acceptable to all stakeholders—ENGOs, indigenous peoples, industry, and government (Clapp & Mortenson, 2011). This organization helped resolve confrontations between ENGOs and industry and encouraged more constructive bargains based on shared knowledge and common areas of agreement. Meanwhile, the Turning Point Initiative depended on recognition by the provincial government that negotiations with indigenous peoples would be "government to government" (G2G), rather than merely stakeholder to government (Davis, 2009). In this regard, a series of court decisions have empowered First Nations, who, from being excluded, have become highly influential, and whose interests, knowledge, and consent must now be explicitly included in forestry decision making.

These institutional process innovations were vital to enable contesting parties to learn and bargain effectively with one another and construct new rules and codes of conduct for the new stakeholder model. Their mission complete in this regard, the JSP and CIT are now disbanded, although the GIS data base remains. Further, as a result of the dialogue inspired by these initiatives, including the G2G negotiations, the final Great Bear Forest Agreement remapped the region in terms of permitted activities and their governance, creating more permanent institutions, in the forms of new rules and organizations—ecosystem-based management, community forests, conservation areas, indigenous forestry firms, and new forms of financing. Will these new developments work out as intended?

Assessing Stakeholder Remapping as an Emergent Form of Good Governance

Do these new stakeholder agreements that have replaced Fordist forestry models, of the kind illustrated by the GBR Agreement or more ambiguously by Tasmania's Conservation Act, constitute good governance? The answer is not straightforward. As long appreciated, global forest types, uses, and governance are highly differentiated within ecosystems and regions as well as among countries. Nevertheless, the idea of global paradigmatic change comprises possibilities for transformation. Moreover, the idea of a stakeholder model informed by ecological science is rooted in the importance of local context. Admittedly, the implications of crisis and conflict for forest governance within stakeholder models may not be important. For example, across Europe, where property rights are well established, institutional innovation in forestry practices is rarely seen as a form of conflict resolution. Elsewhere, however, as Ostrom's (2010) pioneering work has demonstrated for many developing countries, the creation of various forms of polycentric governance has been vital to resolving forest conflicts.

Ostrom (2010, p. 652) systematically analyzed polycentric governance in which problem solving is a major theme and developed a common set of eight "design principles [that promote] long surviving resource institutions" in managing common pool resources.

Locally, the success of stakeholder remapping depends upon the effectiveness of new routines for meeting stakeholder goals for sustained development. These routines are most obviously seen in the establishment of formal organizations and rules, with associated incentives and penalties. Routines and institutions may also be informal and less tangible. Indeed, the authors of a considerable literature suggest that social capital in the form of trust, cooperation and social networking enhances local development (Storper, 1997). From this perspective, brief reference to Ostrom's (2010, p. 652) systematic analysis of polycentric governance as problem solving responses to forestry conflicts around a common set of eight "design principles [that promote] long surviving resource institutions" in managing common pool resources is a good place to begin contemplation of the effectiveness of stakeholder remapping.

As Ostrom argues, polycentric governance encompasses wide-ranging institutional arrangements that are more complex than economic systems controlled by markets and governments. Ostrom's (2010, pp. 653–654) design principles relate to transparency, locally congruent resource appropriation and provision, collective choice arrangements, monitoring of users and resources, sanctions, conflict resolution, recognition of rights, and links to higher levels of governance. Thus, the local resource system should be clearly demarcated, and the boundaries between legitimate users and non-users be defined. Resource appropriation and maintenance rules should make sense in terms of local ecological conditions and the distribution of the benefits and costs of resource utilization should be fair. All stakeholders affected by resource use should be part of the decision making process. Users should monitor resources, and policymakers should graduate sanctions to penalize rule violation and supplement them with low-cost, local mechanisms for conflict resolution. The government should recognize local users' rights to make their own rules. Finally, if part of a larger territory, governance mechanisms are needed to ensure coherence between local and regional decision making. Further, the rules shaping common pool resource use are enormously varied in practice; Ostrom (2010, p. 651), for example, recognizes seven categories of rules pertaining to the role of actors (boundary, position, and choice rules) and potential outcomes (information, aggregation, payoff, and scope rules).

Moreover, in successful polycentric governance design, principles and rules are embedded in social relations in which trust among actors plays a key role (Ostrom, 2010, p. 642). As a central concept of social capital, trust reduces transaction costs by facilitating mutual understanding, cooperation, and the predictability of behavior. Indeed, in their analysis of community forestry projects in BC, McIlveen and Bradshaw (2009) emphasize the strong role of "social cohesion" and "community support" in distinguishing successful projects from failures. Like Markey et al. (2012) in their study of transitioning resource communities in BC, these authors also identified leadership as a key ingredient for local development, but whose availability and capability cannot be guaranteed.

Implications of Conflict Resolution for Trust and Cooperation

In common pool and property resources, problem solving occurs when disputes arise among users with a shared interest in the resource's use. In Ostrom's design principles, the problems that arise are best handled locally by accepted mechanisms. However, more formidable conflicts arise in the context of remapping when stakeholders have diverse motivations and goals for resource ownership and use and where local, low-profile, and low-cost dispute resolution rules are difficult to construct. Indeed, the effectiveness of agreements rooted in deep-seated conflict and mistrust, and where rule setting is politicized and legalistic, must be questioned in regards to promoting sustainable development. Are the new rules appropriate or even permanent? Will investors face increasing transaction costs? Can stakeholder agreements be modified in the light of experience without generating more rancor and mistrust? Are the new agreements less democratic than messy and cumbersome? Is trust and cooperation possible following protest and litigation?

Although finalized only recently, the Great Bear Rainforest Agreement 2016 helps illuminate the ramifications of these questions. This widely lauded agreement demonstrates the trend from top-down hierarchical and distant decision making over forest resources towards more democratic, locally sensitive governance. In particular, the provincial government and industry have both lost influence, whereas ENGOs and most especially First Nations have gained influence. From a social capital perspective, the GBR Agreement has achieved significant progress in developing cooperation, witnessed by the various forms of institutional thickening as well as the various stakeholders' ability to work together. The Agreement also embraces a wider range of values and privileges local interests who have intimate understanding of local resources and can perhaps be better trusted to achieve sustainable practices.

Yet the main actors involved in the GBR Agreement continue to have different values and conflicting mandates. ENGOs, for example, remain as watchdogs over industrial logging and have already claimed violations of the Agreement by local industry. In contrast, ENGOs are reticent to criticize activities by First Nations, even though they were upset (and surprised) by the latter's decision to reject Forest Stewardship Council certification—the supposed Cadillac standard of eco-certification—because of its costs and bureaucracy. Forest resources for commercial use have been allocated among various indigenous groups who appear to be adopting different strategies, ranging from contracting out or hiring specialized companies in ecosystem-based management to hopes of adding value. Whether forest resource allocations are individually large enough to justify investments in processing facilities is an important question, especially if coordination among indigenous groups remains a challenge. Remoteness, difficulties of access, and a small population base mean that both local and export markets are extremely limited.

Within the GBR Agreement, the provincial government reserves the right to approve mining and other resource developments apart from forestry. Any major proposed developments, however, would likely re-ignite opposition and litigation from ENGOs and First Nations who now have the support of recent Canadian Supreme Court precedents. The fear of further litigation is real and in turn indicates higher transaction costs for proposed developments. Beyond perhaps some informal local monitoring and policing by indigenous groups, no locally recognized dispute settlement mechanism seems to exist that can deal with issues expeditiously. Any negotiations for large-scale developments are likely to be messy, at least. Indeed, in stakeholder remapping, such as in the GBR, different interests have distinctive forms of legitimacy, and each group needs to demonstrate some form of achievement to the particular constituencies they serve or wish to address.

From a local development perspective, trust and cooperation are important in developing competitiveness because they reduce the transaction costs of exchange and facilitate the development of localized external economies. Moreover, where human resources are sparse, trust and cooperation are even more important to ensure that available energy is not wasted. But trust and cooperation take time to construct, both informally and formally, and actors must combine them with some form of penalty to prevent opportunism and free riders. From a governance perspective, the failure to develop workable social bargains might well encourage neoliberal impulses in which markets are seen as a more efficient alternative. For local communities and ENGOs, such failure may reinforce conflictual modes of behavior. Even so, from a common property perspective, progress towards clear design principles is evident, for example, in identifying the GBR boundary and the boundaries of ecological sub-regions, the recognition of resource user rights (and benefit streams) among indigenous peoples and principles of local empowerment, and acknowledgement of monitoring functions.

Conclusion

Forest use and governance is highly varied around the world, engaging societies at all levels of development. Driven by the depletion of old-growth forests, rising population pressures, growing industrial demands for resources, and the globalization of markets, forest conflicts have become widespread. Forest conflicts themselves are highly varied, and may involve: arson, violence, and illegal expropriation, impositions of radically different government policies on local economies, localized disputes among competing users, and the paradigmatic remapping of forest use away from a narrow focus on commodity production towards a recognition of environmental and cultural benefits of the forest. In this chapter, we have explored this latter theme.

Remapping is inherently contentious as it involves threats to vested interests and deeply ingrained ideas about resources and development, and it requires alternative patterns of forest use and governance. These proposals have roots in both shareholder and stakeholder models or ways of thinking. If this duality is blurred in practice, we have argued that stakeholder models—akin to Ostrom's concept of polyvalence—potentially provide differentiated, locally informed, and empowering ways to realize the goals of remapping in support of sustainable development. Proponents of shareholder approaches emphasize the public goods and benefits of forests, while recognizing the divergent interests of local and nonlocal actors. Moreover, the creation of stakeholder models on the ground requires innovative thinking, both in terms of bringing divergent parties together and developing new forms of cooperation and trust with new routines and institutions that enact cooperation.

At the same time, stakeholder models need to be validated beyond simple repudiation of shareholder (neoliberal) thinking. That is, stakeholder approaches must be evaluated in terms of their desired goals and mandates. In this regard, future researchers face important challenges. Comparative analyses of the institutional innovations underlying remapping would sharpen understanding of the global-local dynamics shaping initiatives and possibilities regarding new forms of governance. More generally, there is a need to develop frameworks to compare and critically assess whether stakeholder approaches to conflict resolution are working as desired, both in the pioneering regions generating them and in other regions whose inhabitants look to them as models for resource governance. The creators of such frameworks might also incorporate a wider range of resource activities within the theme of remapping, contrasting the spatial footprint and environmental impacts of mining, fishing, and grazing with those of forestry.

References

Affolderbach, J. (2011). Environmental bargains: Power struggles and decision making over British Columbia's and Tasmania's old-growth forests. *Economic Geography, 87,* 181–206. https://doi.org/10.1111/j.1944-8287.2011.01107.x

Affolderbach, J., Clapp, R. A., & Hayter, R. (2012). Environmental bargaining and boundary organizations: Remapping British Columbia's Great Bear Rainforest. *Annals of the Association of American Geographers, 102,* 1391–1408. https://doi.org/10.1080/00045608.2012.706567

Amin, A., & Thrift, N. (1995). Globalisation, institutional "thickness" and the local economy. In P. Healey, S. Cameron, S. Davoudi, S. Graham, & A. Madani-Pour (Eds.), *Managing cities: The new urban context* (pp. 91–108). Chichester: Wiley.

Anderson, T. L., & Leal, D. R. (2015). *Free market environmentalism for the next generation.* New York: Palgrave Macmillan. https://doi.org/10.1057/9781137443397

Bestor, T. C. (1998). Making things clique: Cartels, coalitions, and institutional structure in the Tskujij wholesale seafood market. In M. Fruin (Ed.), *Networks, markets, and the Pacific Rim, Studies in Strategy* (pp. 154–80). Oxford, UK: Oxford University Press.

Bevir, M. (2012). *Governance: A very short introduction*. Oxford, UK: Oxford University Press.

Carron, L. T. (1985). *A history of forestry in Australia*. Canberra, Australia: Australian National University Press.

Clapp, R. A. (1995). Creating competitive advantage: Forest policy as industrial policy in Chile. *Economic Geography, 71,* 273–296. https://doi.org/10.2307/144312

Clapp, R. A. (2001). Tree farming and forest conservation in Chile: Do replacement forests leave any originals behind? *Society & Natural Resources, 14,* 341–356. https://doi.org/10.1080/08941920119176

Clapp, R. A. (2004). Wilderness ethics and political ecology: Remapping the Great Bear Rainforest. *Political Geography, 23,* 839–862. https://doi.org/10.1016/j.polgeo.2004.05.012

Clapp, R. A., Hayter, R., Affolderbach, J., & Guzman, L. (2016). Institutional thickening and innovation: Reflections on the remapping of the Great Bear Rainforest. *Transactions of the Institute of British Geographers, 41,* 244–257. https://doi.org/10.1111/tran.12119

Clapp, R. A., & Mortenson, C. (2011). Adversarial science: Conflict resolution and scientific review in British Columbia's central coast. *Society & Natural Resources, 24,* 902–916. https://doi.org/10.1080/08941921003801505

Coombes, B. (2003). The historicity of institutional trust and the alienation of Maori land for catchment control at Mangatu, New Zealand. *Environment and History, 9,* 333–359. https://doi.org/10.3197/096734003129342872

Dauvergne, P. (1997). *Shadows in the forest: Japan and the politics of timber in Southeast Asia*. Cambridge, MA: The MIT Press.

Davis, H. C., & Hutton, T. A. (1989). The two economies of British Columbia. *BC Studies, 82,* 3–19. https://doi.org/10.14288/bcs.v0i82.1320

Davis, L. (2009). The high stakes of protecting indigenous homelands: Coastal first nations' turning point initiative and environmental groups on the B.C. west coast. *International Journal of Canadian Studies, 39,* 137–159. https://doi.org/10.7202/040827ar

Franklin, J., Berg, D., Thornburgh, D., & Tappeiner, J. (1997). Alternative silvicultural approaches to timber harvesting: Variable retention harvest systems. In K. A. Kohm & J. F. Franklin (Eds.), *Creating a forestry for the 21st century* (pp. 111–139). Washington, D.C.: Island.

Freeman, C., & Louça, F. (2001). *As time goes by: From the industrial revolutions to the information revolution*. Oxford, UK: Oxford University Press.

Gee, H. (2001). *For the forests: A history of the Tasmanian forest campaigns*. Hobart: The Wilderness Society.

Gritten, D., Mola-Yudego, B., Delgado-Matas, C., & Kortelainen, J. (2012). A quantitative review of the representation of forest conflicts across the world: Resource periphery and emerging patterns. *Forest Policy and Economics, 33,* 11–20. https://doi.org/10.1016/j.forpol.2012.06.008

Hayes, S. P., & Glendenning, G. (2005). Human values and forests: Changes in the Great Lakes wildlands. In R. G. Lee & D. R. Field (Eds.), *Communities and forests: Where people meet the land* (pp. 77–95). Corvallis, OR: Oregon State University Press.

Hayter, R. (1976). Corporate strategies and industrial change in the Canadian forest product industries. *Geographical Review, 66,* 209–228. https://doi.org/10.2307/213581

Hayter, R. (2000). *Flexible crossroads: The restructuring of British Columbia's forest economy*. Vancouver, Canada: UBC Press.

Hayter, R. (2003). "The war in the woods": Post-Fordist restructuring, globalization and the contested remapping of British Columbia's forest economy. *Annals of the Association of American Geographers, 93,* 706–729. https://doi.org/10.1111/1467-8306.9303010

Hayter, R., & Barnes, T. J. (2012). Neoliberalism and its geographical limits: Comparative reflections from forest peripheries in the global north. *Economic Geography, 88,* 197–221. https://doi.org/10.1111/j.1944-8287.2011.01143.x

Hayter, R., Barnes, T. J., & Bradshaw, M. J. (2003). Relocating resource peripheries to the core of economic geography's theorizing: Rationale and agenda. *Area, 35,* 15–23. https://doi.org/10.1111/1475-4762.00106

Hayter, R., & Edenhoffer, K. (2016). Evolutionary geography of a mature resource sector: Shakeouts and shakeins in British Columbia's forest industries 1980 to 2008. *Growth and Change, 47,* 497–519. https://doi.org/10.1111/grow.12155

Hirst, P. Q., & Thompson, G. (1996). *Globalization in question: The international economy and possibilities of governance.* Cambridge: Polity.

Jenssen, J. (1989). "Different" but not "exceptional": Canada's permeable Fordism. *Canadian Review of Sociology and Anthropology, 26,* 69–94.

Jessop, B. (1998). *The rise of governance and the risks of failure: The case of economic development. International Social Science Journal, 50,* 29–45. https://doi.org/10.1111/1468-2451.00107

Johnston, R., & Glasmeier, A. (2007). Neo-liberalism, democracy and the state: Temporal and spatial constraints to globalization. *Space and Polity, 11,* 1–23. https://doi.org/10.1080/13562570701406493

Kelly, P. F. (1999). The geographies and politics of globalization. *Progress in Human Geography, 23,* 379–400. https://doi.org/10.1177/030913259902300303

Langston, N. (2005). Resource management as a democratic process: Adaptive management on federal lands. In R. G. Lee & D. R. Field (Eds.), *Communities and forests: Where people meet the land* (pp. 52–76). Corvallis, OR: Oregon State University.

Lee, R. G., & Field, D. R. (2005). Introduction: From scientific forestry to community forestry. In R. G. Lee & D. R. Field (Eds.), *Communities and forests: Where people meet the land* (pp. 1–14). Corvallis, OR: Oregon State University.

Mander, J., & Tauli-Corpuz, V. (2006). *Paradigm wars: Indigenous peoples' resistance to globalization.* San Francisco: Sierra Club.

Marchak, M. P. (1995). *Logging the globe.* Montreal, Canada: McGill-Queens University Press.

Marchak, P., Aycock, S., & Herbert, D. (1999). *Falldown: Forest policy in British Columbia.* Vancouver: David Suzuki Foundation and Ecotrust.

Markey, S., Halseth, G., & Manson, D. (2012). *Investing in place: Economic renewal in northern British Columbia.* Vancouver, Canada: UBC Press.

Mather, A. S. (1992). The forest transition. *Area, 24,* 367–379. Retrieved from https://www.jstor.org/stable/20003181

Mayntz, R. (2003). Governing failures of the problem of governability: Some comments on a theoretical paradigm. In J. Kooiman (Ed.), *Modern governance: New government-society interactions* (pp. 9–20). London: SAGE.

McCarthy, J. (2006). Neoliberalism and the policy of alternatives: Community forestry in British Columbia and the United States. *Annals of the Association of American Geographers, 96,* 84–104. https://doi.org/10.1111/j.1467-8306.2006.00500.x

McIlveen, K., & Bradshaw, B. (2009). Community forestry in British Columbia, Canada: The role of local community support and participation. *Local Environment, 14,* 193–205. https://doi.org/10.1080/13549830802522087

Moran, E. F., & Ostrom, E. (2005). *Seeing the forest and the trees: Human-environment interactions in forest ecosystems.* Cambridge, MA: The MIT Press.

North, D. C. (1990). *Institutions, institutional change and economic performance.* Cambridge, UK: Cambridge University Press.

O'Riordan, T. (2001). *Globalism, localism and identity: Fresh perspectives on the transition to sustainability.* London: Earthscan.

Ostrom, E. (2010). Beyond markets and states: Polycentric governance of complex economic systems. *American Economic Review, 100,* 641–672. https://doi.org/10.1257/aer.100.3.641

Percy, M. (1986). *Forest management and economic growth in British Columbia.* Ottawa: Economic Council of Canada, Supply and Services.

Price, K., Roburn, A., & MacKinnon, A. (2009). Ecosystem-based management in the Great Bear Rainforest. *Forest Ecology and Management, 258,* 495–503. https://doi.org/10.1016/j.foreco.2008.10.010

Rajala, R. A. (1998). *Clearcutting the Pacific rain forest: Production, science and regulation.* Vancouver, Canada: UBC Press.

Raumolin, J. (1985). The impact of forest sector on economic development in Finland and Eastern Canada. *Fennia, 163,* 395–437.

Reiffenstein, T., & Hayter, R. (2006). Domestic timber auctions and flexibly specialized forestry in Japan. *The Canadian Geographer, 50,* 503–525. https://doi.org/10.1111/j.1541-0064.2006.00160.x

Robbins, W. G. (2006). *Hard times in paradise.* Seattle, WA: University of Washington.

Roche, M. M. (1990). Perspectives on the post-1984 restructuring of state forestry in New Zealand. *Environment and Planning A, 22,* 941–999. https://doi.org/10.1068/a220941

Stevens, S. (2014). *Indigenous peoples, national parks, and protected areas: A new paradigm linking conservation, culture, and rights.* Tuscon, AZ: University of Arizona Press.

Storper, M. C. (1997). *The regional world: Territorial development in a global economy.* New York: Guilford.

Totman, C. D. (1989). *The green archipelago: Forestry in preindustrial Japan.* Berkeley, CA: University of California Press.

Tsing, A. L. (2005). *Friction: An ethnography of global connection.* Princeton, NJ: Princeton University Press.

Veblen, T. (1899). *The theory of the leisure class: An economic study of institutions.* New York: Macmillan.

Weiss, G. (2011). The study of innovation in the forest sector: Relevance and research background. In G. Weiss, D. Pettenella, P. Ollonqvist, & B. Slee (Eds.), *Innovation in forestry: Territorial and value chain relationships* (pp. 1–9). Wallingford: CABI. https://doi.org/10.1079/9781845936891.0000

Weller, S., & O'Neill, P. (2014). An argument with neoliberalism: Australia's place in a global imaginary. *Dialogues in Human Geography, 42,* 105–130. https://doi.org/10.1177/2043820614536334

Westoby, J. C. (1989). *Introduction to world forestry: People and their trees.* Oxford: Basil Blackwell.

Widick, R. (2009). *Trouble in the forest: California's redwood timber wars.* Minneapolis, MN: University of Minneapolis Press.

Wilson, J. (1998). *Talk and log: Wilderness politics in British Columbia.* Vancouver, Canada: UBC Press.

Zimmerer, K. S. (Ed.). (2006). *Globalization and new geographies of conservation.* Chicago, IL: University of Chicago.

Chapter 17
The Governance of Global Innovation Systems: Putting Knowledge in Context

Christian Binz and Bernhard Truffer

Broadening the Governance of Innovation Systems

In a globalizing knowledge economy, innovation processes increasingly depend on interaction between distant places (Corpataux, Crevoisier, & Theurillat, 2009). However, most authors of innovation policy literature continue to focus on processes that happen within specific territorial boundaries such as countries or regions. They generally justify their national or regional focus by arguing that science, technology, and innovation policies are typically formulated by national or regional policy makers, or that innovation related governance structures differ substantially between regions and countries. This perspective has been particularly salient in innovation system frameworks, which have become one of the core frameworks to inform innovation policy from a governance perspective (Sharif, 2006). In order to address the challenges of globalization, researchers must thus elaborate a number of conceptual extensions to innovation system approaches. In particular, regionally or nationally embedded innovation processes must be connected to dynamics that emerge from multiscalar actor networks or from the interplay between innovation processes in different spatial subsystems (Martin, 2016; Weber & Truffer, 2017).

C. Binz (✉)
Department of Environmental Social Sciences, Eawag-Swiss Federal Institute of Aquatic Science and Technology, Dübendorf, Switzerland

Centre for Innovation, Research and Competence in the Learning Economy CIRCLE, Lund University, Lund, Sweden
e-mail: christian.binz@eawag.ch; christian.binz@circle.lu.se

B. Truffer
Department of Environmental Social Sciences, Eawag-Swiss Federal Institute of Aquatic Science and Technology, Dübendorf, Switzerland

Copernicus Institute of Sustainable Development, Utrecht University, Utrecht, The Netherlands
e-mail: B.Truffer@uu.nl

J. Glückler et al. (eds.), *Knowledge for Governance*, Knowledge and Space 15,
https://doi.org/10.1007/978-3-030-47150-7_17

The concept of Global Innovation Systems (GIS) was recently developed to address some of these challenges (Binz & Truffer, 2017), building on recent contributions in innovation studies whose authors emphasize actor networks and institutional contexts for innovation that are interrelated across spatial scales (Bunnell & Coe, 2001; Carlsson & Stankiewicz, 1991). Various analytical approaches have conceptualized the increasing importance of international linkages between and beyond territorial innovation systems (for an overview, see, e.g., Carlsson, 2006; Grillitsch & Trippl, 2013). However, other scholars have criticized these concepts for remaining rather vague about how supranational actor networks and institutions influence innovation processes and how they differ between different types of sectors and industries (Binz, Truffer, & Coenen, 2014; Coenen, Benneworth, & Truffer, 2012; Grillitsch & Trippl, 2013).

The GIS framework accordingly explicitly specifies how key system resources for innovation are formed in multiscalar networks. In order to grasp the complex spatial configuration of a GIS, two conceptual extensions are proposed. First, subsystems are defined based on the actors, networks, and institutions that generate key system resources (knowledge, market access, financial investment, and technology legitimacy) that are necessary for innovation processes (see Binz, Truffer, & Coenen, 2016). These subsystems may or may not coincide with territorial boundaries of countries or regions. Second, the overall performance of a GIS depends on whether the relevant subsystems effectively interrelate through so-called structural couplings. These extensions lead to a typology of four ideal-type GIS configurations, distinguishing the industry's dominant innovation modes—STI (science-technology and innovation) versus DUI (doing, using, and interacting) (Jensen, Johnson, Lorenz, & Lundvall, 2007)—and the economic system of valuation in which markets for the innovation are constructed—standardized valuation in global mass markets versus customized valuation in local contexts (Huenteler, Schmidt, Ossenbrink, & Hoffmann, 2016; Jeannerat & Kebir, 2016). Depending on the innovation and valuation mode, the elements of a GIS may be either spatially mobile (with many international spillovers) or rather sticky (with most spillovers constrained to specific regional contexts).

The proposed focus on multiscalar industry dynamics calls for new innovation governance approaches, which require coordination between multiple actors both inside and outside of specific countries or regions. We define innovation governance in line with prior governance and innovation studies as the self-organizing networks, negotiated interorganizational coordination, and decentered, context mediated intersystemic steering that influences the creation, diffusion, and application of novel technologies, products, and services (Benz, Lütz, Schimank, & Simonis, 2007; Carlsson & Stankiewicz, 1991; Jessop, 1998). Proponents of innovation system approaches early on implicitly adopted a governance perspective in that they perceived state actors and their innovation policy interventions as only one—if often very important—actor group in broader agentic constellations that jointly shape the directionality of emerging technologies and industries (Weber & Rohracher, 2012; Weber & Truffer, 2017). Innovation governance in GIS therefore emphasizes the interplay among different actors who may be active in and across distant territorial subsystems, their transnational networks, as well as the multiscalar institutional structures that may promote, hinder, or shape the trajectory of emerging technologies and industries. In this context, conventional national and regional innovation policymakers must reflect on how they intervene in transnational innovation processes, what kind of system resources they mobilize locally or anchor from distant regions, and how actors residing outside their own jurisdictions may be influenced. Furthermore, it also puts in relief what sort of coordinative institutions are needed at a global level that may shape, hinder, or promote new solutions to emerging global challenges. The recent globalization of renewable energy industries highlights the need for such an approach (Binz, Tang, & Huenteler, 2017). It is probably still fair to say that the topic of global innovation system governance has not received sufficient attention (Truffer, 2012).

In the present chapter, we will formulate a first tentative contribution to the problem of global innovation system governances by elaborating on implications of the multiscalar governance constellations that align with the four innovation system configurations proposed by the GIS concept's supporters (Binz & Truffer, 2017). The increasing globalization of innovation activity calls one to understand, anticipate, and integrate effects of spatial spillovers in place-based innovation governance strategies. One of this framework's key implications is that the existing focus on the governance of knowledge creation must be complemented with strategies that target "valuation" processes, as in the construction of new markets, the management of technology legitimacy, or the mobilization of financial investment (Jeannerat & Kebir, 2016).

We will elaborate these arguments as follows. We first review challenges that globalization poses to the existing innovation system literature. In the following section, we introduce the Global Innovation System framework alongside the industrial typology derived from an in-depth assessment of the innovation and valuation dimension. In the section on policy and governance implications of global innovations systems, we elaborate the governance implications for different actors operating at various spatial scales. We then reach a conclusion in the last section and provide a view on how knowledge must be contextualized in order to effectively inform the governance of increasingly globalizing innovation dynamics.

Existing Perspectives on Innovation Systems in Transnational Contexts

Innovation system scholars emphasize that innovation emerges from complex actor networks that combine complementary knowledge stocks and capabilities into new solutions, all in the context of specific institutional settings (Lundvall, 1992). They utilize a system metaphor to emphasize the distributed yet dynamically coordinated nature of many innovation processes. Positive externalities that emerge from the interaction among universities, firms, policy makers, NGOs, and various intermediaries are a key prerequisite to the innovation process, which can however not be steered or controlled by any actor on his or her own (Nelson, 1993).

Scholars have formulated different variants of the IS concept over the years, including a national (NIS) (Lundvall, 1988), sectoral (SIS) (Malerba, 2002), regional (RIS) (Cooke, Gomez Uranga, & Etxebarria, 1997), and technological (TIS) (Carlsson & Stankiewicz, 1991) approach. Although each variant emerged from the same roots in innovation studies and evolutionary economics, their developers also created their own research program with a somewhat distinct epistemology and methodological approach (Coenen & López, 2010). Given these differences, cross-fertilization between the various traditions of IS research has remained surprisingly scant (Weber & Truffer, 2017). This *siloed* thinking in the IS tradition has proven to be particularly problematic when trying to formulate a more internationalized perspective on innovation processes. In an international context, the multiscalar overlaps between regional, national, sectorial, and technological elements move center stage. Yet, the existing literature on *global*, *international*, or *multiscalar* IS has generally reflected a lack of integrative thinking in the IS tradition (Carlsson, 2006; Niosi & Bellon, 1994; Oinas & Malecki, 2002; Pietrobelli & Rabellotti, 2009).

The GIS approach builds on recent developments in the TIS tradition and adds more internationalized perspectives from other IS literatures, whose authors recently emphasized the multiscalar actor networks and institutional contexts that jointly support (or hinder) the formation and diffusion of innovation (Jurowetzki, Lundvall, & Lema, 2015; Oinas & Malecki, 2002). In some cases, the relevant IS structures may be largely reducible to specific territorial contexts, yet in the majority of cases, they depend on actor strategies, networks, and institutional dynamics that coevolve between different parts of the world. The possible combinations of actors, networks, and institutions that support or hinder innovation in GIS are thus almost countless, and alternative configurations of the systems' structure can lead to similar performance characteristics (Edquist, 1997). As the different system elements become more complexly structured internationally, relating not only to system structures but also to key activities, we propose to focus on the formation dynamics around four distinct system resources—knowledge, market access, financial investment, and technology legitimacy (Binz et al., 2016).

Each of these resources emerge from distributed agency in the system and may each depend on actor networks with specific spatial configurations. For example, a market for a novel technology may not preexist and have to be proactively constructed by firms, prosumers, and intermediaries in specific regions (Dewald & Truffer, 2012). Similarly, financial investors may only be willing to invest in an innovation after the related industry has been legitimized to some degree and key performance and quality standards have been defined for the related products. Knowledge may be created in an international network of firms, universities, and private R&D labs, whereas technology legitimacy may depend on the formulation of globally accepted quality standards (like in the ISO process). In this conceptualization, global innovation systems consist of sub-systems that create each of the four necessary system resources and are coupled to each other by multiscalar actor networks and institutional contexts (for a more expansive discussion, see Binz & Truffer, 2017).

Analyzing Global Innovation Systems: Introducing Multiscalar Relationships

To further characterize the spatial configuration of system resource formation processes, we must elaborate two conceptual elements in more detail: (1) subsystems of a GIS and (2) structural couplings between them. In the following, we will elaborate these elements and then propose a heuristic for assessing their spatial configuration.

The notion of subsystems closely relates to the question of *where* system resources form and which actors are able to access them. We define subsystems not in a spatially predefined way, but as the actor networks and institutions involved in the formation of system resources. Subsystem boundaries can correspond to national or regional borders, but they may also develop in networks that transcend these borders. An emblematic example of multiscalar resource formation processes are those created by dispersed communities of practice, as in the open source software field. Here, actors are often spread globally, but still develop shared cultures, knowledge stocks and investment models that outsiders would find hard to copy and access (Binz & Truffer, 2017; Lakhani & von Hippel, 2003). A similar example is innovation in the membrane bioreactor field, which initially emerged from a global R&D network spanning engineers in French transnational water companies and research institutes in various places around the world (Binz et al., 2014).

In a GIS perspective, the successful creation and diffusion of an innovation ultimately depends on how actors combine knowledge, investment, markets, and legitimacy into new configurations that work. The performance of a given GIS thus depends on (1) whether each system resource is created in a subsystem and (2) whether subsystems are coupled to each other in order to form a coherent global innovation trajectory. Structural coupling here relates to the foundational elements of an IS – actors, networks, or institutions (see Bergek et al., 2015). For example, internationally active firms, international conferences, and trade fairs or an industry's professional culture and quality standards are all examples of coupling domains that may integrate knowledge, market, financial, and institutional elements from various GIS subsystems.

Resource formation and coupling dynamics in GIS are accordingly multipolar and fluid. As key system resources are emerging from subsystems with varying geographies, actors in the GIS will in many cases be unable to directly appropriate a dominant share of them in-house or inside a given region or country. Instead, they will have to create strategic alliances and rely on nongeographic types of proximities to access and anchor a full resource portfolio in a given place (Boschma, 2005). Resourceful actors with a global reach (e.g., transnational companies, global donor organizations, or professional associations) are in a superior position to facilitate effective integration of spatially spread subsystems, but integration may also happen in a specific region with very dense inter-organizational networks, or even in a loosely coupled community of traveling technology experts (Larner & Laurie, 2010).

Given this high spatial complexity in any GIS, the question arises of how the generic system configuration may be systematized for different industries. Binz and Truffer (2017) start from the well-established assumption in the sectorial innovation systems literature that differences in an industry's technology characteristics condition the spatial characteristics of the innovation processes (Huenteler et al., 2016; Malerba, 2005). The GIS framework complements this view by emphasizing the coevolution of an industry's knowledge base with capabilities to facilitate institutional and market embedding. The latter concerns in particular processes like market formation, resource mobilization, and technology legitimation, which together define technology *valuation*, as in the ways in which new products, means of production, and infrastructures are connected to relevant demands and symbolic frames of the customer base. Proponents of the GIS framework therefore propose an industry typology that distinguishes two principal components: characteristics of technological innovation (focusing mostly on knowledge related system resources) and product valuation (related to market formation, the mobilization of financial resources, and legitimacy).

On the technological innovation side, industries can be characterized with regard to whether they subscribe more to a science and technology driven (STI) innovation mode or whether they depend on learning by doing, using, and interacting (DUI) (Jensen et al., 2007). The STI mode is more common in industries with science-based, analytical knowledge bases (e.g., biotechnology, pharma, solar PV), whereas the DUI mode characterizes innovation based on engineering-based synthetic knowledge bases (e.g., car manufacturing, machine tools, wind power) (Asheim, Coenen, & Vang, 2007; Herstad, Aslesen, & Ebersberger, 2014; Martin & Moodysson, 2013). STI-based industries rely on scientific principles, which can be codified in models, patents, and reports. Formalized R&D inside the company, tight industry-university linkages and radical technology breakthroughs characterize these fields (Huenteler et al., 2016). In STI-based industries, knowledge can relatively easily get disembodied and exchanged in space. This industry type will accordingly give rise to significant knowledge spillovers beyond regional and national borders (Moodysson & Jonsson, 2007; Schmidt & Huenteler, 2016).

Regarding the characteristics of valuation processes, one may distinguish between *standardized* and *customized* valuation systems (Jeannerat & Kebir, 2016). In the former case, valuation is happening in global mass markets with highly converging user tastes and distribution channels. TV sets, for example, may be produced anywhere and shipped to supermarkets all over the world to cater to very similar user tastes. In customized valuation systems, in contrast, new products will have to be strongly tailored to specialized user preferences that may vary considerably in space. Before an innovation may be sold, the innovation system actors will have to engage in intense user-producer interaction to form new use-patterns, establish socially accepted price-performance relationships, or create reputational capital in the form of brands and labels (Dewald & Truffer, 2011; Fligstein, 2007). It follows that GIS with standardized valuation systems will create considerable spatial spillovers, whereas customized valuation systems require territorially embedded and thus *spatially sticky* interaction processes.

By cross-tabulating these two dimensions, a typology of four ideal-type GIS configurations can be derived (cf. Fig. 17.1). Industries can be roughly positioned in the framework relative to each other, as they often represent different mixtures of innovation and valuation approaches. Also, the position in the typology may depend on the maturation stage of a particular product or industry. These complexities notwithstanding, the four types can be correlated with a characteristic GIS configuration (see Fig. 17.1). The combination of an STI innovation mode and a standardized valuation system will lead to a *footloose* spatial structure of the GIS with various spillovers inside and between each subsystem. We would therefore expect system resource formation processes to depend on substantial interaction at an international level or to shift rapidly from one region to the next. This setup may be exemplified by the solar PV module industry or by home electronics manufacturing. The combination of a customized form of valuation with a DUI technological innovation mode will—in the other extreme—lead to a spatially rather *sticky* GIS configuration, with spillovers largely restricted to specific local/regional contexts. This will ultimately lead to strong regional path dependencies and a high variety of customized product variants. Typical examples may be the biogas industry or the housing construction sector. Industries with a DUI-based innovation mode and standardized valuation system can be described as *production-anchored* GIS types. This setup will create strong and

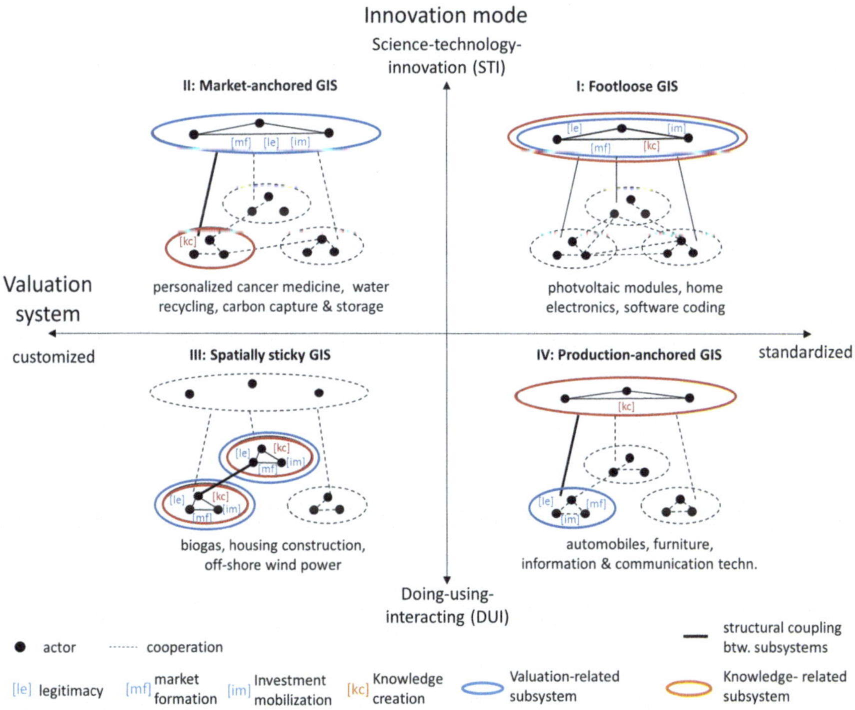

Fig. 17.1 Four ideal-type GIS Configurations. Author's own elaboration, based on Binz and Truffer (2017). Copyright 2017 by Elsevier B.V. Adapted with permission

persistent geographies of industry leaders managing global value chains to serve global markets. Examples are the automobile industry or information and communication technologies (Lee & Lim, 2001). Finally, the combination of an STI knowledge base and highly customized markets will create *market-anchored* GIS structures. This GIS type will often depend on global companies with a strong knowledge base in embedding generic technologies into a variety of local contexts. Examples comprise personalized cancer medicine, but also many civil engineering-based infrastructure services (Moors, Fischer, Boon, Schellen, & Negro, 2017; Yap & Truffer, 2019).

The four ideal-type GIS configurations imply radically different governance approaches for all those actors who attempt to shape the innovation process, comprising multinational companies, universities, as well as policy makers in specific regions or countries. It may also apply to *intermediary* actors like NGOs, civil society groups, or professional associations that have an interest in influencing the directionality of the emerging innovation trajectories. The resulting governance structures of GIS will depend on the interplay between the actors' different strategies. In the following section, we will elaborate some of the implications that the presented typology has for the characteristic governance modes in each GIS type and how they impact strategies of companies, noncommercial, and nongovernmental actors as well as governments at different spatial scales.

Policy and Governance Implications of Global Innovations Systems

The discussion above implies that innovation governance at a national or regional level should not myopically focus on regionally available assets, but closely reflect the targeted industry's GIS configuration. Industries with a spatially sticky GIS can profit most from conventional innovation governance and policy approaches operating predominantly at regional to national scales, while footloose GISs are most directly challenging traditional governance approaches. Market-anchored and production-anchored GISs in turn challenge specific parts of traditional region- or nation-focused innovation governance paradigms.

Innovation Governance in Spatially Sticky GISs

In spatially sticky GISs, system resources as well as innovation and valuation dynamics depend on spatial embedding and profit from dense colocation of the relevant actors. For instance, developing the first wind turbines was only possible in a few territorial clusters in Denmark or Germany, which facilitated dense user-producer-intermediary interaction (Garud & Karnoe, 2003). Similar observations hold for innovation in the biogas industry (Wirth, Markard, Truffer, & Rohracher,

2013). In this GIS type, co-creating the first product designs and market solutions in a given region or country continues to provide a sustained first-mover advantage in later stages of the innovation trajectory.

Effective innovation governance for this GIS type thus depends on a territorially specific *strategic niche management* rationale (Kemp, Truffer, & Harms, 2000). Producers, users, and various intermediary actors will have to be colocated in a given region and supplied with patient capital[1] and a (subsidized) market niche in which experimentation, trial & error, as well as complex learning-by-doing and interacting can occur. In addition, the recombination of preexisting technological capabilities in a gradual, related diversification process will be of key importance (OECD, 2013). The key element providing comparative advantage lies in stimulating continuous interaction between the innovation and valuation side of the industry, thus repeatedly improving the product's main features based on the historically grown industrial specialization and cultural identity of a given region. As such, this approach goes beyond conventional (manufacturing) cluster initiatives, which we would rather relate to the production-anchored GIS type. Structural couplings across different territorial innovation systems are perhaps of lowest importance in this GIS constellation. They may relate to knowledge exchange through mergers and acquisitions or long-term investments in other regions. But essentially, the core system resources are largely available in territories where innovation and valuation actors co-locate.

An emblematic example of the governance challenge for this GIS type is the early wind turbine industry. Garud and Karnoe (2003) observed that the Danish (DUI-based, distributed) *bricolage* approach to developing wind turbines proved more effective than the American (STI-based, centrally controlled) *breakthrough* governance approach. Later studies confirmed that the pioneering wind turbine regions in Denmark retained considerable first-mover advantages, even in later life-cycle stages when strong industrial competitors emerged in India or China and when the industry disintegrated its value chain (Lewis, 2007). Having a tightly integrated innovation governance mode proved a locational asset throughout the industry lifecycle, which was characterized by repeated radical innovations in different parts of the product architecture (blades, gearing, turbine control systems, etc.) (Huenteler et al., 2016).

Innovation Governance in Footloose GISs

The challenges for innovation governance in footloose GIS types directly contrasts the above explanations, as this industry type's innovation system resources emerge in internationally dispersed networks that are hard to locate and retain in any national or regional context. Rather than spatially embedded learning by doing, this

[1] Private or state-based financial investments made without expectation of turning a quick profit.

industry type profits from STI-based innovation processes as well as subsidies and trade policies that enable the quick upscaling of manufacturing capacities, as well as price-competitive production for standardized mass markets.

Conceptually, industries with a footloose GIS type can thus best be governed with an *STI-based free trade* approach: Tax credits, low-interest loans, liberal trade policies, and the creation of centers of excellence in science-based R&D will all support local firms in accessing internationally available system resources and competing in a fierce global price and quality competition. Patenting, trademarks, and technology standardization will be of decisive importance in this type of innovation process with a dynamic and often noncumulative technology trajectory (Castellacci, 2008). Latecomers in this GIS type may profit from anchoring system resources available elsewhere in order to embark on dynamic catch-up processes (Binz & Anadon, 2018). Innovation governance actors in developed economies will thus have to anticipate and deal with latecomers leapfrogging to the technological frontier and fast shifts in the global innovation landscape (Lee & Lim, 2001; Zhu, He, & Zhou, 2017). We would accordingly expect transnational corporations to be a core actor in this GIS type. They are able to serve global markets and shift the production base quickly between regions according to changing production cost differentials. For national or regional governments, this GIS type implies a rather weak role with an inherent risk of suffering from spill-over effects of national policy efforts.

The experience with the solar PV industry in Germany illustrates how this GIS type challenges conventional, nationally delimited innovation governance approaches. The quite ambitious market deployment subsidy (feed-in-tariff) that Germany introduced in 2002 was aimed at creating a mass market that would provide various actors in the German renewable energy sector with a first-mover advantage (Hoppmann, Huenteler, & Girod, 2014). Although the policy proved successful in the wind power and biogas sectors, its frictions with the PV sector's footloose GIS configuration quickly became apparent. Given the ubiquitous international structural couplings and spillovers in this GIS type, the subsidized local mass market induced substantial spillovers to latecomers in other places, in particular to China (Quitzow, 2015). Chinese PV firms were particularly skillful in accessing and anchoring markets, knowledge, investment, and legitimation dynamics from other regions in the GIS into a local industrial path that quickly became globally competitive (Binz & Anadon, 2018). German firms and policy makers were initially not anticipating the disruptive effects of the manifold spatial spillovers to China (Hoppmann et al., 2014), which ultimately led to significant trade disputes with Chinese competitors. The GIS framework suggests that German actors could to some degree have anticipated these global interdependencies and that some sort of international governance scheme would be needed in footloose GISs to regulate the spatially imbalanced costs and benefits of these spatial spillovers.

Innovation Governance in Market-Anchored GIS

Market-anchored GISs may in turn profit most from an innovation governance approach that relies on a *high-quality lead market* rationale: Here, creating pilot applications and successful new market segments for highly demanding customer segments is most decisive for innovation success. In this GIS type, innovation-related system resources may circulate globally whereas the valuation dynamics will depend on institutional embedding in specific local contexts. Market experience and demonstration effects in one place may thus be turned into a *sticky* locational advantage that can be mobilized when exporting the product to new places (Beise & Rennings, 2005).

Conceptually, a governance system that is adapted to this industry type would have to facilitate valuation-side dynamics, for example, in the form of deployment policies, "public procurement for innovation" (Edquist & Zabala-Iturriagagoitia, 2012), and the close coordination between producers and a local customer base with the highest quality requirements (Porter, 1990). For instance, the culturally diverse movie audience in Hollywood is a key resource for the local film industry for testing the reception of their work in various parts of the world. Rather than supporting basic R&D or breakthrough innovation locally, a smart governance design would encourage the local industry to access globally available innovative ideas and optimize their (economic/social/environmental) performance in demanding local application contexts. GIS governance structures will therefore have to facilitate pipelines to global knowledge networks while also supporting a locally embedded actor base that experiments with adaptations to the locally specific institutional settings and preference structures. National and regional policy makers might be able to define specific variations in the design of the solutions that are implemented, while simultaneously having to accept the globally predefined core of the technological paradigms.

Examples that illustrate this governance challenge can be drawn from the membrane bioreactor (MBR) or the personalized cancer medicine industries (Binz & Truffer, 2017; Moors et al., 2017). In the former case, technological innovation emerged from highly internationalized R&D networks that span various EU countries, Asia, and the USA (Binz et al., 2014). Although various technological designs coexisted for a long time, a key question was where first mass-markets would emerge that would steer the technological trajectory in a specific direction and reduce insecurities and manufacturing prices. US, European, and Chinese actors were the first to create niche markets driven by increasingly rigid water quality standards (Yap & Truffer, 2019). Given the geographic and institutional particularities in each region's valuation system, the chosen technological configuration and favored technology providers strongly differed between regions (Yang, Cicek, & Ilg, 2006). The MBR industry is now an oligopoly, dominated by lead firms like GE, Kubota, or Beijing Origin Water that have strongly adapted their valuation strategies to specific market regions. Recently, Chinese firms moved to a leading position based on a targeted valuation-side governance approach. By strongly increasing the quality criteria for surface water and standardizing the favored technology designs (based on criteria from local design institutes), a large market for high-quality MBR systems was created in China that could best be served by a local start-up, Beijing Origin Water. These changes in the valuation system enabled Chinese firms to quickly innovate, gain a dominant market position in the high-quality home market, and subsequently export their products to other latecomer countries (Yap & Truffer, 2019).

Innovation Governance in Production-Anchored GIS

Finally, the case of production-anchored GIS asks for a cluster- and RIS-based governance approach (Tödtling & Trippl, 2005), which creates a highly networked, specialized manufacturing base. Here, competitive regional advantages will not predominantly stem from local niche markets, but rather from cumulative synthetic knowledge stocks and culturally embedded cooperation patterns in manufacturing processes. Innovative products will emerge from collective, experience-based competencies in designing complex technical systems. As such, strengthening local knowledge circulation and recombination at the supply side of innovation is more important than supporting market structures, which can also be accessed globally (Boschma, Coenen, Frenken, & Truffer, 2017).

A governance approach adapted to these challenges would enable informal knowledge exchange in dense industry-supplier-university networks, while also supporting worldwide market access and knowledge exchange. This governance mode may depend on creating intense local knowledge exchange (*local buzz*) (Bathelt, Malmberg, & Maskell, 2004) as well as the creation of local cluster organizations, networking events, or business incubators supporting spin-offs from the incumbent industry and universities. Regional actors would have to jointly create applied science organizations with specialized curricula in the respective technological field and support on-the-job and vocational training. At the same time, the regional/national innovation governance structures would have to support exports into global markets with, for example, image campaigns abroad or by investing into trade infrastructure (roads, ports, airports, etc.) and complementary business service industries (market research or logistics). GIS governance actors will thus have to foster regional/national knowledge and skill specializations in global production networks that are often dominated by multinational lead firms.

The car industry illustrates the challenge of this innovation governance model. In terms of valuation dynamics, user tastes have grown highly standardized and gravitate around a few very similar car models. In addition, the manufacturing and supplier networks for cars are globally standardized and organized in highly fragmented value chains (Dicken, 2015). On the innovation side, the car industry provides an emblematic case of spatially sticky innovation system structures. For almost a century, the decisive innovation and design competencies have remained concentrated in the headquarters of the major OEM companies that are in turn embedded in policy-supported automotive clusters in Germany, France, Japan, Korea, or the USA (Dicken, 2015). The recent ascent of battery-driven and plug-in hybrid vehicles may challenge this production-anchored GIS structure. Newcomers such as Tesla, BYD, or Google cars are disrupting the sector by introducing more STI-based elements related to digitalization and self-driving capabilities. With an increasing shift toward an STI-based and customized GIS configuration, the traditional innovation clusters in the industry are now forced to build new pipelines to partners in regions with IT-based capabilities. Whether and how governance systems in the traditional car manufacturing clusters will successfully transform to react to this challenge remains an open empirical question (Miörner & Trippl, 2018) (Fig. 17.2).

This GIS-based innovation governance frameworks provides two interrelated novel contributions. On the one hand, it helps to assess and predict the prevalence of unintended spatial spillovers from national policy interventions, as in the solar PV case. On the other hand, it may be used to identify and eliminate factors that hinder the fast development and diffusion of innovation at supra-national scales, that is, it could be used to expand Weber and Rohracher's (2012) system failure frameworks with a "global policy coordination failure." In the case of solar PV discussed above, erratic regional and national policy support in various parts of the world led to global overcapacities and trade disputes, which now significantly hamper the GIS's capacity to diffuse the innovation. In this GIS type, a global governance structure would in principle be needed that integrates and coordinates innovation dynamics in various parts of the world. Such a governance structure could construct a more level playing field for all involved parties by mitigating trade

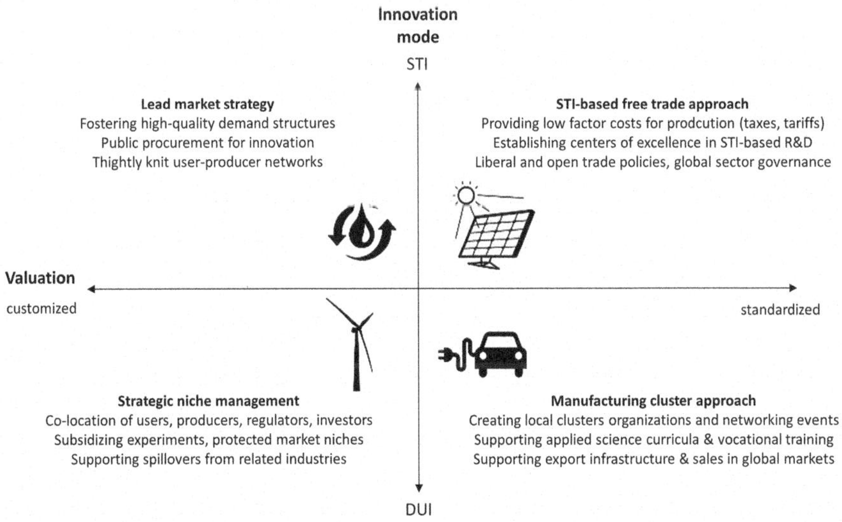

Fig. 17.2 Innovation governance models and policy approaches in each GIS type. Source: Design by authors

disputes and coordinating policy learning and innovation dynamics in various parts of the world.

Policy and Governance Implications: Putting Knowledge in Context

In the present chapter, we have elaborated how innovation governance structures have to be reconsidered under conditions of increasingly globalizing industry dynamics. National and regional competition on industrial leadership grows fiercer and more varied than ever, which asks for new ways of overcoming bottlenecks for innovation success. This is all the more important in the context of industry dynamics that address global challenges, such as in "cleantech" or "green" sectors. The dominant governance approaches to this challenge have so far focused on conditions for leveraging knowledge formation and emphasized intellectual property rights and scientific excellence, combined with regional cluster initiatives or smart specialization strategies. We have extended this view by showing that innovation success in some industries depends on being able to anchor extra-regional system resources and addressing valuation processes, that is, the management of emerging markets, the generation of technology legitimacy, or the leveraging of financial resources. Under these conditions, the portfolio of governance mechanisms extends into more procedural or evolutionary options that favor experimentation, the alignment of production capacities with local institutional structures, or the proactive shaping of new industries in the form of lead market strategies.

A key challenge for governing Global Innovation Systems lies in the fact that policies and industry roadmaps are still mostly formulated, legitimized, and executed in territorially bounded contexts. Yet we have shown that in almost all industry types (except for spatially sticky GISs), a sound understanding of internationalized innovation system structures is a key prerequisite for formulating effective regional or national governance schemes and support policies. Simply focusing on local innovation capabilities (e.g., through R&D support, clusters, RIS, industrial districts, etc.) will not be sufficient. Rather, national or regional governance approaches have to be extended to address actors, institutional contexts, and networks that lie outside their territories. In some extreme cases, supporting the innovative capacity of domestic industries, one might even have to encourage market formation or legitimation processes in distant places (e.g., through cooperation with globally active NGOs and industry associations). In addition, actors may have to consider new governance mechanisms and institutional arrangements that operate at a *global* level in order to counteract barriers for industry formation if pursued through national policies in isolation. This relates to questions of intellectual property rights, free trade arrangements, the definition of international standards, and the like. In the context of Global Innovation Systems, *knowledge for governance* therefore means adapting the governance of knowledge to multiple, multiscalar industrial contexts.

Besides knowledge about producing novel technologies, it is equally important to competently manage social contexts in which these technologies must be embedded and thereby create supportive valuation contexts.

References

Asheim, B. T., Coenen, L., & Vang, J. (2007). Face-to-face, buzz, and knowledge bases: Sociospatial implications for learning, innovation, and innovation policy. *Environment and Planning C: Politics and Space, 25,* 655–670. https://doi.org/10.1068/c0648

Bathelt, H., Malmberg, A., & Maskell, P. (2004). Clusters and knowledge: Local buzz, global pipelines and the process of knowledge creation. *Progress in Human Geography, 28,* 31–56. https://doi.org/10.1191/0309132504ph469oa

Beise, M., & Rennings, K. (2005). Lead markets and regulation: A framework for analyzing the international diffusion of environmental innovations. *Ecological Economics, 52,* 5–17. https://doi.org/10.1016/j.ecolecon.2004.06.007

Benz, A., Lütz, S., Schimank, U., & Simonis, G. (2007). *Handbuch Governance: Theoretische Grundlagen und empirische Anwendungsfelder* [Handbook governance: Theoretical principles and empirical fields of application]. Wiesbaden: Springer VS.

Bergek, A., Hekkert, M., Jacobsson, S., Markard, J., Sanden, B., & Truffer, B. (2015). Technological innovation systems in contexts: Conceptualizing contextual structures and interaction dynamics. *Environmental Innovation and Societal Transitions, 16,* 51–64. https://doi.org/10.1016/j.eist.2015.07.003

Binz, C., & Anadon, L. D. (2018). Unrelated diversification in latecomer contexts: The emergence of the Chinese solar photovoltaics industry. *Environmental Innovation and Societal Transitions, 28,* 14–34. https://doi.org/10.1016/j.eist.2018.03.005

Binz, C., Tang, T., & Huenteler, J. (2017). Spatial lifecycles of cleantech industries: The global development history of solar photovoltaics. *Energy Policy, 101,* 386–402. https://doi.org/10.1016/j.enpol.2016.10.034

Binz, C., & Truffer, B. (2017). Global innovation systems: A conceptual framework for innovation dynamics in transnational contexts. *Research Policy, 64,* 1284–1298. https://doi.org/10.1016/j.respol.2017.05.012

Binz, C., Truffer, B., & Coenen, L. (2014). Why space matters in technological innovation systems: The global knowledge dynamics of membrane bioreactor technology. *Research Policy, 43,* 138–155. https://doi.org/10.1016/j.respol.2013.07.002

Binz, C., Truffer, B., & Coenen, L. (2016). Path creation as a process of resource alignment and anchoring: Industry formation for on-site water recycling in Beijing. *Economic Geography, 92,* 172–200. https://doi.org/10.1080/00130095.2015.1103177

Boschma, R. (2005). Proximity and innovation: A critical assessment. *Regional Studies, 39,* 61–74. https://doi.org/10.1080/0034340052000320887

Boschma, R., Coenen, L., Frenken, K., & Truffer, B. (2017). Towards a theory of regional diversification: Combining insights from evolutionary economic geography and transition studies. *Regional Studies, 51,* 31–45. https://doi.org/10.1080/00343404.2016.1258460

Bunnell, T. G., & Coe, N. M. (2001). Spaces and scales of innovation. *Progress in Human Geography, 25,* 569–589. https://doi.org/10.1191/030913201682688940

Carlsson, B. (2006). Internationalization of innovation systems: A survey of the literature. *Research Policy, 35,* 56–67. https://doi.org/10.1016/j.respol.2005.08.003

Carlsson, B., & Stankiewicz, R. (1991). On the nature, function and composition of technological systems. *Journal of Evolutionary Economics, 1,* 93–118. https://doi.org/10.1007/BF01224915

Castellacci, F. (2008). Technological paradigms, regimes and trajectories: Manufacturing and service industries in a new taxonomy of sectoral patterns of innovation. *Research Policy, 37,* 978–994. https://doi.org/10.1016/j.respol.2008.03.011

Coenen, L., Benneworth, P., & Truffer, B. (2012). Toward a spatial perspective on sustainability transitions. *Research Policy, 41,* 968–979. https://doi.org/10.1016/j.respol.2012.02.014

Coenen, L., & López, F. J. D. (2010). Comparing systems approaches to innovation and technological change for sustainable and competitive economies: An explorative study into conceptual commonalities, differences and complementarities. *Journal of Cleaner Production, 18,* 1149–1160. https://doi.org/10.1016/j.jclepro.2010.04.003

Cooke, P., Gomez Uranga, M., & Etxebarria, G. (1997). Regional innovation systems: Institutional and organisational dimensions. *Research Policy, 26,* 475–491. https://doi.org/10.1016/S0048-7333(97)00025-5

Corpataux, J., Crevoisier, O., & Theurillat, T. (2009). The expansion of the finance industry and its impact on the economy: A territorial approach based on Swiss pension funds. *Economic Geography, 85,* 313–334. https://doi.org/10.1111/j.1944-8287.2009.01035.x

Dewald, U., & Truffer, B. (2011). Market formation in technological innovation systems: Diffusion of photovoltaic applications in Germany. *Industry and Innovation, 18,* 285–300. https://doi.org/10.1080/13662716.2011.561028

Dewald, U., & Truffer, B. (2012). The local sources of market formation: Explaining regional growth differentials in German photovoltaic markets. *European Planning Studies, 20,* 397–420. https://doi.org/10.1080/09654313.2012.651803

Dicken, P. (2015). *Global shift: Mapping the changing contours of the world economy* (7th ed.). New York: The Guilford.

Edquist, C., & Zabala-Iturriagagoitia, J. M. (2012). Public procurement for innovation as mission-oriented innovation policy. *Research Policy, 41,* 1757–1769. https://doi.org/10.1016/j.respol.2012.04.022

Edquist, C. (1997). Systems of innovation approaches: Their emergence and characteristics. In C. Edquist (Ed.), *Systems of innovation: Technologies, institutions and organizations* (pp. 1–35). London: Routledge.

Fligstein, N. (2007). The sociology of markets. *Annual Review of Sociology, 33,* 105–128. https://doi.org/10.1146/annurev.soc.33.040406.131736

Garud, R., & Karnoe, P. (2003). Bricolage versus breakthrough: Distributed and embedded agency in technology entrepreneurship. *Research Policy, 32,* 277–300. https://doi.org/10.1016/S0048-7333(02)00100-2

Grillitsch, M., & Trippl, M. (2013). Combining knowledge from different sources, channels and geographical scales. *European Planning Studies, 22,* 2305–2325. https://doi.org/10.1080/09654313.2013.835793

Herstad, S. J., Aslesen, H. W., & Ebersberger, B. (2014). On industrial knowledge bases, commercial opportunities and global innovation network linkages. *Research Policy, 43,* 495–504. https://doi.org/10.1016/j.respol.2013.08.003

Hoppmann, J., Huenteler, J., & Girod, B. (2014). Compulsive policy-making: The evolution of the German feed-in tariff system for solar photovoltaic power. *Research Policy, 43,* 1422–1441. https://doi.org/10.1016/j.respol.2014.01.014

Huenteler, J., Schmidt, T., Ossenbrink, J., & Hoffmann, V. (2016). Technology life-cycles in the energy sector: Technological characteristics and the role of deployment for innovation. *Technological Forecasting & Social Change, 104,* 102–121. https://doi.org/10.1016/j.techfore.2015.09.022

Jeannerat, H., & Kebir, L. (2016). Knowledge, resources and markets: What economic system of valuation? *Regional Studies, 50,* 274–288. https://doi.org/10.1080/00343404.2014.986718

Jensen, M. B., Johnson, B., Lorenz, E., & Lundvall, B. Å. (2007). Forms of knowledge and modes of innovation. *Research Policy, 36,* 680–693. https://doi.org/10.1016/j.respol.2007.01.006

Jessop, B. (1998). The rise of governance and the risks of failure: The case of economic development. *International Social Science Journal, 50,* 29–45. https://doi.org/10.1111/1468-2451.00107

Jurowetzki, R., Lundvall, B., & Lema, R. (2015). *Combining the global value chain and the innovation system perspectives*. Paper presented at the DRUID Academy conference, Rebild, Aalborg. Retrieved from https://pdfs.semanticscholar.org/e772/e3fc0f320f4d29b4aa6976bd2ff23df0a623.pdf

Kemp, R., Truffer, B., & Harms, S. (2000). Strategic niche management for sustainable mobility. In K. Rennings, O. Hohmeyer, & R. L. Ottinger (Eds.), *Social costs and sustainable mobility: Strategies and experiences in Europe and the United States* (pp. 167–188). ZEW Economic Studies: Vol. 7. Heidelberg: Physica. https://doi.org/10.1007/978-3-642-57669-0

Lakhani, K. R., & von Hippel, E. (2003). How open source software works: "Free" user-to-user assistance. *Research Policy, 32,* 923–943. https://doi.org/10.1016/S0048-7333(02)00095-1

Larner, W., & Laurie, N. (2010). Travelling technocrats, embodied knowledges: Globalising privatisation in telecoms and water. *Geoforum, 41,* 218–226. https://doi.org/10.1016/j.geoforum.2009.11.005

Lee, K., & Lim, C. (2001). Technological regimes, catching-up and leapfrogging: Findings from the Korean industries. *Research Policy, 30,* 459–483. https://doi.org/10.1016/S0048-7333(00)00088-3

Lewis, J. I. (2007). Technology acquisition and innovation in the developing world: Wind turbine development in China and India. *Studies in Comparative International Development, 42,* 208–232. https://doi.org/10.1007/s12116-007-9012-6

Lundvall, B.-Å. (1988). Innovation as an interactive process: From user-producer interaction to the national system of innovation. In G. Dosi, C. Freeman, R. Nelson, G. Silverberg, & L. Soete (Eds.), *Technical change and economic theory* (pp. 349–369). London: Pinter.

Lundvall, B.-Å. (Ed.). (1992). *National systems of innovation: Toward a theory of innovation and interactive learning*. London: Pinter.

Malerba, F. (2002). Sectoral systems of innovation and production. *Research Policy, 31,* 247–264. https://doi.org/10.1016/S0048-7333(01)00139-1

Malerba, F. (2005). Sectoral systems of innovation: A framework for linking innovation to the knowledge base, structure and dynamics of sectors. *Economics of Innovation and New Technology, 14,* 63–82. https://doi.org/10.1080/1043859042000228688

Martin, B. R. (2016). Twenty challenges for innovation studies. *Science and Public Policy, 43,* 432–450. https://doi.org/10.1093/scipol/scv077

Martin, R., & Moodysson, J. (2013). Comparing knowledge bases: On the geography and organization of knowledge sourcing in the regional innovation system of Scania, Sweden. *European Urban and Regional Studies, 20,* 170–187. https://doi.org/10.1177/0969776411427326

Miörner, J., & Trippl, M. (2018). Embracing the future: Path transformation and system reconfiguration for self-driving cars in West Sweden. *European Planning Studies, 27,* 2144–2162. https://doi.org/10.1080/09654313.2019.1652570

Moodysson, J., & Jonsson, O. (2007). Knowledge collaboration and proximity: The spatial organization of biotech innovation projects. *European Urban and Regional Studies, 14,* 115–131. https://doi.org/10.1177/0969776407075556

Moors, E. H., Fischer, P. K., Boon, W. P., Schellen, F., & Negro, S. O. (2017). Institutionalisation of markets: The case of personalised cancer medicine in the Netherlands. *Technological Forecasting and Social Change, 128,* 133–143. https://doi.org/10.1016/j.techfore.2017.11.011

Nelson, R. (1993). *National innovation systems: A comparative analysis*. New York, NY: Oxford University Press.

Niosi, J., & Bellon, B. (1994). The global interdependence of national innovation systems: Evidence, limits, and implications. *Technology in Society, 16,* 173–197. https://doi.org/10.1016/0160-791X(94)90028-0

OECD (Organisation for Economic Cooperation and Development). (2013). Innovation-driven growth in regions: The role of smart specialisation. Paris: OECD. Retrieved from https://www.oecd.org/innovation/inno/smart-specialisation.pdf

Oinas, P., & Malecki, E. J. (2002). The evolution of technologies in time and space: From national and regional to spatial innovation systems. *International Regional Science Review, 25,* 102–131. https://doi.org/10.1177/016001702762039402

Pietrobelli, C., & Rabellotti, R. (2009). The global dimension of innovation systems: Linking innovation systems and global value chains. In B. Lundvall, K. J. Joseph, C. Chaminade, & J. Vang (Eds.), *Handbook of innovation systems and developing countries* (pp. 214–238). Cheltenham: Edward Elgar.

Porter, M. E. (1990). The competitive advantage of nations. *Harvard Business Review, 68*(2), 73–93.

Quitzow, R. (2015). Dynamics of a policy-driven market: The co-evolution of technological innovation systems for solar photovoltaics in China and Germany. *Environmental Innovation and Societal Transitions, 17,* 126–148. https://doi.org/10.1016/j.eist.2014.12.002

Schmidt, T. S., & Huenteler, J. (2016). Anticipating industry localization effects of clean technology deployment policies in developing countries. *Global Environmental Change, 38,* 8–20. https://doi.org/10.1016/j.gloenvcha.2016.02.005

Sharif, N. (2006). Emergence and development of the National Innovation Systems concept. *Research Policy, 35,* 745–766. https://doi.org/10.1016/j.respol.2006.04.001

Tödtling, F., & Trippl, M. (2005). One size fits all? Towards a differentiated regional innovation policy approach. *Research Policy, 34,* 1203–1219. https://doi.org/10.1016/j.respol.2005.01.018

Truffer, B. (2012). The need for a global perspective on sustainability transitions. *Environmental Development, 3,* 182–183. https://doi.org/10.1016/j.envdev.2012.05.010

Weber, K. M., & Rohracher, H. (2012). Legitimizing research, technology and innovation policies for transformative change: Combining insights from innovation systems and multi-level perspective in a comprehensive 'failures' framework. *Research Policy, 41,*1037–1047. https://doi.org/10.1016/j.respol.2011.10.015

Weber, M., & Truffer, B. (2017). Moving innovation systems research to the next level: Towards an integrative agenda. *Oxford Review of Economic Policy, 33,* 101–121. https://doi.org/10.1093/oxrep/grx002

Wirth, S., Markard, J., Truffer, B., & Rohracher, H. (2013). Informal institutions matter: Professional culture and the development of biogas technology. *Environmental Innovation and Societal Transitions, 8,* 20–41. https://doi.org/10.1016/j.eist.2013.06.002

Yang, W., Cicek, N., & Ilg, J. (2006). State-of-the-art of membrane bioreactors: Worldwide research and commercial applications in North America. *Journal of Membrane Science, 270,* 201–211. https://doi.org/10.1016/j.memsci.2005.07.010

Yap, X., & Truffer, B. (2019). Shaping selection environments for industrial catch-up and sustainability transitions: A systemic perspective on endogenizing windows of opportunity. *Research Policy, 48,* 1030–1047. https://doi.org/10.1016/j.respol.2018.10.002

Zhu, S., He, C., & Zhou, Y. (2017). How to jump further and catch up? Path-breaking in an uneven industry space. *Journal of Economic Geography, 17,* 521–545. https://doi.org/10.1093/jeg/lbw047

Chapter 18
Experimentalist Systems in Manufacturing Multinationals: Recursivity and Continuous Learning Through Destabilization

Gary Herrigel

Permanent innovation pressures riddle many contemporary manufacturing sectors with uncertainty. In search of new product ideas and processual improvements to gain advantage over competitors, firms continuously modify and improve both their internal and external relations and processes. Such constant searching and optimization destabilizes routines and relations, creating both possibility and hazard for firms and their stakeholders. Uncertainty, the distrust that current resources and practices will yield future advantage, is both the cause and consequence of these continuous optimization processes. As a result, coping with uncertainty is a core governance challenge in contemporary manufacturing firms and across their supply chains.

Production globalization exacerbates these dynamics in two ways. First, it diffuses routine and relational disruption practices to offshore locations. Manufacturing MNCs compete with one another all over the world on the basis of cost, quality, and innovation. Increasingly, in sectors like automobiles and machinery, product sophistication differs little across the markets in which MNCs actively produce (Horner & Hulme, 2017; Horner & Nadvi, 2017; Sutton, 2012). Consequently, firms confront uncertainty in offshore markets at least as intense as they experience in their home markets. These pressures induce firms to extend their search and optimization practices (and the continuous routine and relational disruptions that they involve) to their offshore locations.

Second, globalized production further generates uncertainty simply through the complexity involved in managing the global interrelations and interdependencies of far-flung and increasingly sophisticated production clusters manufacturing common technologies. Complexity associated with localization pressures—the need to adapt models to local tastes and conform with indigenous standards and regulations—adds to the uncertainty by destabilizing designs and production practice. Further,

G. Herrigel (✉)
Department of Political Science, University of Chicago, Chicago, IL, USA
e-mail: g-herrigel@uchicago.edu

J. Glückler et al. (eds.), *Knowledge for Governance*, Knowledge and Space 15,
https://doi.org/10.1007/978-3-030-47150-7_18

players at the center of an MNC exploit innovations and improvements that are generated both at home and offshore and seek to leverage them for advantage in other locations, including the firm's nominal home location. This in turn further disrupts routines and relations in all the receiving locations, creating uncertainty.

The upshot is that many global manufacturing firms live with the uncertainty of product change and routine and relational disruption throughout all aspects of their global operations, virtually all the time. Indeed, it is not merely something that they respond to; it is an environmental situation that they chronically self-induce in order to leverage innovation from where it occurs to where it can be advantageously deployed.[1]

In this article, I argue that these highly uncertain environmental conditions and internal firm self-disrupting practices are generating a notable shift away from conventionally hierarchical governance architectures in complex manufacturing sectors like the automobile and machinery industries. Continuous innovation and change in products and processes push traditionally separate functionalities, such as conception/execution, design/production, finance/manufacturing, assembler/supplier, and headquarters/local subsidiary into explicit mutual dependence. The content and scope of functionalities and roles, increasingly, are defined and redefined interactively, through the mutual participation of stakeholder parties, rather than through top-down processes of command and control. Roles and identities are too unstable and transient to support rigid hierarchical authority; the imperative for learning and for the diffusion of knowledge throughout the MNC overwhelms the clumsy and slow-moving unilateral pretenses of top-down bureaucratic order.

But if traditional hierarchical bureaucratic governance no longer works, how do firms govern their global operations? What are the principles of effective governance under conditions of uncertainty that can accommodate and foster continuous routine disruption, innovation, and learning? This article uses case study material on German automobile and machinery multinationals to show that at least those MNCs are managing the uncertainties that their practices encounter and generate by developing post-bureaucratic "experimentalist" architectures.[2]

---

[1] All this in addition to the continuous negotiations and struggles between MNC home office interest and the ambitions of local players, dynamics that have been the focus of much literature on MNC Center-Local relations. See, generally, Kristensen and Zeitlin (2004), and for an interesting case, Glückler (2014).

[2] I am not claiming that there is something specifically German about these practices. They are simply German cases. For what it is worth, Netland (e.g., Netland, 2013) and others see the diffusion of self-optimizing CPSs as a general trend. For the material in this paper, I am drawing on a series of research projects I conducted with German collaborators. We conducted over 150 interviews in Germany, Central Europe, and China between 2008 and 2015 in the automobile, auto components, electromechanical, and machinery industries. My aim here is simply to exploit the rich and extensive fieldwork that we did in German companies and describe the experimentalist dynamics that we found there. I am saving the important discussion of national specificities for another day. For a more detailed description of the German case material from which I draw, see Herrigel, Voskamp, and Wittke (2017). For an in-depth study whose authors deal extensively with the implementation of the VW as well as Toyota Production Systems in the BRICs, see Jürgens and Krzywdzinski (2016).

These alternative governance architectures are formal systems of multilevel, interdependent, stakeholder role-, standard- and goal-setting practices that presuppose the provisionality of initial targets (roles, standards, designs, production and cost goals, etc.). Iterative evaluation of frameworks in light of the experience of putting them into practice, at all levels of the MNC, fosters continuous—and formally prescribed—optimization, improvement and learning throughout the organization. Centrally (and collectively or jointly) set goals and standards are modified through the process of attempting to realize them; formally prescribed requirements on the local adapters to defend their changes to central actors both disciplines the local and opens the center to the possibility for the transformation of its own goals and roles.[3] The experimentalist systems are processual and interdependent; they are formal and inclusive; and they use collective self-surveillance to foster organizational and technological learning and innovation.[4]

In a very general way, such architectures instantiate a variant of what the management theorists Maurizio Zollo, Sidney Winter (Winter, 2003; Zollo & Winter, 2002), David Teece (Teece, 2007; Teece, Pisano, & Shuen, 1997), and A. G. L. Romme (1997, 1999; Romme, Zollo, & Berends, 2010) call "dynamic capabilities." According to Zollo and Winter (2002, p. 340): "A dynamic capability is a learned and stable pattern of collective activity through which the organization systematically generates and modifies its operating routines in pursuit of improved effectiveness." It is "exemplified by an organization that adapts its operating processes through a relatively stable activity dedicated to process improvements." Such capabilities arise, they argue, "from learning; they constitute the firm's systematic methods for modifying operating routine."

Dynamic capability theorists use the notion in an extremely general way to apply to architectures that can govern the revision of operating routines under a wide array of environmental conditions. Hence bureaucratic hierarchy and post-bureaucratic experimentalism can both be, so to speak, dynamically capable. Zollo and Winter (2002) suggest that robust dynamic capabilities of the post-bureaucratic sort—in other words, in which deliberative and codified learning processes are integral to

[3] In this sense, the perspective here runs orthogonally to the debate about HQ-subsidiary relations and embeddedness, for instance, the influence of local institutional relations on subsidiary practices (e.g., Becker-Ritterspach & Dörrenbächer, 2009; Dörrenbächer & Geppert, 2011). Under the important background condition of uncertainty, experimentalist practices routinely accept local deviation from central standards, but formally require an explicit justification from local actors. The center is open to local modification because its members want to identify and circulate beneficial or effective aspects of local innovations across all its global subsidiaries performing similar manufacturing projects. For a discussion of the relationship between the perspective here and the global-local literature on MNCs, see Herrigel et al. (2017, Chap. 4).

[4] Indeed, because the experimentalist practices presuppose the existence of deep environmental uncertainty, the analysis here also runs orthogonally to the HQ-subsidiary literature whose authors focus on subsidiary initiative and intra-corporate power relations (e.g., Bouquet & Birkinshaw, 2008; Dörrenbächer & Gammelgaard, 2016; Glückler, 2014). These authors presuppose role positions where the actors know their interests before they act; the analysis here highlights the way in which uncertainty causes precise role identities and interests to emerge out of joint deliberation.

practice—come into being (a) in turbulent and uncertain environments; (b) when there is a learning culture or predisposition in the organization; (c) when less frequently occurring, more heterogeneous, and causally ambiguous tasks are in play.[5] Under these conditions, they argue, purely behavioral or practice-based forms of randomly induced learning are conditioned and modified by more directive, deliberative, and formal learning mechanisms.

I argue here that post-bureaucratic or "experimentalist" architectures, which are generated under the conditions of uncertainty outlined above, are distinctive within this genus of dynamic capabilities, for three reasons.[6] First, they presuppose extensive stakeholder involvement in the establishment and revision of standards and organizational goals. Uncertainty undermines hierarchy, destabilizes roles and functions, and makes stakeholder cooperation across boundaries inescapable. Call this the *principle of transparency and inclusion* (Sabel, 2006; Simon, 2015). Second, they emphasize the provisionality and continuous revision of deliberatively established standards and goals in light of practical experience. Crucially here, where bureaucracy governs by rule and relies on unaccountable, informal, and random discretion to reconcile rule with practice, experimentalism formally demands accountability from discretion (peer review, reason giving) in order to modify and improve, rather than simply reconcile, rule and practice. Call this the *principle of revisable planning or governance by plan rather than by rule* (Simon, 2015).

Third, experimentalist architectures are formally proactive and preemptive rather than reactive about error, mistake, and deviation in processes. Simon (2015, p. 62) says that post-bureaucratic organizations tend "to rely on audits more than complaints, and.... [take] a diagnostic approach to complaints, understanding them not just as evidence of idiosyncratic deviance, but as symptoms of systemic malfunction." Call this the *principle of improvement-oriented self-surveillance.*

Experimentalist governance, then, can be understood as a form of dynamic capabilities in that it involves learning not just at the level of operational routines for the production of given automobiles or machines; it also involves the optimization and recomposition of the learning process itself. But experimentalist architectures are distinctive in their explicit effort to use transparency, inclusion, revisable planning, and improvement-oriented self-surveillance to cast not only routines but also very explicitly the boundaries and inter-relations of roles and functions within the organization into play. By erecting a governance architecture capable of continuously recomposing roles and functions, in addition to routine practices, experimentalist governance induces distinctively self-recomposing circular or recursive learning processes throughout MNC global operations.

The German firms studied here apply this experimentalist version of dynamic capability throughout their organizations, from headquarters to off-shore subsidiary and from top management suite to production-level, shop-floor teams. A central

---

[5] See also Romme et al. (2010).

[6] See Simon (2015) for elucidation of post-bureaucratic principles in administrative law; see Romme et al. (2010) for description of circular learning in dynamic capabilities under conditions of uncertainty.

diffusion mechanism for these practices is the Corporate or Comprehensive (*Ganzheitliche*) Production System (CPS) (Netland, 2013; Netland & Aspelund, 2013; Netland & Ferdows, 2014; Netland & Sanchez, 2014). Inspired by the original Toyota Manufacturing System, such team/stakeholder-driven formal systems have transformed German manufacturing culture today. Their experimentalist character enhances companies' dynamic capability to negotiate uncertainty in their market and technological environments through induced processes of circular learning and the continuous recomposition of roles and function across organizations in response to challenges.

When they are working properly, experimentalist systems foster and diffuse both organizational and technological innovation within companies and across supply chains (Helper, MacDuffie, & Sabel, 2000; Herrigel, 2010; Sabel, 2006; Spear, 2009). As such, they enhance German manufacturing competitiveness and induce continuous producer and regional upgrading in emerging economies (Herrigel, Wittke, & Voskamp, 2013).

Finally, movement towards experimentalism within manufacturing MNCs is neither seamless nor uncomplicated (Netland & Ferdows, 2014). Indeed, there are three characteristic obstacles to the diffusion in practice of experimentalism's recompositional and circular learning dynamics: hierarchical insulation, stakeholder exclusion, and inadequate empowerment resources for participants. Interestingly, these obstacles exist not only *ex ante*, as firms attempt to construct the experimentalist architectures of corporate production systems and implement them throughout their global operations, but they also are continually regenerated by the recompositional dynamics of the CPS's themselves. Revision of commonly agreed-upon frameworks frequently redefines power relations and stakeholders, creating new possibilities for insulation and exclusion.

In order to prevent such obstacles from paralyzing the global process of recursive learning, many experimentalism-oriented German MNCs, using the principle of improvement oriented self-surveillance, introduce an array of destabilization mechanisms to systematically undermine insulation and exclusion strategies within the global firm and reconstitute the deliberative experimentalist learning process. Interestingly, CPSs often contain penalty default mechanisms as part of the experimentalist framework which systematically monitor widespread deliberation processes for possible paralysis and, when finding it, redefine the deliberative terrain to re-start deliberation (and learning) on a new basis.

This article portrays manufacturing MNCs using experimentalist systems as extraordinarily dynamic organizations (embedded in equally dynamic supply chains) characterized by recursive learning and chronic organizational disruption and recomposition over time. A snapshot of these organizations at any instant in time contains complex admixtures of joint problem solving, team-based goal setting, hierarchical insulation strategies, and patterns of stakeholder in/exclusion. Viewed over time, however, as a constantly self-disrupting process, experimentalist multinationals emerge as dynamically recursive learning systems focused on permanent organizational and technological optimization and innovation.

The article proceeds in three sections. The first section outlines the global competitive and strategic conditions that have given rise to experimentalist governance architectures within manufacturing multinationals and their supply chains. The second section describes experimentalist architectures and how they work. It also addresses a number of failed alternative strategies that German firms pursued prior to embracing experimentalist practices. The third section then discusses characteristic obstacles to experimentalism's diffusion, many endogenous to the practice of experimentalism itself, along with a range of strategies and mechanisms that firms deploy to overcome them.

Experimentalism Grows out of Uncertainty Associated with Changing Global Demand and Attendant Production Relocation

From a developed country's manufacturing MNC point view, global opportunities for growth and expansion have shifted notably in the new century (Baldwin, 2016; Horner & Hulme, 2017; Kobrin, 2017; Sutton, 2012). During most of the twentieth century, the largest manufactured goods markets were also the fastest growing ones. For US, German, and Japanese manufacturers, this meant that they targeted the bulk of their exports and FDI efforts at developed (western) European, North American, and North Asian economies. This situation began to change in the new century, however, as emerging market economies, notably the BRICS, became the driving forces of global growth (cf. Horner & Nadvi, 2017; World Bank, 2015).

Significantly, emerging market demand growth has been so rapid, technologically challenging, and quantitatively massive that it cannot be serviced through exports alone. Instead, firms have been forced to expand FDI and service demand in those emerging markets by "producing where they sell." Since 2009, for example, German automobile makers have produced more cars outside of Germany than inside Germany (Heymann, 2014; VDA, 2012, p. 30). China has, by far, become the largest offshore production location for Germans (VDA, 2017). Similar shifts away from export and toward offshore production characterize automobile industry development in Japan, Korea, the USA, and France (VDA, 2017). In most cases, companies develop global models that are then adapted to local market conditions.

Crucially, this shift toward producing where they sell involves significant upgrading of offshore production operations (Herrigel et al., 2013, 2017; Horner & Hulme, 2017; Sutton, 2012). Competition for market share in growing markets such as China is intense and customer sophistication is developing rapidly (Brandt & Thun, 2010). In order to be competitive, FDI manufacturers must pay attention to manufacturing economies and product quality. Moreover, the MNC affiliates must be able to offer products that appeal specifically to the needs and preferences of local customers and that are designed according to host country regulatory norms and standards. This presses manufacturers to upgrade local operations in three areas: production worker skill levels, supply base sophistication, and local R&D, design, and engineering capability (Brandt & Thun, 2010; Herrigel et al., 2013; Sutton, 2012).

These strategic shifts within MNCs have created extensive "internal" networks of interconnected production clusters in different global regions, all focused on trying to manufacture the same or very similar end products. This generates quite distinctive governance challenges within global manufacturing MNCs. Firms need to optimize exports from home locations with global offshore production capacity while simultaneously reconciling constant imperatives for process and product optimization, innovation, cost reduction, and learning, not only within individual plants, but also centrally and locally across vast global organizations. This is not easy: Innovation can increase costs; optimization and cost reduction can undermine learning; too much local autonomy can generate centrifugal pressures weakening the various forms of leverage (learning, knowledge, purchasing) that come with global concern membership; too much central direction can undermine local innovation and organizational capabilities crucial for competitive advantage in foreign markets.[7]

Not only that, global competition is so dynamic that there is never a natural sweet spot in which all of these competing goals and pressures can be stably reconciled or in which a happy equilibrium can be found. New products, technical innovation, competition among suppliers, new local regulations, currency value shifts, organizational learning induced possibility—and much more—all constantly destabilize the ordered practices that firms develop and generate new adjustment and governance challenges. Innovation, cost reduction, and learning are imperatives for all actors throughout MNC operations, yet environmental uncertainty is so great that at any given moment players have no clear sense of what strategy would be most optimal for them to achieve those goals. Forced, nonetheless, to act, I here argue that they turn to the revisable and recursive measures associated with experimentalism.

Global Strategies Governed by Self-Recomposing Experimentalist Learning Architectures Within MNCs and Across Supply Chains

Uncertainty along multiple dimensions is thus a crucial element in the emergence of the new global governance practices within manufacturing MNCs. It is not, however, analytically helpful (nor empirically accurate) to view successful manufacturing

[7] For exploration of these dynamics, see Bouquet and Birkinshaw (2008), Becker-Ritterspach and Dörrenbächer (2009), and Dörrenbächer and Geppert (2011).

MNC governance efforts as "reactions" seeking to "cope" with continuously uncertain and recombinatory practices in their organizations and the environment. Rather, this section shows that many manufacturing MNCs are creating (transnational) governance architectures that *systematically induce* organizational destabilization and recompositional experimentation to foster innovation and learning.[8] This proactive destabilization is accomplished through the workings of formalized "experimentalist" systems, such as corporate production systems (CPSs), Six Sigma programs, and other formal systems of open standards (e.g., ISO certifications). Such systems foster collective self-analysis through formally transparent and inclusive procedures that involve joint goal setting, systematic performance review, prompt problem solving, and organized destabilization of insulation and exclusion strategies.

What Is an Experimentalist Governance Architecture?

I describe the emerging recursive or circular learning-based processes in MNCs as "experimentalist" because they resemble what Charles Sabel, Jonathan Zeitlin, William Simon, and others call "experimentalist governance architectures" in public policy and administrative law contexts (Sabel, 2006; Sabel & Simon, 2011; Sabel & Zeitlin, 2008). Those scholars, in turn, draw theoretically on the American Pragmatic tradition's use of the word experimentation. Experimentation, pragmatically understood, describes the relational, interactive, and social character of identity, goal setting, and action, in which goals and the means adopted to achieve them are continuously modified and optimized through the social action process (cf. Dewey, 1922; Joas, 1996).

In its most abstract analytical form, the experimentalist governance architecture is a formalized four-step recursive process. All actors are aware of the formal rules and obligations that constitute the system. First, there is joint or collective goal setting. Relevant stakeholders (what Dewey called "publics") commonly affected by a given problem openly deliberate about solutions and future goals for their common interactions. Second, local stakeholders then implement/pursue these solutions and goals in their milieu. Application or realization of the common standard in the local environment invariably requires local player discretion: Unanticipated problems emerge, intermediate benchmark goals are not fulfilled, local conditions differ from the stylizations used during the general deliberations, and so forth. The center permits local discretion—deviation from agreed upon practice or standards—in order to facilitate problem solving and adaptation that enables the local organization to achieve the goal target. But these deviations must be transparent (other players must be able to observe or review them). In a third step, the deviation/local change must be defended to the central stakeholder teams who delegated the design/standard/goal. Finally, fourth, successful local experiments are then used to review the effectiveness and desirability

[8]Again, this condition distinguishes the view here from authors of other HQ-subsidiary literatures who take a more structuralist position, attaching specific interests *ex ante* to particular role positions, compare Bouquet and Birkinshaw (2008).

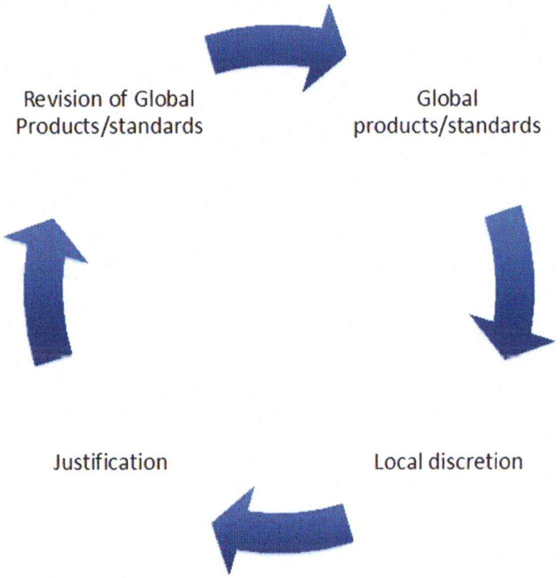

Fig. 18.1 Pragmatist-experimentalist learning architecture. Source: Design by author

of central/common goals and standards. If the local innovation is compelling enough, this can result in modification of the central standard (Fig. 18.1).[9]

The experimentalist governance school applies this framework to workings of public policy and regulation (cf. also Sabel, Herrigel, & Kristensen, 2018), but it is a very fruitful way to understand how many contemporary MNCs are using CPSs and other formal standard-setting and review mechanisms to govern their global operations. CPSs are formal systems that organize group- or stakeholder-based goal setting within firms to achieve product and process innovation, optimization (cost reduction), and learning on a continuous basis.[10] The systems are rooted in team goal-setting/standardization procedures (regular goal setting meetings) and constitute a hierarchical architecture of team-based goal conversations, ultimately linking (through many mediations) the shop floor to the top management.

The conversations are also systematically cross-functional and global. Product teams, customer teams, design and manufacturing teams, and continuous improvement teams all are constituted in multiple locations and form super-ordinate or umbrella teams that engage with one another across markets and geographical space

[9] Compare this inclusive and mutual role-defining four-step circular model to the more hierarchical four-step circular model presented by Zollo and Winter (2002, p. 343): External stimuli and processual feedback produce 1.) generative variation (scanning, recombination) →2.) internal selection (evaluation, legitimation) →3.) replication (knowledge sharing/transfer adaptive variation, problem solving) →4.) retention (enactment, routinization).

[10] See Friedli and Schuh (2012), Sabel (2006), Spear (2009), Westkämper and Zahn (2009) for general discussions; for international case studies of implementation and diffusion, see Clarke (2005), Netland (2013), Netland and Federow (2014), Netland and Aspelund (2013), and Netland and Sanchez (2014).

to identify common goals and standards and compare (and defend) their local experiments.

CPSs have diffused widely among manufacturing MNCs. Many companies brand their CPS (e.g., The Siemens Production System or The Volkswagen Production System). The companies also characteristically provide their own corporate names to the mechanisms of goal setting, self-evaluation (performance review), benchmarking, joint problem solving, and goal revision. Despite this nominal variety, however, all follow the general experimentalist logic outlined above. Many smaller companies also embrace similar formal, experimentalist, team-governed, lean production-based principles, even though most do not attempt to "brand" their system.

The following example, taken from a German Truck and Omnibus transmission producer (*Auto-Getriebe*), illustrates the globally recursive and learning elements of these systems.[11] Joint German design and manufacturing teams developed a new variant of a medium-sized transmission for the global market. They worked out the technical specifications, cost targets, and manufacturing time in an iterative experimentation and exchange process between design and manufacturing engineers, the prototyping workshop, and the home location shop floor. Very early conversations between this product team and a higher-level global strategy team suggested that the transmission would also be produced in China, Russia, India, and other emerging markets. Design and manufacturing teams from these markets were incorporated into the development process and technical specifications, cost targets, and manufacturing cycle parameters (metrics and standards) for those markets were provisionally established.

My colleagues and I followed the transfer of the technology to the Chinese market. German team members, design and manufacturing engineers, as well as skilled workers from the prototype workshop and home location shop floor, traveled to China to assist local engineers and workers with the initial production set up. Local Chinese engineers educated their skeptical German counterparts about the possibilities and limits of the Chinese location. Adjustments along several dimensions were made locally, involving a variety of input material, contour design, machine usage, and cycle time metrics and standards. Engineers from the transmission producer's Chinese design center were called in to assist the collaborating teams with these adjustments. As design and production metrics and standards were altered, the German design office was consulted to approve suggested changes to the original targets. The local production and design teams defended the changes to the central teams. In the process, the central team noticed that the adjustments in the flow of

[11] I draw all examples from a research project I conducted in eastern Europe and China among German automobile and machinery producers and suppliers funded by the Hans Böckler Stiftung. Company names must remain confidential. We conducted over 150 open-ended interviews at multiple firms and subsidiary operations in both industries between 2008 and 2015 and have anonymized specific cases by agreement with the case partners. When I refer to specific firm names in this text, I am drawing on material that is either in the public domain or is publicly available. Compare Herrigel et al. (2017) for full discussion of the study's empirical basis.

manufacturing could be used for the same product in eastern European and Indian production locations. Changes were made to the central design. Production performance both in Germany and in offshore locations was, in this systematic fashion, regularly reviewed, metrics and standards were optimized, and roles and relations recomposed.

All of this iterated transfer and exchange occurred according to team-based CPS procedures. Transnational know-how transfer and experience-driven learning, facilitated by formally prescribed team interactions, were systematic features of this system. It was also deeply recursive, as the central teams learned from the experiments of the local teams even as the latter were learning from the former. Finally, revision of the metrics and standards involved role and rule changes within the organization. The division of labor in design and production was continuously optimized and varied. Stakeholder interests were not aligned by the system, they were continuously changed by the process of metric and standard creation, performance review, and optimization.[12]

Obviously, the key to this system's success is that it is global and extends seamlessly throughout all the operations of a firm. For innovation, optimization, and learning to flow recursively within the MNC, everyone must speak the language of the company's CPS. Practices of joint goal-setting and systematic performance review need to become second nature—a new form of self-disrupting routine (Sabel, 2006). This raises the interesting problem of how such systems are globalized.

Firms deploy a number of different diffusion mechanisms. *Auto-Getriebe*, for example, created teaching units that it called Centers of Competence (CoC). These were especially highly performing functional units (e.g., Transmission Housing Assembly) within specific company production locations that had most successfully implemented the CPS. These units were then given the responsibility to help other units doing the same thing elsewhere adopt the CPS routines and develop the capacity to hit performance targets.

My colleagues and I observed CoC at work in a number of units, including the truck-transmission housing assembly group. In all cases, the most advanced workshops were located in the company's home German location. Teams from, for example, the assembly workshop (including managers, engineers, and line workers) travelled to assembly operations in France, India, and China to assist local teams set up operations. These interactions were, in turn, observed by superordinate "international" teams composed of management, engineering, and shop floor representatives from all truck driveline assembly and logistics operations world-wide. The goals were, on the one hand, to get agreement on product and quality metrics and standards among all truck driveline assembly operations (and among global suppliers), and, on the other hand, to get agreement on the core performance review and problem-solving procedures consistent with the company's CPS. The CoC convened face-to-face international team meetings once a year. In addition, two-hour phone meetings

[12] These dynamics can also generate counter-productive forms of exclusion and hierarchical insulation. I discuss these possibilities and the mechanisms firms are developing to deal with them in the following section.

occurred once every quarter (always at 15:00 German time), in which the callers discussed "red status" (i.e., problem/local deviation) issues and made group decisions.

Crucially, the home country CoC's ambition was not to impose common assembly procedures, materials, or logistical flows on like units across the entire company. Rather, it aimed to construct procedures to achieve agreement on metrics and standards and establish transparent self-optimizing processes of regular performance review. The German location provided the offshore operations with technical advice, demonstrated German procedures, and actively assisted with the industrialization of the offshore locations. But local managers were given much leeway to achieve agreed metrics and standards in locally appropriate ways. Deviations from central practices had to be defended, in particular in discussions within the international team. But if the metrics and standards could be maintained or improved upon, deviations were accepted. Indeed, other operations embraced especially innovative alternative practices, through the information channel of the international team.

Continuous improvement teams (CITs) are an alternative experimentalist governance diffusion mechanism. These teams are especially common in machinery firms. Here, the idea is to create a team of CPS experts (lean production specialists, Six Sigma blackbelts, quality engineers, skilled production workers) travelling throughout the firm's functional areas and all global production locations interacting with functional teams as CPS consultants and service providers. CIT members continually provide advice about CPS procedure, how to implement practices of joint goal setting and systematic performance review (the next section will show that this can be a purposefully disruptive activity). But, significantly, they also enact the CPS with their interlocutors, making suggestions for workflow improvement and socializing teams in joint problem solving. CIT teams routinely help multifunctional production or product teams construct better ergonomic workplace arrangements for machining and assembly (using CIT budgeted resources). CITs are also globally constituted (indeed, in the two largest machinery producers we observed, CITs are the largest global teams) and engage with offshore locations in the same CPS proselytizing and service-providing manner in which it engaged home country teams.

Again, as with the CoCs, the aim of CIT activity is not to impose uniform technical and work practices across all parts of the company. Rather, it is to cultivate a common team-based joint goal setting and systematic performance review practice focused on optimization and learning. Notably, even as they help establish intra- and inter-team communication procedures, the circulation of experienced CIT members diffuses innovation and useful practical innovations throughout the MNCs global operations. As the CIT head at a German Power Drive producer (*SW-Antrieb*) told us:

> We are very careful to ensure that information ... gets transferred.... [W]e train employees, world-wide, in these themes.... [I]deas get discussed and solutions outlined at local units all over the world—we are permanently present, locally. At the same time, our members are constantly traveling between units. We achieve information transfer in this way.

CITs generate organizational learning, establish procedures to sustain it, and help to diffuse it within the global organization.

CoC's and CITs are two of many variants currently diffusing CPSs across MNC firms' global operations. Like the CPSs they are used to construct, these organized practices are disruptive mechanisms. Their aim is to instigate local experimentation for practical improvement of jointly agreed upon central norms, metrics, and procedures. They are not establishing incentive alignments; they are convening discussions to define (and redefine) common goals. The *SW-Antrieb* CIT head describes the character of the process in this way:

> We present our plans for investment and change to the local colleagues in a workshop in their plant. We then get together with the employees and ... look at the existing process and determine what is good and what is bad .... Once we have done all of these things, we work together with the local actors to develop a new production island.... Our job is to see that whatever result is worked out is developed and worked through jointly with the colleagues locally.

The manager then referred to a Korean case where the local players resisted implementing the home company's blueprints in every detail. "We installed a manual conveyor here only three or four years ago," the Koreans told them, "and we don't want to just throw it away." After some detailed back and forth about quality and cost differentials between Korea and Germany, the decision was to keep the manual conveyor. "This was fine with us because the production costs were significantly lower in Korea than in Germany."

The next section will show how this interactive joint problem-solving process plays a crucial role in destabilizing the obstacles to learning that emerge through efforts on the part of managers and worker groups and representatives to insulate or exclude interests from the experimentalist process. In the present context, however, I emphasize again that it is inaccurate to understand the workings of these mechanisms as encounters between distinct home and host country institutional logics.[13] CoC and CIT team actors do not regard the practices of local interlocutors as a foreign logic; they view local player perspectives as potential resources to be leveraged in a continuous optimization and learning process. Similarly, local players regard CoC and CIT players not as hierarchical principals giving them orders out of a foreign, institutionally embedded universe, but as potential resources and partners to help them achieve goals that both have agreed on. Moreover, when insights gained from local deviation prove effective, they are diffused elsewhere in the MNC. In this way, institutional logics do not clash; they give rise to deliberation. Systematic disruption and joint problem solving gives rise to continuous, mutual institutional recomposition.[14]

[13] Once again running orthogonally to the concerns of the embeddedness school of Dörrenbächer and Geppert (2011).

[14] Crucially, when asked in response to the story of module adaptation in Korea above *Is it always simply a matter of deviations from a standard module or can the process also result in a reconstruction (Umbau) of the module (central standard) itself?*, the Power Drive CIT Chief replied: "That happens quite often. It is very explicitly never excluded as a possibility. It is an essential part of the transfer process."

MNCs committed to a CPS logic are very emphatic about the system's distinctively collaborative and experimentalist aspects. Often this is true because many had tried (and failed) to arrange the technology transfer process in a more conventional hierarchical interest alignment way prior to working toward the experimentalist (formally collaborative) architecture. Prior to the CoC's creation, for example, *Auto-Getriebe* tried to manage technology transfer very hierarchically. Products were developed centrally at the home production location. Designs, discrete manufacturing-process instructions, and specific machinery to be deployed were then handed off to the subsidiary location. The subsidiary locals were then expected to implement exactly what had been handed to them, and their incentives were set according to centrally determined output and cost measures.

Invariably, locals ran into trouble getting the central designs and machinery to work in the ways the Germans did: Locals could not get machines to produce error free, costs were out of line, processes ran into unanticipated bottlenecks due to operator unfamiliarity with procedure (or differences in training and competence). New product ramp-up, as a result, chronically took longer than desired. Under that old system, the solution to such problems was to send a team of production and design experts from the home location to the subsidiary, where they would typically spend weeks telling locals exactly how to set up the German system, make the prescribed machinery work properly, and avoid bottlenecks. According to our senior production manager informant, the old system was an endless, and very expensive, cycle. The expert teams were no sooner home than they were called back to address new problems that had emerged. The old system's transfer process was too rigid and the subsidiaries' resources to address ramp-up problems too under-utilized.

CoCs were developed to introduce communication, flexibility, and local discretion into the technology transfer process. As described above, a key evaluative criterion was not the CoC's technological knowledge per se; rather, it was its organizational abilities to excel collaboratively within the MNC organization. Managers at *Auto-Getriebe* were driven to this experimentalist architecture largely because the old hierarchical incentive alignment arrangement was ineffective. Through a benchmarking process, the firm discovered that the Robert Bosch Corporation had implemented these CPS-driven experimentalist architectures for global product management, and the company decided to do so as well.

Endogenously Generated Barriers to Experimentalism's Diffusion Within Manufacturing MNCs and How Improvement-Oriented Self-Surveillance Mechanisms Also Become Destabilization Mechanisms

Market uncertainty, linked and unremitting pressures to innovate and reduce costs, and constantly evolving best-practice models drive the adoption and diffusion of experimentalist CPSs. The systems are attractive under these conditions both because they make organizational practices transparent to the actors engaging in them, and because they help actors see that the endurance of specific practical arrangements is contingent upon good performance.

Naturally, there are many barriers to the diffusion of experimentalist CPS practices. Indeed, even in cases where actors extend the experimentalist logic quite far into their organizations, they often encounter limits to further extension. As a result, organizational optimization, recomposition, and recursive/circular learning dynamics become compromised and blocked. In this section, I will some of the most predominant organizational challenges to the diffusion of experimentalism within German manufacturing MNCs.[15] I will also point to an array of strategies that MNCs pursue to overcome these blockages. The main diffusion barriers observed in our cases are: hierarchical insulation, stakeholder exclusion, and inadequate empowerment resources for participants.

Hierarchical insulation involves efforts on the part of higher level management to remove themselves from the continuing stakeholder joint goal setting and self-review procedures. In such cases, managers foster experimentalist problem solving within the domain that they command, but neither confer across domains with other managers or superiors about their domain's relative performance, nor do they negotiate with peer stakeholders about goal setting. In these cases, principal-agent incentive structures govern top management, whereas experimentalist practices govern the practical domains of design and production. Upward diffusion of experimentalism is blocked by the power desires and egoism of managerial ambition within governance and organizational hierarchies.

Such governance segmentation within companies can lead to suboptimal organizational outcomes. Higher managers pursuing incentives based on results grow impatient with the CPS's process-focused, bottom-up problem-solving procedures. They suspend or circumvent the process to generate results for which they will be immediately rewarded. This can create chaos and incoherence in the design and manufacturing value chain: rigorous problem solving is disrupted, learning blocked, and, ultimately, innovation inhibited. In the worst cases, it frustrates and delegitimizes expansive experimentalism at lower levels. Good-willed innovation is blocked by arbitrary power driven by rarified and incompatible managerial incentive structures (cf. Hafner, 2009).

Hierarchical insulation is a central criticism of CPSs in the critical industrial sociology literature. Sauer (2013), Gerst (2011a, 2011b), Pfeiffer (2007, 2008a, 2008b), and Dörre (2015), to take only the most prominent interventions, all point to hierarchical elements that are imposed on production line workers without negotiation: Financial data targets set without production level participation and human

[15] I draw most of the empirical examples presented from fieldwork on German and US MNCs. For an alternative (complementary) perspective on diffusion obstacles, see Netland and Ferdows (2014).

resource hiring practices are two aspects of hierarchical management insulation that all these authors highlight in particular.[16]

In our interviews, these pathologies were present, but they were by no means uncontested, even by players beyond the shop floor. Indeed, CPSs themselves often "officially" characterize such insulation strategies as typical organizational pathologies that must be combatted. As I discuss below, many firms, following the basic experimentalist principle of improvement oriented self-surveillance, have developed mechanisms to destabilize insulation efforts.

Stakeholder exclusion also engenders governance segmentation within organizations. It involves the implementation of formal self-optimization procedures without involving all relevant stakeholders. Thus, manufacturing and design engineers are included in product development discussions, but purchasing executives are excluded or only brought in after crucial decisions have been made. Alternatively, manufacturing and design engineers as well as purchasing people are included, but key suppliers are left out or brought in late. These used to be classic errors of exclusion in the early days (1990s) of the diffusion of lean practices within industry (Helper, 1991a, 1991b; MacDuffie & Helper, 1999; Schumann, Baethge-Kinsky, Kuhlmann, Kurz, & Neumann, 1994; Springer, 1999). A great deal of progress has been made in this area since then, especially in the supply chain and the product development process (Helper et al., 2000; Herrigel, 2010; Sako, 2006; Whitford, 2005). But the large body of critical industrial sociological literature is evidence of its continued persistence in German plants.

Perhaps the most prevalent exclusionary barriers are those preventing production-line worker participation. Important case studies of CPS introduction in the German automobile industry, such as Pfeiffer's (2007, 2008a, 2008b; Adami, Lang, Pfeiffer, & Rehberg, 2008) analysis of a complex automobile assembly line, point to the continued exclusion of production-level teams' from upgrading and process redesign discussions, at least in some plants. This can occur when management designates its own agent as speaker of an allegedly self-governing production team, or when higher level teams simply rely on plant managers, section supervisors or set-up engineers for information on work team performance. It can also happen when continuous improvement teams unilaterally impose work-flow improvements without interaction with the line workers whose process they are improving. In all these cases, it is usual for workers and supervisors to communicate informally. But without the formal obligation to make their habitual actions transparent, workers can hide information (e.g., about finicky machines), protect favorite routines from alteration and, worst of all, become complacent about opportunities to make their own collective efforts better and more competitive. They withhold or bury basic information about the character of production. The firm (and management) in this way foregoes valuable knowledge of its operations and squanders resources for recomposition, innovation, and competitiveness (Adami et al., 2008; Pfeiffer, 2007, 2008a, 2008b).

[16] Indeed, the Diesel scandal at VW, where subordinates were led to improvise an illegal work-around in order to satisfy unilateral performance pressures from higher level managers, appears to be a model illustration of hierarchical insulation.

The above forms of exclusion are hierarchically imposed, and thus have an affinity with hierarchical insulation. But it is also possible for certain stakeholder groups to self-exclude themselves from experimentalist self-optimization processes. This can occur in a variety of ways. In our German cases, we encountered examples of works council self-exclusion from newly introduced CPSs in their firm. The works councils in these cases viewed the new system as a threat and refused to participate. Abstinence from participation in the CPS, however, proved confusing for the employees in work teams at all levels, because they experienced a management discourse of empowerment and self-organization and a works council/union discourse of property divide, asymmetric interest, and mistrust. Sometimes, self-exclusion backfires for the works council, as the CPS self-optimization process's success undermines the sense of organizational indispensability associated with the works council's role. Other times, the effect is the reverse: Traditional role and identity cultivation prevents the CPS from gaining genuine traction in the firm.

A final internal barrier to experimentalist diffusion is inadequate empowerment resources for participants. Most prominently, this means the lack of adequate skills at the production and lower management levels, such as one finds in emerging market contexts. Many firms that run CPS procedures in broadly inclusive ways in their home market locations find it difficult to implement thoroughly inclusive self-optimization practices in emerging market operations. Language and cultural difference combined with limited skill and educational competences within the available labor pool make it difficult to configure production in a way that engages employees in useful self-optimization.

In part, low wages and structural weakness on the employee side allows management to not try so hard to implement an experimentalist system. Instead, they crassly exploit cost advantages without continuous optimization, or they impose improvements developed elsewhere on a compliant workforce. Self-optimization processes elsewhere in the company—even those higher up in the emerging market operation—in this way carry the inefficiency of very low cost and very manipulable labor.

Inadequate empowerment resources are not, however, only an artifact of power imbalances. Firms often believe that the constraints that available skill and competence pools place on efforts to localize home country production and work practices result in the loss or destruction of useful knowledge, rather than the generation of new or alternative knowledge about familiar technologies and production processes. For example, in their German operations, woodworking machinery and automobile front-end makers from our sample both use highly skilled workers capable of performing a variety of operations. Their experience generates unique product technology knowledge that can be leveraged for innovation in cooperation with engineers and product designers. Such workers in these firm's home locations in *Westfalen* and on the *Schwäbische Alb* are wholesomely incorporated into self-governing, cross-functional teams and in the serial self-review processes in their respective firms' CPS variants.

In China, however, such workers do not exist in the broader Shanghai labor pool where the firms have their production operations. Hence, when the firms produce the same products in Shanghai that they produce in Germany, even when they make significant localization changes in the product design, the local management teams have to devise ways to make the product with far fewer skilled workers. Typically, this means that several more narrowly skilled workers will perform in a sequential and disintegrated way what one highly skilled worker would perform in a synthetic way in Germany. The firms incorporate the disintegrated workers less into the self-optimization procedures of their respective CPSs because they believe that the knowledge such workers could contribute through inclusion is less valuable. In part, this is a violation of the inclusion aspect of the transparency and inclusion principle; but there is also an element of the Chinese employees not being able to engage with the transparency element.

Destabilization Mechanisms

Interestingly, all these obstacles exist not only *ex ante*, as firms attempt to construct corporate production systems and implement them throughout their global operations. They also continually regenerate through the CPS's experimentalist dynamics themselves. The revision of commonly agreed-upon frameworks in light of changes to processes and products introduced at a local level frequently redefines power relations and stakeholders, creating new possibilities for insulation and exclusion at various levels and in different locations within the firm.

For example, many firms find that operations that are highly automated in German operations do not require the same degree of automation in lower-wage locations, such as Poland or China (or Korea, as in the quoted example provided earlier). As a result, local players there deconstruct the home procedures and render production flow into a series of manual operations. In some cases, these innovations actually prove more flexible and productive than the automated operations that they replace and, as a result, cooperating teams in the home operations try to replace automation with the newer manual procedures. In so doing, new worker groups emerge and teams are re-constituted. If the firm is not careful to ensure that the newly emergent groups become integrated into existing team deliberative relations in the plant, the new groups can be excluded (not recognized) as (knowledge-bearing) stakeholders in the production process. In such cases, managers can grab power and monopolize control over crucial levels of knowledge and resource flow within the firm.

In order to prevent such obstacles from paralyzing the global recursive learning process, many MNCs, in line with the principle of improvement-oriented self-surveillance, deploy an array of destabilization mechanisms that systematically undermine insulation and exclusion strategies within the global firm and reconstitute the deliberative experimentalist learning process. In particular, the organizational forms described above as knowledge carriers throughout the MNC's global operations—CoCs, CITs—act as destabilization mechanisms undermining efforts to insulate knowledge and exclude stakeholders. This makes sense, their aim is to manage center and local deliberation in ways that circulate technical and organizational knowledge through the company's transparent formal CPS procedures. They both implement the CPS itself and carry knowledge around the global firm that CPS procedures generate. Local players seeking to exclude stakeholders (e.g., production workers or suppliers) or central actors looking to insulate their own practices from the changes generated by subsidiary actors are targeted by these "third-party" organizations and challenged to defend their efforts. Often, this challenge is enough to initiate inclusion processes: Why are production line workers not involved in team discussions with line leadership staff and application engineers? How will production implementation and run-up problems be dealt with without their input?

It is not only talk, either. Because the CoCs and CITs, in particular, are not supposed to impose solutions on players, but simply to instigate local discussions regarding the implementation of central technologies and metrics (for CoCs) or of possibilities for process improvement in the context of CPS procedures and global best practice (CITs), they have the organizational authority to provoke local actors into defending exclusion or insulation practices. And, as they are globally active, they come to individual central or local conversations with independent knowledge of practices throughout the MNC's operations. They can use this knowledge to insist that specific players contend with best practice within the firm. This use of organizational mission and accumulated practical knowledge to destabilize relations is a crucial dimension of CoC and CIT activity. They do not impose solutions, but rather use their organizational mission to destabilize practices and provoke deliberation about solutions.

But what if, for all that, stalemate or paralysis emerges? Or, what if deliberation proves so contentious and arduous that despite exchange, progress is too long delayed? In such cases, most CPSs provide for a penalty default (Ayres, 2001; Ayres & Gertner, 1989). That is, if local players are locked in dispute or cannot resolve a local problem, a higher order stakeholder team will intervene to redefine the problem that is stymying the local actors in an effort to create better conditions for agreement. Such interventions, moreover, are frequently independent from higher-level team judgment. Rather, the CPS itself establishes penalty default triggers—most often in the form of time frames for decision making, or outer boundary cost or return on investment targets ("gates") for processes. If local deliberation exceeds the allotted time or under- or overshoots cost targets, relevant higher-level teams are automatically called in (often following a logic imposed by Six Sigma instruments) to evaluate the situation.

The CIT head at *SW-Antrieb* (quoted earlier) outlined the procedures *they* follow for improvement project implementation. He highlights both the CIT role as a destabilization mechanism—in particular in identifying stakeholder exclusion and developing strategies for inclusion—and the key role played by penalty default—primarily in the form of strict time frameworks for project implementation.

> We are continuously and permanently improving our processes through the involvement of all those who are affected and participating in them. That is our core approach to secure process sustainability. If you like, you can see it as a mixture of business reengineering and classic Toyota continuous improvement ... broadly lean principles. It is a path of many

> small and tiny steps.... The limiting factor is always the time frame.... Within 12 weeks, we try to organize the process in a new way. The 12 weeks are a basic grid for us. If we see that the spectrum of themes, the degree of effort, or the scope are too large and can't be resolved within 12 weeks, we then start to segment the themes, and create sequences for effort in very transparent ways.

Twelve weeks, in other words, is a penalty default trigger. Projects that cannot be accomplished in that timeframe are redefined to facilitate more possible forms of collective problem solving. The *SW-Antrieb* CIT chief also emphasized that the role of CIT actors is to make all projects inclusive:

> We [the CIT team-gh] in fact rove through the shop floor, observing production and frequently making suggestions for improvement. That is a permanent activity. It frequently happens, however, that the middle management comes to us and says: "I have a concrete problem in this area. Can you try to get people working toward a solution?" A project will be created that is ... interdisciplinary in makeup: That is, all stakeholders (*Beteiligten*) sit at the table and we try to construct a comprehensive/inclusive (*ganzheitliches*) image of optimization. All those affected by the problem participate in the work toward creating a solution. That is a core principle.

The same manager emphasized that CIT teams were needed as destabilizers because the "normal process" of self-optimizing, continuous improvement by specialized multifunctional teams very often tended toward myopia and self-blockage. They inadvertently excluded important players in the value chain:

> No matter where you look, at Toyota or other benchmark companies, not a single company has been able to operate only self-optimizing teams. External destabilizers such as CITs are needed. We are no different. We thought initially that the theme of optimization ... would establish itself and run by itself. That didn't happen.... People do not look at their jobs in the whole context of production. This is what interdisciplinary project work tries to overcome. When a worker is active in a specific area of production, that person is generally very highly specialized.... But in the project teams, suddenly, he has to be concerned with material delivery and preparation as well as where the bevel gears that are made in her station will go to next .... The natural initial response here is to throw up your hands and say—this isn't my job, I don't know anything about this. And the person will avoid looking at the whole production process—not out of ill will, but simply because she sticks to what she knows. Our job is to remind people to see how their activity fits into the whole. We make models of the whole process chain and constantly remind the project participants what the big picture is.

Finally, the CIT head emphasized that it would be counter-productive for his firm's CIT team to impose a solution on the parties engaged in the project. Their job was to convene a conversation among all the relevant stakeholders. The point is to have the stakeholders understand their respective roles and to make all of their actions explicit to one another and to the firm's CIT team. When asked if his firm's CIT imposed solutions on the deliberating project teams, he said:

> If we simply imposed solutions on the projects, they would never actually get implemented. Why? Because every person who worked on the process once it was up and running would try to prove to me that the solution doesn't work. Instead, our idea is ... [that] people have to be at the center in the design of the solution because they are the ones who will be responsible for ensuring that the process yields the quality, safety, and reliability that we need.

And the system only works, he emphasized, if it goes all the way down to the production line workers:

> The project teams have to include machine operators, set-up people, maintenance, logistics, work preparation, work planning, the foreman (depending on how complex the solution looks like it will be), someone from the tool shop, and tool maintenance who will be able to ensure that the proper tools will be delivered to the new process. That whole team works together to come up with a solution for the new production process. In principle, the CIT team indicates only what methods and strategies are required.... It is up to the team members to choose. Of course, the solution has to be better than what it is replacing. That is the standard. The parameters are generated by our company's formal production system, but the solution is generated by the stakeholder problem-solving process...

Conclusion

All of these activities aimed at stakeholder inclusion and transparency creation are permanent and ongoing within successful experimentalist German MNCs. Projects and products set up in one year are reviewed the next in order to ensure that the original designs are having the desired effect—or to see if optimization projects elsewhere in the value chain have created possibilities in the area that were not possible to see when the initial project was undertaken. Such continuous self-surveillance is disruptive. "Daily routines" are made explicit to those enacting them and reflection leads to change or recomposition. As a result, manufacturing MNCs operating on a CPS logic are constantly in flux. Successfully overcoming problems of exclusion or insulation with one project often create new ones. The teams themselves, and the destabilizing CoC and CIT teams, must be vigilant and continue to root out exclusionary/insulating dynamics. Doing this on a permanent basis fosters learning and innovation within the entire global organization.

The upshot of all of this, of course, is quite paradoxical. In the current environment, manufacturing multinationals' commitment to learning, permanent self-optimization, and recomposition constantly creates the possibility for blockage through insulation or exclusion. Indeed, there are no pure examples of a thoroughly inclusive and systematically deliberative recursive learning-oriented manufacturing multinational. Rather, recursive learning organizations have heterogeneous, hybrid, and constantly self-recomposing governance arrangements with varying and highly contingent admixtures of joint problem solving, team-based deliberation, hierarchical insulation, and stakeholder in/exclusion. In this sense, the core emergent institutions in the new multinational are those that disrupt (especially the principles of revisable planning and improvement-oriented self-surveillance), rather than those that govern. Disruption is a permanent process, while specific formal governance arrangements, like the organizational roles they manage, are always ephemeral.

References

Adami, W., Lang, C., Pfeiffer, S., & Rehberg, F. (Eds.). (2008). *Montage braucht Erfahrung: Erfahrungsbasierte Wissensarbeit in der Montage* [Assembly needs experience: Experience based knowledge work in assembly]. Mehring: Rainer Hampp.

Ayres, I. (2001). Default rules for incomplete contracts. Retrieved from https://ianayres.yale.edu/sites/default/files/files/Palgrave.pdf

Ayres, I., & Gertner, R. (1989). Filling gaps in incomplete contracts: An economic theory of default rules. *The Yale Law Journal, 99,* 87–130. https://doi.org/10.2307/796722

Baldwin, R. (2016). *The Great Convergence: Information technology and the new globalization.* Cambridge, MA: Harvard University Press.

Becker-Ritterspach, F., & Dörrenbächer, C. (2009). Intrafirm competition in multinational corporations: Towards a political framework. *Competition and Change, 13,* 199–213. https://doi.org/10.1179/102452909X451332

Bouquet, C., & Birkinshaw, J. (2008). Managing power in the multinational corporation: How low-power actors gain influence. *Journal of Management, 34,* 477–508. https://doi.org/10.1177/0149206308316062

Brandt, L., & Thun, E. (2010). The fight for the middle: Upgrading, competition, and industrial development in China. *World Development, 38,* 1555–1574. https://doi.org/10.1016/j.worlddev.2010.05.003

Clarke, C. (2005). *Automotive production systems and standardization: From Ford to the case of Mercedes-Benz.* Heidelberg: Physica. https://doi.org/10.1007/b138988

Dewey, J. (1922). *Human nature and conduct: An introduction to social psychology.* New York: Henry Holt. https://doi.org/10.1037/14663-000

Dörre, K. (2015). Beyond shareholder value? The impact of capital market-oriented business management on labor relations in Germany. In C. E. Weller (Ed.), *Inequality, uncertainty, and opportunity: The varied and growing role of finance in labor relations* (pp. 85–118). Champaign: Labor and Employment Relations Association.

Dörrenbächer, C., & Gammelgaard, J. (2016). Subsidiary initiative taking in multinational corporations: The relationship between power and issue selling. *Organization Studies, 37,* 1249–1270. https://doi.org/10.1177/0170840616634130

Dörrenbächer, C., & Geppert, M. (2011). *Politics and power in the multinational corporation.* Cambridge, UK: Cambridge University Press. https://doi.org/10.1017/CBO9780511973352

Friedli, T., & Schuh, G. (2012). *Wettbewerbsfähigkeit der Produktion an Hochlohnstandorten* [Competitive production in high wage locations]. Heidelberg: Springer. https://doi.org/10.1007/978-3-642-30276-3

Gerst, D. (2011a). *Intensivierung von Arbeitszeit durch GPS* [Intensification of working time through CPSs] [Presentation].

Gerst, D. (2011b). Sind ganzheitliche Produktionssysteme und Gute Arbeit vereinbar? [Are holistic production systems and good work compatible?]. *Arbeit: Zeitschrift für Arbeitsforschung, Arbeitsgestaltung und Arbeitspolitik, 20,* 246–250. https://doi.org/10.1515/arbeit-2011-0309

Glückler, J. (2014). How controversial innovation succeeds in the periphery: A network perspective of BASF Argentina. *Journal of Economic Geography, 14,* 903–927. https://doi.org/10.1093/jeg/lbu016

Hafner, S. J. (2009). *Sisyphus und Machiavelli bei der Arbeit: Ganzheitliche Produktionssysteme zwischen Mythen und Realitäten* [Sisyphus and Machiavelli at work: Holistic production systems between myths and realities]. Mehring: Rainer Hampp. https://doi.org/10.1688/9783866183261

Helper, S. (1991a). How much has really changed between U.S. automakers and their suppliers? *MIT Sloan Management Review, 32*(4), 15–28.

Helper, S. (1991b). Strategy and irreversibility in supplier relations: The case of the U.S. automobile industry. *Business History Review, 65,* 781–824. https://doi.org/10.2307/3117265

Helper, S., MacDuffie, J. P., & Sabel, C. (2000). Pragmatic collaborations: Advancing knowledge while controlling opportunism. *Industrial and Corporate Change, 9,* 443–488. https://doi.org/10.1093/icc/9.3.443

Herrigel, G. (2010). *Manufacturing possibilities: Creative action and industrial recomposition in the United States, Germany and Japan.* Oxford, UK: Oxford University Press. https://doi.org/10.1093/acprof:oso/9780199557738.001.0001

Herrigel, G., Voskamp, U., & Wittke, V. (2017). *Globale Qualitätsproduktion: Transnationale Produktionssysteme in der Automobilzulieferindustrie und im Maschinenbau* [Global quality production: Transnational production systems in automobile supply chains and in the machinery industry]. Frankfurt am Main & New York: Campus.

Herrigel, G., Wittke, V., & Voskamp, U. (2013). The process of Chinese manufacturing upgrading: Transitioning from unilateral to recursive mutual learning relations. *Global Strategy Journal, 3,* 109–125. https://doi.org/10.1111/j.2042-5805.2012.01046.x

Heymann, E. (2014). Zukunft des Automobilstandorts Deutschland [The future of Germany as an automobile production location]. *Deutsche Bank Research.* Retrieved from https://www.dbresearch.de/PROD/RPS_DE-PROD/PROD0000000000444451/Zukunft_des_Automobilstandorts_Deutschland.pdf

Horner, R., & Hulme, D. (2017). From international to global development: New geographies of 21st century development. *Development and Change, 50,* 347–378. https://doi.org/10.1111/dech.12379

Horner, R., & Nadvi, K. (2017). Global value chains and the rise of the Global South: Unpacking twenty-first century polycentric trade. *Global Networks, 18,* 207–237. https://doi.org/10.1111/glob.12180

Joas, H. (1996). *Die Kreativität des Handelns* [The creativity of action]. Frankfurt: Suhrkamp.

Jürgens, U., & Krzywdzinski, M. (2016). *New worlds of work: Varieties of work in car factories in the BRIC countries.* Oxford, UK: Oxford University Press. https://doi.org/10.1093/acprof:oso/9780198722670.001.0001

Kobrin, S. J. (2017). Bricks and mortar in a borderless world: Globalization, the backlash, and the multinational enterprise. *Global Strategy Journal, 7,* 159–171. https://doi.org/10.1002/gsj.1158

Kristensen, P. H., & Zeitlin, J. (2004). *Local players in global games: The strategic constitution of a multinational corporation.* Oxford, UK: Oxford University Press. https://doi.org/10.1093/acprof:oso/9780199275625.001.0001

MacDuffie, J. P., & Helper, S. (1999). Creating lean suppliers: Diffusing lean production throughout the supply chain. In J. K. Liker, W. M. Fruin, & P. S. Adler (Eds.), *Remade in America: Transplanting and transforming Japanese management systems* (pp. 154–202). New York, NY: Oxford University Press.

Netland, T. (2013). Exploring the phenomenon of company-specific production systems: One-best-way or own-best-way? *International Journal of Production Research, 51,* 1084–1097. https://doi.org/10.1080/00207543.2012.676686

Netland, T., & Aspelund, A. (2013). Company-specific production systems and competitive advantage: A resource-based view on the Volvo production system. *International Journal of Operations & Production Management, 33,* 1511–1531. https://doi.org/10.1108/IJOPM-07-2010-0171

Netland, T., & Ferdows, K. (2014). What to expect from a corporate lean program. *MIT Sloan Management Review, 55*(4), 83–89.

Netland, T., & Sanchez, E. (2014). Effects of a production improvement programme on global quality performance. *The TQM Journal, 26,* 188–201. https://doi.org/10.1108/TQM-03-2012-0023

Pfeiffer, S. (2007). *Montage und Erfahrung: Warum ganzheitliche Produktionssysteme menschliches Arbeitsvermögen brauchen* [Assembly and experience: Why comprehensive production systems need worker resources]. Munich: Rainer Hampp.

Pfeiffer, S. (2008a). Flexible Standardisierung und ganzheitliche Produktionssysteme—erfahrungsförderlich?! [Flexible standardization and comprehensive production systems—the importance of experience]. In W. Adami, C. Lang, S. Pfeiffer, & F. Rehberg (Eds.), *Montage braucht Erfahrung: Erfahrungsbasierte Wissensarbeit in der Montage* (pp. 143–169). Mehring: Rainer Hampp.

Pfeiffer, S. (2008b). Montage, Wissen und Erfahrung: Warum "einfache" Arbeit auch Wissensarbeit ist, warum Erfahrung in flexibler Montage so wichtig ist—und was das alles bildungspolitisch bedeutet [Assembly, nowledge and experience: Why "simple" work is also "knowledge" work, why experience is so important in flexible assembly—and what that implies for training]. In W. Adami, C. Lang, S. Pfeiffer, & F. Rehberg (Eds.), *Montage braucht Erfahrung: Erfahrungsbasierte Wissensarbeit in der Montage* (pp. 14–48). Mehring: Rainer Hampp.

Romme, A. G. L. (1997). Organizational learning, circularity and double-linking. *Management Learning, 28*(2), 149–160. https://doi.org/10.1177/1350507697282004

Romme, A. G. L. (1999). Domination, self-determination and circular organizing. *Organization Studies, 20,* 801–832. https://doi.org/10.1177/0170840699205005

Romme, A. G. L., Zollo, M., & Berends, P. (2010). Dynamic capabilities, deliberate learning and environmental dynamism: A simulation model. *Industrial and Corporate Change, 19,* 1271–1299. https://doi.org/10.1093/icc/dtq031

Sabel, C. F. (2006). A real-time revolution in routines. In C. Heckscher & P. S. Adler (Eds.), *The firm as a collaborative community* (pp. 106–156). New York, NY: Oxford University Press.

Sabel, C. F., Herrigel, G., & Kristensen, P. H. (2018). Regulation under uncertainty: The co-evolution of industry and regulation. *Regulation and Governance, 12,* 371–394. https://doi.org/10.1111/rego.12146

Sabel, C. F., & Simon, W. H. (2011). Minimalism and experimentalism in the administrative state. *Georgetown Law Journal, 100,* 53–93. https://doi.org/10.2139/ssrn.1600898

Sabel, C. F., & Zeitlin, J. (2008). Learning from difference: The new architecture of experimentalist governance in the European Union. *European Law Journal, 14,* 271–327. https://doi.org/10.1111/j.1468-0386.2008.00415.x

Sako, M. (2006). *Shifting boundaries of the firm: Japanese company—Japanese labour*. Oxford, UK: Oxford University Press. https://doi.org/10.1093/acprof:oso/9780199268160.001.0001

Sauer, D. (2013). *Die Organisatorische Revolution: Umbrüche in der Arbeitswelt—Ursachen, Auswirkungen und arbeitspolitische Antworten* [The organizational revolution: Disruption in the world of work—causes, consequences and human resource solutions]. Hamburg: VSA.

Schumann, M., Baethge-Kinsky, V., Kuhlmann, M., Kurz, C., & Neumann, U. (1994). *Trendreport Rationalisierung—Automobilindustrie, Werkzeugmaschinenbau, Chemische Industrie* [Report on rationalization trends in the automobile, machine tool and chemical industries]. Berlin: Edition Sigma.

Simon, W. H. (2015). The organizational premises of administrative law. *Law and Contemporary Problems, 78*(1–2), 61–100. Retrieved from https://scholarship.law.columbia.edu/faculty_scholarship/1827

Spear, S. J. (2009). *Chasing the rabbit: How market leaders outdistance the competition and how great companies can catch up and win.* New York: McGraw-Hill.

Springer, R. (1999). *Rückkehr zum Taylorismus? Arbeitspolitik in der Automobilindustrie am Scheideweg* [Return to Taylorism? A turning point for human resource policy in the automobile industry]. Frankfurt am Main & New York: Campus.

Sutton, J. (2012). *Competing in capabilities: The globalization process*. Oxford, UK: Oxford University Press.

Teece, D. J. (2007). Explicating dynamic capabilities: The nature and microfoundations of (sustainable) enterprise performance. *Strategic Management Journal, 28,* 1319–1350. https://doi.org/10.1002/smj.640

Teece, D. J., Pisano, G., & Shuen, A. (1997). Dynamic capabilities and strategic management. *Strategic Management Journal, 18,* 509–533. https://doi.org/10.1002/(SICI)1097-0266(199708)18:7<509::AID-SMJ882>3.0.CO;2-Z

Verband der Automobilindustrie (VDA). (2012). *Jahresbericht 2012* [Annual report 2012]. Retrieved from https://www.vda.de/dam/vda/publications/Jahresbericht%202012/1341308052_de_437643196.pdf

Verband der Automobilindustrie (VDA). (2017). *Jahresbericht 2017* [Annual report 2017].

Westkämper, E., & Zahn, E. (2009). *Wandlungsfähige Produktionsunternehmen: Das Stuttgarter Unternehmensmodell* [Adaptable manufacturing companies: The Stuttgart Model]. Berlin: Springer. https://doi.org/10.1007/978-3-540-68890-7

Whitford, J. (2005). *The new old economy: Networks, institutions and the organizational transformation of American manufacturing*. Oxford, UK: Oxford University Press. https://doi.org/10.1093/acprof:oso/9780199286010.001.0001

Winter, S. G. (2003). Understanding dynamic capabilities. *Strategic Management Journal, 24,* 991–995. https://doi.org/10.1002/smj.318

World Bank. (2015). *Global economic prospects: Having fiscal space and using it.* Washington, D.C.: World Bank Group. Retrieved from https://www.worldbank.org/content/dam/Worldbank/GEP/GEP2015a/pdfs/GEP15a_web_full.pdf

Zollo, M., & Winter, S. G. (2002). Deliberate learning and the evolution of dynamic capabilities. *Organization Science, 13,* 339–351. https://doi.org/10.1287/orsc.13.3.339.2780

Chapter 19
Networks as Facilitators of Innovation in Technology-Based Industries: The Case of Flat Glass

Nebahat Tokatli

To what extent do networks facilitate innovation in technology-based industries? So far, most scholars of this question have focused on just a few high-technology industries. Consider, for example, the work of Powell and his collaborators, Hagedoorn and his collaborators, and Stuart and his collaborators on industries such as computers, semiconductors, pharmaceuticals, and biotechnology.[1] Their overall conclusion is that collaborations within networks (which are made up of firms as well as institutions such as universities, research institutes, and venture capital) are absolutely necessary components of the *virtuous cycles* that networks and innovation now constitute in these industries: Networks facilitate innovation, and innovative outputs then attract further collaborative ties (see Powell & Grodal, 2005, p. 67). Simply put, the networks are the *locus of innovation* in these industries: For example, in biotechnology, the most important force behind innovation is reported to be the "structure of its networks" and the "rules governing these networks"—not "money," not "market power," not even the "sheer force of novel ideas" (Powell & Grodal, 2005, p. 59; Powell, White, Koput, & Owen-Smith, 2005, p. 1187).

It follows from this conclusion that if one wishes to thoroughly understand the most significant opportunities for innovation in any particular high-technology industry (or, perhaps, in any industry for that matter), one should consider focusing on that industry's networks. By placing networks at the heart of their study of

[1] Here, I basically refer to a number of publications especially including Owen-Smith and Powell (2004), Powell et al. (2005), Powell, Packalen, and Whittington (2012), Whittington, Owen-Smith, and Powell (2009), Powell and Owen-Smith (2012), Powell and Standholtz (2012), Powell and Padgett (2012), Hagedoorn and Schakenraad (1994), Stuart and Podolny (1996), Stuart (2000), and Robinson and Stuart (2007).

N. Tokatli (✉)
Milano School of Policy, Management, and Environment, The New School, New York, NY, USA
e-mail: tokatlin@newschool.edu

J. Glückler et al. (eds.), *Knowledge for Governance*, Knowledge and Space 15,
https://doi.org/10.1007/978-3-030-47150-7_19

innovation, the above mentioned researchers make the observations that, first, innovative firms and other institutions are now compelled to be part of networks, and, secondly, that the network's *structure* (architecture or topology) and the manner in which they are *governed* (institutionalized and managed outside governmental rules and regulations, except for antitrust regulations and such) matter a great deal (Powell et al., 2005, p. 1187). Here it is important to point out that as they give prominence to the role that networks play in mobilizing resources for innovation (as scientific and technological knowledge is transferred from university laboratories or R&D departments to industry and from firm to firm, and converted into commercially successful products and processes), they are basically interested in mature and diverse industries with large and complex networks. Consequently, their work, for the most part, is able to accommodate formal network approaches with their particular vocabulary (namely, *nodes*, *lines*, and their chartable *patterns*) and reliance on mathematical modelling.

Now that researchers have all this work at their disposal, it would be interesting to go beyond high-technology industries such as biotechnology, and look into some other *technology-based* industries whose products are developed by the application of some degree of technological expertise (without the entire industries necessarily being labelled as *high-technology*), and whose networks are typically smaller and less complex than those belonging to biotechnology and similar industries.[2] In this chapter, I consider one of these industries, namely the flat glass industry, with two purposes in mind—one being the primary purpose of the chapter, and the other a secondary and somewhat provisional one pointing to further research. My primary purpose is to present a reasonably complete picture of the flat glass industry around a research question that concerns the extent to which this industry's networks facilitated innovation in the past and continue to do so now. My secondary purpose is to explore whether or not a different sort of network thinking is required for this particular industry—different from the thinking that the students of high-technology industries subscribe to as they study, say, biotechnology.

Concerning the primary purpose, I rely on historical accounts such as those of Harris (1975), Barker (1977), and Uusitalo and his collaborators, together with contemporary information available on the Internet as well as in industry publications.[3] What is common in all these accounts is the observation that the innovative firms in the flat glass industry have not been, as Teece (2016, p. 11) would put it, "islands"—at least not during the last few centuries. I elaborate on this observation here, but then argue that each time period requires a slightly different understanding of the network concept as the flat-glass industry is studied over a long period of time. This is because most innovations in this industry have required technical collaborations

[2] I here use the term *technology-based* as somewhat of a catch-all phrase to refer to all industries whose products are developed by the application of some degree of technological expertise. I am thus not using the terms *technology-based* and *high- technology* industries interchangeably—a point that I will elaborate on later.

[3] By "Uusitalo and his collaborators," I mean the authors of the following publications: Uusitalo (1997, 2014a, 2014b), Uusitalo and Mikkola (2010), and Uusitalo and Möller (2015).

between firms, but the contexts in which these have taken place have changed from period to period. For example, in the eighteenth century, technical collaborations occurred between firms located on both sides of the English Channel, and in the nineteenth century between firms on both sides of the Atlantic Ocean. And then, in the twentieth century, thanks to a breakthrough innovation, the flat-glass industry became basically a "one technology, one license, one product" industry dominated by the British flat-glass firm Pilkington. During this period, Pilkington made sure that its licensees all over the world were legally bound to share with it whatever technologies they might develop in the future—thereby changing the role that networks played in this industry with respect to innovations.

Today, there are six firms meeting most of the global demand for flat glass; and as far as the primary processing of flat glass (that is the basic manufacture of flat glass) goes, collaboration within networks still refers to provisional and transitional steps taken by firms in order to enter new markets, open doors for mergers and joint ventures, spread risks, or minimize the costs of research and development. However, when it comes to the secondary processing of flat glass (i.e., the further processing of flat glass by laminating, toughening, coating, or silvering the product), there are signs that collaboration within networks might be starting to get closer to that of high-technology industries—one sign being a newly emerging idea of establishing consortia for major research activities, including a discussion as to who should take a leading role in such efforts. Perhaps, we are entering into a period where exciting scholarly research possibilities will arise from the larger and more complex networks that might emerge from the now maturing industry.

If there really is a possibility of large and complex networks emerging in the world of flat glass (at least with respect to the secondary processing of flat glass), then it is time to consider the sort of network thinking that might be useful concerning the contemporary flat glass industry. Here, we have some choices to make. After all, there is a spectrum of network thinking at our disposal with one end represented by Uusitalo and his collaborators, and the other by Powell and his collaborators. The former group utilizes the network concept basically as, what Marsden (1990, p. 436) calls, a "sensitizing metaphor," and refrains from developing a formal network approach[4]; whereas at the other end of the spectrum, the latter group utilizes a more formal representation which leads to a particular vocabulary—namely, "nodes," "lines," and their chartable "patterns." Consequently, whereas Uusitalo and his collaborators become interested in distribution of market power and cogency of innovative ideas among firms, Powell and his collaborators let "money," "market power," and the "sheer force of novel ideas" take the back seat to "structure of the networks" and the "rules governing these networks" (Powell et al., 2005, p. 1187).

Concerning this chapter's second purpose (i.e., exploring whether or not a different sort of network thinking is needed for this particular industry), I cautiously

[4] For example, consider the extent to which Uusitalo and Möller (2015) use the words "network" (or, in their words, "net") and "industry" interchangeably. More specifically, by the "two ... industrial networks" of the past, they mean "the sheet glass industry and the plate glass industry" (p. 1379).

consider Coward's (2018, p. 454) critique of the network thinking that Powell and his collaborators subscribe to—one such criticism being that there is "no natural outside to the network" in this particular network thinking, which Coward (2018, p. 453) calls the "unbounded-ness" problem: "Networks are, in principle, infinitely extensible" in this understanding—something which leads to "fantasies of precision"—as if it is possible to have an accurate understanding of how networks work and what disrupts their workings. There really are opportunities to reflect on Coward's (2018) concerns here. Consider, for example, the footnote by Powell et al. (2005, p. 1191), in which they inform the reader that they had at one point conducted an analysis on a biotechnology network with "250,000 nodes"—that is, 250,000 firms and institutions such as universities, research institutes, and venture capital that were simultaneously sourcing and receiving knowledge. Before elaborating on this and other issues, in the next section, I will first consider some conceptual challenges arising from the particularities of the flat-glass industry. It should here suffice to point out that the idea is certainly not to contest network thinking in its entirety, but rather to open up a discussion of the many conceptual ways of situating networks within industrial studies.

Some Conceptual Challenges

The flat-glass industry is not perfectly comparable to the high-technology industries from which the research question originated. First of all, the pace of scientific and technological development in high-technology industries such as computers, semiconductors, pharmaceuticals, and biotechnology is now such that single firms no longer have all the necessary skills to keep up with progress and come up with significant innovations (Powell & Grodal, 2005; Teece, 2016). In fact, collaboration within networks in these industries is now more than a provisional and transitional step taken by firms in order to enter new markets, spread risks, or reduce the unnecessary duplication of research costs and efforts. Rather, they are absolutely necessary components that facilitate innovation. Once innovation occurs, innovative outputs attract further collaborative ties (see Powell & Grodal, 2005, p. 67). However, most of what is known about networks' role in innovation comes from the study of only a few industries.

Obviously, the particularities of the flat-glass industry matter. For example, as mentioned before, the flat-glass industry is not a high-technology industry in the sense that, for example, biotechnology is a high-technology industry. Simply put, innovation in the flat-glass industry is not as cumulative as in some other industries, and innovation's assets are not as dispersed (locally and globally) as they are elsewhere. On the other hand, in some areas such as "high resolution flat panel displays," this particular industry is not significantly dissimilar to the above-mentioned industries either (see Stuart, 2000, p. 793). Here, it is vital to remember not only that there are many technology intensive firms in industries that are not regarded as high-technology industries per se, but also that not all firms belonging to high-technology industries are necessarily research-intensive and technologically dynamic (Keeble & Wilkinson, 1999).

Another reason why the flat-glass industry might not be perfectly comparable to the high-technology industries from which my research question originated is that industries such as biotechnology happen to be especially mature and diverse with unusually large and complex networks. Obviously, not all technology-based industries have such large networks. In addition, because history plays a role, it might not be possible to understand what compels contemporary firms to be part of today's networks without a thorough understanding of what compelled the firms to network with others in the past (including pairing up with only one other firm or institution). All this points to the importance of developing a dynamic understanding of the networks in a variety of technology-based industries—including those with networks not as cohesive and complicated as, say, those belonging to biotechnology.

I must here make a few more conceptual points. First, by giving priority to the networks and the rules governing them in the above-mentioned studies on high-technology I certainly do not mean to deny the role of what Powell et al. (2005, p. 1187) call "the sheer force of novel ideas." Rather, I mean to emphasize the association between such ideas and the networks. Burt's (2004) fundamental observation that novel ideas are better expressed, kept, and evaluated as valuable in networked contexts than in isolated contexts can serve as an example: Here, although Burt (2004) hypothesizes that this is the case in all settings, there are good reasons to believe that it is perhaps especially the case in technology-based industries. Just as scientists are supposedly "stimulated to their best ideas by people outside their own discipline," firms are stimulated to their best innovative performances when they are placed in contact with firms and institutions that are dissimilar to themselves (Burt, 2004, p. 59).

Secondly, I do not deny the importance of money and market power in the above-mentioned studies; rather, I mean to emphasize the association between these factors and the networks. Money and market power are important resources, and "resource rich" firms are more capable of altering their positions by reconfiguring their networks; still, what matters the most is being networked. Simply put, *disconnectedness* is a "liability" (Powell et al., 2005, pp. 1137–1138). In fact, economists now sometimes wonder whether a firm is "too connected" to fail, just as they once used to wonder whether a firm was "too big" to fail. This focus makes especial sense in high-technology industries, where the fast pace of scientific and technological developments leads to insurmountable disadvantages for isolated firms (including even the most moneyed and powerful ones). Simply put, in such industries, leaving aside the fact that no single firm can master and control all the competencies required for innovation, no single firm can even absorb the available resources (say, the $17 billion grant money offered to biotechnology by the federal government in the United States in the year 2000 only—Powell et al., 2005, p. 1142).

Obviously, it is relatively easy to see how and why the "network" concept comes into play in the context of innovation in high-technology industries: The term "research consortium" has been commonplace in Japan and Europe for a long time, and the Unites States caught up with the idea in the late 1980s after the National

Cooperative Research Act of 1984 stipulated that joint R&D ventures between competitive firms "must not be held illegal per se" (Lee & Lee, 1991). At that time, perhaps only "50–100 firms worldwide" had excellent R&D capabilities whose research collaboration with each other, universities, and governments could turn otherwise daunting research projects into possibilities (Teece, 2016, p. 30). The early well-known research collaboration experiences in the United States included the Sematech consortium (founded in 1987), whose network of innovators came from 14 domestic semiconductor manufacturers and the U.S. government. The consortium was created to search for new semiconductor manufacturing processes that would make the United States competitive with Japan—fundamentally because Japan's dominance was causing concern within the American defense establishment, whose military systems relied heavily on sophisticated electronics (Lee & Lee, 1991).

Today there are a "hundred plus strong technology firms … and hundreds of new enterprises" in almost every technology-based industry worldwide with excellent R&D capabilities that have research collaborations with public institutions and universities in the U.S., Europe, and Asia (Teece, 2016, p. 30). Consequently, today's well-known research consortia are often multinational in nature—the Human Genome Project (launched in 1990), whose two multinational rival networks published their findings between 2001 and 2006, being an example.

The literature provides some information on the structure and governance of the networks in these well-known examples of research consortia—the most challenging governance issue being that of making sure that the research collaboration is far enough removed from commercialization that firms can cooperate in the laboratory, though still remain able to compete in the marketplace (Lee & Lee, 1991). This issue is connected to the potential antitrust violations that have always been a concern: The National Cooperative Research Act protects "pure research activity" and allows commercial activity only as far as it is "reasonably required" for research (ibid.).

However, there is now a widespread belief that the idea of "pure research activity" is unrealistic. Today, excellent R&D capabilities (both basic and applied research) must go hand in hand with excellent commercial capabilities (which require research into products' commercialization and marketing)—see Hage and Hollingsworth (2000). Also, what Teece (2016, p. 31) calls "multi-invention contexts" are now the norm—contexts in which individual products and processes "draw on multiple internal and external sources of technology (patented or unpatented)." For example, innovations in laser and computer technology can now be developed much more effectively in collaboration than separately, and further possibilities appear when innovations in optical fiber are also added to the mix (Teece, 2016, p. 31). In fact, in many industries, innovation is now all about combining multiple innovations from different industries—Apple's iPod being an example (Pitelis & Teece, 2010).

Consequently, today's networks now comprise multiple forms of research cooperation including, on the one hand, research consortia, industrial parks, and such, and, on the other, what Teece (2016, p. 27) calls "naked licensing." In between, there are multiple options including joint ventures and strategic alliances—either equity-based or non-equity-based (see Robinson & Stuart, 2007); there are mergers and acquisitions; and there are corporate venture capital investments whose influences on innovation performance have recently received special attention (see, e.g., Phelps, Heidl, & Wadhwa, 2012; Wadhwa & Kotha, 2006; Wadhwa, Phelps, & Kotha, 2016). As a result, empirical research into the manner in which networks influence knowledge creation (and, thus, innovations) often produces conflicting results. For example, Phelps et al. (2012) review article on the subject comprises empirical results from the literature they describe as "conflicting/inconsistent/mixed" at least 35 times. A single empirical study can sometimes include different conclusions about different industries, just as what is found in one region is not always found in another region—see Baptista (2000) and Rowley, Behrens, and Krackhardt (2000) for industry cases, and Beaudy and Breschi (2003) for the regional cases.

Among other factors, the degree to which knowledge sharing is institutionalized as a professional practice in a particular industry makes a significant difference; whether or not a high degree of uncertainty surrounds a particular industry matters; and whether or not the industry's competitive demands require the exploration of new ideas (rather than the exploitation of what is already known) is a factor. In addition to the industries and regions, the firms themselves also matter—for example, see Stuart (2000) and Robinson and Stuart (2007) for the degree to which network effects can be different for publicly and privately held firms, for well endowed and not so well endowed firms, and for large and technologically sophisticated firms and small and technologically unsophisticated firms (see also Beaudry & Breschi, 2003). Finally, the particular products and processes matter as well: For example, whereas some products require tight links between those doing basic research and those doing product development research (as well as research on manufacturing processes, research on quality control, and research about the commercialization and marketing of products), many products do not need such tight links (Hage & Hollingsworth, 2000).

In summary, the literature's contributors offer many conjectures concerning the structures and governance rules of high-technology industries that might be studied in an actual, not necessarily high-technology, industry context. For example, one of the many conjectures in stated in the literature is: "Equity joint ventures facilitate knowledge transfer better than other governance modes and lead to increased organizational knowledge creation," and "can mitigate the unintended leakage of partner knowledge that is unrelated to the partnership" (Phelps et al., 2012, p. 1134).

As mentioned before, the actual industry that is of primary interest here is the flat-glass industry. If one is to investigate the degree to which networks facilitate innovation in the flat-glass industry or consider conjectures around this question (conjectures similar to those mentioned above), one must clearly first develop some basic understanding of the past and present particularities of the flat-glass industry.

The Flat-Glass Industry—A Brief History

I here use term "flat-glass industry" to refer to the manufacturing of the glass found in windows, cars, appliances, electronic devices, solar panels, and so forth. In his book entitled *The Substance of Civilization*, Sass (1998) writes that flat glass was as technology-based "ten thousand years ago" as it is today. Although this statement is certainly correct on a fundamental level, it requires elaboration: Even though the product might have stayed "forever the same," the process of manufacturing glass has, of course, changed over history—a history that I will discuss in this section.

Any historic account of glass must include the fact that although "glass was glass," there were, for a long time, important differences between *sheet* glass (ordinary window glass), and *plate* glass (more sophisticated window glass, and, since the 1920s, automobile glass). Sheet glass was expected to have some imperfections (such as little pits, tiny bubbles, and slight distortions), and thus could be manufactured relatively inexpensively. On the other hand, plate glass was expected to be ground and polished for perfection, and thus could only be produced on a large scale at capital-intensive and innovation-heavy plants. Until the 1980s, these two types of glass were manufactured by two different industries. This was not because the plate glass manufacturers did not manufacture any sheet glass, but because they never really excelled in the manufacturing of this relatively inexpensive kind of glass.

Anyone offering an historic account of glass must also pay attention to the degree to which, as early as the eighteenth century, glass manufacturing firms (especially those which manufactured the relatively more expensive plate glass) learned from each other and collaborated. For example, as Harris (1975) discusses in detail, in the eighteenth century, the British Ravenhead learned the technique of casting plate glass from the French Saint-Gobain, whereas Saint-Gobain learned the technique of using coal in glass furnaces from its British competitor. Barker's (1977) excellent account of the glass industry also contains information as to the manner in which British firms such as Pilkington, Chances, and Hartleys and the French Saint-Gobain continued learning from each other during the nineteenth century—a time period when the competition coming from cheap Belgian glass became significant. Barker (1977) also offers similar information on the beginning of the twentieth century—a time period that he calls the "new age of industrial diplomacy" (p. 197). This period (1900–1914) was the time when glass manufacturers from continental Europe formed "conventions" to determine glass prices—a point whose importance will become clearer later. This time period was also when Saint-Gobain decided to take a stake in the British Chances, and when North American influences were being felt in Britain and Europe.

In the 1920s, there was intense industrial diplomacy between the major plate glass manufacturing firms such as the British Pilkington; the U.S. based firms Pittsburg Plate Glass, Guardian, Libbey-Owens-Ford, and Ford Motor; the French Saint-Gobain; and the Japanese Asahi Glass. While these plate glass manufacturers were producing their *plate* glass (distortion-free glass not only for architectural uses but also for more sophisticated applications such as automobile glass), the sheet-glass manufacturers such as the German firms Deutsche Libbey-Owens Gesellschaft (Delog) and Deutsche Tafelglas (Detag) were manufacturing ordinary window glass somewhat inexpensively. This was possible because, as mentioned above, ordinary glass was expected to contain occasional flaws and distortions, and this allowed manufacturers to dispense with grinding and polishing the product.

During the 1930s, the plate-glass manufacturers improved the quality of their glass, thanks to an important technical development that had first been commercialized in the late 1920s: the Pittsburgh technique of grinding and polishing. The Pittsburgh technique replaced the Fourcault technique that glass manufacturers had used to grind and polish since the mid-1910s. Meanwhile, sheet-glass manufacturers such as the German Deutsche Tafelglas (Detag) were also steadily producing better and better quality sheet glass. In the 1940s and 1950s, as the quality gap between the expensive plate glass and the inexpensive sheet glass kept narrowing, the sheet-glass industry started to seriously threaten the plate-glass industry. This was because, as mentioned before, although plate glass manufacturers such as Pilkington and Pittsburgh Plate Glass had been excelling in the manufacturing of expensive plate glass, they were not so successful in the production of the cheaper sheet glass (Barker, 1977; Uusitalo & Möller, 2015).

It is worth repeating here how intimate and interconnected the firms within the plate-glass industry (including Pilkington and Pittsburgh Plate Glass) were just before and during this especially competitive period. In the 1940s and 1950s, the plate-glass manufacturers had shares in each other's companies, and exclusive (or non-exclusive) cross-licensing (or sub-licensing) arrangements with each other. Moreover, they routinely engaged in fidelity agreements and quota arrangements among themselves, which required that there were always some "diplomatic activities" going on between the plate glass manufacturing firms—something which reveals quite a lot about the historic particularities of the industry (Barker, 1977, p. 368). The British firm Pilkington's "hand in the world of industrial diplomacy" was especially strong (p. 357)—"diplomatic activities" being perhaps a euphemism for cartel-like practices, something to which I will return later.

Then in 1959, just when the competition between plate-glass manufacturers and sheet-glass manufacturers had become especially fierce, Pilkington's innovation of *float glass* came to the rescue of the plate-glass industry. This technique was a game changer for two reasons: First, it brought tremendous cost and efficiency improvements to the plate-glass manufacturers; and, secondly, the licensing policy that Pilkington designed and implemented for this new technology turned out to be consequential.

I cannot over-emphasize the first point: The new technique of manufacturing *float glass* (in which sand, soda ash, limestone, dolomite, alumina are all melted, then poured across the surface of a bath of molten tin, and then spread and flattened before being drawn horizontally in a continuous ribbon) was unquestionably superior to any previous technique for two basic reasons. First of all, the float-glass process finally enabled firms to design processes that could operate continuously: that is, 365 days per year, throughout the working life of the float lines (which is between 10 and 15 years). Linking "all the islands of automation into one continuous process" led to significant cost and efficiency-related improvements. In fact, the technique turned glassmaking "into a highly efficient and automated industry": It reduced the length of the production line by more than half, increased the volume of finished glass by 15–25%, eliminated the use of costly abrasives, lessened waste, and finally diminished labor costs by 80% and energy costs by 50% (Utterback, 1994, pp. 112–113). Secondly, the float technique eliminated the processes of grinding and polishing and produced glass with such an incredible flatness that no amount of grinding and polishing could match it (Pilkington 1969, cited in Utterback, 1994, p. 115). The overall result was a dramatic decline in glass prices (Teece, 2000, p. 225).

Pilkington's float-glass technique was also a game changer in the industry because of Pilkington's licensing strategy: With the help of Alastair Pilkington himself (the mechanical engineer who developed the process), the firm decided on a policy of initially licensing the process only to the existing major plate-glass manufacturers. Today's observers have different interpretations here. Uusitalo and Möller (2015) write that the previous Pittsburgh and Fourcault techniques had been liberally licensed to all firms that could afford it, but that Pilkington chose a different path and left out an entire industry: that of the sheet-glass industry. Here, they point to the fact that, in 1960, the German sheet-glass manufacturer Deutsche Tafelglas (Detag) applied for a license from Pilkington and was refused, and consider this as a sign of Pilkington's illiberal licensing policy (Uusitalo & Möller, 2015).[5] Their conclusion is that Pilkington's decision of which firms would be allowed to acquire the license ended up determining today's winners and losers (the German and Scandinavian sheet-glass manufacturers were wiped out after their applications for a Pilkington license were rejected), and redrawing the geography of flat-glass manufacturing (through the territorial restrictions that came with the license itself).

Teece's (2000) interpretation is completely different: Pilkington's float process replaced only the grinding and polishing stages of the overall glass-production process, and it was for this reason that Pilkington initially restricted the license to those major plate-glass manufacturers that were already skilled at the other stages of production and had the other marketing and distribution capabilities necessary to rapidly commercialize the technique. Furthermore, Teece (2000, p. 228) claims that otherwise "the terms and conditions employed by Pilkington to license the process innovation were generally as liberal or more liberal than those found in agreements for other glass technologies," such as the Pittsburgh and Fourcault techniques.

In 1962, it was Pittsburg Plate Glass (PPG), Pilkington's licensor of the Pittsburg technique since 1929, who (after lengthy negotiations over the terms) became the first flat-glass manufacturer to be licensed to use the float process. During the rest of the 1960s and 1970s, the remaining plate manufacturers worldwide acquired the license one by one. Pilkington invested in a number of countries (Canada in 1967 and Mexico in 1968) before becoming a public company in 1970, but especially

[5] Uusitalo and Möller (2015) believe that the fate of the German Deutsche Tafelglas (Detag) was sealed in 1960 when Detag's application for a license from Pilkington was refused.

accelerated its efforts to invest abroad after going public (again Canada in 1970, Australia in 1974, Sweden in 1976, and South Africa in 1977). The U.S. firm Guardian acquired the Pilkington technique in 1971, and in 1976, 21 non-British firms were paying float royalties to Pilkington (Barker, 1977). Thanks to Pilkington's own investments and those of its licensees, float glass replaced plate glass in a short period of time (Neuman, 1996).[6] In the 1980s, float glass then also replaced sheet glass, thus allowing the industry to converge into a single industry, classified by a single four-digit code—the Standard Industrial Classification (SIC) code (Uusitalo, 2014a, 2014b). In the process, a number of German, Scandinavian, and Belgian firms (and, in fact, almost the entire German sheet-glass industry) disappeared.

The plate-glass manufacturers who acquired the license agreed to a number of conditions: They would use the technology to construct and operate float-glass plants in a limited number of countries (a territorial restriction), observe Pilkington's sub-licensing rules, and report and share with Pilkington all future technological improvements that they might develop (Neuman, 1996). Consequently, Pilkington enjoyed an unquestionable power over the industry and, according to an estimate by Teece, Grindley, and Sherry (cited in Teece, 2000), made $5.3 billion (in 1992 prices) from its float innovation. However, as early as the 1980s, there were also a few signs suggesting that Pilkington would not be able to retain its licensing privileges forever. For example, in 1983, following a lawsuit by Pilkington, the U.S. firm Guardian was legally released from the majority of its obligations under its licensing agreements with Pilkington. From then on, Guardian was able to construct float glass plants outside its territory. And then in 1994, when over 90% of flat glass worldwide was being manufactured under the Pilkington license, the U.S. Department of Justice challenged Pilkington and alleged that the firm was continuing to impose restrictions on its licensees even after the expiration of its licenses. Following this lawsuit, a consent decree eliminated all of the limitations Pilkington had imposed on its U.S. licensees: from then on, they were allowed to use the float technology free of charge and sublicense any third party anywhere in the world. The decree also provided, in effect, a similar "safe harbor" for any other American individual or firm who was not a Pilkington float-glass licensee to use any float technology in its possession without any liability to Pilkington (Neuman, 1996).

Intense intra-industry competition followed the decree. The new century's first decade was a time period in which there was an increase in demand for safety glass, stricter regulations concerning energy efficiency, capacity increases, and declining prices. Herold and Paha (2016) write that it was under these conditions that, between 2004 and 2005, the top four firms in Europe initiated the establishment of a cartel. In fact, the European Commission claimed in 2008 that these cartel practices had started even earlier: namely, between 1998 and 2003, when the firms' representatives regularly met and decided on the allocation of glass supplies and the division of market shares. That same year, in 2008, the European Commission fined the

[6] See, for example, how Gobain-Pont-a-Mousson's then newest factory became outdated before it even went into production (Utterback, 1994, p. 121, Footnote 18).

French Saint-Gobain, the British unit of Japanese Nippon Sheet Glass and Asahi and others a record 1.4 billion euros ($1.77 billion) for price-fixing during 1998–2003 (Kanter, 2008). The French Saint-Gobain was fined the largest amount (896 million euros, i.e., $1.1 billion), because it was a repeat offender when it came to engaging in cartel practices. The firm had been fined twice in the 1980s for cartel pricing in Belgium, Italy, Luxembourg, and the Netherlands. Nippon Sheet Glass's British unit was fined 370 million euros ($470 million), and Asahi was fined 113.5 million euros ($144 million). Although these high fines indicated that cartel practices were risky, they proved not to be crippling (see Stephan, 2009).

Today's Flat Glass Industry

Today, six firms meet most of the global demand for float glass: Namely, NSG (Nippon Sheet Glass, Japan—which acquired Pilkington in 2006), AGC (Asahi Glass Co, Japan), Saint-Gobain (France), Guardian (US), Taiwan Glass (Taiwan), and Şişecam (Turkey). Of these six, NSG, AGC, and Saint-Gobain represent the old guard, whereas Guardian is a relatively newer firm (established in 1932 and acquiring the Pilkington technique in 1971) that has been a powerful actor for some time. Şişecam is a genuine newcomer (established in 1935, acquiring the Pilkington technique in 1977, which has become a regional power in Europe during the last few years). Taiwan Glass is the youngest of all (established in 1964 and acquiring the Pilkington technique in 1974) and, like Şişecam, is more regional than global, with float lines only in China and Taiwan.

These firms are remarkably similar to each other: They all are multi-product/multi-owned brand firms operating geographically separated production facilities in multiple countries. The first four (namely, NSG, AGC, Saint-Gobain, and Guardian) share most of the global demand almost equally, whereas the other two (namely, Şişecam and Taiwan Glass) fulfill regional roles. Among the countries in which these six firms operate float lines, Germany and the BRIC countries (namely Brazil, Russia, India, and China) are at the top of the list, with float lines operated by four or five of these six firms. As mentioned before, each of these six firms reports exactly the same turning point in their respective histories: namely, the year when they acquired the Pilkington license.

There seems to be an overall disappointment in the industry today that, when it comes to the primary processing of glass, the industry has not experienced any innovation that can be considered as consequential as Pilkington's float-glass process.[7] In fact, not only have no industry-shattering innovations occurred since the introduction of the Pilkington technique, but even the most incremental innovations have been few and far between. According to Thomsen (2013), this last point

[7] Since the publication of the Pilkington license, thousands of licenses related to float glass have been published in the world (Nascimento, 2014). However, unlike the Pilkington license, they have not been breakthrough licenses.

makes the industry one with a "slow clock speed," and, in fact, a "dinosaur." Obviously, industry insiders are not exactly satisfied with recent innovations—such as Corning's *fusion drawing technique* in which the pristine surfaces of the glass are not touched by molten tin—something which results in a remarkably flat and uniformly thick glass that can withstand a high degree of chemical strengthening, providing it with unusual optical clarity, touch sensitivity, and damage resistance, and thus making it even more suitable for use as electronic device display glass.[8]

However, when it comes to glass's secondary processing (i.e., the further processing of flat glass by, among others, laminating, toughening, coating, or silvering the product), there are more technological developments to be discussed with respect to architectural glass, automotive glass, electronic display glass, and solar glass. For example, in the area of architectural glass, which fills an overwhelming share of the global demand for flat glass, there are possibilities for further innovations with respect to the coating technology[9] and the fenestration of flat glass.[10] The most recent trends seem to be towards increased energy performance (increased stringency, better tuning of the building envelope, and optimal use of daylighting), increased building performance (acoustics optimization and such), and better aesthetics/design (larger glass units, and larger ranges of decorative glass).[11]

In order to be used in the automotive industry, glass needs to be appropriately toughened (thermally or chemically strengthened), laminated, and bent specifically for use in automobile windshields (and perhaps occasionally turned into

[8] Corning based its *fusion drawing* technique on James Franklin Hyde's 1932 discovery of a high purity fused silica that only became useful over 80 years after its conception. Corning now uses the fusion drawing technique in the production of its *Gorilla Glass* (a cover glass used in devices such as tablets, notebook personal computers, and cell phones).

[9] The coating technology enhances the performance of architectural glass by turning it into glass with low-emissivity and solar control, glass with scratch resistance qualities, and even self-cleaning glass. There are basically two alternatives: pyrolytic coating (hard coating—used since the 1970s) is produced through chemical vapors deposited onto the glass during the glass production. A newer technique which is called sputtered coating (soft coating—used since the 1980s) is applied off-line independently of the float manufacturing process by depositing layers of thin metallic and oxide coatings onto the surface of precut glass. In sputtered coating, the primary material has for some time been silver: Today single-silver, double-silver, and triple-silver layers are all commercially viable, although there are now concerns about moving towards four-silver layers because it looks as if this might be a point of diminishing returns for firms when the weight and cost of the product are taken into account.

[10] The fenestration of flat glass refers to the design and placing of windows in building including double-glazing (insulating). There are a number of alternative double-glazing techniques: warm-edge spacers, gas filled glazing (the gases being argon, krypton, and xenon), aerogel filled glazing, vacuum glazing, and, more recently, glazing via the application of an electric current (electrochromic glass).

[11] In the absence of industry shattering innovations, a number of observers believe that what would be best for the industry is a substantial increase in today's window to wall ratios—preferably 60% glass on the surface of buildings (instead of today's 40%). Obviously, this is a policy-related issue.

bullet-proof safety glass). Here, innovation researchers focus on making windshields especially thin in order to make vehicles lighter and more energy efficient, in addition to giving automotive glass scratch resistance and noise-reducing qualities. The trend seems to be going away from the traditional soda-lime-silica windshield glass to high-alumina, high-strength windshield glass. Devices such as televisions, cell phones, laptop computers, tablet computers, and notebook computers require specially manufactured cover glass with a specific thickness, optical clarity, touch sensitivity, and damage resistance. Finally, there is the area of solar photovoltaic and solar thermal power generation that requires glass to be used as plates, front electrodes, and condensing heliostats.

In summary, in the area of the primary processing of its product, the flat-glass industry is not at all similar to well-studied high-technology industries (such as computers, semiconductors, pharmaceuticals, and biotechnology). In addition, collaborations between the industry's top six firms continue to be along the lines of licensing, joint-ventures, and such, with Saint-Gobain's recent collaborations with Şişecam being examples. In 2008, the French and Turkish firms agreed to combine forces to first open a flat-glass factory in Egypt in 2010 and then a flat-glass, mirror, and coated-glass manufacturing facility in the Alabuga Special Economic Zone of the Republic of Tatarstan in Russia in 2014. The industry also continues to be characterized by mergers and acquisitions. For example, in 2012, the Turkish firm Şişecam acquired the Romanian GlassCorp (that was formerly known as Geamuri SA), reportedly for the purpose of capacity increases. This acquisition was followed by a 2014 investment of 65 million euros in GlassCorp. In 2013, Şişecam also acquired an over 50% stake in HNG Float Glass Limited in India and acquired Fritz Holding (one of the world's technology leaders in encapsulated automotive glass), reportedly for the purposes of capacity increases and market access: Fritz had facilities in Germany, Slovakia, and Hungary and manufactured automotive glass for clients such as Mercedes-Benz, BMW, Porsche, Audi, and Lamborghini. Finally, in 2016, Şişecam acquired the Sangalli Vetro Porto Nogaro facility at a price of 84.7 million euros ($91.81 million). During the same year, in 2016, the owner of Pittsburgh Plate Glass (PPG) sold the company to a Mexican manufacturer for $750 million.

However, things might be different when it comes to the product's secondary processing: Here there are signs that collaboration might now refer to something more than it does in the area of glass's primary processing. For one thing, the flat glass firms now seem to be aware of the possibility that networks might not only reduce the unnecessary duplication of their research efforts, but also make otherwise daunting research projects possible. For example, Beerkens, Bange, and Durán (2008, p. 368) mention two industry meetings that took place in Switzerland less than a decade ago during which "the first steps to define and organize large scale projects in the glass society with the aim of developing breakthrough technologies in glass products and glass production" were discussed. More recently, an industry insider has stated that the lead firms want "there to be linkages amongst all the people who work in glass research" (a research director at Corning, cited in Corning, 2016). At least there is now talk about who "should take a leading role in defining, initializing, and organizing consortia for major research activities, addressing innovative glass products" (Beerkens et al., 2008, p. 368). A new period is perhaps dawning in which exciting scholarly research possibilities will arise from the newly emerging, large, and complex networks of a now-maturing industry.

Conclusion: Looking at the Flat-Glass Industry from a Network Perspective

To what extent do networks facilitate innovation in technology-based industries? Scholars who have studied a number of high-technology industries such as today's biotechnology industry answer this question with certainty: Networks facilitate innovation to such an extent that the primary force behind innovation is now the "structure" of these networks and the "rules governing these networks"—not "money," not "market power," not even the "sheer force of novel ideas" (Powell & Grodal, 2005, p. 59; Powell et al., 2005, p. 1187). Consequently, they allocate most of their research energy into developing formal network approaches (known for their particular vocabulary of *nodes*, *lines*, and their chartable *patterns*, and for their reliance on mathematical modelling), and conduct analyses of large and complex networks—sometimes large enough to have as many as "250,000 nodes" (see Powell et al., 2005, p. 1191). The conviction is that if researchers thoroughly understand the structure and governance of the networks through which thousands (if not hundreds of thousands) of firms and institutions such as universities, research institutes, and venture capital simultaneously source and receive knowledge from each other, then they can even make predictions about these high-technology industries' future innovations. This approach has certainly not been without its critics. For example, consider Coward (2018, p. 453), who is concerned, among other things, about the "fantasies of precision" that this sort of thinking might lead to as networks expand and researchers unavoidably abandon more and more of the substantive contents of the nodes whose numbers keep increasing.

Here, with the above mentioned criticism of Coward (2018) in mind, I have looked into the flat-glass industry—an industry in which a number of research intensive and technologically dynamic firms are now reasonably networked, as opposed to being isolated like "islands," as Teece (2016, p. 11) would put it. However, the flat-glass industry does not make the official lists of high-technology industries; the assets of innovation might not yet be as dispersed (locally and globally) as they are elsewhere; and its networks are not as large and complex as, say, in biotechnology. Consequently, it is understandable that the flat-glass industry has not yet been subject to any formal network approaches in a manner similar to some other technology-based industries.

However, throughout their history (or at least during the last few centuries) the innovative flat-glass firms seem to have been stimulated to their best innovative performances when they established contact with other firms. The accounts of how, say, the British Ravenhead learned the technique of casting plate-glass from the French Saint-Gobain, whereas Saint-Gobain learned the technique of using coal in glass furnaces from its British competitor, can serve as an example. Thus, the observations around the case of the flat-glass industry are mostly in accordance with Burt's (2004) fundamental observation that novel ideas are better expressed, kept, and considered valuable in networked contexts than in isolated contexts.

On the others hand, it is difficult to go as far as imagining "money," "market power," and the "sheer force of novel ideas" taking the back seat to the "structure of its networks" and the "rules governing these networks" in the flat-glass industry, even though this might be the case in biotechnology (see the related claims of Powell et al., 2005, p. 1187; and of Powell & Grodal, 2005, p. 59). The most breakthrough innovation in the flat-glass industry, namely the innovation of the float technique, was fundamentally a "single firm" innovation (the single firm being the British Pilkington) that resulted in the industry turning into a "one technology, one license, one product" industry for decades. After this innovation, Pilkington made sure to legally obligate its licensees worldwide to share with Pilkington whatever technologies they might develop in the future—thereby suggesting the possibility that in this industry, instead of networks facilitating this breakthrough innovation, the innovation itself facilitated the configuration of today's networks. More specifically, I am here referring to the networks of Pilkington's licensees who found themselves compelled to share all innovative ideas they might develop in the future with Pilkington. Obviously, the idea behind Pilkington's licensing policy was making sure that Pilkington would not miss out on any innovation that might appear anywhere in the world.[12]

What is perhaps most interesting here is that Pilkington offered its float-glass licenses only to the existing major plate-glass manufacturers, thereby wiping out the sheet-glass manufacturers in Germany, Scandinavia, and elsewhere, and redrawing the geography of flat-glass manufacturing (through the territorial restrictions that came with the license itself). However, no consensus as yet exists as to the nature and consequences of the manner in which Pilkington governed its innovation. It is here worth mentioning Teece's (2000, 2016, p. 19) assertion that by settling for "naked licensing," Pilkington failed to fully benefit from its innovation: It should have created a "true network" around its game-changing innovation in the manner that today's high-technology firms do: "Pilkington [was] unprepared and unable to implement the technology by itself on a worldwide basis … [Thus] widespread licensing seemed the best alternative" (Teece, 2016, p. 8).

Interestingly, Pilkington's network-creating policies were also shrouded in controversy: For example, Uusitalo and Möller (2015) observed that the previous Pittsburgh and Fourcault techniques had been liberally licensed to all the firms that could afford it, but that Pilkington chose a different path and excluded an entire industry. On the other hand, Teece (2000, p. 228) claims the opposite: "The terms

[12] It is interesting to note that since the publication of the Pilkington license, no license has been related to a breakthrough innovation, perhaps rendering Pilkington's expectations a little too optimistic (see Nascimento, 2014).

and conditions employed by Pilkington to license the process innovation were generally as liberal or more liberal than those found in agreements for other glass technologies," such as the Pittsburgh and Fourcault techniques. This certainly would be a good starting point for a thorough analysis of the flat-glass industry following this breakthrough innovation. More specifically, I suggest further research on the flat-glass industry similar to the somewhat Schumpeterian analysis that Powell and his collaborators conducted on the emergence and growth of the biotechnology industry in the Unites States in the pre-1988 period. I also suggest that future researchers should consider the question of what constitutes a "true network" (Provan & Kenis, p. 231; Teece, 2000, 2016, p. 19).

Obviously, my above suggestions for further work refer to the primary processing of flat glass. There is even more room for further work concerning the secondary processing of the product (i.e., the further processing of flat glass by laminating, toughening, coating, or silvering the product with the purpose of turning glass into more sophisticated products such as high resolution flat panel displays). When it comes to flat glass's secondary processing, innovation's assets are now much more dispersed (locally and globally) than they were before. Also, contexts in which individual products and processes draw on multiple internal and external sources of technology are now becoming more and more pervasive in glass's secondary processing. For example, innovations in laser technology (such as those around ultrashort pulse lasers) and innovations in glass can now be developed much more effectively in collaboration than separately. Similarly, when innovations in optical fiber and glass are brought together, new near-to-eye display technologies become possible. It is now possible to pick up from where Uusitalo and his collaborators left off and study the contemporary manner in which flat-glass firms feel compelled to be part of networks, together with the network structures and rules that seem to be emerging.

References

Baptista, R. (2000). Do innovations diffuse faster within geographical clusters? *International Journal of Industrial Organization, 18,* 515–535. https://doi.org/10.1016/S0167-7187(99)00045-4

Barker, T. C. (1977). *The glassmakers: Pilkington: The rise of an international company, 1826–1976.* London: Weidenfeld and Nicolson. https://doi.org/10.1017/S002205070008726X

Beaudry, C., & Breschi, S. (2003). Are firms in clusters really more innovative? *Economics of Innovation and New Technology, 12,* 325–342. https://doi.org/10.1080/10438590290020197

Beerkens, R., Bange, K., & Durán, A. (2008). Glass: The challenge for the 21st century. *Boletin de la Sociedad Española de Cerámica y Vidrio, 47,* 366–370.

Burt, R. S. (2004). Structural holes and good ideas. *American Journal of Sociology, 110,* 349–399. https://doi.org/10.1086/421787

Corning. (2016). Inside Corning's Glass Summit. Retrieved from https://www.corning.com/emea/de/innovation/culture-of-innovation/glass-summit-feature-2016.html

Coward, M. (2018). Against network thinking: A critique of pathological sovereignty. *European Journal of International Relations, 24,* 440–463. https://doi.org/10.1177/1354066117705704

Hage, J., & Hollingsworth, J. R. (2000). A strategy for the analysis of idea innovation networks and institutions. *Organization Studies, 21,* 971–1004. https://doi.org/10.1177/0170840600215006

Hagedoorn, J., & Schakenraad, J. (1994). The effect of strategic technology alliances on company performance. *Strategic Management Journal, 15,* 291–309. https://doi.org/10.1002/smj.4250150404

Harris, J. R. (1975). Saint-Gobain and Ravenhead. In B. M. Ratcliffe (Ed.), *Great Britain and her World 1750–1914: Essays in honour of W. O. Henderson* (pp. 27–70). Manchester, UK: Manchester University Press.

Herold, D., & Paha, J. (2016). Predicting cartel formation. Retrieved from https://ssrn.com/abstract=2740528. https://doi.org/10.2139/ssrn.2740528

Kanter, J. (2008, November 12). Glass makers are fined $1.7 billion in Europe's war on price fixing. *The New York Times*. Retrieved from https://www.nytimes.com/2008/11/13/business/worldbusiness/13glass.html

Keeble, D., & Wilkinson, F. (1999). Collective learning and knowledge development in the evolution of regional clusters of high technology SMEs in Europe. *Regional Studies, 33,* 295–303. https://doi.org/10.1080/00343409950081167

Lee, M. K., & Lee, M. K. (1991). High technology consortia: A panacea for America's technological competitiveness problems? *High Technology Law Journal, 6,* 335–362.

Marsden, P. V. (1990). Network data and measurement. *Annual Review of Sociology, 16,* 435–463. https://doi.org/10.1146/annurev.so.16.080190.002251

Nascimento, M. L. F. (2014). Brief history of the flat glass patent: Sixty years of the float process. *World Patent Information, 38,* 50–56. https://doi.org/10.1016/j.wpi.2014.04.006

Neuman, J. N. (1996). Through a glass darkly: The case against Pilkington plc. under the new U.S. Department of Justice international enforcement policy. *Northwestern Journal of International Law & Business, 16,* 284–317.

Owen-Smith, J., & Powell, W. W. (2004). Knowledge networks as channels and conduits: The effect of spillovers in the Boston biotechnology community. *Organization Science, 15,* 5–21. https://doi.org/10.1287/orsc.1030.0054

Phelps, C., Heidl, R., & Wadhwa, A. (2012). Knowledge, networks, and knowledge networks: A review and research agenda. *Journal of Management, 38,* 1115–1166. https://doi.org/10.1177/0149206311432640

Pitelis, C. N., & Teece, D. J. (2010). Cross-border market co-creation, dynamic capabilities and the entrepreneurial theory of the multinational enterprise. *Industrial and Corporate Change, 19,* 1247–1270. https://doi.org/10.1093/icc/dtq030

Powell, W. W., & Grodal, S. (2005). Networks of innovators. In J. Fagerberg, D. C. Mowery, & R. R. Nelson (Eds.), *The Oxford handbook of innovation* (pp. 56–85). Oxford, UK: Oxford University Press.

Powell, W. W., & Owen-Smith, J. (2012). An open elite: Arbiters, catalysts, or gatekeepers in the dynamics of industry evolution? In J. F. Padgett & W. W. Powell (Eds.), *The emergence of organizations and markets* (pp. 466–495). Princeton, NJ: Princeton University Press.

Powell, W., Packalen, K., & Whittington, K. (2012). Organizational and institutional genesis: The emergence of high-tech clusters in life sciences. In J. F. Padgett & W. W. Powell (Eds.), *The emergence of organizations and markets* (pp. 434–465). Princeton, NJ: Princeton University Press.

Powell, W. W., & Padgett, J. F. (2012). Coda: Reflections on the study of multiple networks. In J. F. Padgett & W. W. Powell (Eds.), *The emergence of organizations and markets* (pp. 566–570). Princeton, NJ: Princeton University Press.

Powell, W. W., & Standholtz, K. (2012). Chance, nécessité, et naïveté: Ingredients to create a new organizational form. In J. F. Padgett & W. W. Powell (Eds.), *The emergence of organizations and markets* (pp. 379–433). Princeton, NJ: Princeton University Press.

Powell, W. W., White, D. R., Koput, K. W., & Owen-Smith, J. (2005). Networks dynamics and field evolution: The growth of interorganizational collaboration in the life sciences. *American Journal of Sociology, 110,* 1132–1205. https://doi.org/10.1086/421508

Provan, K. G., & Kenis, P. (2007). Modes of network governance: Structure, management, and effectiveness. *Journal of Public Administration Research and Theory, 18,* 229–252. https://doi.org/10.1093/jopart/mum015

Robinson, D. T., & Stuart, T. E. (2007). Network effects in the governance of strategic alliances. *The Journal of Law, Economics, and Organization, 23*(1), 242–273. https://doi.org/10.1093/jleo/ewm010

Rowley, T., Behrens, D., & Krackhardt, D. (2000). Redundant governance structures: An analysis of structural and relational embeddedness in the steel and semiconductor industries. *Strategic Management Journal, 21,* 369–386. https://doi.org/10.1002/(SICI)1097-0266(200003)21:3<369::AID-SMJ93>3.0.CO;2-M

Sass, S. L. (1998). *The substance of civilization: Materials and human history from the stone age to the age of silicon.* New York: Arcade.

Stephan, A. (2009). Hear no evil, see no evil: Why antitrust compliance programs may be ineffective at preventing cartels (ESRC CCP Working Paper No. 09-09). https://doi.org/10.2139/ssrn.1432340

Stuart, T. E. (2000). Interorganizational alliances and the performance of firms: A study of growth and innovation rates in a high-technology industry. *Strategic Management Journal, 21,* 791–811. https://doi.org/10.1002/1097-0266(200008)21:8<791::AID-SMJ121>3.0.CO;2-K

Stuart, T. E., & Podolny, J. M. (1996). Local search and the evolution of technological capabilities. *Strategic Management Journal, 17,* 21–38. https://doi.org/10.1002/smj.4250171004

Teece, D. J. (2000). *Managing intellectual capital: Organizational, strategic, and policy dimensions.* Oxford, UK: Oxford University Press. https://doi.org/10.1093/0198295421.001.0001

Teece, D. J. (2016). Profiting from innovation in the digital economy: Standards, complementary assets, and business models in the wireless world (Working Paper Series No. 16). Tusher Center for the Management of Intellectual Capital. Retrieved from https://businessinnovation.berkeley.edu/wp-content/uploads/2014/07/Tusher-Center-Working-Paper-No.-16.pdf

Thomsen, S. (2013). *Step change innovation for glass.* Keynote Presentation at the Glass Performance Days, Tampere.

Utterback, J. M. (1994). *Mastering the dynamics of innovation: How companies can seize opportunities in the face of technological change.* Boston, MA: Harvard Business School Press.

Uusitalo, O. (1997). Developments in the flat glass industry in Scandinavia, 1910–1990: The impact of technological change. *Scandinavian Economic History Review, 45,* 276–295. https://doi.org/10.1080/03585522.1997.10414672

Uusitalo, O. (2014a). Understanding the inter-linkage of innovations: The Finnish float glass and value added industries. *Intellectual Economics, 8*(2), 92–111. https://doi.org/10.13165/IE-14-8-2-07

Uusitalo, O. (2014b). *Float glass innovation in the flat glass industry.* Cham: Springer International.

Uusitalo, O., & Mikkola, T. (2010). Revisiting the case of float glass: Understanding the industrial revolution through the design envelope. *European Journal of Innovation Management, 13,* 24–45. https://doi.org/10.1108/14601061011013212

Uusitalo, O., & Möller, K. (2015). *Historical determination of a business network: The case of the flat glass industry.* Paper presented at 12th IMP Conference, Karlsruhe.

Wadhwa, A., & Kotha, S. (2006). Knowledge creation through external venturing: Evidence from the telecommunications equipment manufacturing industry. *Academy of Management Journal, 49,* 819–835. https://doi.org/10.5465/amj.2006.22083132

Wadhwa, A., Phelps, C., & Kotha, S. (2016). Corporate venture capital portfolios and firm innovation. *Journal of Business Venturing, 31,* 95–112. https://doi.org/10.1016/j.jbusvent.2015.04.006

Whittington, K. B., Owen-Smith, J., & Powell, W. W. (2009). Networks, propinquity, and innovation in knowledge-intensive industries. *Administrative Science Quarterly, 54,* 90–122. https://doi.org/10.2189/asqu.2009.54.1.90

The Klaus Tschira Foundation

The Klaus Tschira Foundation was created in 1995 by the physicist Klaus Tschira (1940–2015). It is one of Europe's largest privately funded non-profit foundations. The foundation promotes the advancement of natural sciences, mathematics, and computer science and strives to raise appreciation of these fields. The focal points of the foundation are "Natural Sciences—Right from the Beginning," "Research," and "Science Communication." The involvement of the Klaus Tschira Foundation begins in kindergartens and continues in primary and secondary schools, universities, and research facilities. The foundation champions new methods in the transfer of scientific knowledge, and supports both the development and intelligible presentation of research findings. The Klaus Tschira Foundation pursues its objectives by conducting projects of its own but also awards subsidies after approval of applications. To foster and sustain work on selected topics, the foundation has also founded its own affiliates. Klaus Tschira's commitment to this objective was honored in 1999 with the "Deutscher Stifterpreis," the award conferred by the National Association of German Foundations.

The Klaus Tschira Foundation is located in Heidelberg and has its head office in the Villa Bosch, once the residence of Carl Bosch, a Nobel laureate in chemistry.

www.klaus-tschira-stiftung.de

J. Glückler et al. (eds.), *Knowledge for Governance*, Knowledge and Space 15,
https://doi.org/10.1007/978-3-030-47150-7

Fig. 1 Participants of the symposium "Knowledge for Governance" at the Studio Villa Bosch in Heidelberg, Germany. © Johannes Glückler, Heidelberg

Fig. 2 Villa Bosch, the head office of the Klaus Tschira Foundation, Heidelberg, Germany. © Peter Meusburger, Heidelberg

Index

J. Glückler et al. (eds.), *Knowledge for Governance*, Knowledge and Space 15,
https://doi.org/10.1007/978-3-030-47150-7

The manufacturer's authorised representative in the EU is Springer Nature Customer Service Centre GmbH, Europaplatz 3, 69115 Heidelberg, Germany. If you have any concerns regarding our products, please contact ProductSafety@springernature.com

Printed and bound by CPI Group (UK) Ltd, Croydon, CR0 4YY
05/10/2026
02228127-0015